Soil Algae

Arti Mishra • Ajit Varma
Editors

Soil Algae

Morphology, Ecology and Biotechnological Applications

Springer

Editors
Arti Mishra
Department of Botany
Hansraj College, University of Delhi
New Delhi, Delhi, India

Ajit Varma
Amity Institute of Microbial Technology
Amity University
Noida, Uttar Pradesh, India

ISBN 978-981-97-9872-8 ISBN 978-981-97-9873-5 (eBook)
https://doi.org/10.1007/978-981-97-9873-5

This Springer imprint is published by the registered company Springer Nature Singapore Pte Ltd.
The registered company address is: 152 Beach Road, #21-01/04 Gateway East, Singapore 189721, Singapore

Preface

This book covers all aspects of soil algal species, which improves our understanding of the algal flora that exist in the terrestrial environment. This book deals with the updated knowledge about algal species that are present, particularly in the soil. Also, how we can cultivate them in laboratory conditions for further detailed study about their morphology, physiology, and applications in various fields. So, by keeping such views in mind, this book is designed to enhance the knowledge about the algae present in the soil.

Different chapters covered in this book range from ecology to physiology to molecular biology and their applications. In addition, conventional and novel strategies for isolation, enumeration, and cultivation of soil microalgae and cyanobacteria are discussed in detail. Ecological significance of soil algae and their biotechnological importance are key features of this book. Furthermore, symbiosis of soil algae, bioremediation, and enzymes from soil algae are also included. In this book, I have attempted to fill the gap in knowledge in this field.

This book will assist the researchers working in the different aspects of phycology. This book is also handy to researchers of molecular biology, agriculture, and biotechnology and also to scientists working on soil biology, agronomy, and crop research. Also, the book serves as additional reading material for undergraduate and graduate students of microbiology, agriculture, molecular biology, and biotechnology. Researchers from biotechnological and biofertilizer companies will also find this to be a useful read.

New Delhi, Delhi, India — Arti Mishra
Noida, Uttar Pradesh, India — Ajit Varma

Contents

About the Editors

Arti Mishra is an Assistant Professor at Hansraj College, University of Delhi, India. She is working in the area of environmental microbiology, phycology, bioremediation, and plant microbe interaction. Dr. Mishra has a number of research papers and review articles to her credit in the journals of international repute. In addition, Dr. Mishra has been recognized as a member of the Biotech Research Society (BRSI), India. She has been awarded Dr. D. S. Kothari postdoctoral fellowship from UGC in 2016 for postdoctoral research work at School of Environmental Sciences, Jawaharlal Nehru University, New Delhi, India. In 2023, she received Carl Trygger fellowship for doing postdoctoral work at Umeå Plant Science Center, Department of Plant Physiology, Umeå University, Umeå, Sweden. Dr. Mishra received her PhD in Botany from APS University, Rewa, Madhya Pradesh, India, and MS in Biology from Linköping University, Sweden.

Ajit Varma has completed his PhD at the age of 25 years from Allahabad University, and is Former Professor, School of Life Sciences, Jawaharlal Nehru University, India. Presently, he is the Emeritus Professor at Amity Institute of Microbial Technology; Former Pro-Vice Chancellor, Ritnand Balved Education Foundation; and Former Vice Chairman, Amity Science, Technology & Innovation Foundation Amity University, Uttar Pradesh, India. He has published more than 314 papers in reputed journals and has been serving as an editor-in-chief of Soil Biology Series, Springer Verlag Germany. Dr. Varma is the Fellow of Alexander-von-Humboldt Society, Germany, elected Fellow of National Academy Agricultural Sciences, and Fellow of Microbiology Society of India.

Microbial Interactions in Soil Algae

S. Murugesan, P. Priya, and V. Sivamurugan

Abstract The present chapter encapsulates interactions of soil algae with various microbes. The nature and benefits of algae-microbial interactions are discussed briefly. The advantages and applications of algae-microbes mutual synergism are emphasized in this chapter. The interaction between soil algae-microorganism-plants resulted in the availability of nutrients, conditioning of soil, biofertilizers for plants, and improved crop production and plant defensive metabolites and pigments. The roles of microbial interaction with soil algae are to maintain soil fertilization and reduce the use of chemical fertilizers and pollutants in the soil. The chapter is intended to provide an overview of the benefits arising out of soil algae-microorganism-plants interaction.

Keywords Algae · Biofertilizers · Alternative fuel · Microbes · Nutrients · Mutualism

1 Introduction

Plant growth, production, and fruit quality have all been improved using microalgae species (Guo et al. 2020). Microalgae synthesize and accumulate huge amounts of lipids, proteins, and carbohydrates in the cell under certain growth circumstances.

S. Murugesan
PG and Research Department of Botany, Pachaiyappa's College, Chennai, Tamil Nadu, India

P. Priya
PG and Research Department of Zoology, Pachaiyappa's College, Chennai, Tamil Nadu, India

V. Sivamurugan (✉)
PG and Research Department of Chemistry, Pachaiyappa's College, Chennai, Tamil Nadu, India

A. Mishra, A. Varma (eds.), *Soil Algae*, https://doi.org/10.1007/978-981-97-9873-5_1

Microalgae have a higher lipid content than other biofuels, ranging from 20 to 50% of dry cell weight but occasionally reaching 70% (Alvarez et al. 2021; Sun et al. 2019). Ramanan et al. (2016) reviewed investigations on algae-bacteria interactions and elaborated types and beneficial outcomes of the interactions. Mutual support between algae-microorganism established through interaction benefited for nutrient exchange and growth (Cirri and Pohnert 2019). Crop development and plant protection are achieved by a symbiotic relationship between plant microorganisms and soil algae (Lee and Ryu 2021).

The use of algae or other plant extracts can help with growth and stress conversion. Algal extracts, protein hydrolysates, humic and fulvic acid, and other composite substances have higher nutritional characteristics, all of which promote growth and stress tolerance. Many plant biostimulants are administered to the rhizosphere to enhance nutrient uptake, but many also protect against environmental stresses such as dehydration, salt, and exposure to extremely low temperatures. The sole purpose of these rhizosphere microalgal communities is to convert atmospheric nitrogen to ammonia, which is then used by plants to produce bioavailable nitrogen and stimulate plant growth (Fig. 1).

Certain algal and cyanobacterial organisms, as well as extracts from them, have been found to have soil-beneficial properties. Only a few of their well-known functions include the ability to (i) convert atmospheric nitrogen into nitrogen compounds that plants may easily absorb; (ii) provide important nutrients for plants; and (iii) improve the physical and chemical environment.

To deal with selective pressures working within their natural environment, colonies of diverse individuals living in the rhizosphere or any other habitat exhibit a variety of morphological, physiological, and biochemical adaptations (Galhano et al. 2011). Plant gibberellins and microalgae/cyanobacteria gibberellins, for example, have similar roles in terms of growth promotion, colour induction, and protein accumulation (Han et al. 2018). Gibberellins from *Chlorella ellipsoidea*, *Chlorella pyrenoidosa*, *Chlorella vulgaris*, and *Microcystis aeruginosa* have been linked to larger cell sizes and faster growth rates in microalgae and cyanobacteria (González-Garcinuño et al. 2016). Phytohormones derived from microalgae and cyanobacteria emphasize the additional benefits of employing these species (or their extracts) in the agricultural sector, particularly those related to plant growth and protection in both biotic and abiotic conditions. Plant growth factors include metabolites such as phytohormones, as well as other amino acids, vitamins (such as vitamin B12 and biotin), polysaccharides, and polypeptides (Sharma et al. 2012). Kumar et al. (2013) and Priya et al. (2015) have both documented the implementation of plant protection strategies against microalgae and cyanobacteria. As a result, as indicated in the literature, the use of microalgae and/or cyanobacteria in food plants might elicit plant defences not only through the activity of their polysaccharides but also through the activity of these biologically active chemicals.

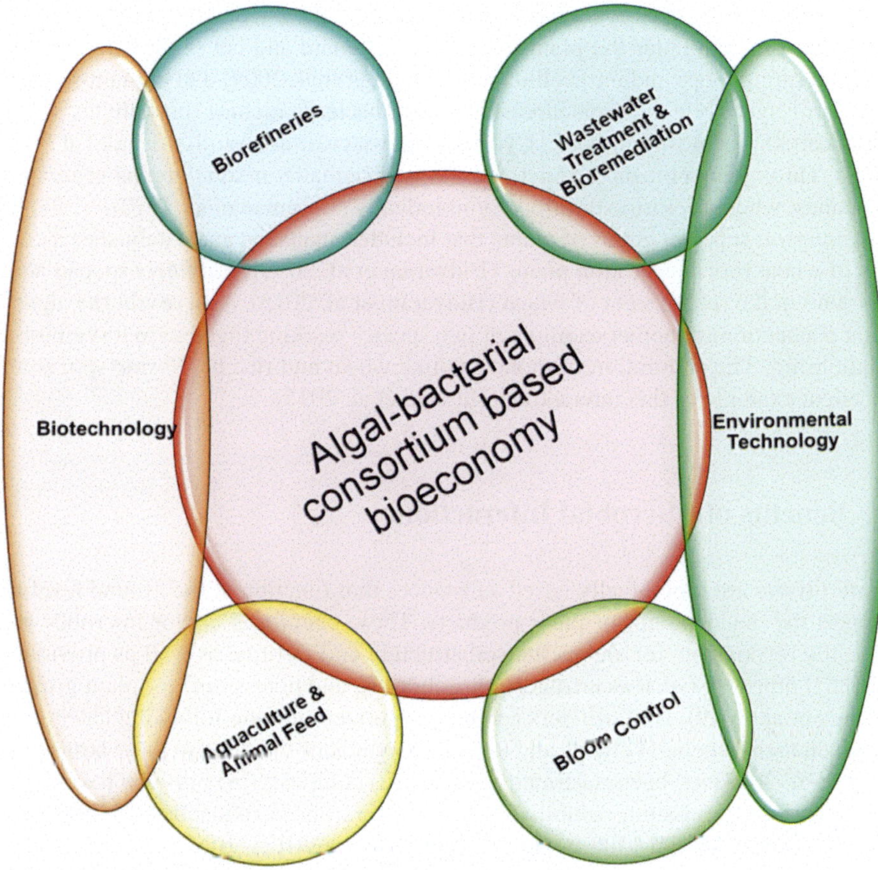

Fig. 1 Applications of algae-bacterial interactions (Credit: Ramanan et al. 2016, This article attributed by CC BY-NC-ND 4.0)

2 Types of Microbes-Algae Interactions

A symbiotic microbe interacts with algae to improve stress tolerance and the synthesis of secondary metabolites (Yao et al. 2019). Phytohormones are signalling cells in plants that regulate a variety of cellular functions. Indeed, phytohormones (e.g., auxins, abscisic acid, cytokinins, ethylene, and gibberellins) have been found in a variety of algal species, including brown, red, and green algae, cyanobacteria, and diatoms (Han et al. 2018). In higher plants, auxins, especially indole-3-acetic acid, regulate growth and development (Lu and Xu 2015).

From mosses to angiosperms, several organisms have symbiotic relationships with algae (Santi et al. 2013). Algae cover the plant surface with cells from within the plant tissues, such as other microorganisms in symbiotic and mutualistic interactions with plants, to interact with them. Plant-algae interactions can be seen in prokaryotic algae like cyanobacteria (Treves et al. 2016; Zhu et al. 2018). Cyanobacteria can enter the plant through the stomata and cover the intercellular space, forming loops and intracellular coils (Krings et al. 2009). Other plant species that form symbiotic relationships with cyanobacteria include bryophytes (e.g., Anthoceros), gymnosperms (e.g., Cycads), and water ferns (Azolla) (Lindblad et al. 1990). Through symbiosis, Gunnera develops systematic, unique organs generated by glands, whereas Nostoc develops symbiotically (Bergman et al. 1992).

Anabaena spp. is a genus of plants that includes anabaena and establishes a colony of wheat roots and cotton plants (Bidyarani et al. 2015). *Calothrix* sp. can also be found in the root system of wheat (Bidyarani et al. 2015). As a result, the algae-plant connection is another example of two species working together in a symbiotic relationship. The restoration of monocots like wheat and rice by *Nostoc* spp. is an excellent example of this interaction (Hussain et al. 2015).

3 Benefits of Microbial Interactions

Biofertilizers are biologically based substances that function at the ground level to support the development of plant products. These chemicals and/or microbes are typically responsible for enhancing soil structure and fertility, as well as providing essential nutrients (such as nitrogen, phosphorous, and potassium) for plant growth (Ronga et al. 2019). Biofertilizers can be categorized into the following categories based on their beneficial effects and the microorganisms that bind to these structures (Win et al. 2018): (i) bacteria-friendly plants; (ii) compost; (iii) nitrogen fixers; (iv) phosphate- and potassium-solubilizing biofertilizers; and (v) phosphorus-mobilizing biofertilizers (Carvajal-Muñoz and Carmona-Garcia 2012). Biofertilizers provide various advantages in agriculture, according to Carvajal-Muñoz and Carmona-Garcia (Carvajal-Muñoz and Carmona-Garcia 2012): (i) higher crop yield per unit area and time; (ii) reduced pressure; (iii) control and protection of sufficient soil structures and fertility; (iv) low risk of soil and water pollution; and (v) plant protection against disease-causing pests.

The usage of microalgae and cyanobacteria (and/or their extracts) can directly boost plant growth and development by increasing plant growth and development (e.g., shoot growth and root length, leaf growth, and high nutrient content). This development is due to the action of microalgae/cyanobacterial metabolites, which can induce a variety of metabolic assignments such as respiration, photosynthesis, nucleic acid synthesis, chlorophyll production, and ion uptake (Górka et al. 2018). Soil is home to a variety of terrestrial cyanobacteria, which benefit from soil fertility by fixing atmospheric nitrogen (N), binding soil particles, and assisting in moisture retention and erosion prevention. The most important factors for plant growth that regenerate the effects of these pests are an increase in essential soil microelements needed for plant growth and ion uptake, increased N content in topsoil, and production of plant-growth compounds such as phytohormones and other plant growth regulators (PGRs) such as amino acids, sugars, and vitamins (Misra and Kaushik 1989; Karthikeyan 2006; Obana et al. 2007).

Plant production is aided by biostimulants, which work directly on the plant. Górka et al. (2018) and Rouphael and Colla (2020) attribute these substances to increased respiration, photosynthetic activity, nucleic acid synthesis, and ion uptake, which improves plant metabolism and, as a result, plant development. These compounds and/or microbes can help plants grow faster, boost their resistance to harsh environmental circumstances, and make better use of their resources (Campobenedetto 2021). Polysaccharides, pigment, phenolic compounds, proteins and (bioactive) peptides, phytohormones, and micro- and macronutrients are among the bioactive substances found in seaweed extracts (El Boukhari 2020). Biostimulants' stimulating activity can be detected in both positive and negative environments, implying that these substances can help improve plant resilience and stress tolerance (Dias et al. 2016). As a result, biostimulants have the following advantages in agriculture (Dias et al. 2016): (i) higher yields; (ii) better nutrient efficiency; and (iii) better crop quality. It can be difficult to define and organize a large range of chemicals (e.g., polysaccharides, phenolic compounds, hormone-like compounds, and vitamins) discovered in their biostimulating regenerative plant function.

Plant protection from biotic and abiotic factors is critical for agricultural crop production. Although certain plants have immunity and resistance, external protection agents must be utilized in food plants in order to achieve the future product. Given the wide range of bioactive chemicals found in microalgae and cyanobacteria, using these microorganisms (or extracts) can provide enough plant protection in both biotic and abiotic environments.

The soil algal community is one of the most vulnerable biotic components, and it plays a critical role in the natural stability of soil systems (Megharaj et al. 2000). They are also important members of the global ecosystem, as they are important members of the soil's autotrophic biotic community. Climate, soil types, and local vegetation all influence the structure of the algal community (Bohlen et al. 2001). Because they mix natural changes with the seasons in a few weeks, algae colonies are heavily influenced by environmental influences. Land use has a greater impact on algal/cyanobacterial community structure than physico-chemical soil limits. The food sources of microarthropods as a paradigm for functional roles for algae/cyanobacteria diversity. Because algae and cyanobacteria are major producers around the world and serve a variety of roles in soil fertility and composition, there is a strong link between biodiversity and algae consumers. Organic C, organic N, and organic P, as well as moisture, are absorbed by algal species in the earth, which Na^+ clay into Ca^{2+} clay. Organic nitrogen and nitrogen-added cyanobacteria bind to soil particles, allowing for greater penetration and permeability (Singh 1961).

3.1 Plant Growth

Different human microbiota fragments are abundant in the rhizosphere of plants. Cyanobacteria-based symbiosis, or creatures with roots from a range of plants, is one such parasite. Other free species (*Anabaena*, *Calothrix*, *Oscillatoria*, *Tolypothrix*, and *Scytonema*, among others) can be found living in terrestrial soil systems, creating healthy soil for a variety of plant kingdom members, from Bryophytes to angiosperms (Adams et al. 2013). Other species of the genus *Nostoc* have been discovered to form endosymbiotic interactions with coralloid roots of Gymnosperms (Chang et al. 2019). They remove the membrane to protect their thalloid bodies from shrinkage and mechanical stress.

In terms of cyanobacteria's physiological and biochemical characteristics in relation to life, their ability to divide into various cells performing specific functions within the same filament is critical. Plant cells and photosynthetic cells comprise the major body of the thalloid under ideal conditions. In the worst-case scenario, certain filament cells separate to generate kinetes, which are weather-resistant compounds (Sukenik et al. 2015). Furthermore, many cyanobacterial species produce a large number of thick-walled cells known as heterocysts, with 9–15 cells forming at regular intervals throughout the filament (Kumar et al. 2010). These specialized cells contain oxygen-sensitive nitrogenase (an enzyme molecule that regulates nitrogen fixation) and serve as the most suited anaerobic component for nitrogen fixation.

It's worth noting that free-living cyanobacteria like *Nostoc* can give massive amounts of nitrogen to plant roots after death, which is followed by filament lysis and the formation of heterocyst-rich ammonia cells (Meeks and Elhai 2002). Because of their metabolic evolution, cyanobacteria obtain nutrition predominantly through photosynthesis, allowing them to be self-sufficient in food production and the primary source of oxygen evolution (Hamilton et al. 2016). With encouraging results, powerful structures that encourage cyanobacterial inoculant plant development can be employed commercially as Biofertilizers (Chittora et al. 2020).

Gantar et al. (1993) recovered heterocystous nitrogen-fixing cyanobacteria from plant roots and soil, including *Nostoc*, *Anabaena*, and *Cylindrospermum*. *Anabaena* loose loam of root wool and a strong root colony in the area were discovered when wheat seedlings were examined. *Nostoc* species live as endosymbionts in the flowering plant (angiosperm) *Gunnera*, in addition to living freely in the root zone of the plant (Lindblad et al. 1990). Gunnera's glands are strikingly similar to the crown glands generated by *Agrobacterium tumefaciens*. Cyanobacteria alter nitrogen fixation and release, heterocyst growth, and symbiosis compatibility during symbiosis. The nodules created by *Rhizobium* spp. and the gall produced by *A. tumefaciens* are similar to the *Nostoc* glands in the stem of the *Gunnera* plant (Rasmussen et al. 1996). These findings suggest that the Rhizobium-legume symbiosis differs from the *Gunnera-Nostoc* relationship. Non-legume plants have been studied for their ability to extend structures such as nodules and fix nitrogen. Chan and Kennedy (1989) used 2, 4-dichlorophenoxy acetic acid (2, 4-D), a synthetic auxin-based substance, to successfully produce nodule-like structures dubbed after para-nodule; however, they were unable to repair nitrogen utilizes nitrogenase includes germs.

Anabaena isolate treatment of rice fields resulted in a significant increase in seed germination and the growth of plant structures such as plant length, root length, and fresh and dry leaves on stems and roots (Saadatnia and Riahi 2009). Improvements in soil structure, such as humidity and porosity, were also noted. Aqueous extracts of *Gracilaria corticata* and *Enteromorpha flexuosa* were mixed into maize and sunflower seeds in a study by Omar et al. (2014). Although extracts from both species resulted in root and shoot growth as well as dry weights, the effects of extracted *E. flexuosa* on portable plants were the most evident. Elarroussia et al. (2016) investigated the effect of employing a solution of polysaccharides isolated from *S. platensis* in tomato and pepper plant culture, finding that both tomato and pepper plants grew larger.

Auxins discovered in *C. vulgaris*, *D. salina*, *Nannochloropsis oceanica*, *Scenedesmus obliquus*, and *Scenedesmus quadricauda* operate as growth regulators in microalgae, supporting a rise in cell size, chlorophyll, and protein content, as well as cell division, in high biomass production (Kozlova et al. 2017). Abscisic acid is frequently linked to cellular responses to environmental stresses such as excessive salt levels, drought, and food inadequacy (Han et al. 2018). This chemical is particularly effective in modulating cell response to salt, osmotic, oxidative, drought, and nutritional stress in microalgae and cyanobacteria (Yoshida et al. 2004; Duan et al. 2012). The abscisic acid pathway is responsible for a number of processes, including the formation of carotenoids, which are commonly used as antioxidants in the face of oxidative stress (Mousavi et al. 2016). These molecules, which are dependent on cytokinin, regulate a variety of functions in higher plants, including cell division, growth; biogenesis, chloroplasts, and seed sleep regulation, among others (Han et al. 2018). Aside from these tasks, cytokinins also help plants cope with harsh environmental circumstances (Han et al. 2018). Furthermore, the D-cytokinin generated by *D. salina*, kinetin, creates a pigment that protects the microalga's β-carotene image (Mousavi et al. 2016). Ethylene has a wide range of effects on plants, from influencing development and ageing to enhancing plant tolerance to biotic and abiotic stress factors (Han et al. 2018). The role of ethylene in microalgae and cyanobacteria is not fully understood. Other research has found that this chemical stimulates growth in *Chlamydomonas reinhardtii*, *Chlorella* sp., and *Spirogyra platensis* (Ju et al. 2015).

3.2 Crop Production

When combined with *Calothrix elenkinii* in the genus, injections with *A. laxa* boost coriander seed germination and promote root growth and germination in coriander, cumin, and fennel (Kumar et al. 2013). When planted on the roots and leaves of the *Perilla* plant, onions, lettuce, Chinese cabbage, seedlings, turnips, and spinach, eukaryotic *Chlorella* spp. promote cyanobacteria-like growth (Kim et al. 2012, 2018).

At the cellular level, laboratory research has shown that certain quotes have an interesting ability to activate plant-immune response systems that include Organized Acquired Objections.

Seaweed extracts, employed in very small amounts as leaf sprays or soil fertilizers, have been demonstrated to improve yields in a variety of crops, including grape wine, sugarcane, strawberries, and vegetables, according to field testing (Arioli et al. 2015; Shukla et al. 2019). Plant tolerance to abiotic stress such as drought, cold, and salt has been established in studies utilizing Arabidopsis, soybeans, and tomato plants (Santaniello et al. 2017; Goni et al. 2018; Jithesh et al. 2019). Laboratory studies have demonstrated that particular statements have an interesting capacity to trigger plant-protection response mechanisms such as Organized Acquired Resistance at the molecular level (Goni et al. 2018; Islam et al. 2020). The extraction of seaweed stimulated early seed germination and establishment, increased root growth, increased chlorophyll of leaves (Jannin et al. 2013), improved plant performance and tomato yield (Kumari et al. 2011), and increased resistance to pressure of biotic/abiotic (González et al. 2013). In addition, seaweed also affects the physical, chemical, and biological properties of the soil. The extraction of marine vegetation can incorporate the iron ion of the soil to produce colloids and prevent the formation of compacted soils (Halpern et al. 2015).

Microalgae and cyanobacteria are major sources of organic matter that can help farmers increase their yield. Because of the extensive range of substances found in microalgae/cyanobacteria, as well as their numerous functions in plant production, they can be employed as pure components or natural fertilizers. Furthermore, different metabolites contribute to plant production at different levels: (i) other extracts and microorganisms that improve soil quality, (ii) others that promote plant growth by protecting them from biotic and abiotic stress, and (iii) other metabolites that act specifically to promote plant growth.

3.3 Nitrogen Fixation

Cyanobacteria consumption enhances nitrogen availability in the soil, which is one of the most important factors in plant growth (Win et al. 2018; Baweja et al. 2019). Cyanobacteria showed favourable effects on agricultural soils (for various crop production) that have already been established (Renuka et al. 2016). According to Jha and Prasad (2006), this inoculum boosts grain and grass yields while also increasing soil nitrogen concentration. Similarly, *Anabaena variabilis* species utilized in rice field. According to Singh and Datta (2007), the rice types employed encourage rice crop development (e.g., increase in crop length and leaf length, as well as improved seed, grain, and grass production). Innok et al. (2009) observed that employing *Nostoc* sp. plant cells in the rice field increased grain yield when compared to chemical-based nitrogen fertilizers. Osman et al. (2010) found a considerable rise in the percentage of germination, photosynthetic pigments, and growth limits in soils meant for the development of peanut plants by cyanobacterial species *Nostoc entophytum* and *Oscillatoria angustissima*. The investigators found this rise was linked to both cyanobacteria in N_2 remodelling activities. Swarnalakshmi et al. (2013) discovered that inoculating a wheat plant with a biofilm made up of the cyanobacterium *Anabaena torulosa* greatly boosted nitrogen availability in the soil. These bacteria also help with nitrogen digestion, reducing the demand for chemical nitrogen fertilizers. Finally, cyanobacteria are well-known for their ability to create soil layers that limit nitrogen depletion and excretion, lowering the danger of nitrogen pollution (Renuka et al. 2018).

3.4 Nutrient Absorption in Soils

Essential nutrients are required for high plant growth; instead of utilizing traditional chemical fertilizers, microalgal/cyanobacterial biomass can be employed as a source of these nutrients (Ronga et al. 2019). Other essential microelements for plant growth and development (such as potassium, magnesium, sulphur, and iron) can also be found in microalgal/cyanobacterial biomass. These nutrients are frequently implicated in the redox response, and they are crucial for plant consumption. As a result, using microalgae and cyanobacteria in the soil can boost macro- and micronutrient availability in plant growth. These increases in nutritional availability have already been shown in the scientific literature (Manjunath et al. 2016).

According to Manjunath et al. (2016), the application of cyanobacterial-bacterial biofilms and consortia in okra plants resulted in an increase in zinc and iron content in the soil, as well as root growth and weight. Microalgae and cyanobacteria can also be employed as a biological supply for high-yielding plants, but their concentration must be high enough to be successful (Renuka et al. 2018). Extracellular polymeric compounds are usually released by these insects, which increase the amount of organic carbon in the soil and promote the growth of bacteria. Finally, changing species and pastures can change the amount of microalgal and cyanobacterial biomass in the agricultural sector, which increases soil carbon content.

The use of microalgae and/or cyanobacteria to raise soil carbon content has been documented previously in the literature (Karthikeyan et al. 2007). The updated consortium boosted the soil's organic carbon content while also increasing the availability of micronutrients (especially zinc, iron, copper, and manganese). As a result of the enhanced soil condition, grain yields have increased and production quality has improved, as they now have grain with a higher nutritional content.

3.5 Physical and Chemical Amendments to the Soil

Microalgae and cyanobacteria are particularly significant in soil damage restoration because they can (i) change the pH of the soil; (ii) regulate salt levels in the soil or provide crop protection in high-salt locations; and (iii) remove heavy metals and other contaminants from the ground (Sharma et al. 2012). Several studies have demonstrated that using microalgal/cyanobacterial biomass to control pH causes an increase or reduction in soil pH. However, it is critical to emphasize the importance of this biomass in managing the acidic/basic state of various soils (Chatterjee et al. 2017). In terms of soil salinity, multiple studies have found that plants grow better in soil with high salinity (Sharma et al. 2012). This progress could be attributable to (i) better plant tolerance to high salt levels or (ii) decreased soil salinity (due to improved soil water storage capacity) (which can be achieved with gibberellic acid, phytohormone commonly found in microalgae and cyanobacteria).

Microalgae and cyanobacteria can detect elements like iron, zinc, copper, and manganese in the environment; heavy metals like cadmium, lead, and chromium; and hydrocarbons and other chemicals (Hakimi 2020). Without delivering vital nutrients or regenerating the soil structure, microalgae and cyanobacteria can increase soil properties such as concentration, porosity, penetration, air permeability, and moisture (Sharma et al. 2012; Kumar et al. 2015a, b). Polysaccharides generated by many microalgal/cyanobacterial species are usually linked to the development of these features (Sharma et al. 2012). When these polysaccharides are removed from the soil, they generally form a sticky coating and become gelatinous, which adds to soil compaction and erosion prevention (Baweja et al. 2019). These polysaccharides are also critical for maintaining soil permeability and water retention capacity, which is necessary for controlling soil temperature, pH, and salinity (Sharma et al. 2012; Chatzissavvidis and Therios 2014).

3.6 Plant Protection

Protection from Biotic Substances

The presence of insects, nematodes, bacteria, and fungi can have a negative impact on plant output (Hamed et al. 2018). Polysaccharides, which are recognized as reflecting compounds present in the cell walls of pathogenic organisms and create multiple protective reactions, are extensively used to guard against these harmful insects (Chanda et al. 2019). Common immunological responses that lead to the formation of secondary metabolites with antimicrobial and fungicidal qualities include signalling genetic expression and stimulation of particular biosynthetic pathways (e.g., phenolic compounds, terpenoids, and other substances) (Renuka et al. 2018). Because these insects are a key source of polysaccharides, microalgae and cyanobacteria may enhance these immune systems when used on agricultural plants (Pardee et al. 2004).

Farid et al. (2018) have demonstrated that algal polysaccharide activity may play a role in the discovery of plant defence mechanisms. The authors conclude that (i) polysaccharides extracted from *C. vulgaris ne-C. sorokiniana* stimulated 1,3-endoglucanase activity in tomato plants; (ii) polysaccharides extracted from *C. sorokiniana* increased phenylalanine ammonia-lyase activity; and (iii) polysaccharides extracted from *C. vulgaris no-C. reinhardtii* causes increased peroxidase activity. Several metabolites with antioxidant, antibacterial, and cytotoxic activities are typically found in microalgal/cyanobacterial biomass, despite the fact that numerous plant protection strategies can be developed with microalgal/cyanobacterial polysaccharides (Hamed et al. 2018). Methanol extracts from the brown sea *Sargassum wightii* have been proven to be effective against *Xanthomonas oryzae*, a frequent bacterial blight of rice, according to Arunkumar et al. (2005).

Abiotic Factors Defence

Plants subjected to severe environmental circumstances (such as extreme cold and humidity, excessive salinity, and drought) can have a negative impact on agricultural yields and biological development (Bulgari et al. 2019). Because the phospholipidic layer that generates these pores is fragile, it may interfere with the metabolic activity of higher plants and damage cell membranes at very high temperatures. High temperatures can damage cell membranes and alter plant protein and enzyme synthesis and function (Bulgari et al. 2019). Osmotic stress is caused by high levels of salt in plants, which impacts plant growth (salted plants often have lower chlorophyll content than plants that are not stressed) and, as a result, plant metabolism and growth (Bulgari et al. 2019).

Plants are also subjected to osmotic stress because of drought. However, it has an impact on the plants' ability to exchange gas, which has a significant impact on photosynthetic and respiratory levels, as well as plant yields (Bulgari et al. 2019). Although biochemical compounds have been shown to stimulate plant reactions and metabolic processes, the use of secondary metabolites produced by microalgae and cyanobacteria has annoyed the interest of researchers all over the world due to their versatility and potential to attract a wide range of plant reactions and metabolic processes. Algal-based chemicals have been employed as biostimulants in food plants successfully under a variety of circumstances, including temperature, salt, and drought (Haggag et al. 2018). In drought-tolerant bean plants, using *Ulva rigida* and *Fucus spiralis* seaweed promotes drought tolerance and growth (Mansori et al. 2014).

Promoting Direct Growth

Apart from their beneficial effects on soil quality and crop protection, microalgae and cyanobacteria (and/or their extracts) can directly boost plant growth and development by improving plant levels and characteristics (e.g., increased shoots and root lengths, increased leaf area, and higher nutrient content). This is handled by the activity of microalgal/cyanobacterial metabolites, which can stimulate a variety of metabolic activities such as respiration, photosynthesis, nucleic acid synthesis, chlorophyll production, and ion uptake (Górka et al. 2018).

Anabaena isolation treatment of rice fields increased seed germination and the growth of plant structures such as plant height, root length, new and dried leaf weights, stems, and roots (Saadatnia and Riahi 2009). There were also improvements in soil structure, such as humidity and porosity. Elarroussia et al. (2016) evaluated the effect of using a polysaccharide extract obtained from *S. platensis* in tomato and pepper culture and discovered that both tomato and pepper plants grew larger. Root weight, size, and number of nodes all rose in the treatment plants.

Improved Tolerance Against Abiotic Depression
Using algae-derived compounds could aid in improving abiotic stress resistance. Rice stems and dry weight are protected against salt stress (5 g/ml NaCl) by extracellular products of the cyanobacterium *Scytonema hofmanni* (Rodríguez et al. 2006). Under high salt pressure, tomato plants sprayed with 1% *Dunaliella salina* hydrolysate had higher dry shoot weight, dry root weight, and chlorophyll a and b concentrations than untreated plants (Arroussi et al. 2018). When fava beans are grown, their drought resistance improves as well. When fava bean plants are treated with *C. vulgaris*, they become more drought resistant (Li et al. 2014). Plant tolerance of abiotic stress caused by microalgae treatment is strongly connected to the generation of active oxygen (ROS) and antioxidant activity (Li et al. 2014; Arroussi et al. 2018).

By increasing the frequency of stomata closure and fluid retention, traditional *C. vulgaris* therapy decreases respiration and enhances drought tolerance in bean plants (Li et al. 2014). However, more research into the precise molecular process and algal choice will be required to increase the plant's abiotic pressure tolerance.

The Impact of Seaweed Quotes on Plant Tolerance to Biotic Pressures
The rise of infectious and disease-resistant pests in giant plants has increased as a result of changing climate and excessive use of chemical pesticides, drastically lowering agriculture's effects (Ayliffe and Lagudah 2004; Anderson et al. 2004). Seaweed extracts have been demonstrated to reduce nematode infestation in plants such as *Arabidopsis thaliana* (Beckett and van Staden 1989; Crouch and van Staden 1993), sunflowers (Vernieri et al. 2006), and tomatoes (Beckett and van Staden 1989; Crouch and van Staden 1993; Mancuso et al. 2006). However, because seaweed extracts did not exhibit distinct nematocidal capabilities, this nematocidal activity is essentially part of the plant's immune response, probably with cytokinin: a modulation of auxin dose (De Waele et al. 1988). Furthermore, extracts of *Sargassum wightii* and *Padina pavonica* have demonstrated significant antibacterial activity against the red cotton spot (*Dysdercusc ingulatus*), a serious disease that affects cotton plants (Asaraja and Sahayaraj 2013). In the combination treatment with *Sargassum* spp., *A. nodosum*, and *Laminaria* spp., inflammation of the greenfly aphid (*Aphis gossypii*) and serpentine leafminer (*Liriomy zatrifolii*) dramatically reduced cotton (Gençsoylu 2016). In addition, extracts from marine plants have been able to significantly reduce the effects of borer, aphids, and thrips on sugarcane, thus preventing significant economic losses (Carolina and Helena 2019).

Seaweed extracts have been demonstrated to be effective in reducing the severity of viroid and plant virus symptoms (Vera et al. 2011). When previously performed with carrageenan polysaccharide from seaweed, the symptoms of tomato chlorotic dwarf viroid were considerably reduced. Similarly, when tobacco is treated with sulphated galactans, a prominent component of marine chemicals, the severity of the tobacco mosaic virus (TMV) is considerably reduced (Vera et al. 2011).

The Impact of Seaweed Quotes on Crop Tolerance in Abiotic and Environmental Depression
Surprisingly, seaweed-treated plants including *A. nodosum* and *Sargassum* spp. were able to survive the negative impacts of these living stressors. *Kappaphycus alvarezzi* produces treatment with different wheat under salt and drought pressure, causing plants with increased root lengths, improved chlorophyll content and carotenoids, and a water content of tissues. Exfoliation also reduced electrolyte leakage and lipid peroxidation, decreased the Na^+/K^+ ratio, and raised Ca content, all of which reduced ionic variability. Furthermore, osmoprotectants such as proline, amino acids, and complete proteins have accumulated in portable wheat plants (Patel et al. 2018). Seaweed plants also promote frost tolerance in barley (Dalal et al. 2019) and *A. thaliana* (Ganesan et al. 2015) for the growth of winter hardiness when treated with sprays that release seaweed. The extraction of marine vegetation that has reduced the effects of severe droughts, colds, and salts has been shown to mediate the development of root morphology, the formation of non-structural carbohydrates that improve energy retention, metabolic stability, and water repair, as well (Dalal et al. 2019).

3.7 Ecological Balance

Soil algae are a type of colony that does not rely on the organic matter (carbon) in the soil. They produce a lot of secretory polysaccharides that help with topsoil and nitrogen fixation (Goyal 1997). The most significant non-aquatic organisms in microalgae are soil habitats, where these insects play a role in soil formation and sustainability (Zenova et al. 1995; Metting and Rayburn 1979). Microalgae also contribute to the energy and matter sources (Kuzyakhmetov 1998). Green and blue microalgae in the top-soil can form important resources for soil ecosystems (Starks et al. 1981) and agriculture (Ruble and Davis 1988). The generation of organic matter from non-living elements is one of microalgae's major advantages (Alexander 1977). Microalgae create active organic substances such as enzymes and ions that can affect other portions of soil communities, including plants, in addition to supplying food for other microbes, nematodes, and animals (Zenova et al. 1995).

Functional roles for algae/cyanobacteria diversity: microarthropods' food sources as a role model. There is active interaction between biodiversity and algae consumers because algae and cyanobacteria are important producers in the world and play various roles in maintaining soil fertility and composition. According to Neustupa and Skaloud (2010) diversity is most prevalent in open areas where light is a major influence, but observed by Ray and Binoy Thomas (2012) canopy cover composed of high vegetation and natural and chemical properties of the affected soil formation of algal value in forest area. The positive effects of nutrients, moisture, and soil compaction have been observed in the growth of soil algae (El-Gamal et al. 2008). Algae of temperate soils are well explored (El-Gamal et al. 2008; Mansour and Shaaban 2010) and those of some tropical soils are also being tested more recently. In tropical soils, Řeháková et al. (2011) observed the influence of altitude and vegetation on small algal communities in Ladakh, while the collection of Cyanobacteria in the Thar desert soil as an activity of agroecological conditions was reported by Bhatnagar et al. (2008). Tropical wet soils are rich in algae, unlike cool soils where algae diversity is low (Ray and Binoy Thomas 2012).

Research into the major biogeographic distribution patterns of living communities is important in classifying microbial diversity. Previous biogeographical studies have focused on rivers (Shi et al. 2020) and plants in wastewater treatment plants (Zhang et al. 2019) as research sites to determine microbial diversity and diversity in communities. Recently, there has been a growing need for more information on lake ecosystems in biogeography—a method that studies the formation of microbial communities in the best possible way (Liu et al. 2019). Many evidence suggests that microbial biogeography research of urban lakes is important in revealing the biological process.

Despite the fact that many papers and textbooks acknowledge the significant role of benthos in the overall production of the lake ecosystem, research into the causes of increasing lake production in the late twentieth and early twenty-first centuries focuses more on phytoplankton than on benthic algal assemblages (Wetzel 1996; Winder et al. 2009; Yvon-Durocher et al. 2010). However, there are growing reports of benthic algal blooms in historically oligotrophic lakes around the world (Vadeboncoeur et al. 2020). Despite their relevance, these benthic compounds have received little research worldwide, making it difficult to discern between supporting mechanisms that interact with other ecosystems (Vadeboncoeur et al. 2002).

Although algae are typically thought to live only in "free" water, they have long been a source of essential biodiversity above ground (John et al. 2011; Foets et al. 2020). When compared to pure autotrophic conditions, most green algae may grow heterotrophically in the dark while utilizing an external source of organic carbon (Fan et al. 2012; Bell 2013). The photosynthetic and heterotrophic combination of algae colonies can compact soil from top to bottom. Algae are actively building the earth's carbon and nitrogen cycles in the soil (Elbert et al. 2012). Furthermore, the soil may contain algal cysts that are dormant or non-motile seeds (aplanospores), acting as a repository for organisms that require specific circumstances to thrive. This latter environment is difficult to assess because many microbial species are not cultured, and determining life cycles is highly difficult. Non-liquid water systems, particularly ice and icebergs, can be used by other species of algae (Lu et al. 2016; Hoham and Remias 2019).

Natural and chemical treatments for recycling alkaline soils, such as adding gypsum, sulphur, or over-irrigation, do not eradicate the soluble salt and sodium transformed into the soil. Singh (1961) suggested that cyanobacteria may replenish alkaline soils because they form a thick stratum on the earth's surface during the rainy and winter months. They can dissolve microbial nutrients (Fritsch 1945) and dissolve dissolved nodules of carbonates by the release of oxalic acid (Fritsch 1945; Singh 1961). Improvements have been made in soil compactness through reducing pH and increasing conductivity of saline in alkaline soils promoted by cyanobacteria (Kaushik and Subhashini, 1985).

In the past, even today, soil microbiologists and microbial ecologists use microbes to restore salt-affected soil to increase plant growth and production and improve soil fertility in sustainable agriculture (Jha et al. 1987; Jan et al. 2017). In a state of high salinity, it provides oxygen to the roots of rice plants. For the first time, cyanobacteria have been reported to be effective in rehabilitating saline-affected soils by producing a thick stratum on the surface of salty soils and preserving and improving organic matter, N, P, and K (Jan et al. 2017). Cyanobacteria have important features that can help regenerate soils, such as nitrogen fixation, extracellular polymeric production (EPS), and increased organic carbon content (Muñoz-Rojas et al. 2018; Abinandan et al. 2019). Cyanobacteria play an important role in the ecology of earth's ecosystems because they contribute to soil fertility through the production of organic carbon (Kabirov and Gaisina 2009) and because they facilitate the processing of atmospheric nitrogen (Pankratova 2006).

3.8 Biofertilizers

Biofertilizer is a natural fertilizer that increases the availability of nutrients in plants by containing microbial inoculants such as bacteria, algae, and fungi, either alone or in combination. Beneficial bacteria and fungi in biofertilizers improve soil chemical and biological properties, phosphate solutions, and agricultural yield (El-Habbasha et al. 2007). Biofertilizers, rather than chemical fertilizers, provide farmers with economic and environmental benefits such as soil health and fertility. Biofertilizers provide nutrients by utilizing natural processes such as nitrogen fixation, phosphorus dissolution, and the use of a combination of growth-promoting chemicals to promote plant growth. Chemical fertilizers and insecticides are predicted to be used less with biofertilizers. Bacteria in the biofertilizer (*Azotobacter*, blue-green algae, *Rhizobium*, *Azospirillum*) repair the soil's natural nutrient cycle and build up organic matter.

Biofertilizers are composed of microorganisms that assure enough nutrient delivery to host plants, as well as proper development and physiological management. Only those pesticides with specialized activities to increase plant growth and reproduction are employed in the preparation of biofertilizer, which are made from organic microorganisms. For plant production, the microbe transforms complicated nutrients into simple nutrients. If the biofertilizer is utilized correctly, plant yields can rise by 20–30%. Plants can also be protected from soil-borne illnesses by using biofertilizers. The necessity for biofertilizers has developed primarily due to two factors. First, since increasing fertilizer use leads to higher crop yields; second, because increased chemical fertilizer use damages soil texture and causes other environmental issues. As a result, using biofertilizers is both cost-effective and environmentally friendly.

Cyanobacteria as a Biofertilizer

Cyanobacteria are a rapidly expanding microbe that contributes to long-term agricultural development (Singh et al. 2016a, b; Singh et al. 2017). Cyanobacteria are useful in the production of affordable and environmentally friendly biofertilizers. They can help plants handle nitrogen deficiencies, improve soil penetration, and water retention capacity, as well as provide vitamin B12 (Paumann et al. 2005). The most effective nitrogen-fixing cyanobacteria are *Nostoc linkia*; *Anabaena variabilis*, *Aulosira fertilizer*, *Calothrix* sp., *Tolypothrix* sp., and *Scytonema* sp. are present in the rice field (Prasad and Prasad 2001). *Anabaena* and *Nostoc* are soil and rock-dwelling plants that fix up to 20–25 kg of nitrogen per hectare. *Anabaena* can produce up to 60 kg of nitrogen per hectare per season while also enriching the soil with organic matter (Moore 1969).

Cyanobacteria may grow, develop, and manufacture significant natural products without the need for a host. In the rice paddy field, the *Azolla-Anabaena* symbiosis of nitrogen fixation and nutrient enrichment is an example. They indicate cell wall lignin rupture and the release of phenolic chemicals, which has resulted in a large biological flood (Malliga et al. 1996a). According to Song et al. (2005), cyanobacteria serve an important role in maintaining and improving soil fertility, acting as a natural fertilizer for biofertilizers. Blue algae's main functions are to (a) create a hollow soil and manufacture sticky compounds. (b) Phytohormones (auxin, gibberellins, etc.), vitamins, and amino acids are released. (c) By virtue of their gel structure, they improve groundwater holding capacity (Roger and Reynaud 1982). (d) Increased soil biomass following death and decay (Saadatnia and Riahi 2009). (e) Salt depletion in the soil (Saadatnia and Riahi 2009). (f) Prevents weed growth (Saadatnia and Riahi 2009). (g) Soil phosphate acquisition by the release of organic acids (Wilson 2006). (h) Bioremediation (active absorption of heavy metals in the microbiological environment) (Ibraheem 2007a, b).

To support crop production, the use of chemical fertilizers has increased many barriers. Tons of chemical fertilizers are used annually to make the Nations dependent on food production but also to damage the environment and cause harmful effects on biodiversity. These fertilizers are easily accessed by rainwater, which affects living organisms that include the growth of permanent microorganisms. Excessive use of chemical fertilizers in agriculture is expensive and has various adverse effects on the soil. BGA has unique environmental properties and acts as an excellent substitute for chemical fertilizers. Biofertilizers can reduce environmental disturbances. There are costly, environmentally friendly costs, and when needed in bulk they can be produced on the farm itself. Another unifying point is that after 3–4 years of continuous use, there is no need for biofertilization because parental claims are sufficient for growth and replication. About NPK it was found to be better than farm compost and other types of compost.

Cyanobacteria have a significant function in soil fertility maintenance and formation, resulting in higher rice growth and production (Song et al. 2005): (1) the release of growth-promoting substances such as hormones (auxin, gibberellin), vitamins, and amino acids (Roger and Reynaud 1982; Rodríguez et al. 2006); (2) the enhancement of soil porosity by a group of cyanobacteria with a filamentous structure and the production of adhesive substances (Sahu et al. 2012); (3) increased water capture capacity through their gel structure (Roger and Reynaud 1982); (4) the increase in soil biomass as a result of death and decay (Saadatnia and Riahi 2009); (5) the depletion of salts in the soil (Al-Sherif et al. 2015); (6) the inhibition of weed growth (Al-Sherif et al. 2015); and (7) the increase in soil phosphate due to the release of organic acids (Al-Sherif et al. 2015; Wilson 2006). Many additional crops, including barley, oats, tomatoes, seedlings, cotton, sugarcane, corn, peppers, and lettuce, have been shown to benefit from cyanobacterial inoculation (Thajuddin and Subramanian 2005). According to Malliga et al. (1996b), *Anabaena azollae*, when employed as a biofertilizer, demonstrates lignolysis and releases phenolic chemicals, causing considerable bodily harm. This research examines the use of coir waste as a biofertilizer carrier for cyanobacteria. In the fields of biofertilizer, energy production, human nutrition, animal feed, polysaccharides, biochemistry, medicine, and environmental change, cyanobacteria serve a variety of fascinating roles. Cyanobacteria offer cheap nitrogen to plants without improving crop yields by fertilizing the soil. The "algalization" of BGA biofertilizer in rice helps to produce an agroecosystem setting that promotes economic efficiency in paddy cultivation while reducing energy-intensive inputs. There is currently insufficient data to distinguish the use of small algae as a biofertilizer, yet reports suggest small algae could be a viable alternative to commercial or organic fertilizers. Biofertilizer is used in sustainable farming to boost the nutrient content of the soil, resulting in increased productivity. Rice production has been shown to be affected by nitrogen deficit in numerous ways, including slower growth and development, decreased crop production, leaf transplantation, and poorer yields (Lian et al. 2005). The investment strategy for most rehabilitation programmes is less than 50% (Fageria 2013). Furthermore, due to variations, leaching, denitrification, and soil erosion, a significant amount of nitrogen is lost (Fageria et al. 2015). Algae, which can be found practically everywhere on the planet, is one of the most visible organisms on the planet and can be used in nutrition as food additives, agriculture as biofertilizers, soil preparation, wastewater purification, and as a source of biofuel.

Marine Plants as Biofertilizers

Seaweeds have been utilized as a soil supplement to boost agricultural output in coastal locations and repair alkaline soils, where nutrient deficits have been common, since ancient times (Craigie 2011). Seaweed was fertilized with grass, peat, or other natural waste and combined with sand or gravel (Craigie 2011). Seaweed is widely used in agriculture because of its purported benefits as bioactive organisms and biological and organic fertilizers (Nelson and Van Staden 1984). The utilization of micronutrients from various oceans has recently resulted in an increase in plant biomass, resulting in high levels of zinc (Tuhy et al. 2015). A new generation of seaweed liquid fertilizer has been developed as a natural organic fertilizer rich in nutrients, with benefits such as faster seed formation, higher crop yields, and greater resistance to multi-crop pathogens. Sathya et al. (2010), Hurtado et al. (2009), Srijaya et al. (2010), Shahbazi et al. (2015), and Ciepiela et al. (2016) have effectively employed liquid manure obtained from seaweed extraction, initially introduced by Milton in 1952, as a fertilizer in horticulture and agriculture.

Macroalgae are employed as soil stabilizers as well as biofertilizers (Bhardwaj et al. 2014; Arioli et al. 2015). The new kelp (*Macrocystis integrifolia*) has a very positive effect on soil, according to Temple and Bomke (1988), with good plant growth and nutrient response. *Fucus*, *Laminaria*, *Ascophyllum*, *Sargassum*, and other seaweed fertilizers are perishable, non-toxic, non-polluting, and safe for humans, farm animals, and birds (Dhargalkar and Pereira 2005). Macroalgae mixes create significant quantities of organisms and various nutrients, according to Nedzarek and Rakusa-Suszczewski (2004), particularly those rich in NH_4^+, NO_3^-, NO_2^-, and phosphate. In addition, foliar infusion promoted okra growth (Abbasi et al. 2010), while the release of *Sargassum myriocystum* (Kalaivanan and Venkatesalu 2012) stimulated *Vigna mungo* growth including amylase activity in barley (Rayorath et al. 2008a, b). *Ulva fasciata* helped in seed germination, growth restrictions, pigment and carbohydrate content of wheat variety (Shahbazi et al. 2015), rice, and maize growth using seaweed milk (Singh et al. 2015a, b, 2016a, b), and root development and sprout length with increasing leaf number (Singh et al. 2015a, b, 2016a, b). *Vigna* sp. has recently been promoted to full development utilizing various macroalgae as biofertilizers (Reddy et al. 2016). Binumol (2016) and Gopalakrishnan (2016) were discovered. The regulation of phytohormone biosynthesis and accumulation in Arabidopsis has been affected by a commercial extract from *Ascophyllum nodosom* (Wally et al. 2012). In addition, extracts from the brown macroalga *Aqcophylum nodosum* were used to induce cytokinin-like activity in *Arabidopis thaliana* (Khan et al. 2011). Many additional seaweeds were investigated as biofertilizer sources (Zamani et al. 2013; Jayasinghe et al. 2016).

3.9 Bioremediation

Pollution issues have gotten worse because of rapid population development, urbanization, and industrialization, and they're wreaking havoc on ecosystems (Koul and Taak 2018). Every year, hundreds of thousands of tonnes of waste (solid and liquid) are produced all over the world. By 2050, global garbage production is anticipated to increase from 2.01 billion tonnes to 3.40 billion tonnes. Only a small portion is recycled, while the remainder is thrown away or left untreated, posing health and environmental risks to people and the environment (soil, water, air, heat, visual pollution, and so on). As a result, it's critical to find long-term, low-cost medicines that require minimum infrastructure and inputs.

The use of algae (micro or macroalgae) to remove or pollute biotransforms such as heavy metals, fertilizers, and other poisons (including xenobiotics) appears to be a continuous process (Upadhyay et al. 2018). Algae are a major source of carbon dioxide and can help to reduce carbon emissions (pollution caused by carbon dioxide, global warming, and the consequences of global warming) (Ding et al. 1995). Microalgae are considered productive in treating various types of domestic and industrial wastewater due to: (i) low demand, (ii) low energy requirement, (iii) low residual biodiversity, (iv) mud formation is relatively low, and aerobic medical technology, (v) effective extraction of heavy metals (HMs) and emerging organic contaminants (EOCs), (vi) use of contaminated water for growing microalgae can significantly reduce the need for clean water (90%), and increase the cost of nutrients (Gupta et al. 2019).

Phosphorus, nitrogen, ammonium, and heavy metals can all be released by microalgae cells when they are exposed to contaminated water (Abdel-Raouf et al. 2012). In microalgae cultivation, the creation and formation of growth media has a substantial impact on operational expenses and is a major issue for design, according to the economic cycle of the life cycle (Lam and Lee 2011). As a result, it's important to hunt for low-cost sources of nutrients, and wastewater having the correct nutrients can be a good option. This lowers the cost of producing microalgae biomass for biofuels, animal feed, and essential oils. At the same time, microalgae biomass produced following heavy metal treatment in polluted water can be used as a potential stock for charcoal, coal, biofuel, and biogas production (Poo et al. 2018).

Using molecular and functional genomic techniques, researchers have evolved algal species bioremediation of wastewaters by improving photosynthetic efficiency, flexibility (resistance to extreme weather), and the ability to release toxins (Lutzu et al. 2020). Phycoremediation offers various advantages over other physiochemical techniques because to its improved nutrient dose (reverse osmosis, exchange ion and electro-dialysis, dialysis, membrane-mediated separation, carbon-mediated activation adsorption, and chemical oxidation or reduction) (Hussain et al. 2018, 2021). Lower operational costs and algal biomass compatibility with N and P, which eliminates the need to manage mud and remove toxins from the water before they penetrate into the water, are just a few of the benefits. This is a natural process since algal biomass can be reused as a fertilizer after nutrients have been removed, resulting in no secondary pollution (Liu et al. 2016).

As a result, heavy-metal water treatment has sparked the curiosity of scientists, biologists, and politicians all around the world. As a result, to protect the environment and public health, such toxins must be removed from industrial effluent and aquatic ecosystems in the future. Heavy metals have been removed using a variety of processes, including adsorption, nanofiltration, ultrafiltration, reversible osmosis, electrochemical treatment methods, and biological therapies (Kumar et al. 2015b).

The blue-green algae and green algae are still being studied for their heavy metal potential in microalgae. Meanwhile, while there is little data on diatoms' ability to absorb heavy metals, diatoms are abundant and readily available in aquatic habitats. Heavy metals gathered by microalgal cells will play a role in cell carbon metabolism. The synthetic carbon molecules could make biofuels (Piccini et al. 2019). The presence of metals like Fe boosts the yield of biodiesel generation in microalgae. It encourages the formation of carbon molecules in tiny cells, which are required for microalgae-based biodiesel generation (Piccini et al. 2019).

As a result, growing algae in dirty water has the added benefit of decreasing heat emissions, treating wastewater, and producing algal biomass all at the same time. This biomass can be used to provide protein and dietary supplements (animal and human feed), bioenergy resources (biogas and biofuels), bio-ore for heavy metals, medicines, cosmetics, and other essential components (Pittman et al. 2011; Sahu et al. 2013). Nutrient removal strategies, such as phosphorus or nitrogen removal, can promote eutrophication in surface water. To remove tiny particles in minor activities, techniques such as soil erosion, filtration, and pond storage are used. To remove solvents and oxidize organic material in wastewater, multiple primary and secondary treatment plants have been installed in various places.

The potential for microbial growth in the circular economy is expanding, especially if the biomass generated can be used and commercialized in a few industries (e.g., food, textiles, pharmaceuticals, bioplastic, biofuel, biofertilizer, and cosmetics) (Fazal et al. 2018). Harvested microalgae biomass can create neutral fats, high-value pigs, fish and animal feed, biofertilizers, bioplastics, and carbon dioxide cyclically by incorporating the concept of biorefinery.

The concept of a waste-free economy is depends on cost-effective process efficiency. However, the cost of efficacy trading remains an issue, and it is critical to entice investors to higher-level manufacturing, thus technology is continuously being developed through R&D (Enamala et al. 2018). The current study focuses on four strategic planning areas: feedstock, process, product, and market acquisition (Koreiviene and Paskauskas 2019). Bioremediation approaches tend to be more competitive and advantageous than other conventional physicochemical treatments since they are effective, ecologically benign, and affordable (Chibueze et al. 2016).

Phycoremediation can take many forms, depending on the HM and its speciation, the type of microalgae utilized (living or non-living), and the functional condition of the microalgae (Gonzalez-dkila 1995). Living microalgae's ability to extract and eliminate HM toxins is the product of evolutionary mechanisms that have evolved over centuries (Monteiro et al. 2012). Because of the enormous demand for energy and the high operational costs of harvesting systems, ways to simplify this process were researched (Pal et al. 2019). As a result, novel microalgae cultivation methods have emerged and been integrated into wastewater treatment, such as the way of avoiding alginate beads and the construction of bioreactors employing microalgae biofilm (Gao et al. 2019). In addition, the biological interaction between microalgae and bacteria or fungi, as well as that microalgae has the ability to float naturally under unfavourable conditions, are being studied in order to enhance microalgae biomass output (Siddeeg et al. 2020) and reduce yields (Lavrinovics and Juhna 2018). Phycoremediation processes will be able to be applied on a wider scale with more study into genetic engineering, pre-treatment, energy prevention measures, and a combination of other physicochemical techniques.

Algae are also used as bioindicators, which maintain a constant eye on pollution levels and assist limit pollutants within the area's bounds (García-Seoane et al. 2018). Algal-based biofilms have been discovered as a viable alternative to increase algae biomass output. In addition, algae have been employed as an anode in MFCs to create power and energy (Shukla and Kumar 2018). Although there are great opportunities for directing the efficient use of photosynthetic algae in the control of waste and productive products in various areas, this approach poses a number of challenges, including the environment required for algae species to grow and the time required to dispose of waste in a polluted environment (Vassilev and Vassileva 2016). To deploy phycoremediation technology successfully in the future, we must first endeavour to reduce the quantity of land, energy, nutrients, and water consumed.

Technology

1. Algae can solve multiple problems at once, which is difficult with traditional chemical treatment. Because, for example, algae can be used for bioremediation, and biomass can be used to address a power problem and generate a new bioactive product in an environmentally acceptable way.
2. Algae can be employed based on the requirements of the hour. In some cases, an algae-based photobioreactor system can be used in batch mode, semi-continuous mode, or continuous mode.
3. Algal biomass contains the most important commercial products in the form of secondary metabolites such as astaxanthin, β-carotene, and eicosahexanoic acid.
4. Algal technology is a reliable and non-disruptive method, especially when used in conjunction with other industrial processes. An algae-based system requires adequate nourishment, water, and land.
5. Phycoremediation technology is a cost-effective and practical way of removing dangerous chemicals and energy from the environment.
6. The ability of algae to absorb CO_2 is the most likely answer to the global warming problem. Exhaust oxygen can assist with environmental clean-up.
7. Phycoremediation technology can handle a wide range of quality and quantity of commodities delivered as wastewater, as well as varying levels of contaminants.
8. The functions of phytotechnology are quite basic, and their usefulness does not necessitate the use of expert professionals.
9. Because of its local friendliness, the method is easy and merely removes unwanted substances from the region of contamination without disrupting the environment.
10. Finally, phycoremediation technology can be further evaluated in order to uncover other benefits of this process so that it can be successfully applied in the industrial setting.

More research is needed to determine whether such a vast amount of algal biomass can be used in large-scale industrial applications. Testing algal-based innovations in the industry, as well as educational preparations for greater awareness and discussion among industries, researchers, and students, is required so that this process can be used in the most advantageous way for all human development while avoiding environmental harm.

3.10 Removal of Nutrients and Contaminants

The approach of purifying wastewater using microorganisms such as high-density algae or high-rate algal ponds (HRAPs) has garnered a lot of attention in recent years because of the finding of algal biomass, which can be utilized as fertilizer, protein, or biofuel, and high-quality pollutants (Craggs et al. 2012). HRAP is a shallow reactor that fosters a symbiotic relationship between microalgae and bacteria. In such environments, heterotrophic bacteria that consume the oxygen produced by microalgal photosynthesis cause harm to the living animal (Passos et al. 2014). Only a few studies have focused on the removal of organic pollutants such as phenolic compounds, surfactants, alloys, biocides, and polycyclic aromatic hydrocarbons, despite the fact that microalgae wastewater purification systems have been studied since the 1950s for their ability to remove nutrients, heavy metals, bacteria, and helminthic eggs (Abdel-Raouf et al. 2012). Indeed, the usefulness of HRAPs in addressing environmental concerns about emerging organic contaminants (EOCs) has gotten little attention.

The efficiency of HRAPs in removing EOCs from contaminated water has yet to be determined. Traditional mud-wired wastewater treatment facilities (WWTPs) have been extensively investigated for the removal of EOCs. There has only been one study of HRAPs' ability to remove tetracyclines, and it was done on a lab scale with contaminated water (de Godos et al. 2012). Other research on small-scale pollutants like PAHs, biocides (e.g., organotin chemicals), surfactants, and phenolic compounds suggests that microalgae-based wastewater treatments could remove both abiotic and biotic pollutants (sorption, volatile). Biotic process includes biodegradation, microalgae absorption, or metabolization and abiotic process emphasize on photodegradation activities (Abargues et al. 2013). Some recent investigations have shown the potential of some algae to biodegrade phenolic chemicals in both light and darkness.

Microalgae can trigger the breakdown of many complex pesticides by interacting with heterotrophic bacteria (the complexity of the pesticide depends on the chemical composition) (Chen and Wang 2020). Water pollution is one of the most serious issues the environment has ever faced (Jin et al. 2012). Green algae have been shown to minimize the number of pesticides such as prometryne, according to research (Kumar and Sharma 2013). Small algae production has four primary steps: (i) land selection, (ii) growing, (iii) harvesting, and (iv) extraction (Munoz and Guieysse 2006).

Aside from the pollution mentioned above, agricultural pesticides, green sewage, and urban and industrial pollution all contribute to water contamination (Barakat 2004). As a result, pesticide residues in water, plants, and grass may be ingested by predators and end up in the tissues. As a result, pesticide preparation, particularly bioremediation with microalgae, is critical. The metabolic deterioration of pesticides is influenced by chemical parameters such as cell mass, active groups, and toxins (Priyadarshani et al. 2011). Algae appear to be better at digesting low-molecular-weight chemical substances than bigger molecules (Heredia-Arroyo et al. 2011).

This offers the appearance that pesticides will be removed from dirty water using a filter made of dead algal biomass. Employing dead biomass instead of living biomass provides the advantages of product stability and no risk of harming the intended cells. Aksu and Dönmez (2006) found that dead biomass can be recycled. Biosorbed pollutants can be washed into algal biomass and safely treated (Friesen-Pankratz et al. 2003), and the biomass filter itself can then be reused (Smith and Pepich 2007).

4 Future Perspectives

Overall plant growth, such as seed germination, pigment production, and root growth, as well as nutritional content, is influenced by plant interactions with microbes. The plant-microbe symbiotic relationship allows the plant to thrive in a variety of stressful situations. The role of microbial interaction could aid in maintaining healthy soil conditions such as nitrogen fixation, salinity, ecological balance, soil erosion, and nutrient seepage. Soil algae aided in the breakdown of organic and inorganic contaminants. Soil algae, for example, can boost a plant's defence system by stimulating immunological responses and promoting antioxidant, fungicidal, and bactericidal properties. By boosting nitrogen availability to the plants, microorganisms such as cyanobacteria can be used as a replacement for commercial nitrogen fertilizers, lowering production costs by up to 25–40%.

As a critical parameter, plants' macro- and micronutrients as well as organic matter can be made more available by using microalgae and cyanobacteria in the soil. Microalgae and cyanobacteria, like photosynthetic microorganisms, take in nitrogen and phosphorus from various sources and store it in their biomass. Plant output is considerably improved by the development of these soil structures, which encourage root growth and increase microbial activity. As a result, microalgae and cyanobacteria have become popular for soil repair and fertility enhancement. Aside from pests and illnesses, soil algae-based biofertilizers may help plants withstand different environmental stresses such as drought, high temperatures, salinity, and cold climes, all of which can impair crop production.

Biogeography of algae communities can help us understand more about how algae communities form in various lake settings around the world. However, little is known about how the algal community reflects local diversity or how the local community influences the algal community's formation. We could solve the country's fertilizer storage problem if we employ blue-green algae (BGA) as an excellent fertilizer. Nitrogen is one of the most important fertilizers in the plant (Sainju et al. 2019). Although it makes up 70% of our universe, plants cannot use it directly from nature. BGA is a natural way of absorbing nutrients and increasing soil fertility.

5 Conclusions

An important aspect of recognizing soil algae-plant interaction is increasing crop productivity. Introducing new mechanisms could aid in the fine-tuning of soil fertility. Biofertilizers made from soil algae might replace chemically synthesized fertilizers while also preserving the natural state of the soil. Using soil algae as growth promoters, plants created secondary metabolites or plant defences such as carotenoids, alkaloids, and lipids. SAR is a whole-plant defence system that is influenced by microorganisms that live in a symbiotic relationship with plants. For better crop production, plant protection, and stress tolerance, further research into the mechanistic study of soil algae-microorganism-plant interaction is required. Microorganisms help the soil by regulating salinity and pH, and the availability of nutrients, as well as removing organic/inorganic contaminants.

References

Abargues MR, Ferrer J, Bouzas A, Seco A (2013) Removal and fate of endocrine disruptors chemicals under lab-scale postreatment stage. Removal assessment using light, oxygen and microalgae. Bioresour Technol 149:142–148

Abbasi FF, Baloch MA, Aia-ulhassan WKH, Shah AN, Rajpar I (2010) Growth and yield of okra under foliar application of some new multinutrient fertilizer products. Pak J Agric Eng Vet Sci 26:11–18

Abdel-Raouf N, Al-Homaidan AA, Ibraheem IBM (2012) Microalgae and wastewater treatment. Saudi J Biol Sci 19(3):257–275

Abinandan S, Subashchandrabose SR, Venkateswarlu K, Megharaj M (2019) Soil microalgae and cyanobacteria: the biotechnological potential in the maintenance of soil fertility and health. Crit Rev Biotechnol 39:981–998

Adams DG, Bergman B, Nierzwicki-Bauer SA, Duggan PS, Rai AN, Schüßler A (2013) Cyanobacterial–plant symbioses. In: Rosenberg E, DeLong EF, Stackebrandt E, Lory S, Thompson F (eds) The prokaryotes: prokaryotic biology and symbiotic associations. Springer, Berlin, pp 359–400

Aksu Z, Dönmez G (2006) Binary biosorption of cadmium (II) and nickel (II) onto dried Chlorella vulgaris: co-ion effect on mono-component isotherm parameters. Process Biochem 41:860–868

Alexander M (1977) Introduction to soil microbiology. Wiley, New York, p 467

Al-Sherif EA, Abd El-Hameed MS, Mahmoud HS, Ahmed (2015) Am-Eur J Agric Environ Sci 15:794e799

Alvarez AL, Weyers SL, Goemann HM, Peyton BM, Gardner RD (2021) Microalgae, soil and plants: a critical review of microalgae as renewable resources for agriculture. Algal Res 54:102200. https://doi.org/10.1016/j.algal.2021.102200

Anderson PK, Cunningham AA, Patel NG, Morales FJ, Epstein PR, Daszak P (2004) Emerging infectious diseases of plants: pathogen pollution, climate change and agrotechnology drivers. Trends Ecol Evol 19:535–544

Arioli T, Mattner S, Winberg P (2015) Applications of seaweed extracts in Australian agriculture: past, present and future. J Appl Phycol 27:2007–2015

Arroussi HE, Benhima R, Elbaouchi A, Sijilmassi B, Mernissi NE, Aafsar A (2018) Dunaliella salina exopolysaccharides: a promising biostimulant for salt stress tolerance in tomato (Solanum lycopersicum). J Appl Phycol 30:2929–2941

Arunkumar K, Selvapalam N, Rengasamy R (2005) The antibacterial compound sulphoglycerolipid 1–0 palmitoyl-3-0(60-sulpho--quinovopyranosyl)-glycerol from Sargassum wightii Greville (Phaeophyceae). Bot Mar 48:441–445

Asaraja A, Sahayaraj K (2013) Screening of insecticidal activity of Brown macroalgal extracts against Dysdercuscingulatus (Fab.)(Hemiptera: Pyrrhocoridae). J Biopest 6:193–203

Ayliffe MA, Lagudah ES (2004) Molecular genetics of disease resistance in cereals. Ann Bot 94:765–773. https://doi.org/10.1093/aob/mch207

Barakat AO (2004) Assessment of persistent toxic substances in the environment of Egypt. Environ Int 30:309–322

Baweja P, Kumar S, Kumar G (2019) Organic fertilizer from algae: a novel approach towards sustainable agriculture. In: Giri B, Prasad R, Wu Q-S, Varma A (eds) Biofertilizers for sustainable agriculture and environment. Springer, Cham, pp 353–370

Beckett RP, van Staden J (1989) The effect of seaweed concentrate on the growth and yield of potassium stressed wheat. Plant Soil 53:187–189

Bell G (2013) Experimental evolution of heterotrophy in a green alga. Evolution 67:468–476

Bergman B, Johansson C, Söderbäck E (1992) The *Nostoc–Gunnera* symbiosis. New Phytol 122:379–400

Bhardwaj D, Wahid Ansari M, Kumar RS, Tuteja N (2014) Biofertilizers function as key player in sustainable agriculture by improving soil fertility, plant tolerance and crop productivity. Microb Cell Factories 13:13–66

Bhatnagar A, Makandar MB, Garg MK, Bhatnagar M (2008) Community structure and diversity of cyanobacteria and green algae in the soils of Thar Desert (India). J Arid Environ 72:73–834

Bidyarani N, Prasanna R, Chawla G, Babu S, Singh R (2015) Deciphering the factors associated with the colonization of rice plants by cyanobacteria. J Basic Microbiol 55:407–419

Bohlen PJ, Groffman PM, Driscoll CT, Fahey TJ, Siccama TG (2001) Plant-soil-microbial interactions in a northern hardwood forest. Ecology 82:965–978

Bulgari R, Franzoni G, Ferrante A (2019) Biostimulants application in horticultural crops under abiotic stress conditions. Agronomy 9:306

Campobenedetto C (2021) The application of a biostimulant based on tannins affects root architecture and improves tolerance to salinity in tomato plants. Sci Rep Nat Publ Group UK 11:354

Carolina FVA, Helena GCL (2019) Biostimulants and their role in improving plant growth under abiotic stresses. In: Biostimulants in plant science. IntechOpen, London

Carvajal-Muñoz J, Carmona-Garcia C (2012) Benefits and limitations of biofertilization in agricultural practices. Livest Res Rural Dev 24:1–8

Chan YT, Kennedy IR (1989) IR possible N_2-fixing root nodules induced in non-legumes. Agric Sci 2:57–59

Chanda MJ, Merghoub N, Hicham EA (2019) Microalgae polysaccharides: the new sustainable bioactive products for the development of plant bio-stimulants? World J Microbiol Biotechnol 35:177

Chang ACG, Jun D, Li N (2019) Perspectives on endosymbiosis in coralloid roots: association of cycads and cyanobacteria. Front Microbiol 10:1888

Chatterjee A, Singh S, Agrawal C, Yadav S, Rai R, Rai L (2017) Role of algae as a biofertilizer. In: Algal green chemistry. Elsevier BV, Amsterdam, pp 189–200

Chatzissavvidis C, Therios I (2014) Role of algae in agriculture. In: Pomin VH (ed) Seaweeds: agricultural uses, biological and antioxidant agents. Nova Science Publishers, Hauppauge, pp 1–37

Chen H, Wang Q (2020) Microalgae-based nitrogen bioremediation. Algal Res 46:101775. https://doi.org/10.1016/j.algal.2019.101775

Chibueze C, Chioma A, Chikere B (2016) Bioremediation techniques—classification based on site of application: principles, advantages, limitations and prospects

Chittora D, Meena M, Barupal T, Swapnil P, Sharma K (2020) Cyanobacteria as a source of biofertilizers for sustainable agriculture. Biochem Biophys Rep 22:100737

Ciepiela GA, Godlewska A, Jankowska J (2016) The effect of seaweed Ecklonia maxima extract and mineral nitrogen on fodder grass chemical composition. Environ Sci Pollut Res 23:2301–2307

Cirri E, Pohnert G (2019) Algae−bacteria interactions that balance the planktonic microbiome. New Phytol 223:100–106. https://doi.org/10.1111/nph.15765

Craggs R, Sutherland D, Campbell H (2012) Hectare-scale demonstration of high rate algal ponds for enhanced wastewater treatment and biofuel production. J Appl Phycol 24:329–337

Craigie JS (2011) Seaweed extract stimuli in plant science and agriculture. J Appl Phycol 23:371–339

Crouch IJ, van Staden J (1993) Evidence for the presence of plant growth regulators in commercial seaweed products. Plant Growth Regul 6:345–388

Dalal A, Bourstein R, Haish N, Shenhar I, Wallach R, Moshelion M (2019) Dynamic physiological phenotyping of drought-stressed pepper plants treated with "productivity-enhancing" and "survivability-enhancing" biostimulants. Front Plant Sci 10:905. https://doi.org/10.3389/fpls.2019.00905

de Godos I, Muñóz R, Guieysse B (2012) Tetracycline removal during wastewater treatment in high-rate algal ponds. J Hazard Mater 229-230:229–230

De Waele D, Mcdonald H, De Waele E (1988) Influence of seaweed concentrate on the reproduction of Pratylenchus zeae (Nematoda) on maize. Nematologica 34:71–77

Dhargalkar VK, Pereira N (2005) Seaweed: promising plant of the millennium. Sci Cult 71:60–66

Dias GA, Rocha RHC, Araújo JL, De Lima JF, Guedes WA (2016) Growth, yield, and postharvest quality in eggplant produced under different foliar fertiliser (Spirulina platensis) treatments. Semin Cienc Agrar 37:3893–3902

Ding GT, Mohd Yasin NH, Takriff MS, Kamarudin KF, Salihon J, Yaakob Z, MohdGonzalez-dkila M (1995) The role of phytoplankton cells on the control of heavy metal concentration in seawater. Mar Chem 48:215–236

Duan X, Ren GY, Liu LL, Zhu WX (2012) Salt-induced osmotic stress for lipid overproduction in batch culture of Chlorella vulgaris. Afr J Biotechnol 11:7072–7078

El Boukhari MEM (2020) Trends in seaweed extract based biostimulants: manufacturing process and beneficial effect on soil-plant systems. Plan Theory 9:359

Elarroussia H, Elmernissia N, Benhimaa R, El Kadmiria IM, Bendaou N, Smouni A, Wahbya I (2016) Microalgae polysaccharides a promising plant growth biostimulant. J Algal Biomass Util 7:55–63

Elbert W, Weber B, Burrows S, Steinkamp J, Budel B, Andreae MO (2012) Contribution of cryptogamic covers to the global cycles of carbon and nitrogen. Nat Geosci 5:459–462

El-Gamal AD, Ghanem NAE, El-Ayouty EY, Shehata EF (2008) Studies on soil algal flora in Kafr El-Sheikh governorate. Egypt Egypt J Phycol 9:1–23

El-Habbasha SF, Hozayn M, Khalafallah MA (2007) Integration effect between phosphorus levels and biofertilizers on quality and quantity yield of faba bean (*Vicia faba* L.) in newly cultivated sandy soils. Res J Agric Biol Sci 3(6):966–971

Enamala MK, Enamala S, Chavali M, Donepudi J, Yadavalli R, Kolapalli B, Aradhyula TV, Velpuri J, Kuppam C (2018) Production of biofuels from microalgae – a review on cultivation, harvesting, lipid extraction, and numerous applications of microalgae. Renew Sust Energ Rev 94:49–68

Fageria NK (2013) The role of plant roots in crop production. CRC Press, Boca Raton

Fageria NK, Baligar VC, Heinemann AB, Carvalho MBS (2015) Nitrogen uptake and use efficiency in rice. In: Rakshit A, Bahadur Singh H, Sen A (eds) Nutrient use efficiency: from basics to advances. Springer, India, p 285e296

Fan J, Huang J, Li Y, Han F, Wang J, Li X (2012) Sequential heterotrophy-dilution-photoinduction cultivation for efficient microalgal biomass and lipid production. Bioresour Technol 112:206–211

Farid R, Mutale-Joan C, Redouane B, Najib EM, Abderahime A, Laila S, Hicham EA (2018) Effect of microalgae polysaccharides on biochemical and metabolomics pathways related to plant defense in Solanum lycopersicum. Appl Biochem Biotechnol 188:225–240

Fazal T, Mushtaq A, Rehman F, Ullah Khan A, Rashid N, Farooq W, Rehman MSU, Xu J (2018) Bioremediation of textile wastewater and successive biodiesel production using microalgae. Renew Sust Energ Rev 82:3107–3126

Foets J, Wetzel CE, Teuling AJ, Pfister L (2020) Temporal and spatial variability of terrestrial diatoms at the catchment scale: controls on productivity and comparison with other soil algae. PeerJ 8:e9198

Friesen-Pankratz BB, Doebel CC, Farenhorst AA, Gordon Goldsborough L (2003) Interactions between algae (Selenastrum capricornutum) and pesticides: implications for managing constructed wetlands for pesticide removal. J Environ Sci Health B 38:147–155

Fritsch FE (1945) The structure and reproduction of algae–II. Cambridge University Press, London, pp 658–767

Galhano V, De Figueiredo DR, Alves A, Correia A, Pereira MJ, Gomes-Laranjo J, Peixoto F (2011) Morphological, biochemical and molecular characterization of Anabaena, Aphanizomenon and Nostoc strains (cyanobacteria, Nostocales) isolated from Portuguese freshwater habitats. Hydrobiologia 663:187–203

Ganesan S, Vadivel K, Jayaraman J (2015) Sustainable crop disease management using natural products. CABI, Wallingford

Gantar M, Kerby N, Rowell P (1993) Colonization of wheat (Triticum vulgare L.) by N2-fixing cyanobacteria: III. The role of a hormogonia-promoting factor. New Phytol 124:505–513

Gao F, Cui W, Xu JP, Li C, Jin WH, Yang HL (2019) Lipid accumulation properties of Chlorella vulgaris and Scenedesmus obliquus in membrane photobioreactor (MPBR) fed with secondary effluent from municipal wastewater treatment plant. Renew Energy 136:671–676

García-Seoane R, Fernández J, Villares R, Aboal J (2018) Use of macroalgae to biomonitor pollutants in coastal waters: optimization of the methodology. Ecol Indic 84:710–726

Gençsoylu I (2016) Effect of seaweeds and organic foliar fertilizers on the cotton pests, predators, yield and fiber quality in cotton. Adnan Menderes Üniv. Ziraat Fakültesi Derg

Goni O, Quille P, O'Connell S (2018) Ascophyllum nodosum extract biostimulants and their role in enhancing tolerance to drought stress in tomato plants. Plant Physiol Biochem 126:63–73

González A, Castro J, Vera J, Moenne A (2013) Seaweed oligosaccharides stimulate plant growth by enhancing carbon and nitrogen assimilation, basal metabolism, and cell division. J Plant Growth Regul 32:443–448

Gonzalez-dkila M (1995) The role of phytoplankton cells on the control of heavy metal concentration in seawater. Mar Chem 48:215–236

González-Garcinuño Á, Sánchez-Álvarez JM, Galán MA, Del Valle EMM (2016) Understanding and optimizing the addition of phytohormones in the culture of microalgae for lipid production. Biotechnol Prog 32:1203–1211

Gopalakrishnan CN, Binumol T (2016) Preliminary studies on the effect of bioactive substances of Hypnea musciformis (Wulf) Lamour on the growth of seedlings in green gram, Vigna radiata L. J Phytology 8:1–6

Górka B, Korzeniowska K, Lipok J, Wieczorek PP (2018) The biomass of algae and algal extracts in agricultural production. In: Algae biomass: characteristics and applications. Springer, Berlin, pp 103–114

Goyal SK (1997) Algae and the soil environment. Phykos 36:1–13

Guo S, Wang P, Wang X, Zou M, Liu C, Hao J (2020) Microalgae as biofertilizer in modern agriculture. In: Alam M, Xu JL, Wang Z (eds) Microalgae biotechnology for food, health and high value products. Springer, Singapore, pp 397–411

Gupta S, Pawar SB, Pandey R (2019) Current practices and challenges in using microalgae for treatment of nutrient rich wastewater from agro-based industries. Sci Total Environ 687:1107–1126

Haggag W, Hoballah M, Ali R (2018) Applications of nano biotechnological microalgae product for improve wheat productivity in semaiaird areas. Int J Agric Technol 14:675–692

Hakimi NIN (2020) Phycoremediation of palm oil mill effluent (POME) and CO_2 fixation by tool for heavy metal remediation. Ecotoxicol Environ Saf 113:329–352

Halpern M, Bar-Tal A, Ofek M, Minz D, Muller T, Yermiyahu U (2015) The use of biostimulants for enhancing nutrient uptake. Adv Agron 130:141–174

Hamed SM, El-Rhman AAA, Abdel-Raouf N, Ibraheem IB (2018) Role of marine macroalgae in plant protection & improvement for sustainable agriculture technology. Beni Suef Univ J Basic Appl Sci 7:104–110

Hamilton TL, Bryant DA, Macalady JL (2016) The role of biology in planetary evolution: cyanobacterial primary production in low-oxygen proterozoic oceans. Environ Microbiol 18:325–340

Han X, Zeng H, Bartocci P, Fantozzi F, Yan Y (2018) Phytohormones and effects on growth and metabolites of microalgae: a review. Fermentation 4:25

Heredia-Arroyo T, Wei W, Ruan R, Hu B (2011) Mixotrophic cultivation of Chlorella vulgaris and its potential application for the oil accumulation from non-sugar materials. Biomass Bioenergy 35:2245–2253

Hoham RW, Remias D (2019) Snow and glacial algae: a review. J Phycol 56:264–282

Hurtado AQ, Yunque DA, Tibudos K, Crithley AT (2009) Use of Acadian marine plant extract powder from Ascophyllum nodosumin tissue culture of Kappaphycus varieties. J Appl Phycol 21:633–639

Hussain A, Shah ST, Rahman H, Irshad M, Iqbal A (2015) Effect of IAA on in vitro growth and colonization of Nostoc in plant roots. Front Plant Sci 6:46

Hussain F, Shah SZ, Ahmad H, Abubshait SA, Abubshait HA, Laref A, Iqbal M, Koul B, Taak P (2018) Biotechnological strategies for effective remediation of polluted soils. Springer. https://doi.org/10.1007/978-981-13-2420-8

Hussain F, Shah SZ, Ahmad H, Abubshait SA, Abubshait HA, Laref A, Manikandan A, Kusuma HS, Iqbal M (2021) Microalgae an ecofriendly and sustainable wastewater treatment option: biomass application in biofuel and bio-fertilizer production. A review. Renew Sustain Energy Rev 137:110603. https://doi.org/10.1016/j.rser.2020.110603

Ibraheem IBM (2007a) Cyanobacteria as alternative biological conditioners for bioremediation in agriculture, ecosystem, and environmental sustainability. Front Microbiol 7:529

Ibraheem IBM (2007b) Cyanobacteria as alternative biological conditioners for bioremediation of barren soil. Egypt J Phycol 8:99–116

Innok S, Chunleuchanon S, Boonkerd N, Teaumroong N (2009) Cyanobacterial akinete induction and its application as biofertilizer for rice cultivation. Environ Boil Fishes 21:737–744

Islam M, Gan H, Ziemann M, Hussain H, Arioli T, Cahill D (2020) Phaeophyceaean (brown algal) extracts activate plant defense systems in Arabidopsis thaliana challenged with Phytophthora cinnamomi. Front Plant Sci 11:852

Jan Z, Ali S, Sultan T, Wasiullah, Ahmad W (2017) The role of cyanobacteria in availability of major plant nutrients and soil organic matter to rice crop under saline soil condition. Sarhad J Agric 33(4):566–572

Jannin L, Arkoun M, Etienne P, Laîné P, Goux D, Garnica M, Fuentes M, Francisco SS, Baigorri R, Cruz F, Houdusse F, Garcia-Mina JM, Yvin JC, Ourry A (2013) Brassica napus growth is promoted by Ascophyllum nodosum (L.) Le Jol. Seaweed extract: microarray analysis and physiological characterization of N, C, and S metabolisms. J Plant Growth Regul 32:31–52

Jayasinghe PS, Pahalawattaarachchi V, Ranaweera KKDS (2016) Fertilizer on plant growth of Capsicum annum. Discovery 52:723–734

Jha MN, Prasad AN (2006) Efficacy of new inexpensive cyanobacterial biofertilizer including its shelf-life. World J Microbiol Biotechnol 22:73–79

Jha MN, Venkataraman GS, Kaushik BD (1987) Response of Westiellopsis prolifica and Anabaena sp. to salt stress. Mircen J 3:99–103

Jin ZP, Luo K, Zhang S, Zheng Q, Yang H (2012) Bioaccumulation and catabolism of prometryne in green algae. Chemosphere 87:278–284. https://doi.org/10.1016/j.chemosphere.2011.12.071

Jithesh M, Shukla S, Kant P, Joshi J, Critchley T, Prithiviraj B (2019) Physiological and transcriptomics analyses reveal that Ascophyllum nodosum extracts induce salinity tolerance in

Arabidopsis by regulating the expression of stress responsive genes. J Plant Growth Regul 38:463–478

John D, Whitton B, Brook A (2011) The freshwater algal Flora of the British Isles. An identification guide to freshwater and terrestrial algae, 2nd edn. Cambridge University Press, Cambridge, MA

Ju C, Van De Poel B, Cooper ED, Thierer JH, Gibbons TR, Delwiche CF, Chang C (2015) Conservation of ethylene as a plant hormone over 450 million years of evolution. Nat Plants 1:14004

Kabirov RR, Gaisina LA (2009) Parameters of the productivity of soil algae in terrestrial ecosystems. Eurasian Soil Sci 42:1374–1379

Kalaivanan C, Venkatesalu V (2012) Utilization of seaweed Sargassum myriocystum extracts as a stimulant of seedlings of Vigna mungo (L.) Hepper. Span J Agric Res 10:466–470

Karthikeyan N (2006) Characterization of cyanobacteria from the rhizoshere of wheat. M.Sc thesis, Division of Microbiology, Post Graduate School, Indian Agricultural. Research Institute, New Delhi, India

Karthikeyan N, Prasanna R, Nain L, Kaushik BD (2007) Evaluating the potential of plant growth promoting cyanobacteria as inoculants for wheat. Eur J Soil Biol 43:23–30

Kaushik BD, Subhashini D (1985) Amelioration of salt affected soils with blue-green algae: improvements in soil properties. Proc Ind Natl Sci Acad 51:380–389

Khan K, David H, Alan TC, Balakrishnan P (2011) Bioassay to detect Ascophyllum nodosum extract-induced cytokinin-like activity in Arabidopsis thaliana. J Appl Phycol 23:409–414

Kim MJ, Shim CK, Kim YK, Jee HJ, Park JH, Hong SJ (2012) Effect of green algae, Chlorella fusca on improving seed germination and seedling growth in vegetable crop and minor cereal. Kor J Org Agric 2012:491–492

Kim MJ, Shim CK, Kim YK, Ko BG, Park JH, Hwang SG (2018) Effect of biostimulator Chlorella fusca on improving growth and qualities of Chinese chives and spinach in organic farm. Plant Pathol J 34:567

Koreiviene J, Paskauskas R (2019) EU project of LIFE programme 'algae service for LIFE' develops ecologicaly sustainable bioproducts from freshwater cyanobacteria and macroalgae biomass. Botanica 25:176–185

Koul B, Taak P (2018) Biotechnological strategies for effective remediation of polluted soils. Springer. https://doi.org/10.1007/978-981-13-2420-8

Kozlova TA, Hardy BP, Krishna P, Levin DB (2017) Effect of phytohormones on growth and accumulation of pigments and fatty acids in the microalgae Scenedesmus quadricauda. Algal Res 27:325–334

Krings M, Hass H, Kerp H, Taylor TN, Agerer R, Dotzler N (2009) Endophytic cyanobacteria in a 400-million-yr-old land plant: a scenario for the origin of a symbiosis? Rev Palaeobot Palynol 153:62–69

Kumar M, Sharma MP (2013) Production methodology of biodiesel from microalgae. Int J Appl Eng Res 8:1825–1832

Kumar K, Mella-Herrera RA, Golden JW (2010) Cyanobacterial heterocysts. Cold Spring Harb Perspect Biol 2:a000315

Kumar M, Prasanna R, Bidyarani N, Babu S, Mishra BK, Kumar A, Adak A, Jauhari S, Yadav K, Singh R (2013) Evaluating the plant growth promoting ability of thermotolerant bacteria and cyanobacteria and their interactions with seed spice crops. Sci Hortic 164:94–101

Kumar KS, Dahms HU, Won EJ, Lee JS, Shin KH (2015a) Microalgae – a promising locally isolated microalgae: *Chlorella sorokiniana* UKM2, *Coelastrella* sp. UKM4 and *Chlorella pyrenoidosa* UKM7. J Water Process Eng 35

Kumar KS, Dahms HU, Won EJ, Lee JS, Shin KH (2015b) Microalgae – a promising tool for heavy metal remediation. Ecotoxicol Environ Saf 113:329–352

Kumari R, Kaur I, Bhatnagar AK (2011) Effect of aqueous extract of Sargassum johnstonii Setchell & Gardner on growth, yield and quality of Lycopersicon esculentum Mill. J Appl Phycol 23:623–633

Kuzyakhmetov GG (1998) Algological evaluation of the toxicity of copper compounds in grey forest soil and leached chernozem. Eurasian Soil Sci 31(8):877–882

Lam MK, Lee KT (2011) Renewable and sustainable bioenergies production from palm oil mill effluent (POME): Win-win strategies toward better environmental protection. Biotechnol Adv 29(1):124–141

Lavrinovics A, Juhna T (2018) Review on challenges and limitations for algae-based waste water treatment. Constr Sci 20:17–25

Lee S-M, Ryu C-M (2021) Algae as new kids in the beneficial plant microbiome. Front Plant Sci 12:91. https://doi.org/10.3389/fpls.2021.599742

Li Y, Xu SS, Gao J, Pan S, Wang GX (2014) Chlorella induces stomatal closure via NADPH oxidase-dependent ROS production and its effects on instantaneous water use efficiency in Vicia faba. PLoS One 9:e93290

Lian XM, Xing YZ, Yan H, Xu CG, Li XH, Zhang QF (2005) QTLs for low nitrogen tolerance at seedling stage identified using a recombinant inbred line population derived from an elite rice hybrid. Theor Appl Genet 112:85e96

Lindblad P, Bergman B, Rai AN (1990) The cycad-cyanobacterial symbiosis. In: Rai AN (ed) CRC handbook of symbiotic cyanobacteria. CRC Press, Boca Raton, pp 137–159

Liu C, Subashchandrabose S, Ming H, Xiao B, Naidu R, Megharaj M (2016) Phycoremediation of dairy and winery wastewater using Diplosphaera sp. MM1. J Appl Phycol 28(6):3331–3341

Liu J, Zheng Y, Lin H, Wang X, Li M, Liu Y, Yu M, Zhao M, Pedentchouk N, Lea-Smith DJ, Todd JD, Magill CR, Zhang WJ, Zhou S, Song D, Zhong H, Xin Y, Yu M, Tian J, Zhang X-H (2019) Proliferation of hydrocarbon-degrading microbes at the bottom of the Mariana Trench. Microbiome 7:47

Lu Y, Xu J (2015) Phytohormones in microalgae: a new opportunity for microalgal biotechnology? Trends Plant Sci 20:273–282

Lu N, Chen JH, Wei D, Chen F, Chen G (2016) Global metabolic regulation of the snow alga Chlamydomonas nivalis in response to nitrate or phosphate deprivation by a metabolome profile analysis. Int J Mol Sci 2016:17

Lutzu GA, Ciurli A, Chiellini C, Di Caprio F, Concas A, Dunford NT (2020) Latest developments in wastewater treatment and biopolymer production by microalgae. J Environ Chem Eng:104926

Malliga P, Uma L, Subramanian G (1996a) Lignolytic activity of the cyanobacterium microbiology of tropical soils and plant productivity. Springer, Dordrecht, pp 147–168

Malliga P, Uma L, Subramanian G (1996b) Lignolytic activity of the cyanobacterium Anabaena azollae ML2 and the value of coir waste as a carrier for BGA biofertilizer. Microbios 86:175–183

Mancuso S, Azzarello E, Mugnai S, Briand X (2006) Marine bioactive substances (IPA extract) improve foliar ion uptake and water stress tolerance in potted Vitis vinifera plants. Adv Hortic Sci 20:156–161

Manjunath M, Kanchan A, Ranjan K, Venkatachalam S, Prasanna R, Ramakrishnan B, Hossain F, Nain L, Shivay YS, Rai AB (2016) Beneficial cyanobacteria and eubacteria synergistically enhance bioavailability of soil nutrients and yield of okra. Heliyon 2:e00066

Mansori M, Chernane H, Latique S, Benaliat A, Hsissou D, El Kaoua M (2014) Seaweed extract effect on water deficit and antioxidative mechanisms in bean plants (Phaseolus vulgaris L.). Environ Boil Fishes 27:1689–1698

Mansour HA, Shaaban AS (2010) Algae of soil surface layer of Wadi Al-Hitan protective area (world heritage site), El-Fayum depression. Egypt J Am Sci 6:243–255

Meeks JC, Elhai J (2002) Regulation of cellular differentiation in filamentous cyanobacteria in freeliving and plant-associated symbiotic growth states. Microbiol Mol Biol Rev 66:94–121

Megharaj M, Singleton I, McClure NC, Naidu R (2000) Influence of petroleum hydrocarbon contamination on microalgae and microbial activities in a long-term contaminated soil. Arch Environ Contam Toxicol 38(4):439–445

Metting B, Rayburn W (1979) Algal communities and soil microenvironments in an Eastern Washington silt loam. Soil Sci 127(2):74–78

Milton RF (1952) Improvements in or relating to horticultural and agricultural fertilizers. Br Patent 664:989

Misra S, Kaushik BD (1989) Growth promoting substances of cyanobacteria II. Detection of amino acids, sugars and auxins. Proc Indian Natl Sci Acad B55:499–504

Monteiro CM, Castro PML, Malcata FX, Instituto I, Carlos A, Campos O, Maia C, Pedro PAS (2012) Metal uptake by Microalgae: underlying mechanisms and practical applications. Biotechnol Prog 28(2):299–311

Moore AW (1969) *Azolla*: biology and agronomic significance. Bot Rev 35(1)

Mousavi P, Montazeri-Najafabady N, Abolhasanzadeh Z, Mohagheghzadeh A, Hamidi M, Niazi A, Morowvat MH, Ghasemi Y (2016) Investigating the effects of phytohormones on growth and beta-carotene production in a naturally isolates stain of Dunaliella salina. J Appl Pharm Sci 6:164–171

Munoz R, Guieysse B (2006) Algal–bacterial processes for the treatment of hazardous contaminants: a review. Water Res 40:2799–2815

Muñoz-Rojas M, Román JR, Roncero-Ramos B, Erickson TE, Merritt DJ, Aguila-Carricondo P, Cantón Y (2018) Cyanobacteria inoculation enhances carbon sequestration in soil substrates used in dryland restoration. Sci Total Environ 636:1149–1154

Nedzarek A, Rakusa-Suszczewski S (2004) Decomposition of macroalgae and the release of nutrients into Admiralty Bay, King George Island, Antarctica. Polar Biosci 17:26–35

Nelson WR, Van Staden J (1984) The effect of seaweed concentrate on wheat culms. J Plant Physiol 115:433–437

Neustupa J, Skaloud P (2010) Diversity of subaerial algae and cyanobacteria growing on bark and wood in the lowland tropical forests of Singapore. Plant Ecol Evol 143(1):51–62

Obana S, Miyamoto K, Morita S, Ohmori M, Inubushi K (2007) Effect of Nostoc sp. on soil characteristics, plant growth and nutrient uptake. J Appl Phycol 19:641–646

Omar HH, Abdullatif BM, Al-Kazan MM, El-Gendy AM (2014) Various applications of seaweed improves growth and biochemical constituents of Zea mays L. and Helianthus annuus L. J Plant Nutr 38:28–40

Osman MEH, El-Sheekh MM, El-Naggar AH, Gheda SF (2010) Effect of two species of cyanobacteria as biofertilizers on some metabolic activities, growth, and yield of pea plant. Biol Fertil Soils 46:861–875

Pal P, Chew KW, Yen HW, Lim JW, Lam MK, Show PL (2019) Cultivation of oily microalgae for the production of third-generation biofuels. Sustain For 11:5424

Pankratova EM (2006) Functioning of cyanobacteria in soil ecosystems. Eurasian Soil Sci 39:S118–S127

Pardee I, Ellis K, Bouthillier P, Towers M, French CJ (2004) Plant virus inhibitors from marine algae. Can J Bot 82:304–309

Passos F, Mariona HM, García J, Ferrer I (2014) Long-term anaerobic digestion of microalgae grown in HRAP for wastewater treatment. Effect of microwave pretreatment. Water Res 49:351–359

Patel K, Agarwal P, Agarwal PK (2018) Kappaphycus alvarezii sap mitigates abiotic-induced stress in Triticum durum by modulating metabolic coordination and improves growth and yield. J Appl Phycol 30:2659–2673. https://doi.org/10.1007/s10811-018-1423-4

Paumann M, Regelsberger G, Obinger C, Peschek GA (2005) The bioenergetic role of dioxygen and the terminal oxidase(s) in cyanobacteria. BiochimBiophys Acta 1707:231–253. https://doi.org/10.1016/j.bbabio.2004.12.007

Piccini M, Raikova S, Allen MJ, Chuck CJ (2019) A synergistic use of microalgae and macroalgae for heavy metal bioremediation and bioenergy production through hydrothermal liquefaction. Sustain Energy Fuels 3(1):292–301

Pittman JK, Dean AP, Osundeko O (2011) The potential of sustainable algal biofuel production using wastewater resources. Bioresour Technol 102:17–25

Poo K, Son E, Chang J, Ren X, Choi Y, Chae K (2018) Biochars derived from wasted marine macro-algae (Saccharina japonica and Sargassum fusiforme) and their potential for heavy metal removal in aqueous solution. J Environ Manag 206:364–372
Prasad RC, Prasad BN (2001) Cyanobacteria as a source biofertilizer for sustainable agriculture in Nepal. J Plant Sci Bot Orient 1:127–133
Priya H, Prasanna R, Ramakrishnan B, Bidyarani N, Babu S, Thapa S, Renuka N (2015) Influence of cyanobacterial inoculation on the culturable microbiome and growth of rice. Microbiol Res 171:78–89
Priyadarshani I, Sahu D, Rath B (2011) Microalgal bioremediation: current practices and perspectives. J Biochem Technol 3:299–304
Ramanan R, Kim BH, Cho DH, Oh HM, Kim HS (2016) Algae-bacteria interactions: evolution, ecology and emerging applications. Biotechnol Adv 34:14–29. https://doi.org/10.1016/j.biotechadv.2015.12.003
Rasmussen U, Johansson C, Renglin A, Petersson C, Bergman B (1996) A molecular characterization of the Gunnera–Nostoc symbiosis: comparison with Rhizobium–and Agrobacterium–plant interactions. New Phytol 133:391–398
Ray JG, Binoy Thomas T (2012) Ecology and diversity of green-algae of tropical Oxic Dystrustepts soils in relation to different soil parameters and vegetation. Res J Soil Biol 4(3):42–68
Rayorath P, Khan K, Ravishankar P, Shawna LM, Roumiana S, Simon DH, Alan TC, Balakrishan P (2008a) Extract of the brown seaweed Ascophyllum nodosum induce gibberellic acid (GA3)-independent amylase activity in barley. J Plant Growth Regul 27:370–379
Rayorath P, Khan W, Palanisamy R, Mackinnon SL, Stefanova R, Hankins SD, Critchley AT, Prithiviraj B (2008b) Extracts of the Brown seaweed Ascophyllum nodosum induce gibberellic acid (GA3)-independent amylase activity in barley. J Plant Growth Regul 32:123–128
Reddy AS, Rao PV, Sateesh BJ, Ramana MV (2016) Impact of seaweed liquid fertilizers on productivity of blackgram [Vigna mungo (L.) Hepper]. Int J Curr Res Biosci Plant Biol 3:88–92
Řeháková K, Chlumská Z, Doleža J (2011) Soil cyanobacterial and microalgal diversity in dry mountains of Ladakh, NW Himalaya, as related to site, altitude, and vegetation. Microb Ecol 62:337–346
Renuka N, Prasanna R, Sood A, Ahluwalia AS, Bansal R, Babu S, Singh R, Shivay YS, Nain L (2016) Exploring the efficacy of wastewater-grown microalgal biomass as a biofertilizer for wheat. Environ Sci Pollut Res 23:6608–6620
Renuka N, Guldhe A, Prasanna R, Singh P, Bux F (2018) Microalgae as multi-functional options in modern agriculture: current trends, prospects and challenges. Biotechnol Adv 36:1255–1273
Rodríguez AA, Stella AM, Storni MM, Zulpa G, Zaccaro MC (2006) Effects of cyanobacterial extracellular products and gibberellic acid on salinity tolerance in Oryza sativa L. Saline Syst 2:7. https://doi.org/10.1186/1746-1448-2-7
Roger PA, Reynaud PA (1982) Free-living blue-green algae in tropical soils. In: Dommergues YR, Diem HG (eds) Microbiology of tropical soils and plant productivity. Developments in plant and soil sciences, vol 5. Springer, Dordrecht. https://doi.org/10.1007/978-94-009-7529-3_5
Ronga D, Biazzi E, Parati K, Carminati D, Carminati E, Tava A (2019) Microalgal biostimulants and biofertilisers in crop productions. Agronomy 9:192
Rouphael Y, Colla G (2020) Editorial: biostimulants in agriculture. Front Plant Sci 11:40
Ruble RW, Davis JS (1988) Soil algae from fallow potato fields in South Florida (USA) marl. Nova Hedwigia 47(3–4):403–414
Saadatnia H, Riahi H (2009) Cyanobacteria from paddy fields in Iran as a biofertilizer in rice plants. Plant Soil Environ 55:207–212
Sahu D, Priyadarshani I, Rath B (2012) Cyanobacteria das potential biofertilizer. CIBTech J Microbiol 1:20e26
Sahu AK, Siljudalen J, Trydal T, Rusten B (2013) Utilization of wastewater nutrients for microalgae growth for anaerobic digestion. J Environ Manag 122:113–120

Sainju UM, Ghimire R, Pradhan GP (2019) Nitrogen fertilization I: impact on crop, soil, and environment, September 6th 2019. Available: https://www.intechopen.com/books/nitrogen-fixation/nitrogen-fertilizationi-impact-on-crop-soil-and-environment

Santaniello A, Scartazza A, Gresta F, Loreti E, Biasone A, Tommaso D, Piaggesi A, Perata P (2017) Ascophyllum nodosum seaweed extract alleviates drought stress in Arabidopsis by affecting photosynthetic performance and related gene expression. Front Plant Sci 8:1362

Santi S, De Marco F, Polizzotto R, Grisan S, Musetti R (2013) Recovery from stolbur disease in grapevine involves changes in sugar transport and metabolism. Front Plant Sci 4:171

Sathya B, Indu H, Seenivasan R, Geetha S (2010) Influence of seaweed liquid fertilizer on the growth and biochemical composition of the legume crop Cajanus cajan (L.) Mill sp. J Phytology 2:50–63

Shahbazi F, Nejad SM, Salimi A, Gilani A (2015) Effect of seaweed extracts on the growth and biochemical constituents of wheat. Int J Agric Crop Sci 8:283–287

Sharma R, Khokhar M, Jat R, Khandelwal S (2012) Role of algae and cyanobacteria in sustainable agriculture system. Wud Pecker J Agric Res 1:381–388

Shukla M, Kumar S (2018) Algal growth in photosynthetic algal microbial fuel cell and its subsequent utilization for biofuels. Renew Sust Energ Rev 82:402–414

Shukla P, Mantin E, Adil M, Bajpai S, Critchley A, Prithiviraj B (2019) Ascophyllum nodosum-based biostimulants: sustainable applications in agriculture for the stimulation of plant growth, stress tolerance, and disease management. Front Plant Sci 10:655

Siddeeg SM, Tahoon MA, Ben Rebah F (2020) Agro-industrial waste materials and wastewater as growth media for microbial bioflocculants production: a review. Mater Res Express:7

Singh RN (1961) Role of blue-green algae in nitrogen economy of Indian agriculture. ICAR, New Delhi, p 175

Singh S, Datta P (2007) Outdoor evaluation of herbicide resistant strains of Anabaena variabilis as biofertilizer for rice plants. Plant Soil 296:95–102

Singh RP, Kumari P, Reddy CR (2015a) Antimicrobial compounds from seaweed-associated bacteria and fungi. Appl Microbiol Biotechnol 99:1571–1586

Singh SK, Thakur R, Singh MK, Singh CS, Pal SK (2015b) Effect of fertilizer level and seaweed sap on productivity and profitability of rice (Oryza sativa). Indian J Agron 60:420–425

Singh S, Singh MK, Pal SK, Trivedi K, Yesuraj D, Singh CS, Anand VKG, Chandramohan M, Patidar R, Kubavat D, Zodape ST, Ghosh A (2016a) Sustainable enhancement in yield and quality of rain-fed maize through Gracilaria edulis and Kappaphycus alvarezii seaweed sap. J Appl Phycol 28:2099–2112

Singh JS, Kumar A, Rai AN, Singh DP (2016b) Cyanobacteria: a precious bio-resource in agriculture, ecosystem, and environmental sustainability. Front Microbiol 7:529

Singh R, Parihar P, Singh M, Bajguz A, Kumar J, Singh S, Singh VP, Prasad SM (2017) Uncovering potential applications of cyanobacteria and algal metabolites in biology, agriculture and medicine: current status and future prospects. Front Microbiol 8:515

Smith GA, Pepich BV (2007) EPA method 536: determination of triazine pesticides and their degradates in drinking water by liquid chromatography electrospray ionization tandem mass spectrometry (LC/ESI-MS/MS)

Song T, Mårtensson L, Eriksson T, Zheng WU (2005) Rasmussen biodiversity and seasonal variation of the cyanobacterial assemblage in a rice paddy field in Fujian. China FEMS Microbiol Ecol 54(1):131–140

Srijaya TC, Pradeep PJ, Chtterji A (2010) Effect of seaweed extract as an organic fertilizer on the growth enhancement of blackmustard plant. J Coast Environ 1:137–150S

Starks TL, Shubert LE, Trainor FR (1981) Ecology of soil algae: a review. Phycologia 20(1):65–80

Sukenik A, Maldener I, Delhaye T, Viner-Mozzini Y, Sela D, Bormans M (2015) Carbon assimilation and accumulation of cyanophycin during the development of dormant cells (akinetes) in the cyanobacterium Aphanizomenon ovalisporum. Front Microbiol 6:1067

Sun J, Xiong X, Wang M, Du H, Li J, Zhou D, Zuo J (2019) Microalgae biodiesel production in China: a preliminary economic analysis renew. Sustain Energy Rev 104:296–306

Swarnalakshmi K, Prasanna R, Kumar A, Pattnaik S, Chakravarty K, Shivay YS, Singh R, Saxena AK (2013) Evaluating the influence of novel cyanobacterial biofilmed biofertilizers on soil fertility and plant nutrition in wheat. Eur J Soil Biol 55:107–116
Temple WD, Bomke AA (1988) Effects of kelp (Macrocystis integrifolia) on soil chemical properties and crop response. Plant Soil 105:213–222
Thajuddin N, Subramanian G (2005) Cyanobacterial biodiversity and potential applications in biotechnology. Curr Sci 89(2005):47e57
Shi T, Li M, Wei G, Liu J, Gao Z (2020) Distribution patterns of microeukaryotic community between sediment and water of the Yellow River Estuary. Curr Microbiol 77:1496–1505
Treves H, Raanan H, Kedem I, Murik O, Keren N, Zer H (2016) The mechanisms whereby the green alga Chlorella ohadii, isolated from desert soil crust, exhibits unparalleled photodamage resistance. New Phytol 210:1229–1243
Tuhy Ł, Samoraj M, Baśladyńska S, Chojnacka K (2015) New micronutrient fertilizer biocomponents based on seaweed biomass. Pol J Environ Stud 24:2213–2222
Upadhyay AK, Singh R, Singh DP (2018) Phycotechnological approaches toward wastewater management. Emerging and eco-friendly approaches for waste management. 423–435. World J Microbiol Biotechnol 32:1–18
Vadeboncoeur YM, Zanden JV, Lodge DM (2002) Putting the Lake back together: reintegrating benthic pathways into lake food web models: lake ecologists tend to focus their research on pelagic energy pathways, but, from algae to fish, benthic organisms form an integral part of lake food webs. Bioscience 52:44–54
Vadeboncoeur YM, Moore MV, Chandra S, Stewart SD (2020) Littoral greening of clear lakes: the mystery of benthic filamentous algal blooms. Limnol Oceanogr Bull 29:57–58
Vassilev SV, Vassileva CG (2016) Composition, properties and challenges of algae biomass for biofuel application: an overview. Fuel 181:1–33
Vera J, Castro J, Gonzalez A, Moenne A (2011) Seaweed polysaccharides and derived oligosaccharides stimulate defense responses and protection against pathogens in plants. Mar Drugs 44:2514–2525
Vernieri P, Borghesi E, Tognoni F, Serra G, Ferrante A, Piaggesi A (2006) Use of biostimulants for reducing nutrient solution concentration in floating system. Acta Hortic 21:123–134
Wally OS, Critchley AT, Hiltz D, Craigie JS, Han X, Zaharia LI, Abrams SR, Prithiviraj B (2012) Regulation of phytohormone biosynthesis and accumulation in Arabidopsis following treatment with commercial extract from the marine macroalga Ascophyllum nodosum. J Plant Growth Regul 32:324–339
Wetzel RG (1996) Benthic algae and nutrient cycling in lentic freshwater ecosystems. In: Stevenson R, Bothwell M, Lowe R, Thorp JH (eds) Algal ecology, 1st edn. Elsevier, Amsterdam, pp 641–667
Wilson LT (2006) Cyanobacteria: a potential nitrogen source in rice fields. Texas Rice 6:9–10
Win TT, Barone GD, Secundo F, Fu P (2018) Algal biofertilizers and plant growth stimulants for sustainable agriculture. Ind Biotechnol 14:203–211
Winder M, Reuter JE, Schladow SG (2009) Lake warming favours small-sized planktonic diatom species. Proc R Soc B Biol Sci 276:427–435
Yao S, Lyu S, An Y, Lu J, Gjermansen C, Schramm A (2019) Microalgae–bacteria symbiosis in microalgal growth and biofuel production: a review. J Appl Microbiol 126:359–368. https://doi.org/10.1111/jam.14095
Yoshida K, Igarashi E, Wakatsuki E, Miyamoto K, Hirata K (2004) Mitigation of osmotic and salt stresses by abscisic acid through reduction of stress-derived oxidative damage in Chlamydomonas reinhardtii. Plant Sci 167:1335–1341
Yvon-Durocher G, Jones JI, Trimmer M, Woodward G, Montoya JM (2010) Warming alters the metabolic balance of ecosystems. Philos Trans R Soc B Biol Sci 365:2117–2126
Zamani S, Khorasaninejad S, Kashefi B (2013) The important role of seaweeds for some characters of plants. Int J Agric Crop Sci 5:1789–1793

Zenova GM, Shtina EA, Dedysh SN (1995) Ecological relations of algae in biocenoses. Mikrobiologiya 64(2):121–133

Zhang B, Ning D, Yang Y, Van Nostrand JD, Zhou J, Wen X (2019) Biodegradability of wastewater determines microbial assembly mechanisms in full-scale wastewater treatment plants. Water Res 169:115276

Zhu H, Li S, Hu Z, Liu G (2018) Molecular characterization of eukaryotic algal communities in the tropical phyllosphere based on real-time sequencing of the 18S rDNA gene. BMC Plant Biol 18:1–14

Soil Algae Symbioses: Raising the Curtain of an Ancient Play

Maya P. Stoyneva-Gärtner, Georg Gärtner, Blagoy Uzunov, Miroslav Androv, and Kristian Ivanov

Abstract This review provides a synthesis of the terrestrial symbiotic associations of soil algae, and the term cyanosymbiosis is coined as operational term for those in which cyanoprokaryotes (cyanobacteria) are involved. All types of associations (endo- and ectobiotic, intra- and extracellular, horizontally and vertically transmitted, bi- and tri-membered, including algae as mediators in complex guilds, natural and artificial, immobile and mobile) are presented with special attention to algal diversity and specificity of interactions with the host. Summary of peculiarities of symbioses of pro- and eukaryotic algae is provided before their more detailed description according to the hosts. The search for common features shows that the widespread presence among symbionts of resistant cell wall or protective mucilage sheaths around the cells and the ability to produce dormant cells (akinetes, aplanospores, and cysts) is a favorable feature for the vertical transfer of symbionts. The major difference between symbioses of pro- and eukaryotic algae, according to the available descriptions, is in their active or passive behavior: the active infection of host by the motile hormogonia of prokaryotic algae, which usually occurs in most of cases of endobiotic symbioses, contrasts with the "capture" by the host of eukaryotic algae, which remain "passive" during the infection or simply "land" on the surface of the host. Based on our expertise in algology and lichenology, we hypothesize that zoospores or planogametes of eukaryotic algae (commonly considered primordial features) may have played the role of active infectants in ancient times when symbioses emerged and propose to explore this topic more actively together with the possibilities for signal transmissions between the partners. Other specific traits of symbiotic algae along with some unresolved problems and perspectives for

---

M. P. Stoyneva-Gärtner (✉) · B. Uzunov · M. Androv · K. Ivanov
Department of Botany, Faculty of Biology, Sofia University "St. Kliment Ohridski", Sofia, Bulgaria
e-mail: mstoyneva@uni-sofia.bg; buzunov@uni-sofia.bg; androv@uni-sofia.bg; kristianri@uni-sofia.bg

G. Gärtner
Institut für Botanik der Universität Innsbruck, Innsbruck, Austria
e-mail: georg.gaertner@uibc.ac.at

A. Mishra, A. Varma (eds.), *Soil Algae*, https://doi.org/10.1007/978-981-97-9873-5_2

future research are also noted. Although it is a brief overview, this review mentions to novel methods and approaches in fundamental scientific research, but also those used for symbiosis formation with regard to the development of modern agriculture and biotechnology with the search for higher crops and specific natural products. When data allows, fossil symbioses are mentioned and attempt is made to outline the driving forces that have led to the establishment and maintenance of the special partnerships of algae with an extraordinary wide range of organisms, such as bacteria, other algae, fungi, land plants, and animals. In total, 89 algal genera from six phyla are discussed as real or potential participants of symbiotic interactions. This distribution of terrestrial symbionts reflects the general biodiversity of soil algae and their availability, which can be taken as other important "background" factors for establishment of symbioses. In this regard, absence of widespread streptophyte soil green algae from terrestrial symbioses allows to hypothesize also that evolutionary origin and age may be among the selectivity factors, assuming that ancient organisms had some very specific traits, which provided them with advantages to be chosen as partners during the establishment of the earliest symbioses.

Keywords Cyanobacteria · Cyanosymbioses · Driving forces · Fossil symbioses · Symbiotic interactions · Terrestrial symbioses

1 Introduction

Close partnerships between different organisms during their long life on Earth led to the development of complex associations with balanced relations between the participants (e.g., Paracer and Ahmadjian 2000). It is widely known that Heinrich Anton de Bary (1831–1888), a German surgeon and biologist, coined in 1878 the term "symbiosis" for "living together of unlike organisms," but it is to recall that almost year earlier, in 1877, another German botanist and mycologist Albert Bernhard Frank (1839–1900) also developed the idea and proposed the term "symbiotismus" for coexisting organisms. Both terms have their origin in Greek language (*sym*—together, *biosis*—living), and both, in a common way, did not stress on the types of relations (beneficial, harmful, or neutral), underlining only the coexistence of dissimilar organisms. After more than a century debates, modern biology has not adopted a single definition but has abandoned the severe restrictions proposed by several authors for the types of interactions between the partners, practically accepting the original ideas of Frank (1877) and de Bary (1878, 1879). The broadly accepted definitions of symbiosis as "living together in physical contact of organisms of different species" (Reisser 1992, p. xii) and as "association between two or more different organisms" (Paracer and Ahmadjian 2000, p. 3) were enlarged by Rai (2018, p. 1) with two important features, and thus, symbiosis was defined as "permanent living together of two dissimilar organisms involving exchange of metabolites between the symbionts." The need for consistency was also noted by

Paracer and Ahmadjian (2000, p. 3), who understood "permanent" as "the organisms never being separated or long-lasting."

According to the dependence of organisms from each other, symbioses can be obligate, which means that at least one of the partners is so highly adapted that it completely depends on the other for its survival, or facultative (optional), in which both partners can be free-living, independently of each other (for historical and terminological details see Martin and Schwab 2012; Mishra et al. 2018). Regarding physical aspects, symbioses can be divided into endo- (with one symbiont inhabiting the tissues of the other) and ecto- (with one symbiont living on the surface of the other) types (for details see Martin and Schwab 2012). Regarding the mode of transmission, symbioses with de novo acquired partner are horizontal (environmental), and those with partner transferred through the germ-line or off-springs are vertical (hereditary), but combination of both types is also known (e.g., Hirsh 2004). Considering the geological timing, symbioses are fossil and extant. By time, the evolutionary trend of symbioses is to achieve closer relations between partners, ultimately turning them from facultative to obligate (Lewin 1995; Usher et al. 2007), and the high specificity of partners is considered to be an important result of evolutionary longer adaptive relations (Paracer and Ahmadjian 2000). Therefore, symbioses are recognized as benchmark biological systems to understand better the evolutionary history, and interest in them is gradually progressing (Graham et al. 2018; Bouchard et al. 2020). In addition, currently, there is a surge of research focused on the diverse metabolites produced in these "natural factories" in relation to rising demand for high-value compounds and natural products for biotechnology and different industries related to human life (Lutzu and Dunford 2018; Singh et al. 2020).

Algae, defined broadly by Lewin (1995) as "photosynthetic plants that are neither bryophytes or tracheophytes," represent a unique group of morphologically versatile micro- and macro-organisms, which, due to their pro- or eukaryotic cytological organization, belong to two main domains, Bacteria (Eubacteria) and Eukarya. Algae differ also in their ecology and occur in various environments, including aquatic habitats, soil, which comprises a suite of algal niches in its surface and deep layers (Metting 1981; Johansen and Shubert 2001; Gärtner et al. 2021). It is most likely that soil algae are quite ancient, related to Precambrian times when first prokaryotic blue-green algae occupied the terrestrial habitats and, later, were followed by the eukaryotic algae from other taxonomic groups (Johansen and Shubert 2001). Some fossil discoveries in a shallow subtidal region of China allowed tentatively to suppose that the transfer of aquatic blue-green algae to the land was enhanced by fungi through developing a lichen-like symbiosis (Yuan et al. 2005; Usher et al. 2007). However, the question remains open since algal symbionts in these fossils were not certainly identified (Usher et al. 2007).

We can only speculate about the selection pressures that led to the maintenance and improvement of symbiotic organisms, but the paramount seems to be the need of ancient land organisms to overcome the harsh environmental and poor-soils conditions (including competition for nutrients and living space) and to avoid parasitism or grazing by herbivores. This explains why oxygenic photoautotrophs, such as algae, and especially those, which had the additional ability to convert the atmospheric gaseous nitrogen (N) to mineral forms, were readily recruited by different lineages of fungi and land plants to form diverse symbiotic associations (e.g., Usher et al. 2007). The increased ability of nutrients that symbiont provided to the host, especially in cases when the host had not the relevant metabolic capacity, made the association more productive (Larkum et al. 2003). In addition, the large amount of versatile biologically active secondary metabolites, including growth hormones, protective sunscreens, antioxidant, allelopathic, antibiotic and antifeedant compounds, produced by soil algae, is gaining emerging recognition (e.g., Hrouzek et al. 2011; Singh et al. 2020; Stoyneva-Gärtner et al. 2020, 2022; Gärtner et al. 2021). It is widely accepted that the harbored algae by these substances may elicit the host's defense against pests, diseases, or grazing, and enhance its growth (Graham et al. 2018; Galanakis 2021). However, since some of the so-called secondary metabolites, such as algal toxins, have appeared long before the potential grazers, their role is still disputable with increasing evidence for their function in primary metabolism with importance in cell-cell communication and in protection from oxidative stress (Zilliges et al. 2011; Liaimer et al. 2016; Gärtner et al. 2021). In return for the profits provided by the "tenant," the host ensures its symbiotic algae a more stable protected environment (e.g., Rai et al. 2000, 2002; Larkum et al. 2003), with less fluctuating physical environmental conditions (such as temperature, drought and insolation) compared to the soil surface, which allows to avoid grazing, and, in addition, provides a reduced carbon. In the same time, the symbiont inside the host has to adapt to a different environment in terms of pH, oxygen content, light regime, moistness, and eventual host immune reaction as well (Lewin 1995; Usher et al. 2007). In this adaptation, most probably, advantageous is the one "equipped" with a protective mucilaginous sheath or resistant cell wall (Usher et al. 2007), which are features of many algae. But according to the same authors, it is also true that many hosts recognize the "invaders" and even stimulate them. Although little is known on the exact biochemical, molecular, and physiological relations between the partners in most of these associations (Larkum et al. 2003), a lot of evidence for chemical signaling between them was accumulated.

It is supposed that cyanoprokaryotes were driving forces for the evolution of their hosts (Usher et al. 2007) and first symbiotic associations with blue-green algae appeared about 500 MA (Rai et al. 2002). In support of this view, explaining the colonization of nutrient-poor ancient soils and advantageous trait that symbiosis provides to this process can be interpreted the current tracheophyte recolonization of deglaciated areas in Tierra del Fuego (southern South America), initiated by *Gunnera magellanica*, which lives in partnership with the blue-green *Nostoc* (Fernández-Martínez et al. 2013).

Currently, plant ecology is increasingly recognizing plants as meta-organisms that possess distinct microbiomes and enter into symbiotic relations with them, the detailed mechanisms of which are still unclear (Ho et al. 2017). Over the last decade, interest in bacteria as an additional, third components of different symbiotic associations have grown, and their role is increasingly being evaluated with new insights into complex interactions but also in their potential for new biotechnological applications (e.g., Grube and Berg 2009; Lutzu and Dunford 2018). However, their detailed analysis is beyond the scope of the current review, and data on microbiomes from such underutilized niches are only briefly mentioned for completeness.

This review is focused on soil algae as partners in versatile symbioses which occur in nature or are artificially achieved for agricultural or biotechnological purposes. It is organized in two parts: first oriented towards symbioses of prokaryotic blue-green algae and second describes the associations of eukaryotic algae. This separation is based on their different functions and related traits because if all algae have the capacity to perform photosynthesis and to fix carbon, only blue-green prokaryotes are able to convert atmospheric N into forms available for other organisms.

2 Results

2.1 Symbioses of Soil Cyanoprokaryota (Soil Cyanosymbioses)

Prokaryotic algae, Cyanoprokaryota (blue-green algae/cyanobacteria/oxyphotobacteria), are the most ancient inhabitants of the Earth, which occurred more than 3.5 billion years ago and currently are among the most quantitatively important organisms on our planet (Car and Whitton 1982; Whitton 2012). About two billion years ago, they played a key role in the evolution of life on the Earth not only due to the oxygenation of the atmosphere, but also as progenitors of plastids through their incorporation into a eukaryotic organism (Bergman et al. 1996; Raven and Allen 2003; Whitton 2012). In addition, these algae have the unprecedented capacity to be the only oxyphototrophs, primary sustainable producers, capable of N fixation through the nitrogenase enzyme complex (e.g., Whitton 2012; Liang et al. 2018), which prized them as wanted partners, involved in many and different (extra- and intracellular) symbioses with aquatic and terrestrial bacteria, eukaryotic algae (diatoms, dinoflagellates, etc.), land plants, fungi, invertebrates, and vertebrates (Lewin and Robinson 1979; Reisser 1992; Paracer and Ahmadjian 2000; Usher et al. 2007; Adams and Duggan 2008; Rai 2018; Singh et al. 2020). In these lineages, however, the partners are more restricted, represented by a limited number of genera (Pawlowski 2009). The terrestrial symbioses of cyanoprokaryotes involve bacteria, eukaryotic algae, fungi, land plants, invertebrates, and vertebrates (e.g., Rai 2018) (Fig. 1). From here on, for the purpose of this review, we shall briefly refer these associations as "cyanosymbioses" along with the widely accepted term "cyanobiont" for the cyanoprokaryote partner.

Although some authors express hesitations about the benefits for cyanobionts to enter symbiotic interactions (e.g., Liaimer et al. 2016), we concur with the prevailing view that the host provision of secure space, moisture, nutrients, shelter from herbivores, parasites, intense sunlight, and temperature extremes provides advantageous profits to the symbiotic algae. Despite the remarkable range of main hosts lineages, some common features of terrestrial cyanosymbioses can be outlined:

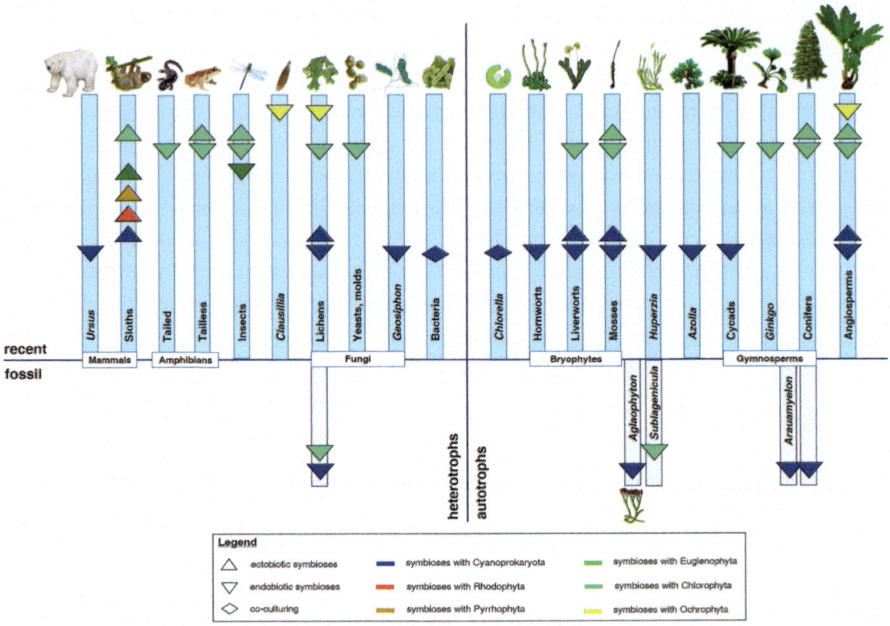

Fig. 1 Symbioses of soil algae according to their types (endobiotic and ectobiotic), hosts (autotrophic and heterotrophic), and geological time (fossil and recent)

1. Symbiotic interactions, and particularly the mutualistic ones, commonly occur in nutrient-poor environments allowing the host to colonize them (e.g., Boucher et al. 1982; Rai 1990; Rasmussen and Johansson 2002; Whitton and Potts 2002; Pereira 2017).
2. Cyanoprokaryotes can function as hosts (cyanosymbioses with bacteria and some fungi), or as cyanobionts in all other cyanosymbioses, in which they occur as intra- or extracellular endophytes, or as epiphytes as well.
3. All structures, inhabited by cyanobionts, preexist in the absence of cyanobiont, and this is different with the symbioses known between other N-fixing bacteria and plants, such as legume nodules in *Rhizobia*-legume symbiosis (Meeks 1998). The sites where cyanobionts reside include the roots of cycads, the stem glands of *Gunnera*, the leaf cavities of *Azolla*, domatia or ventral auricles within the thallus of hornworts and liverworts, the bladders of *Geosiphon*, etc. (Adams et al. 2013).
4. Hosts, in most cases, increase the surface contact with the cyanobionts by production of hairs, ingrowths of cell walls, and protrusions and extensions of contacting cells to facilitate the exchange of nutrients (e.g., Duckett and Prasad 1974; Duckett et al. 1977; Rodgers and Stewart 1977; Ogle and Brown 1997; Paracer and Ahmadjian 2000; McCurdy et al. 2008).
5. Cyanobionts (except those in cycads, and possibly *in Gunnera*, Bergman et al. 1992; Lindblad and Costa 2002) commonly provide N compounds (mainly in the form of ammonia NH_3) that can be assimilated by the host in contrast with

the stable N molecule, which is largely available in the atmosphere, but inaccessible to most organisms and thus unusable (Leigh 2002). These N compounds are excreted as a result of the so-called N fixation, a process of the conversion of atmospheric gaseous N, conducted using enzymes named nitrogenases (Leigh 2002). The organisms containing such nitrogenases, which are able to realize the process of biological N fixation, are known as N-fixing organisms, or diazotrophs, and the designation *nif* (derived from *ni*trogen *f*ixation) is used to indicate genes, necessary for this sustainable for life biological process (Leigh 2002). Nitrogenase has three types: molybdenum (Mo) nitrogenase, vanadium (V) nitrogenase, and iron-only (Fe) nitrogenase, among which the most widely spread is the Mo type (Peters and Szilagyi 2006);

6. In order to increase the level of N fixation of the cyanobiont, host may (a) suppress important N assimilating enzymes of the partner, such as glutamine synthetase (GS) and glutamine oxoglutarate aminotransferase (GOGAT, known also as glutamate synthase), thus making impossible the utilization of the produced NH_3 (Rai et al. 1986; Bergman et al. 1992; Paracer and Ahmadjian 2000; Issa et al. 2014), or (b) excretes chemical signals to stimulate the N fixation (Adams and Duggan 2008). Experiments showed that the activity of NH_3 assimilating enzyme glutamate dehydrogenase (GDH) in its both forms (NADH- and NADPH-dependent) was primarily associated with the hosts, indicating that N fixed by the cyanobiont was transferred to the host, where it was subsequently metabolized through the GS-GOGAT pathway (Rai 2018). In many cases, the host ensures microaerobic and dark conditions, which promote the N-fixing process. For example, the heterotrophic N fixation was proved physiologically in *Nostoc* from bryophytes, which receives organic carbon from the host and corroborates with the absence of phycobilisomes and high heterocyst frequencies in its filaments (Duckett et al. 1977), and in hornworts and *Azolla*, the rise of nitrogenase activity along with photosystem inactivity was noted (Mishra et al. 2018).
7. Cyanobionts are able to stimulate cell (mitotic) division of the host (Bergman et al. 1996). Most cyanobionts produce carbohydrate-rich arabinogalactan proteins, which act as signaling molecules and can also enhance the growth and development of the host (Bergman et al. 1996; Chang et al. 2019).
8. Almost all cyanobionts, except those in some lichens and in *Azolla*, are recruited de novo by the host, as it will be described in the text below for each particular case. This need continually to find new cyanobionts is related with their great diversity, indicating a flexibility in choice, which was supposed to be an evolutionary and ecological advantage (West and Adams 1997).
9. The symbiotic competence seems to be a trait of filamentous unbranched heterocytous algae (Bergman et al. 1996), with *Nostoc* (Fig. 2) and *Desmonostoc* being most commonly reported cyanobionts with a wide range of strains (Bergman et al. 1992; Svenning et al. 2005; Adams and Duggan 2008; Papaefthimiou et al. 2008; Cuddy et al. 2012; Warshan et al. 2018; Pecundo et al. 2021). These genera bear a number of unique morphologically and functionally distinct heterocytes (heterocysts)—transparent cells of peculiar ultra-

Fig. 2 *Nostoc*—the most often reported symbiotic blue-green alga: (**a**) habitat on the asphalt roadside together with the moss *Syntrichia ruralis*; (**b**) comparison of dry and wetted material; (**c**) closer view to wetted thalli and mosses; (**d**) wetted *Nostoc* in a natural size

structure and almost anaerobic internal conditions, usually accepted as responsible for the N fixation (Komárek 2013). They can have other role in the life of cyanoprokaryotes ("pointing" the place where akinetes are formed, or places of simple fragmentation of algal trichomes), and heterocyte germlings into new trichomes through formation of endospores were often reported for both cyanobionts (e.g., Smith 1950; Caiola 1980) and free-living algae (Geitler 1932; Komárek 2013). Besides *Nostoc* and *Desmonostoc*, other heterocytous non-ramified (*Anabaena*, *Calothrix*, *Cylindrospermum*, *Dolichospermum Raphidiopsis*, *Trichormus*) or ramified (*Dichothrix*, *Fischerella*, *Hyphomorpha*, *Mastigocladu*s, *Microchaete*, *Microcoleus*, *Rhizonema*, *Scytonem*a, *Stigonem*a, *Tolypothrix*) genera can enter in these relations, as well as some non-heterocytous ("*Archaeothrix*-type," *Leptolyngbya*, *Lyngbya*, *Oscillatoria*, *Phormidium*, *Plectonema*) or coccal genera (*Anacystis*, *Aphanocapsa*, *Asterocapsa*, *Chlorogloeopsis*, *Chroococcus*, *Chroococcidiopsis*, *Cyanosarcina*, *Entophysalis*, *Gloeocapsa*, *Gloeothece*, *Hyella*, *Microcystis*, *Myxosarcina*, and *Synechocystis*) have been recorded as cyanobionts. Although the taxonomy of cyanobionts in many cases is controversial and not completely resolved, the paradigm is that one and the same strain is able to infect/colonize a wide range of phylogenetically different hosts, thus suggesting a general lack of host specialization, except perhaps for the more specific cluster of strains related with *Azolla* (Svenning et al. 2005; Papaefthimiou et al. 2008; Cuddy et al. 2012; Liaimer et al. 2016; Bell-Doyon et al. 2020).

10. Heterocytous cyanobionts (except those in *Geosiphon*—Dworkin et al. 2006) have more heterocytes in comparison with their free-living counterparts, where they comprise 5–6% of the total number of cells (Bergman et al. 1992; Meeks and Elhai 2002). For example, in bryophytes, they reach 40–50% (Duckett et al.

1977; Bergman et al. 1992); in *Azolla*, they consist about 25–40% (Peters 1991); in cycads, they range from 15% to 46% (Bergman et al. 1992; Caiola 1980) to 65–80% (Chang et al. 2019); and in *Gunnera*, they reach up to 80% (Rasmussen and Johansson 2002). This three- to tenfold increase is thought to be relevant to higher rates of N fixation (Meeks and Elhai 2002) and is believed to be caused by signaling from the host (Adams and Duggan 2008). Although heterocytes are ultimately related with N fixation, we would like to recall that for many filamentous and coccal cyanoprokaryotes that do not form such specific cells, N fixation was experimentally proved (Stewart 1977a, b; Gallon et al. 1991; Bergman et al. 1997; Berrendero et al. 2016);

11. In order to keep the cyanobiont biomass in profitable amount, host has to regulate the symbiont growth and, indeed, reduction of cyanobiont growth and CO_2 production (Adams and Duggan 2008) was detected along with decreased (sometimes to undetectable levels) photosynthetic capacity (Paracer and Ahmadjian 2000; Mishra et al. 2018). In some cases, such reduction is due to the internal location of cyanobionts inside the hosts (e.g., hornworts, cycads, *Gunnera*), where they are not reached by light. In experiments, freshly isolated cycad cyanobionts did not show carbon dioxide (CO_2) fixation neither in dark nor in light conditions (Lindblad et al. 1987). Exceptions from cases of photosynthetic inactivity are cyanobionts that have to provide fixed carbon to non-photosynthetic hosts such as bacteria and fungi (Lindblad et al. 1987; Larkum et al. 2003; Issa et al. 2014). Wagner (1997) summarized data on *Azolla* symbiotic system and showed that despite the decreased levels of mRNA for the main photosynthetic enzyme RuBisCO (ribulose-1,5-bisphosphate carboxylase/oxygenase), it could photosynthesize at high rates, which is explainable with the concerted work of the light-harvesting pigments of both partners. These pigments are complementary and allow the complex symbiotic system to use a wide range of waves from the light spectrum. In different species of cycads from botanical gardens of Italy, many vegetative cells of *Nostoc* cyanobionts lack thylakoids and have homogenous cytoplasm (Caiola 1980). Endosymbiont still can take CO_2 from the host respiration and from the surrounding environment, which has to be transported to it by the host (Larkum et al. 2003). The presence of specific carbon concentration mechanisms (CCMs) that contain RuBisCO to enable efficient photosynthesis and enhance growth in CO_2-limiting environments was proved for many endosymbionts (Larkum et al. 2003; Ettl and Gärtner 2014; Zhan et al. 2018). For example, in lichens, cyanobionts possess carboxysomes which contain RuBisCO and carbonic anhydrase (or carbonate dehydratase) that catalyzes the interconversion of CO_2 to carbonic acid (H_2CO_3), while most of the eukaryotic algae have their functional analogues, pyrenoids (Fig. 3), in their plastids to harbor RuBisCO (e.g., Larkum et al. 2003; Mitchell et al. 2017; Atkinson et al. 2020). These CCM structures are with different functions and can serve as metabolic hubs that also store reserve ch products, but most of the carbon is transferred to the host. In lichens, cyanobionts secrete glucose (it is transferred to the mycobiont, which converses it to mannitol that is kept as a carbon reserve since it cannot be utilized by the

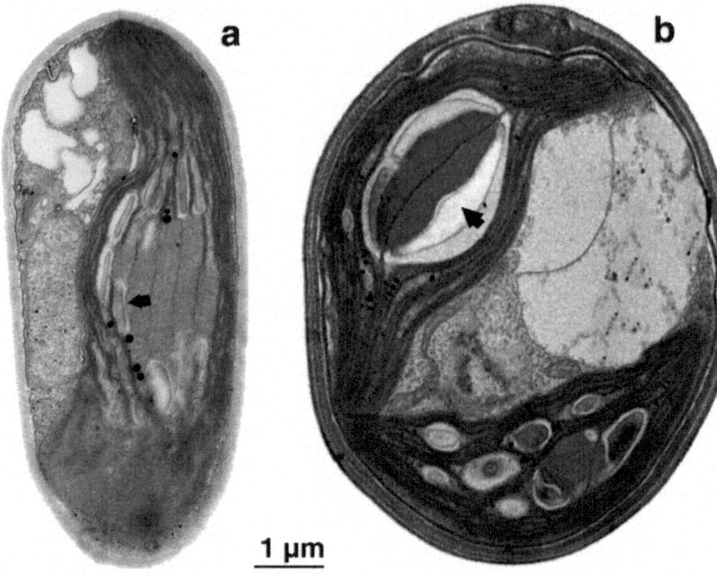

Fig. 3 Ultrastructure of algal cells by TEM with visible pyrenoids (starch indicated by arrows); (**a**) *Stichococcus bacillaris*; (**b**) *Chlorella vulgaris*

cyanobiont), but eukaryotic photobionts transfer polyol, sorbitol, erythryol, or ribitrol (Larkum et al. 2003; Rai 2018). Decreased photosynthesis of cyanobionts is explainable with the fact that many blue-green algae can be facultative heterotrophs and can develop in darkness inside the host plants (Issa et al. 2014). In support of this, evidence was collected that host provides fixed carbon to the cyanobiont in the form of sucrose (Bergman et al. 1992). For example, in a cyanobiont of *Blasia*, the CO_2 fixation was measured to be 12% less than that in the free-living algae, which was compensated by sugars derived from the host (Adams and Duggan 2008).

12. In most endophytic plant and fungal cyanosymbioses, the vegetative cells of cyanobionts were found to differ from the free-living cells being larger, more rounded and with easier breaking connections, which leads to observation of more single cells or their short chains on microscopic slides after pressure of the cover slip, especially on fresh isolated material (Paracer and Ahmadjian 2000; Rai 2018).
13. Most researchers of symbioses pointed hormogonia as the main agents of infection in majority of the cyanosymbioses, and ability to produce hormogonia was stressed among the key factors for symbiotic competence of the cyanobiont strains (e.g., West and Adams 1997; Duggan et al. 2007; Issa et al. 2014). Hormogonia are short, mobile, gliding trichomes formed for propagation and dispersal, which are delimited by so-called "separation disks" (double concave disks of gelatinous material, named also "necridic bands") derived mostly after the formation of so-called necridic cells, a result of programmed cell death (for

details see Geitler 1932; Smith 1950; Komárek 2013; Stoyneva-Gärtner and Uzunov 2017; Aguilera et al. 2021). Here, we have boldly to underline that hormogonia are considered unique for Cyanoprokaryota and are commonly formed by almost all filamentous cyanoprokaryotes, independently of their heterocytous or non-heterocytous organization (Fig. 4) and of their symbiotic competence, with the process being enhanced, if not completely driven, by unfavorable environmental conditions. Gliding motility is an important feature of hormogonia, but gliding is known also for vegetative non-heterocytous trichomes of some genera (e.g., *Oscillatoria*, *Phormidium*) (e.g., Smith 1950). In blue-green algae that are able to glide, additional specific cell layers containing pili (fimbria) have been discovered and pilus-like structures cover the surface of hormogonia cells of *N. punctiforme*, but they have not been found on the vegetative cells of parental filaments (e.g., Dick and Stewart 1980; Duggan et al. 2007; Schuergers and Wilde 2015). Contrasting opinions on the importance of pili for symbiotic competence were expressed: Johansson and Bergman (1994) found fimbria in both infective and non-infective strains of *Nostoc-Gunnera* symbioses, but Dick and Stewart (1980) and Duggan et al. (2007) noted the role of pili in establishment of the *Nostoc-Peltigera* and *Nostoc-Blasia* symbioses, respectively. Moreover, Duggan et al. (2007) demonstrated that inactivation of

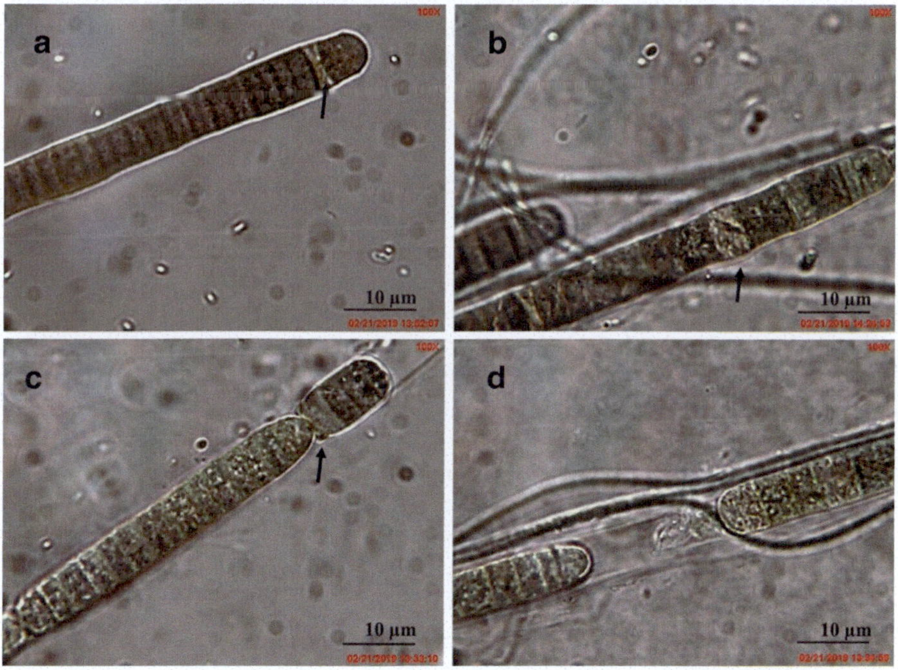

Fig. 4 Stages of formation of short gliding trichomes, hormogonia, in a non-heterocytous filamentous blue-green alga (**a–d**). Necridic discs (separation discs), delimitting hormogonia, are indicated by arrows

the pili encoding genes led to reduction of surface piliation and subsequent reduction of symbiotic competency.

14. Hormogonia are actively motile, but the process of infection/colonization is thought to be initiated or at least facilitated by the host. It is believed that host releases hormogonia-inducing factors (HIF) and chemical compounds (chemoattractants) to entice and guide the hormogonia of the potential cyanobiont to the proper host sites (Meeks et al. 2001; Rai et al. 2002; Meeks and Elhai 2002; Meeks 2003; Elmerich and Newton 2007; Duggan et al. 2007; Adams and Duggan 2008; Issa et al. 2014; Khayatan et al. 2017; Hashidoko et al. 2019, etc.). Therefore, sensitivity of hormogonia to such substances may be important for differences in the symbiotic competence of cyanoprokaryotes (e.g., West and Adams 1997). Production of HIF, which significantly promoted hormogonia formation in different strains of *Nostoc*, was proved for hornworts, liverworts, cycads, and *Gunnera*, and among them some monosaccharides (from *Gunnera*) and diacylglycerols (from cycads) were identified (Campbell and Meeks 1989; Rasmussen et al. 1994; Bergman et al. 1996; West and Adams 1997; Meeks 1998; Adams and Duggan 2008; Adams et al. 2013; Nilsson et al. 2006; Khamar et al. 2010; Chang et al. 2019; Hashidoko et al. 2019). For example, an infection frequency 50 times higher in comparison with that of the wild type was exhibited by a mutant of the cyanobiont strain *Nostoc* sp. 29133, which had increased sensitivity to the HIF of *Anthoceros* (Cohen et al. 1994). Of special interest are studies showing that hosts (*Anthoceros*, *Blasia*, cycads, *Gunnera*, etc.) excrete exudates to induce hormogonia formation in conditions of N starvation or strong N limitation, and vice versa; HIF was not released in experimental culture media with excess N (Campbell and Meeks 1989; Meeks and Elhai 2002; Rai et al. 2002; Meeks 2003; Adams and Duggan 2008; Khamar et al. 2010; Christman et al. 2011; Adams et al. 2013; Hashidoko et al. 2019). Excretion by an N-starved *Blasia* of a potent chemoattractant for the hormogonia of a symbiotic *Nostoc* was demonstrated by Knight and Adams (1996). Data on genetic bases for hormogonia production and heterocyst formation can be found in Christman et al. (2011) who revealed that more than 1762 and 1062 genes were involved. Earlier studies of cyanobionts of *Anthoceros* demonstrated significant change in gene expression patterns during the formation of hormogonia, induced by N starvation or by HIF (Campbell et al. 2008; Christman et al. 2011).
15. Partner recognition may also be an important factor in effective invasion and occurs through pre-invasion signalling. Lectins are thought to be involved in processes of recognition and attraction in many cyanosymbioses (e.g., *Azolla*, *Geosiphon*, *Gunnera*, lichens) (Reisser 1992; Johansson and Bergman 1994; Rasmussen et al. 1994; Knight and Adams 1996; West and Adams 1997; Dworkin et al. 2006; Rai 2018).
16. After the infection or surface colonization, realized by hormogonia, host must stop excretion of HIF and, conversely, start to produce other specific chemical cues that suppress hormogonia formation, enhance the vegetative growth of

trichomes or filaments, and in cases of heterocytous algae, allow differentiation of heterocytes (Warshan et al. 2017; Chang et al. 2019). Such signals, which hinder HIF activities, are united under the name hormogonium-repressing factors (HRF). Some relevant genes (e.g., *hrm*U and *hrm*A), the expression of which was controlled by the host and also some open reading frames (ORFs, such as *hrm*I, *hrm*R, *hrm*K and *hrm*E) related with a set of transcriptional inhibitor proteins, were discovered as affecting the transcription of genes, which induce the formation of hormogonia in studies of *Nostoc-Anthoceros* symbioses (Cohen and Meeks 1997; Meeks et al. 1999; Campbell et al. 2003; Adams et al. 2013; Chang et al. 2019). In *Gunnera*, for example, soluble sugars (sucrose, glucose, and fructose) accumulated in gland tissue were presumed to be HRF factors that also enhance vegetative growth (Khamar et al. 2010). In some cases, depending on their extracellular concentration, the same substances can serve either as a chemoattractant or as a HRF (Liaimer et al. 2015). They can also be autogenic, produced by the cyanobiont. Evidence on the multifaceted character of the specific peptides nostopeptolides, released by *Nostoc*, was provided by Liaimer et al. (2011, 2015) and was also applicable to the similar nostocyclopeptides (Liaimer et al. 2016). In cycad cyanosymbioses, phenolics were found in the cortical layers of the host and in the mucilage, surrounding the cyanobionts, presuming their role in signaling (Caiola 1980; Obukowicz et al. 1981; Lobakova et al. 2004; Chang et al. 2019).

2.1.1 Cyanosymbioses with Bacteria

These symbioses include the naturally co-occurring filamentous N-fixing soil cyanoprokaryotes and non-photosynthetic bacteria (Fig. 1), such as pseudomonads, which usually adhere to the algal mucilaginous sheaths and are commonly concentrated around the heterocytes or their connections with vegetative cells (Car and Whitton 1982; Paracer and Ahmadjian 2000; Rai 2018) but also, according to our observations, can develop on the peculiar, thick-walled dormant and reproductive algal cells, known as akinetes (Smith 1950; Komárek 2013). All these symbioses, in which blue-green algae are the hosts, are considered mutualistic due to beneficial effects for both partners: bacteria use the N compounds excreted by cyanoprokaryotes and, in the same time, consume oxygen with subsequent release of CO_2, the increase of which stimulates the growth of algal filaments, while the decreased oxygen levels enhance the process of N fixation (Paracer and Ahmadjian 2000).

In the field of biotechnology, successful experiments with significantly increased methane yield have been obtained by co-culturing non-heterocytous filamentous terrestrial cyanoprokaryote genus *Oscillatoria* with different plankton microorganisms (Yeung et al. 2017). Consortia and biofilms formed by blue-green N-fixing algae (*Anabaena*, *Calothrix*, and *Nostoc*) with non-photosynthetic bacteria (plant-growth promoting rhizobacteria, phosphor solubilizers, etc.) successfully applied in pot and greenhouse experiments for improving soil and plant parameters of chickpea, corn, cotton, rice, and wheat for agricultural needs have been comprehensively reviewed by Alvarez et al. (2021).

2.1.2 Cyanosymbioses with Eukaryotic Algae

Currently, an emerging approach to maximize the effectiveness of indoor or outdoor cultivation systems appears to be the use of auto-heterotrophic symbiotic model of lichen symbioses, which leads to increased crop yields, better stability, easier maintenance, and reduced costs (Singh et al. 2020). These new biotechnological symbioses include cyanoprokaryotes and eukaryotic algae from different habitats and sometimes are enriched by bacteria (Singh et al. 2020). Experiments with a native co-culture of the soil non-heterocytous filamentous cyanoprokaryote genus *Leptolyngbya* and green coccal alga *Chlorella* (Fig. 1) indicated the great biotechnological potential of such consortia in regard to obtaining high-value products, although the exact mechanisms of their metabolic interactions are not yet clear (Silaban et al. 2014; Lutzu and Dunford 2018). In this exact case, the lipid content and biomass were higher, when cultivated mixotrophically in comparison with autotrophic growth, with the following respectful values: 18.2% vs 8.7% for the lipid content and 134 vs 66 g m^{-3} d^{-1} for the biomass productivity (Silaban et al. 2014).

2.1.3 Cyanosymbioses with Fungi (Lichens and *Geosiphon*)

2.1.3.1 Extracellular Cyanosymbioses with Fungi (Lichens)

The discovery in the South Chinese Doushantuo Formation (551–635 MA) of the earliest fossil filamentous hyphae surrounding unidentified to species level, coccal algal cells suggested that associations resembling modern lichens have been developed in aquatic habitats before the colonization of land by tracheophytes (Yuan et al. 2005) (Fig. 1). Much better preserved thalli from the Early Devonian (415 MA), consisting of septate hyphae of a presumably ascomycetous fungus and containing blue-green algae, led to the description of the fossil species *Cyanolichenomycites devonicushas* and proved once again that lichenification is an ancient event (Honegger et al. 2013). All collected evidence suggests that partnership between fungi and algae occurred several times, with repeated losses and independently in different groups (Lutzoni et al. 2001, 2004; Nash III 2008; Nelsen et al. 2009). In these associations, named lichens, dikaryotic fungi are commonly termed as mycobionts and algae—as photobionts or phycobionts.

Today, a large number of lichen-forming species (ca. 20,000) exists in association with just a few (ca. 150) photobionts (Pardo-De la Hoz et al. 2018) (Fig. 1). Some bipartite lichens contain Cyanoprokaryota as sole, primary photobionts, which usually form a layer below the upper cortex of the thallus, and are commonly named cyanolichens (e.g., Pawlowski 2009). But in some bipartite lichens, the cyanobiont is leading in the shaping of the thallus structure, and such lichens with a typical mucilaginous outfit are named gelatinous (jelly-like) lichens (Nash III 2008). More than 1500 species are known as cyanolichens (approx. 10% of existing lichens—Nash III 2008; Honegger et al. 2013), which are important for N input into terrestrial ecosystems (Hobara et al. 2006; Nash III 1996, 2008; Graham et al. 2018; Singh et al. 2017). Although rarer (in 4% of extant lichens—Honegger et al. 2013), blue-green algae can occur also as a third partner, an accessory photobiont in lichens which already have eukaryotic green algae as primary photobionts (Nash III 2008; Pawlowski 2009). In these lichens, known as tripartite, cyanobionts are not essential for thallus morphology and develop in specific small, discrete cephalodia, which can be external (visible as dark spots on the surface, e.g., *Peltigera*—Fig. 5) or, less commonly, are internal, immersed in lichen tissues (*Lobaria*) (Stewart et al. 1983; Nash III 1996, 2008). Exceptionally, both photobionts form distinct layers in the thallus (*Solorina crocea*), with cyanobiont (*Nostoc*) layer positioned beneath the layer of the green alga *Coccomyxa* (Rai 2018). Usually, there is a single cyanobiont per thallus, but development of different cyanobionts in different cephalodia of the same thallus has been reported (Rai 2018). Such cases are known from lichens *Lecidea* and *Pelobotrya* containing both coccal *Gloeocapsa* and filamentous heterocytous *Fischerella* (Rai 2018). Some lichens can be tripartite or bipartite, depending on the ecological conditions, as it is well exemplified by *Stereocaulon*, which in alkaline or neutral habitats develops cephalodia, but is bipartite in acid sites (Stewart et al. 1983). The role of cyanobionts in bi- and tripartite lichens is different, although in both cases they provide N to the mycobiont, but in bipartite lichens, they also have to fix CO_2 (Stewart et al. 1983). However, the intensity of processes in bi- and tripartite lichens is different: usually the cyanobionts of tripartite lichens show higher level of N fixation than the bipartite cyanobionts, while the green photobiont

Fig. 5 *Peltigera aphtosa*—foliose lichen thallus with cephalodia, containing blue-green algae on the upper surface (visible as dark spots)

is more photosynthetically active (Pawlowski 2009). In addition, cyanobiont produces the UV-protecting pigment scytonemin and mycosporine-like amino acids (MAAs) that are important for occupying some open habitats at higher latitudes and altitudes (Pathak et al. 2019; Singh et al. 2020).

Some bipartite lichens with green photobionts are associated with "free-living" cyanoprokaryotes, which develop on the upper surface of the thalli and supply the hosts with fixed nitrogen. Sometimes these "free-living" blue-green algae are covered by lichen hyphae, forming so-called pseudocephalodia (Poelt and Mayrhofer 1988; Pawlowski 2009).

Extremely interesting are the very rare photosymbiodemes, formed by tripartite lichens, in which the blue-green and green algae alter as primary photobionts in different parts of the same thallus (Pawlowski 2009). In some cases, such thalli may separate into two different morphotypes, which even can live independently (Pawlowski 2009). At present state-of-art photosymbiodemes are known only in some species of *Lobaria*, *Nephroma*, *Peltigera*, *Pseudocyphellaria*, and *Sticta* (Pawlowski 2009).

One of the usual cyanobionts in lichens is *Nostoc*, probably due to its stress-resistant features (Graham et al. 2018), and many studies addressed its identity and phylogenetic relationships revealing a considerable diversity of the strains (e.g., O'Brien et al. 2005, 2013; Dworkin et al. 2006; Otálora et al. 2013; Fedrowitz et al. 2011; Rikkinen 2013; Magain and Sérusiaux 2014; Miadlikowska et al. 2014; Magain et al. 2016; Pardo-De la Hoz et al. 2018). Phylogenetic analyses of a worldwide dataset, based on *rbc*LX genetic region, revealed that *Nostoc* cyanobionts are different from free-living strains in surrounding soils in a given site but are close to other symbiotic strains of the genus, some of which even from other continents (Joneson and O'Brien 2017). Based on these data, it was supposed that different variables, but not the physical proximity, drive the symbiotic lifestyles within given *Nostoc* lineages (Joneson and O'Brien 2017). It was supposed that factors driving lichen symbiotic associations vary at spatial and temporal scales, and symbiont selectivity operates from mycobiont toward cyanobiont, and vice versa (Pardo-De la Hoz et al. 2018). The high mycobiont selectivity for its partner was shown, with only a few strains of *Nostoc* having symbiotic competence to associate with a certain fungal species (Pawlowski 2009). It was hypothesized also that bipartite and tripartite lichens have different specificity patterns toward *Nostoc* with tripartite lichens being associated with less *Nostoc* lineages (Elvebakk et al. 2008; Pardo-De la Hoz et al. 2018). Difference in cyanobionts in bi- and tripartite lichens of the same genus has been demonstrated: all bipartite *Nephroma* species contain one group of *Nostoc* strains, different from *Nostoc* in tripartite *Nephroma* species (Fedrowitz et al. 2012). In the same time, a common cyanobiont can be shared between different co-occurring lichen populations (formed by different fungal genera and even families) in the so-called photobiont-mediated guilds (Rikkinen et al. 2002; Pawlowski 2009). This is explained by the fact that many lichens share common environment with a common pool of free-living cyanoprokaryotes (Dworkin et al. 2006).

Along with *Nostoc*, many other cyanoprokaryote genera occur as cyanobionts in ascomycetous lichens, such as heterocytous *Calothrix*, *Dichothrix*, *Fischerella*, *Hyphomorpha*, *Mastigocladus*, *Scytonema*, and *Stigonema*, as well as the *Aphanocapsa*, *Chroococcus*, *Chroococcidiopsis*, *Cyanosarcina*, *Entophysalis*, *Gloeocapsa*, *Gloeothece*, *Hyella*, *Myxosarcina*, and *Synechocystis* (Dworkin et al. 2006; Nash III 2008). By contrast, only three genera, *Chroococcus*, *Rhizonema*, and *Scytonema*, occur in basiodiomycetous lichens (Dworkin et al. 2006; Nash III 2008; Lücking et al. 2014). However, identification of some of these cyanobionts may be disputable (e.g., Rai 2018), and further investigations, based on polyphasic approach, are necessary.

Lichen cyanobionts have been proved to produce toxins, similar to those excreted by free-living blue-green algae (for details see Kaasalainen et al. 2012; Gärtner et al. 2021; Ivanov et al. 2021; Stoyneva-Gärtner et al. 2025, this volume), but the importance of these toxic substances for symbiont competence in case of lichens needs to be investigated. In regard to symbiotic specificity, the mode of lichen reproduction can be presumed as its important driving factor (Pardo-De la Hoz et al. 2018). Sexually reproducing species (based on the sexual reproduction of the mycobiont) require the horizontal perpetuation of symbiosis, by new acquisition of a "fresh" cyanoprokaryote from the environment (Dworkin et al. 2006; Pardo-De la Hoz et al. 2018). This likely results in high specificity (Paulsrud et al. 1998; O'Brien et al. 2009, 2013; Otálora et al. 2013; Dal Grande et al. 2012; Magain et al. 2017). Vegetatively reproducing lichens (through specific propagules, such as soredia, isidia, phyllidia, or hormocystangia, which contain algal cells along with the fungal cells) promote direct, vertical transmission of cyanobiont to the next generation through its re-acquirement from the preexisting thallus (Dworkin et al. 2006; Rai 2018). In these cases, each lichen species may harbor lower species diversity, and therefore, such lichens may represent evolutionary "dead ends" (Del-Prado et al. 2016). The close relation between the lichen propagation mode and genetic diversity of mycobionts and cyanobionts was demonstrated in a comparative study of two species of the same genus, which differ in having or not sexual reproduction (Otálora et al. 2013). Among them, the sexual bipartite *Degelia plumbea* exhibited much higher genetic diversity of both partners in comparison with the asexual *Degelia atlantica*, limiting its potentially occupied ecological niches. Similar results on importance of reproduction mode were obtained in the comparative study of tripartite *Peltigera* lichens from sections *Chloropeltigera* and *Peltidea*, in which photobionts were *Nostoc* and the green coccal alga *Coccomyxa*, based on a worldwide phylogenetic framework (Pardo-De la Hoz et al. 2018). In both sections, *Coccomyxa* was represented by the same two species, but differences were found in the diversity of mycobionts and of *Nostoc*: eight fungal species (five of which new) in association with nine *Nostoc* phylogroups in *Chloropeltigera* and eight fungal species (including three newly delimited) with only four *Nostoc* phylogroups in *Peltidea*, suggesting more prevalent horizontal transmission of the cyanobiont in *Chloropeltigera*, and more common vertical transmission in *Peltidea*. These differences impacted also the distribution: all *Chloropeltigera* species are generalists, while species of *Peltidea* are more specialized. All lichens of both sections, strictly associated with a single species of *Coccomyxa* (*C. subellipsoidea*), have constrained distributions, indicating that factors shaping geographic ranges of tripartite *Peltigera* include the availability of the green alga in addition to the specificity of interactions between the partners, to the symbiont distribution and to the constraints imposed by climatic factors (Pardo-De la Hoz et al. 2018).

In some cases, lichens and bryophytes share similar habitats, having far more complex and complicated interactions that presumed before, with cases when bryophytes provide a substratum for the juvenile stages of lichens (Cornejo and Scheidegger 2016). Moreover, both bryophytes and lichens can associate with the same cyanoprokaryotes, thus making the blue-green algae an interaction point between these two different organisms. Investigation of the epiphytic on the balsam fir tree, small, leafy liverwort *Frullania asagrayana*, which is a substrate for the foliose cyanolichen *Erioderma pedicellatum* (boreal felt lichen), demonstrated such mediating role of blue-greens in a complex, tri-membered partnership (Cornejo and Scheidegger 2016; Cornejo et al. 2016). *Frullania*, a type of leafy liverworts, has the so-called "water sacs," formed between their lateral (upper) and lower (ventral) leaves, which have a small opening to the environment and retain water (Düll and Düll-Wunder 2008). The water sacs of the *F. asagrayana* contain blue-green algae (considered by some authors as epiphytic), which according to DNA barcoding analyses belong to the filamentous heterocytous genus *Rhizonema* (Cornejo and Scheidegger 2016). According to Maas and Yetman (2002); lichenization between the germinating ascospore of *Erioderma* mycobiont and *Rhizonema* can occur only in the water sacs of the liverwort, where the juvenile lichen thallus is formed and needs about 5–10 years to reach a visible size.

Modern approaches, which use DNA/RNA-based and microscopic techniques (e.g., FISH and confocal laser scanning microscopy), have revealed that diverse microbial communities are associated with lichens (Grube and Berg 2009). The harbored bacteria are diverse, but majority of them belongs to Alphaproteobacteria and many form epiphytic biofilm-like structures. Nonetheless that their interactions with lichens are not fully understood, it was supposed that bacteria may have more than a single function and through producing of bioactive substances (e.g., hormones, antibiotics), complementing the N fixation or exhibiting lytic activities, may support a complex functional network in a profitable way (Grube and Berg 2009).

2.1.3.2 Intracellular Cyanosymbioses with Fungi (*Geosiphon pyriforme*)

The intracellular symbiosis between the glomeromycote *Geosiphon pyriformis* and *Nostoc* is unique, being the only known syncyanosis (endocyanosis) of a fungus (Geitler 1959; Kluge et al. 1994; Schüßler et al. 1994; Schüßler and Kluge 2001; Alexopolous et al. 2004) (Fig. 1). Discovered by the Austrian botanist Friedrich (Fritz) von Wettstein (1895–1945), this association has so far only been found in a single location, where it inhabits the surface layer of moist loamy or silty (with quartz particles) soils with a low pH (3–4) and stands heavy inundation caused by heavy rain or dew (Mollenhauer 1988; Reisser 1992). These soils are nutrient-deficient, mainly poor in phosphates (Schüßler et al. 1994), and this, together with all other peculiarities of this unfavorable habitat, to great extent explains the need for joint life of both partners: a need for additional feeding of the fungus through incorporation of a photosynthetic and N-fixing partner, but also a necessity for the alga to overcome the unfavorable pH and intermittent moistness, as well as to be supplied water, phosphate, CO_2, and other inorganic nutrients (Reisser 1992; Dworkin et al. 2006). Current studies revealed that phosphorus deficit provokes strong stress on the cyanobiont and, in this way, can promote *Geosiphon* symbiosis (Dworkin et al. 2006).

Considering the taxonomy of symbionts, it is to recall that first the host was thought to be a green heterotrophic siphonal alga, which contains chitin in the cell walls (von Wettstein 1915). Later, Knapp (1933) identified it as a fungus with aseptate mycelium and classified the association as an intracellular phycomycetal lichen (Schüßler et al. 1994). Today it is clear that the fungal partner belongs to the phylum Glomeromycota and, according to spore morphology (confirmed as a reasonable diagnostic feature by sequence analysis of rRNA small subunit), resembles another glomeromycote genus *Glomus* (Schüßler 2002; Malar et al. 2021). Glomeromycote fungi, forming obligatory symbiotic relations with roots of land plants, known as arbuscular mycorrhizae, are believed to have originated before the land plants, about 500 million years ago (Heckman et al. 2001; Schüßler 2002; Malar et al. 2021). In this way, the carbon supply from a cyanobiont was advantageous for the archaesporous *Geosiphon* and, most probably, its symbiosis with cyanoprokaryotes appeared quite early, long before the land plants evolved, and can be seen as an important step in land colonization (Schüßler 2002; Dworkin et al. 2006). The most striking peculiarity of *Geosiphon* in comparison with the other species of Glomeromycota is the transformation of part of the plagiotropic coenocyte hyphae into pear-shaped overground multinucleate vesicles (bladders), each longer than 1–2 mm (Reisser 1992). The usual cyanobiont is *Nostoc*, with species *N. punctiforme* being the most common one (Kluge et al. 1994). The result of *Nostoc* incorporation is that bladders become photosynthetically active and N-fixing and easily visible over the soil surface (Knapp 1933; Mollenhauer 1988; Kluge et al. 1991; Reisser 1992; Dworkin et al. 2006). Their ultrastructure was investigated by Schnepf (1964), who showed that *Nostoc* filaments are enclosed in the single, sack-like peripheral compartment of the uppermost part of each bladder, separated by a membrane. This structure, known as phagosome or symbiosome, occupies 2/3 bladder part above the centrally positioned numerous vacuoles and lipid droplets in the basal portion, which is cyanobiont-free (Kluge et al. 1991, 1992, 1994; Dworkin et al. 2006). During the growth of the bladder, the cells of cyanobiont multiply and significantly increase in volume (up to eight times), but the number of heterocytes and ultrastructure of all cells remain similar to this in filaments of free-living *Nostoc* (Dworkin et al. 2006). Therefore, it seems that the primary function of the *Geosiphon* cyanobiont is the photosynthesis, and not the N fixation (Dworkin et al. 2006). This function is most likely to be attributed to other prokaryotic cells, harbored by *Geosiphon* and commonly abbreviated as BLOs (bacteria-like organisms) (Dworkin et al. 2006). Despite that in *Geosiphon* they are not enclosed by host membrane, similar BLOs are commonly found in other mycorrhizal fungi of Glomeromycota and are thought to be horizontally transferred ancient inhabitants (Dworkin et al. 2006).

It is possible to culture *Geosiphon pyriforme* in the laboratory, at least for some months (Mollenhauer and Mollenhauer 1988; Reisser 1992; Schüßler and Wolf 2005). The symbiosis has been considered facultative (Usher et al. 2007) because it is obligate for the fungus, but not for the cyanobiont (Schüßler 2012), which has to be incorporated anew whenever a vesicle is formed, thereby providing the possibility of different bladders containing different strains (Reisser 1992). This incorporation of the cyanobiont happens just below the growing tip of each hypha, when the fungus "meets" the algal colony (Reisser 1992). At contact, part of fungal wall breaks, the outcoming cytoplasm protrusion intakes the algae and subsequently is covered by a new wall (Reisser 1992). The occasional character of incorporation is confirmed by the fact that, although occasionally, *Geosiphon* can uptake other phototrophs and in a single case ingested green alga *Coccomyxa*, forming a new bladder (Reisser 1992).

Since this fungus has been found together with bryophytes (mosses, liverwort *Blasia* and hornwort *Anthoceros*) to found peculiar synusia in ephemeral habitats (Mollenhauer 1988; Reisser 1992), it can share cyanobiont with them (Dworkin et al. 2006). Therefore, it is possible to suppose that future studies will reveal more complicated relations between more partners. Possible relations of *Geosiphon* with vascular plants to form arbuscular mycorrhizae were also speculated (Dworkin et al. 2006). *Geosiphon* was recognized as an important model for research on endocyanoses and endosymbiosis (Rai et al. 2002); but due to its close resemblance to *Glomus*, it also allowed to speculate the probability of hypothetical association of a *Glomus*-like fungus with a green alga to be a step during the evolution of the land plants (Gehrig et al. 1996). Currently, the genome of *Geosiphon* was investigated as a model of ancestral arbuscular mycorrhizal fungi, and a large proportion of transposable elements was identified along with evidence for horizontal transfer of genes between fungus and *Nostoc* (Malar et al. 2021). Moreover, for *Geosiphon*, the absence of thiamine metabolisms and synthesis of fatty acids were demonstrated, whereas specific mating type (MAT) locus and a meiosis-specific gene were detected (Malar et al. 2021).

2.1.4 Cyanosymbioses with Bryophytes

2.1.4.1 Endocyanosymbioses with Bryophytes

Bryophytes (hornworts, liverworts, and mosses) diverged from extant plants about 400 MA ago, with hornworts considered to be the earliest derived lineage (Li et al. 2020; Frangedakis et al. 2021). Despite the poor fossil records preclude from precise placing of hornworts within the evolutionary timescale (Villarreal and Renzaglia 2015), they are accepted as an ancient group, sister to vascular plants, and currently its members are widespread in both tropical and temperate regions, where they occupy moist but nitrogen-poor substrates (e.g., Desirò et al. 2013). We believe that these habitat characteristics explain the need to recruit N-fixing algae (which have been shown to supply the modern bryophytes with N in the form of NH_3, Bergman et al. 1992) and that evolutionary origins of all these symbioses may trace back to pre-Permian times (Rikkinen and Virtanen 2008). Hornworts are considered to be among the oldest land plants that developed cyanosymbioses, some of which exhibit similarity with associations of cyanoprokaryotes with the Devonian land plant *Aglaophyton* (Villarreal and Renzaglia 2015).

The extant bryophyte cyanosymbioses comprise the permanent partnership between terrestrial hornworts from genera *Anthoceros*, *Dendroceros*, *Notothylas*, and *Phaeoceros* and liverworts from genera *Blasia*, *Cavicularia*, *Pellia*, *Riccardia*, and *Riccia* with cyanoprokaryotes, mostly from genus *Nostoc* (Duckett et al. 1977; Bergman et al. 1992; Costa et al. 2001; Meeks 2003; Mishra et al. 2018) (Fig. 1). These symbioses were first noted by the German botanist Wilhelm Friedrich Benedikt Hofmeister (1824–1877) (Hofmeister 1862). The interest to them was provoked by the partially pioneering character of bryophytes, which usually grow on poor soils of disturbed sites and other habitats unfavorable for more evolved land plants (Reisser 1992). In 1955, for first time, the N-fixing capacity of these symbioses was demonstrated using ^{15}N in *Blasia pusilla* (Bond and Scott 1955). Since then, a lot of evidence was accumulated that bryophyte cyanosymbioses are a key source for N input in terrestrial ecosystems in nutrient-limited regions (Rai 2018; Pi et al. 2018). Although the benefits for cyanobionts have been questioned, the cyanosymbioses with hornworts and liverworts were noted as the most ancient among the N-fixing symbioses (Sprent and Raven 1985; Rai 2018). However, taking into account the high number of all bryophyte species, it seems that cyanosymbioses, and endosymbiosis in particular, are exceedingly rare (Rai 2018).

In hornworts, filaments of *Nostoc* inhabit the specific "slime cavities" (mucilage-filled cavities) of the thalloid gametophytes, situated at their lower (ventral) surface, while in thalloid liverworts they are housed also on the ventral side but in hemispherical auricules, named domatia that develop from the slime-producing hairs (Duckett et al. 1977; Renzaglia 1982; Renzaglia et al. 2000; Costa et al. 2001; Adams and Duggan 2008; Bouchard et al. 2020). The infection in hornworts happens through the two-celled mucilage clefts (stomata, slime pores) on the thalli epidermis (Frangedakis et al. 2021). These clefts are ephemeral, and after infection, the number of epidermal cells increases along with their enlargement, closing the cleft opening to the cavity. In this way, infection by other algae is prevented, but also lower oxygen conditions are ensured to enhance the process of N fixation (Goffinet and Shaw 2009; Mishra et al. 2018; Frangedakis et al. 2021). In liverworts, it is also the excreted mucilage that attracts the *Nostoc* hormogonia, which become "trapped" in the auricle, which subsequently grows and closes off (Renzaglia 1982; Renzaglia et al. 2000). Due to the presence of numerous mucilage clefts on the thalli surfaces, different *Nostoc* hormogonia or young colonies can enter (Renzaglia 1978; Rai et al. 2002; Villarreal and Renzaglia 2006), and this explains the cyanobiont diversity, described below in the text. The origin of the mucilage was not clarified, and it was supposed that it is produced or by specific penetrating cells ("slime papillae") of the host, the formation of which is thought to be induced by the cyanobiont or by the cyanobiont itself (for details see Duckett et al. 1977). The last opinion is supported by the fact that *Nostoc* produces mucilage envelopes around each trichome, forming filaments, which are embedded in the common slime of the colony (Komárek 2013). The abovementioned "slime papillae" penetrate the slimy colonies of the cyanobiont and are always multicellular but differ in hornworts and liverworts: in *Anthoceros*, they are thin-walled cells with numerous vacuoles, which occasionally may contain intracellular filaments of the cyanobiont, while in *Blasia* they are thick-walled and usually develop labyrinthine-like wall ingrowths (Duckett et al. 1977; Rodgers and Stewart 1977), comparable only with those in *Azolla* (Duckett and Prasad 1974). These ingrowths increase the absorption area, thus enhancing the partner interconnection, and in the interpretation of Duckett et al. (1977), their absence in hornworts can be caused by higher nutrient content of *Anthoceros* habitats or by its less harmonious relationship with algae.

The endophytic algal life is restricted to gametophytes (which dominate in the life-cycle of bryophytes), and this explains the need for a constant spring recruitment of new bryophyte generations by cyanobionts mainly from surrounding soil, but also from nearby bryophytes or from lichens containing cyanoprokaryotes (cyanolichens) (Costa et al. 2001; Bouchard et al. 2020). The influence of genetic mutations on the infection capacity and effectiveness of *Blasia* cyanobionts was demonstrated using the *cya*C complex gene, which is widespread among cyanoprokaryotes and encodes the enzyme adenylate cyclase (Adams and Duggan 2008). Outgrowth of *Nostoc* onto *Anthoceros* gametophyte can also cause additional infections of growing bryophytes, as was demonstrated by Enderlin and Meeks (1983). An exception from this rule was reported by Costa et al. (2001), who discovered that, by contrast to hornworts, *Blasia* contained *Nostoc* cyanobionts in one type of its asexual propagules, specifically in apical gemmae, which were completely developed before their detachment from the main thalli. Later, the same mode of propagation was recorded from *Cavicularia densa* (Rikkinen and Virtanen 2008).

Once, *Calothrix* was identified as a cyanobiont in the hornwort *Phaeoceros laevis* (West and Adams 1997), but in almost all bryophytes the cyanoprokaryote partner belongs to *Nostoc*. In single gametophytes of *Anthoceros* and *Blasia*, different *Nostoc* strains were identified according to morphological, ultrastructural, and biochemical features (phycoerythrin content), suggesting the non-specificity of the algal-bryophyte association (Duckett et al. 1977; Enderlin and Meeks 1983; Rai 2018). The non-specificity was also proved by experiments, in which isolates from *Anthoceros* may infect *Blasia*, and vice versa, with the same results for isolates from *Cavicularia*, cycads, and *Gunnera* (for details see Duckett et al. 1977). In addition, comparative studies of liverworts showed no geographical dependence, revealing similarity between the dominant *Nostoc* genotype in *Cavicularia densa* from Japan to this in *Blasia pusilla* from Finland (Rikkinen and Virtanen 2008). The few 16S RNA and PCR-based investigations proved the cyanobiont diversity from individual thalli of *Blasia pusilla* in the same location (Liaimer et al. 2016) and demonstrated great genetic diversity of cyanobiont *Nostoc* strains in different cavities within a single gametophyte of *Anthoceros fusiformis*, *Phaeoceros laevis*, and of *Blasia pusilla* (West and Adams 1997; Costa et al. 2001; Rai et al. 2002). However, in each single cavity, only one *Nostoc* strain develops (Costa et al. 2001; Liaimer et al. 2016) and gametophytes containing a single cyanobiont strain were also recorded (Costa et al. 2001). Moreover, with a single exception, there was a clear sequence dissimilarity between symbiotic and free-living strains (West and Adams 1997; Liaimer et al. 2016). It was demonstrated also that cyanobionts do not occur free-living in large numbers at a distance from the host, but they can survive epiphytically on its surface (West and Adams 1997). The persistence of the same *Blasia* cyanobiont genotypes through several years in the same location has been recorded (Costa et al. 2001). Together with differences between free-living and symbiotic genotypes, this persistence was interestingly related by Liaimer et al. (2016) with possibility symbiotic strains to "hide" in bryophytes; in cases when being free-living in the soil, they have been allelochemically outcompeted. In this way, Liaimer et al. (2016) used allelopathic pressure to explain in part the genotypic differences between symbiotic and free-living algae.

The potential of symbiotic *Nostoc* strains to produce cytotoxic and cytostatic compounds against mammalian cell lines was demonstrated: in a set of 82 *Nostoc* terrestrial isolates, ten were symbiotic, among which 60% had such capacity (Hrouzek et al. 2011). In a further investigation of 800 *Nostoc* isolates from four thalli of *Blasia*, microcystins and nodularins along with some nostocyclopeptides and nosperin, as well as some novel aeruginosins and banyaside-like compounds (for details on these cyanotoxins see Gärtner et al. 2021; Stoyneva-Gärtner et al. 2025, this volume) together with yet unidentified cytotoxic substances, were detected (Liaimer et al. 2016). This implies that host cannot discriminate the toxigenic strains and even more, microcystin producers dominated among isolated cyanobiont strains but were not found in *Nostoc* that was free-living in the soil around the studied liverworts (Liaimer et al. 2016). Moreover, the microcystin-producers were acquired more often in comparison with the other strains (Liaimer et al. 2016). A gene cluster for nostocyclopeptides, identified in cyanobionts of *Blasia*, allowed to suppose that, due to their ability to regulate hormogonia motility, these nostocyclopeptides played role in establishment of the symbiosis (Warshan et al. 2018). Regarding the unidentified cytotoxic compounds, it is to note that extracts from four cyanobiont isolates were shown to inhibit cell growth of metastatic human melanoma A2068 cell line and two of them inhibited MRC5 fibroblasts (for details see Liaimer et al. 2016).

Although with different success, reconstruction of bryophyte cyanosymbioses was accomplished in the lab, with a range of symbiotic and free-living *Nostoc* strains infecting *Anthoceros* (Enderlin and Meeks 1983), *Blasia* (Rodgers and Stewart 1977), and *Phaeoceros* (West and Adams 1997). A successful experimental reconstitution of symbiosis between 70 strains of *Nostoc*, *Calothrix*, and *Chlorogloeopsis* with *Blasia* and *Phaeoceros* was achieved (West and Adams 1997). According to modern transcriptomic studies, about 40 genes (transporters, receptor kinases, and transcription factors) are possibly involved in hornwort cyanosymbioses (Li et al. 2020). They include the special epidermal pattering factor-like (EPF-like) gene (from the large EPF-genes family encoding plant-specific secretory peptides), which may be responsible for the formation of mucilage clefts and their functioning (Li et al. 2020; Frangedakis et al. 2021).

A new association between *Nostoc* and the hornwort *Leurosporoceros dussii* was described from two sites in central Panama (Villarreal and Renzaglia 2006; Bouchard et al. 2020). This neotropical association is novel among extant symbioses according to the strikingly different arrangement of cyanobionts, visible from the surface as long and dichotomously branched strands, extending along the main thallus axis (Villarreal and Renzaglia 2006; Adams and Duggan 2008). In all investigated gametophytes, regardless of their age, algal filaments inhabit peculiar schizogenous (originated through cell-separation) channels filled with mucilage, which occupy 20–50% of the thallus, forming an integrated endophytic network (Villarreal and Renzaglia 2006; Bouchard et al. 2020). Since the elongation and branching of the strands is synchronized with the apical growth of the thallus, a single invasion is sufficient for development of these strands (Villarreal and Renzaglia 2006). Lack of cyanobiont specificity was supposed for this tropical association after two different cyanobiont (*Nostoc*) morphotypes were found (Villarreal and Renzaglia 2006). Further genetic studies identified six *Nostoc* genomes that formed a polyphyletic group (Bouchard et al. 2020). There, some of the *Nostoc* strains were distributed among four clades, different from those found in symbioses, and it seemed that *Nostoc* diversity depended on the soil type (Bouchard et al. 2020). The lack of cyanobiont specificity was interpreted as surprising, since the ventral clefs typical of all hornworts are only formed during the sporling-rosette stage in this tropical species and do not develop further after algal infection. In this way they prevent any further invasion, but also provide conditions to evolve partner specificity (Bouchard et al. 2020). However, following the descriptions of Villarreal and Renzaglia (2006), the lack of specificity is explainable by the early infections, because each rosette-sporeling contains numerous lobes, each of them bears many mucilage clefts, and individual potential cyanobiont enters into each of these clefts independently of the other hormogonia or young colonies. *Leurosporoceros-Nostoc* symbiosis is also interesting because the cyanobionts contain the unusual for tropics V nitrogenase, which possibly originates from rich in V volcano soils of the region (Nelson et al. 2019). In addition, peculiarity of this symbiosis is related with eventual fungal inhibition from the cyanobiont, since *Leiosporoceros* is usually considered to be a fungus-free hornwort (Desirò et al. 2013). This presumption was based on some earlier indications that, in cases of abundant cyanobionts, bryophytes seemingly do not host fungi (Desirò et al. 2013).

Another newly supposed endophytic association is between *Microcoleus* (*M. vaginatus*) and *Mnium cuspidatum*, collected from a *Larix* forest of Guandi Mountain, Chanxi Province of northern China (Zhang et al. 2014) (Fig. 1). Despite provided morphological descriptions and phylogenetic data, authors only presume its endophytic character because it is impossible to remove algae by washing of the thalli (Zhang et al. 2014). If further studies prove the endophytic life of *Microcoleus*, this will be the first recognized endophytic cyanosymbiosis with a moss.

2.1.4.2 Ectocyanosymbioses with Bryophytes

Liverworts *Cryptochila*, *Frullania*, *Marchantia*, and *Porella* are known to have epiphytic associations with cyanoprokaryotes (Rai 1990; Adams and Duggan 2008; Arróniz-Crespo et al. 2014; Cornejo and Scheidegger 2016; Dworkin et al. 2006) (Fig. 1), the first of them detected on *Porella navicularis*, endemic for North America (Dalton and Chatfield 1985; Rai 2018). In comparative studies of N fixation at glacier foreland of Cordillera Darwin (Tierra del Fuego), only one of six investigated liverworts, the terricolous *Cryptochila grandiflora*, exhibited high N fixation rates (Arróniz-Crespo et al. 2014).

There is growing evidence on epiphytic associations between N-fixing blue-greens and different bryophytes and their contribution to N accumulation, which helps to promote soil development (Jordan et al. 1978; Stewart et al. 2011; Solheim et al. 2004; Arróniz-Crespo et al. 2014; Liaimer et al. 2016). Epiphytic blue-greens and N fixation were demonstrated for the mosses *Ceratodon purpureus* and *Wessia controversa* (for details see Rai 2018). In the last decade, more attention is paid to the extracellular and epiphytic symbiosis of pleurocarpous feathermosses and *Nostoc* (DeLuca et al. 2002; Ininbergs et al. 2011; Bay et al. 2013; Warshan et al. 2017; Mishra et al. 2018). Although all benefits for the host were not fully clarified, it was supposed to be mutualistic, since the moss tissue became enriched in N over time (Bay et al. 2013), and moss, in turn, provides sulfur to the cyanobiont (Warshan et al. 2017). Moreover, for first time, nitric oxide and aliphatic sulfonates were suggested as eventually involved in establishment of this peculiar ectosymbiosis, and 32 gene families were identified only among cyanobionts (Warshan et al. 2017). The N fixation was demonstrated in cases of *Nostoc* associated with the hylocomian red-stemmed feather moss *Pleurozium schreberi* (DeLuca et al. 2002; Zackrisson et al. 2004; Bay et al. 2013) and hylocomian stair-step moss *Hylocomium splendens* (Solheim et al. 2002, 2004; Bay et al. 2013) (Fig. 1). Six genera were identified morphologically and genetically as epiphytes of *H. splendens* and *Bryum pseudotriquetrum*: *Asterocapsa*, *Calothrix*, *Gloeocapsa*, *Microchaete*, *Nostoc*, and *Stigonema* (Solheim et al. 2004; Gentili et al. 2005; Liu and Rousk 2021). For three of them, *Calothrix*, *Nostoc*, and *Stigonema*, epiphytic on *H. splendens*, physiological differences in temperature optima and relevant changes in N fixation levels seemingly were significantly important for the moss colonization (Gentili et al. 2005). Interestingly, in an experiment with *Pleurozium schreberi*, *Hylocomium splendens*, and two acrocarpous mosses, *Dicranum polysetum* and *Polytrichum commune*, it was shown that all four species induced hormogonia differentiation, but only feather mosses were subsequently colonized (Bay et al. 2013). This result corresponds to the situation in nature, in which *D. polysetum* and *P. commune* also do not form associations with *Nostoc* (Bay et al. 2013), but *Pleurozium* bears algal epiphytes (Scheirer and Dolan 1983). In addition, evidence was provided that colonization on *Pleurozium schreberi* was enhanced during the moss reproductive phase but was impaired by elevated N (Bay et al. 2013).

High rates of N fixation were measured for the terricolous mosses *Ditrichum cylindrocarpum*, *Racomitrium laevigatum*, and *R. lanuginosum* and saxicolous

Acroschisma wilsonii, *Racomitrium didymium*, and *R. subcrispipilum* (Arróniz-Crespo et al. 2014). Relations between the N fixation and stages of ecological succession were supposed for *R. didymium* and *A. wilsonii*, which demonstrated highest rates at longer deglaciated sites of Tierra del Fuego (Arróniz-Crespo et al. 2014).

Epiphytic association between *Anabaena* and *Nostoc* and bonfire moss (common cord-moss) *Funaria hygrometrica* was identified during studies of colonization of Iceland island Surtsey, with demonstration of increased growth and total N contents of the moss (Rodgers and Henriksson 1976). Later, the same algae were revealed by epifluorescent microscopic technique as developing on the leaves, stem, and in the rhizosphere of *F. hygrometrica* (Scheirer and Brasell 1984). In these and other studies of epiphytic moss algae, morphological peculiarities of their surface in relation to retaining of cyanoprokaryote epiphytes were not described (Solheim et al. 2004). By contrast, using different techniques, including confocal laser microscopy, transmission (TEM), and scanning electron microscopy (SEM), special adaptations were found on *Calliergon richardsonii* and *Sanionia uncinata* (Syn. *Drepanocladus uncinatus*) to harbor blue-green algal epiphytes (Solheim et al. 2004). By this study, it was shown that epiphytes of *C. richardsonii* (*Nostoc*) were located on the inner side of the leaves facing the stem, or between the stems and leaves, and some of them were on stems but very close to the leaf bases, whereas the outer side of the leaves was never inhabited. Thus, moss protects the epiphytic algae from direct strong insolation and desiccation (Solheim et al. 2004) and from drastic temperature variations. In addition, authors stressed that exactly the site between the leaf base and the stem is the best place on moss to be colonized since the polysaccharide matrix located there was thought to stabilize the association and to enhance the metabolite transfer between the partners. In *S. uncinata*, the tubular leaves curled around the colonies of *Nostoc*, forming specific grooves in which algae are almost completely enclosed. In this way, the tubular leaf is suitable to maintain high humidity even during the dry summer periods (Solheim et al. 2004). In addition, a colonization by blue-green algae of dead leaf cells was observed, but most probably they serve only as refuge places because no contact with surrounding living cells was found (Solheim et al. 2004). The dependence of the colonization of mosses on morphological features was demonstrated also in a current study by Liu and Rousk (2021). They found higher abundance of epiphytes on shoots with smaller leaves, on shoots with a high frequency of leaves, and, for first time, on the filamentous appendages on *Hylocomium splendens* stems, named paraphyllia. In support of the importance of morphology is the lack of epiphytic cyanoprokaryotes, stressed for the moss *Dendroligotrichum squamosum* and explained by the unfavorable conditions for colonization on their rigid, water repellent leaves (Arróniz-Crespo et al. 2014). In a search for specific moss traits that facilitate epiphytes and regulate N fixation, Liu and Rousk (2021) pointed out also the key importance of the moss hydration rate.

Series of investigations in high-latitude and high-altitude regions revealed important epiphytic development of different species of typical heterocytous *Calothrix*, *Microchaete*, *Nostoc* and *Stigonema*, but also of non-heterocytous *Leptolyngbya*, *Lyngbya*, *Microcoleus*, *Oscillatoria*, and *Phormidium*, and coccal *Asterocapsa*, *Chroococcus*, *Gloeocapsa*, and *Microcystis* in moss communities and their significance as main N source in such pristine areas (Jordan et al. 1978; Madhusoodanan and Dominic 1996; Solheim et al. 2004; Gentili et al. 2005; Arróniz-Crespo et al. 2014; Zhang et al. 2014; Liu and Rousk 2021). Firstly, it was believed that association between feathermosses and *Nostoc* is characterized by low specificity (DeLuca et al. 2002), but later host specificity of *Pleurozium schreberi* and *Hylocomium splendens* in these epiphytic relations was outlined by Bay et al. (2013), supporting the earlier conclusion of Ininbergs et al. (2011). In a study of community composition and diversity of *nifH* phylotypes associated with each of these two feathermoss species, Ininbergs et al. (2011) supposed that the diazotrophic epiphytic communities were determined by the moss species identity, and not by the external environmental conditions. Six different *Nostoc* sequences, forming two sister lineages, were isolated from *P. schreberi* and *H. splendens*, collected from boreal regions (Warshan et al. 2017, 2018). In a comparative study of Warshan et al. (2018), *Nostoc* strains that participate in both epi- and endophytic symbioses (the last being intra- or extracellular) with a broad host-range, form a monophyletic clade, which emerged before the origin of land plants about 600 MA ago. By contrast, *Nostoc* strains restricted only to extracellular symbioses were polyphyletic in origin (Warshan et al. 2018).

2.1.5 Cyanosymbiosis with *Aglaophyton*

Aglaophyton is a fossil sporophyte land plant genus closely related to Rhyniophyta, discovered in Rhynie chert flora (Krings et al. 2009). Only in four axis segments from two different chert blocks, the prostrate axes of *Aglaophyton major* were found to contain endophytic filamentous cyanoprokaryotes of *Archaeothrix*-type (Taylor and Krings 2005; Krings et al. 2009) (Fig. 1). Most probably, the colonization starts through the stomata of the outer cortex and then algal filaments continue to grow in substomatal chambers, subsequently extending throughout the whole intercellular cortex system. To date, this association represents the oldest fossil evidence for cyanoprokaryote-land plants symbiosis, traced back to Lower Devonian, about 410 MA ago (Taylor and Krings 2005; Krings et al. 2009). It has to be noted also that dense aggregations of *Archaeothrix*-type were observed on the surface of some wounded sites of the prostrate axes of *Aglaophyton* (Taylor and Krings 2005), and multiple infection based on accidental events, related with temporary habitat flooding, was presumed (Krings et al. 2009). Of special interest is that cyanoprokaryote filaments were also observed intracellularly, where they coiled, resembling the mycorrhizal hyphae, which form arbuscles in the same plant (Taylor and Krings 2005; Krings et al. 2009). Krings et al. (2009) underlined the lack of heterocytes in the material and supposed that blue-green algae found belong to non-heterocytous order Oscillatoriales, questioning the mutualistic character of the symbiosis. Yet questions on the algal role in this partnership have to be answered, but the possibility host to get N compounds firstly was not excluded (Taylor and Krings 2005). We find a support for this presumption in the fact that the N fixation is not restricted only to heterocytous forms (e.g., Stewart 1977a, b; Gallon et al. 1991; Bergman et al. 1997; Berrendero et al. 2016). Even more interesting is the suggested scenario that the discovered fossils provide evidence for a complex, tripartite consortium, in which blue-green algae have role of "helpers" in the establishment of the mycorrhiza, or they have entered in specific relations with the mycorrhizal hyphae, which enhance the N fixation (for details see Taylor and Krings 2005; Krings et al. 2009).

2.1.6 Cyanosymbiosis with Lycophytes

A symbiotic association of blue-green algae with the extant lycophyte *Huperzia serrata* was discovered (Guo et al. 2012) (Fig. 1). Cyanobionts were isolated from the sporangium of *Huperzia*, but studies were oriented mostly towards their physiological reactions to light, and by series of experiment, it was demonstrated that optimal for the growth was blue or green light with intensity of 500 lx, whereas higher applied intensities (>2000 lx) strongly inhibited the growth (Guo et al. 2012). Extant lycophytes (with classification ranging from order to phylum in different systems) comprise a relatively small number of species and are small vascular plants, but during the Paleozoic Carboniferous period (358.9–298.9 MA), they reached maximum biodiversity and huge forests of their tree-like forms dominated the landscape, contributing to coal deposits, which gave the name of the period (Willis and McElwain 2002; Raven et al. 2005). Lycophytes are one of the oldest lineages of vascular plants, with earliest fossils dated from the Silurian, about 425 MA (Willis and McElwain 2002; Raven et al. 2005).

2.1.7 Cyanosymbioses with Pteridophytes ("Greenhouse" Symbiosis with *Azolla*)

Filamentous heterocytous cyanoprokaryotes form symbiosis with the small heterosporous water pteridophyte *Azolla* (Fig. 1), which is widespread mainly in tropical, subtropical, and temperate regions and contains seven species (Feyisa et al. 2013; Pereira 2017; Pratte and Thiel 2021). This symbiosis is one of the best known due to its broad application in Asian agricultural tradition as a green manure, which could be traced back at least to eleventh century and its additional applications as animal and fish feed, a human food as leafy vegetable and medicine, a water purifier, a phytoremediator, a basis for production of hydrogen fuel, etc. (Hamdi 1982; Whitton and Roger 1989; Pereira 2017), with numerous papers written on the topic. However, negative effects on local biodiversity were detected during *Azolla* invasions and some biological control measures were examined in order to reduce the negative ecological impacts (Dworkin et al. 2006). Currently, this peculiar symbiosis is accepted in a broader sense, as a complex, at least tri-membered synergistic symbiont system, as it will be described below.

The etymology of the fern name, given in 1783 by the French naturalist Jean-Baptiste Pierre Antoine de Monet, chevalier de Lamarck (1744–1829), is commonly explained by the combination of Greek words *azo* and *ollyo* (to dry and to kill) to point its inability to survive drought (Lumpkin 1993). However, more probably, it is related with a local word, used by indigenous tribes in Argentina or Chile, where *Azolla* was firstly collected by the French botanists Philibert Commerson (1727–1773) and Jeanne Baret (1740–1807), during their participation in the expedition, leaded by the French admiral and explorer Louis-Antoine, Comte de Bougainville (1729–1811) (for details see Carrapiço 2010).

Although *Azolla* is known as a freshwater free-floating fern, we describe it among the terrestrial symbiosis because it has amphibial mode of life and can develop on damp soils of the shores of the waterbodies after water withdrawal (Draganov and Stoyneva 1992; Stoyneva 1995). On soils, mostly the red form of *Azolla* was found (Stoyneva 1991), which was considered more resistant to lower temperatures in comparison to two other known forms, green and purple (Tuan and Thuyet 1979). These our observations on *Azolla* (mainly *A. filiculoides*, but also *A. caroliniana*) in the adjacent to the Danube waterbodies, and mainly in the Lake Sreburna, in the end of twentieth century, corroborate with current noting of two ecotypes of *Azolla*, a free floating and a semi-aquatic (Carrapiço 2010). Although "more rare, can be found in sand wet banks, when the water level goes down and where *Azolla* is fixed to the substrate by the root system. …. and turning reddish very quickly due to the anthocyanins production and accumulation in the vacuoles of the epidermal cells of the leaves" (Carrapiço 2010, p. 236).

Fossil records of *Azolla* date back to Eocene (ca. 48.5 MA) of the Cainozoic Era (Collinson 2001; Brinkhuis et al. 2006; Collinson et al. 2009). Its abundant remains in middle Eocene deposits suggested a mass development that lasted for about a million years, which brought to abrupt reduction of CO_2 with enormous impact on global climate and biogeochemical cycles (Brinkhuis et al. 2006; Bujak 2007; Carrapiço 2010). Up to now, no direct correspondence between the fern evolution and the complexity of the symbiosis has been found since *Azolla*, being phylogenetically lower, exhibits more complex and more evolved association characteristics than the other tracheophyte cyanosymbioses (Carrapiço 2010).

Symbiotic algae live in the fern in two topographically different sites, first of which is at its apex, under the overarching of the young leaves, where a small part of the filaments is in constant close connection with the apical meristem of the fern (Becking 1978). Second site is represented by the specific extracellular ellipsoidal leaf compartments, so-called cavities, each formed by infolding of the upper (dorsal) aerial lobe of the bilobed leave, and covered by multicellular hairs (Duckett et al. 1975; Becking 1978; Paracer and Ahmadjian 2000). Set of studies revealed that the center of the cavity (possibly containing gas or liquid) is empty and cyanobionts inhabit only the narrow peripheral layer, which is enclosed by two (internal and external) envelopes and is filled with mucilage (Carrapiço 1991, 2010). Yet the origin of this mucilage is questionable. Discussing whether it is produced by the cavity hairs, Duckett et al. (1975) noted the lack of mucilage in cyanobiont-free *Azolla* in addition to the results from ultrastructural investigation of the hairs, which revealed the absence of strongly developed endoplasmic reticulum and large vesicles of Golgi bodies, typical for secretory cells. Therefore, the authors tentatively guessed that the slime within the cavities may be derived from the cyanobiont. We support this presumption because it is well known that *Trichormus* (*Nostoc*/*Anabaena*), commonly identified as *Azolla* cyanobiont, is capable of producing its own mucilage envelopes (Komárek 2013). However, it has to be noted that along with cyanobionts in the cavities may grow other organisms, mainly non-photosynthetic bacteria, some of which were also supposed to excrete mucilage (e.g., Grilli 1964; Rai 1990, 2018; Wagner 1997; Peters and Meeks 1989; Carrapiço 1991, 2010; Paracer and Ahmadjian 2000; Pawlowski 2009; Pereira 2017; Mishra et al. 2018). Currently, about 60 bacterial isolates from nine genera have been cultured from *Azolla* (Pratte and Thiel 2021). The respiration of bacteria decreases the oxygen level in the cavities, enhancing the N fixation by the algae and some of them, such as *Arthrobacter*, can release the phytohormone auxin (Pawlowski 2009). The real function and relations with all bacteria still need to be clarified, but there were opinions that *Arthrobacter* occurs consistently in all generations of *Azolla* and is the third partner (bactobiont; Pawlowski 2009) in this symbiosis (Carrapiço 1991; Wagner 1997; Pawlowski 2009; Pereira 2017). Therefore, currently, some authors consider this complex symbiont system as a superorganism in ecological terms (Carrapiço 2010).

The hairs, embedded in the peripheral mucilage of the leaf cavities, are of different structure, branched and unbranched, sometimes anastomosing and having filiform wall ingrowths (Duckett et al. 1975). They are supposed to be significant for the metabolite transfer between the symbiotic partners because there is no direct connection between the leaf cavity and the fern vascular system (Duckett and Prasad 1974; Duckett et al. 1975; Rai 1990; Pereira and Carrapiço 2007; Carrapiço 2010). However, it has to be stressed that such hairs have been observed by electron microscopy also in cyanobiont absence (Duckett et al. 1975). In the vacuoles of the simple, unbranched hairs diverse bioactive metabolites, such as polysaccharides, lipids, polyphenols (including tannins), and alkaloids, were identified (Pereira and Carrapiço 2007). Their role in the symbiosis has to be understood, but it was supposed that they may be an important chemical "communication language" or may participate in selection and control of microorganisms in the cavities and, in this way, in the symbiosis establishment and maintenance (Pereira and Carrapiço 2007).

The "*Anabaena-Azolla*" association, first noted by the German botanist Eduard Strasburger (1844–1912) in 1873, was classified as mutualistic by De Bary (1878) and by Danish botanist Eugenius Warming (1841–1924) in his book "Plantesamfund" (Ecology of Plants), published in 1895 (Warming 1895). It was long considered that host functions as a "greenhouse" providing moist and protected environment together with some nutrients to the cyanobiont, which, in turn, supplies it with NH_3 and growth hormones (Hamdi 1982; Bergman et al. 1992; Wagner 1997; Feyisa et al. 2013; Pereira 2017). This providing of N happens because the fern inhibits cyanobionts' NH_3-assimilating enzymes (i.e., GS); therefore, alga cannot utilize it and excretes the produced NH_3, which is furtherly assimilated by the fern (Wagner 1997). For example, the host accounted for at least 90% of the total GS and for 80% of NADH-dependent GDH activities, and the intracellular amounts of NH_3 of symbiotic *Azolla* are five times greater when compared with cyanobiont-free fern (Rai 1990). In addition to NH_3, the cyanobiont releases glutamine, glutamate, and glutamate derivative (Bergman et al. 1992). However, it has to be noted that, as experimentally proved, 12–20% of the fixed N, pumped out as NH_3, can be released in the water environment and it is known that cyanobiont contributes 15–20% of the association chlorophyll (Hamdi 1982). Despite the maintaining of the chlorophyll (chl *a*) and other photosynthetic pigments, such as carotenoids and phycobiliproteins, the photosynthetic rate of cyanobiont is low (Pereira and Vasconcelos 2014). An extra sheath was seen around the heterocysts of the cyanobiont, which possibly helps in maintenance of oxygen-free microenvironment and enhances N fixation (Pereira and Carrapiço 2016). The complexity of the relationships between the symbiotic partners currently is seen as a new level of biological organization, taken as a natural microcosm, or a microecosystem (Carrapiço 2002, 2010). Earlier, the *Azolla* symbiotic system was compared with a natural photobioreactor (Shi and Hall 1988), which went through millions of years of evolution and partners increased some of their physiological and metabolic activities (Carrapiço 2010). In addition, the conferred features and capabilities allowed the symbiotic organism to go functionally and metabolically beyond the abilities of each individual partner and to achieve a principally new and complex level of organization (Carrapiço 2010), suggesting the great importance of symbiosis among the driving forces of evolution.

As it was mentioned above, a part of algal filaments is always attached tightly to the fern meristem (Becking 1978), and in this manner, algae are incorporated in each newly developed leaf during the vegetation season (Wagner 1997). Overwintering is ensured by the female fern sexual structures—the megasporocarps, each covered by a thick, double-layered indusium. Megasporocarp produces one megasporangium with a single megaspore, which bears a few algal filaments with akinetes (vegetatively developed enlarged and thick-walled dormant cells), or only akinetes of the alga, and in this way cyanobiont is inherited in the next generation (Wagner 1997; Paracer and Ahmadjian 2000; Rai 2018). During maturation of the megasporocarp, indusium accumulates tannins in its distal part and this is believed to be related with its function to protect the cyanobiont colony (Rai 2018). In the megasporocarp, below the indusium, along with the distally positioned cyanobiont, bacteria were observed and supposed to be the third partner in this complex symbiotic system (Carrapiço 1991, 2010; Rai 2018). During the embryogenesis, the akinetes germinate into new filaments that spread around the cotyledonary leaf and become associated with the apical meristem of the new plant (Becking 1978; Peters and Perkins 1993; Pereira and Vasconcelos 2014). By this vertical transmission of cyanobiont from host to its offspring, without the need of de novo infection and lack of autonomous growth, this symbiosis is unique among other extant symbioses of cyanoprokaryotes and plants (Ran et al. 2010; Pereira 2017). It is considered to be obligate and most authors note that its successful re-establishment has not been reported (e.g., Becking 1978; Hamdi 1982; Singh 1998; Feyisa et al. 2013; Warshan et al. 2018; Pratte and Thiel 2021). However, Lin et al. (1989) accomplished this task by placing of indusium that contains cyanobiont onto the cyanobiont-free megasporocarp and confirmed the recovery of the symbiotic system using SEM, monoclonal antibody test (McAb-C16), and N fixation capacity (ARA). Moreover, the laboratory-grown cyanobiont was observed to enter in the leaf cavities of the sporophyte (Lin et al. 1989). This successful experiment was immediately cited in 1990 by Rai (1990, 2018). The recognition between the partners seems to be facilitated by lectins, which have found in the fern, in cyanobionts, but also in bactobionts (Dworkin et al. 2006).

The taxonomy of cyanobionts yet is not completely resolved (Komárek 2013; Pereira and Vasconcelos 2014). Historically, they were firstly identified as *Anabaena azollae* (Strasburger 1873), then were supposed to be *Nostoc* sp. (Peters 1991), *Anabaena variabilis* status *azollae* (Fjerdingstad 1976), and *Trichormus azollae* (Komárek 2013). However, according to the genetic sequencing, the *Azolla* cyanobionts, previously assigned to *Nostoc*, differed from all other symbiotic *Nostoc* strains from other hosts and clustered with the freshwater planktic *Raphidiopsis raciborskii* (Syn. *Cylindrospermopsis raciborskii*), benthic *Anabaena cylindrica* 133, and *A. oscillarioides* Hindák 1984/43 (Papaefthimiou et al. 2008). The sequencing of the complete genome of the strain *Trichormus* (*Nostoc*) *azollae* 0708 revealed total length of 5.4 Mb in this obligate cyanobiont, almost twice less in comparison with length of 9.0 Mb in the facultative cyanobiont *Nostoc punctiforme* PCC 73102 (Ran et al. 2010). A large proportion of pseudogenes (>31%) combined with a high frequency of transposal elements (ca. 600) was identified, suggesting genome erosion (Ran et al. 2010). This first evidence for genome degradation of the cyanobiont, in accordance with the lack of autonomous growth, presumed initial phase of a transition from a free-living, independent organism to a N-fixing plant unit (Ran et al. 2010).

2.1.8 Cyanosymbioses with Gymnosperms

These cyanosymbioses include cycads, araucarians, and conifers, the last two known only from Mesozoic fossil findings (Fig. 1).

2.1.8.1 Cyanosymbioses with Cycads

Cycads are gymnosperms that appeared in the Pensylvanian period (325–280 MA) of the late Paleozoic and become widespread in the Mesozoic (Jones 2002; Lindblad and Costa 2002; Kipp et al. 2020). Nowadays they are "living fossils" represented by a few (9) genera and less (ca. 160) species, the small populations of which are distributed mainly in the tropical and subtropical regions of both hemispheres (Paracer and Ahmadjian 2000; Jones 2002; Lindblad and Costa 2002; Baulina and Lobakova 2003; Costa et al. 2004; Chang et al. 2019). The specific endophytic association between blue-green algae ("*Anabaena*") and roots of *Cycas revoluta* was first described in 1872 by German botanist and philosopher Johannes Reinke (1849–1931). By time it was clarified that all cycadean species can form symbiosis with Cyanoprokaryota (Fig. 6) and since both organisms are ancient, their partnership may be traced back to the late Paleozoic and early history of cycads (Bergman et al. 1992; Usher et al. 2007; Chang et al. 2019; Kipp et al. 2020). It is believed that

Fig. 6 Different species of cycads with coralloid roots, containing blue-green algal symbionts. White arrows point the "algal layers", visible as dark green rings in transversal cuttings inside the roots

the symbiosis with diazotrophs allowed cycads to survive the unfertile conditions of barren soils and harsh dry climate (Zheng et al. 2018), relieving them from N limitation (Nelson et al. 2019). As far as cyanoprokaryotes in the soil are concerned, it was assumed that, although they are able to overcome various inimical conditions, they prefer a stable environment, a shelter, as a survival strategy to avoid desiccation, but also to prevent them from being grazed (Adams et al. 2013; Chang et al. 2019). This symbiosis is considered mutualistic and cycads are the only extant gymnosperms known to have evolve it (Rai et al. 2000). Due to limited available light, cyanobionts live heterotrophically, relying completely on the carbon supplied by their hosts (Lindblad 2009). Photosynthetic activity of cyanobionts in *Cycas circinalis*, *C. revoluta*, *Ceratozamia mexicana*, and *Encephalartos villosus* was partially or completely suppressed; nevertheless endophytic algae kept the green color and maintained complete plastid apparatus, associated pigments, and enzyme levels compared with their free-living counterparts (Lindblad et al. 1985, 1987; Gehringer et al. 2010; Baulina and Lobakova 2003; Adams et al. 2013; Chang et al. 2019). However, according to some authors, cyanobionts may supply carbon to the hosting partner along with the N compounds (Bell-Doyon et al. 2020). In cycads, contrary to almost all other cyanosymbioses, glutamine and citrulline move from roots to the stem instead of NH_3 (Lindblad and Bergman 1986; Bergman et al. 1992; Lindblad and Costa 2002). Much rarer, only in some species, N is transferred by the xylem sap as glutamine and glutamic acid (Lindblad and Costa 2002).

Some non-photosynthetic bacteria reside with algae in the coralloid roots and recent studies reveal their diverse bacteriome (e.g., Grobbelaar et al. 1987; Chang et al. 1988; Suárez-Moo et al. 2019; Zheng et al. 2018; Gutiérrez-García et al. 2018; Bell-Doyon et al. 2020; Pecundo et al. 2021). Controversial data were provided by Caiola (1980), who did not find any other endophytic organisms in a comprehensive study of cycad endosymbiosis from Italian botanical gardens. In any case, up-to-now records show that most bacteria develop in limited populations and this was explained by presumed production of secondary growth inhibitory metabolites by the host alone or in a concerted communication with the cyanobiont (Caiola 1980; Obukowicz et al. 1981; Adams et al. 2013; Chang et al. 2019). Currently, by genome mining, Cruz-Morales et al. (2017) identified novel, unique biosynthetic gene clusters in this cyanosymbiosis, which can help to understand coevolution and to discover pathways responsible for the synthesis of different natural substances.

Cyanobionts develop intercellularly in darker micro-aerial conditions of special nodules of peculiar negatively geotropic cycad roots, named coralloid roots (formerly "pneumatophores") (Wittmann et al. 1965; Yamada et al. 2012). They arise from the "normal," lateral cycad roots and contain numerous dichotomously branched and engorged structures, which are visible as slightly protruding above the soil surface near to the stems (Chang et al. 1988; Thajuddin et al. 2010). There, cyanobionts are generally restricted to a single intercellular layer, "algal zone" ("cyanobacterial zone"), but intracellular occurrence of algae is also known (Duckett et al. 1975, 1977; Wittmann et al. 1965; Nathanielsz and Staff 1975a, b; Obukowicz et al. 1981; Zimmerman and Bergman 1990; Rai et al. 2002; Chang et al. 2019; Bell-Doyon et al. 2020). The algal zone, easily visible in a transversal root section even with a naked eye as a dark green ring, is surrounded by a secondary cortex, which does not exist in normal roots, and is full of slime (Chang et al. 1988, 2019; Baulina and Lobakova 2003; Zheng et al. 2018; Bell-Doyon et al. 2020). The algal zone is transversed by elongated cells of the transformed epidermis of the cycad, which not only interconnect both cortical layers, but possibly ensure the metabolite transfer between the partners (Wittmann et al. 1965; Lindblad et al. 1985; Lindblad and Costa 2002; Costa et al. 2004). In addition, in the host cells contacting with cyanobionts, by TEM were observed cell wall ingrowths, typical for transfer cells (Lindblad et al. 1985; McCurdy et al. 2008).

Although it had been demonstrated that initially the cyanobiont is absent in the roots and colonization occurs through hormogonia, details on the exact mode, place, and moment of penetration still need to be determined (Lindblad and Costa 2002; Costa et al. 2004; Yamada et al. 2012). Usually, the recruitment happens de novo from the surrounding environment at any stage of the root development, and this invasion, likewise all other cyanosymbioses, is initiated by a HIF/chemoattractant secreted by the coralloid roots (Costa et al. 2004; Cuddy et al. 2012; Zheng et al. 2018). However, cycads differ from other symbioses by forming of young, apogeotropic papillose roots, named pre-coralloid roots (Ahern and Staff 1994; Nathanielsz and Staff 1975a; Chang et al. 2019). Only after these young roots reach maturity, the dermal tissue on some places disrupts, providing numerous "gates" through which the blue-green algae from the neighboring soil can enter the root (Chang et al. 2019). Once algae appear inside, the morphology and growth of the young root change irreversibly, and it turns into a typical engorged dichotomously divided coralloid root (Wittmann et al. 1965; Lindblad and Costa 2002; Costa et al. 2004; Chang et al. 2019). Before infection, all cortical cells of cycad roots contain high amounts of phenolics, but afterwards, these compounds could be found mainly in the outer cortex cells (Chang et al. 1988). As mentioned before, phenolic substances were presumed as important in establishment of the symbiosis due to their participation in signaling, but their role as antimicrobial compounds and source of other secondary metabolites (Obukowicz et al. 1981) has also to be considered. Although many questions remain answered, it has been supposed that cycads developed so peculiar roots in order to facilitate symbiosis and this idea is supported by the fact that cyanoprokaryotes do not show affinity to enter into a symbiotic relationship with the other two types of cycad roots (primary and lateral) (Chang et al. 2019).

The usually identified cycad cyanobiont is *Nostoc*, and genetic studies, based on 16S RNA sequences, almost similarly demonstrated the grouping of *Nostoc* strains in a monophyletic clade with two main sub-clades, first of which is composed of strains, which develop only epiphytically on lichens and bryophytes, and second contains both symbiotic and free-living strains, with *N. punctiforme* PCC 73120 used as its referent strain (Svenning et al. 2005; Thajuddin et al. 2010). Controversial opinions exist about the presence/absence of biogeographical and phylogeographical structure of *Nostoc* genotypes (e.g., Costa et al. 2004; Yamada et al. 2012; Bell-Doyon et al. 2020). However, contrary to other mostly northern temperate cyanoprokaryotes, which contain both the conventional Mo and rarer V nitrogenases (Nelson et al. 2019), in tropical *Nostoc* from epiphytic endemic *Zamia nana*, only Mo nitrogenase was detected (Bell-Doyon et al. 2020).

The whole-genome sequence of *Nostoc cycadae* WK-1 (formerly WUEB-01), obtained from *Cycas revoluta* planted in the campus of Wakayama University, Japan, was identified (Kanesaki et al. 2018). According to 16S rRNA sequence, WK-1 showed the highest similarity (99.6%) with *Nostoc* sp. PCC 7107, free-living in a shallow pond, but there was a slight difference in the genome size and the number of coding sequences (CDSs) of both strains: 6.77 Mb total length and 5465 CDSs in WK-1 vs 6.33 Mb and 5317 CDSs of PCC 7107 (Kanesaki et al. 2018). The next-generation sequencing techniques were applied to investigate the composition of coralloid, but also of other two types of so-called regular roots of *Cycas bifida* from Vietnam (Zheng et al. 2018). It has to be noted that cyanoprokaryotes were found in all root types but dominated in the coralloid roots, being represented by *Nostoc* and *Dolichospermum* (Syn. *Anabaena* p.p.) (Zheng et al. 2018). In addition, the same study presumed inhibition of any other endophytes by abundant cyanobionts in cycad roots.

Long before these studies, taxonomic efforts revealed a palette of cyanoprokaryotes, which develop in coralloid roots, among which were identified the species of heterocytous *Anabaena*, *Nostoc* (sometimes described as "*Anabaena* or *Nostoc*") and *Calothrix* (Reinke 1872; Grobbelaar et al. 1987; Wittmann et al. 1965; Gehringer et al. 2010; Cuddy et al. 2012; Bell-Doyon et al. 2020). According to Komárek (2013), endobionts of cycads, described as *Anabaena cycadearum* and *Anabaena cycadae*, are united as *Trichormus cycadearum*. *Anabaena* and *Calothrix* are less frequent cyanobionts in cycads (Lindblad et al. 1985; Lindblad and Costa 2002; Yamada et al. 2012; Chang et al. 2019). *Anabaena* was reported from *Cycas revoluta* (Obukowicz et al. 1981; Zhu 1982) and *Calothrix*, first morphologically identified in *Encephalartos hildenbrandtii* (Grobbelaar et al. 1986), was later confirmed genetically using PCR amplification and 16S rRNA gene sequence analysis as endosymbiont in three species of *Macrozamia* (viz. *M. fearnsidei*, *M. parcifolia*, and *M. pauli-gulielmi*) (Gehringer et al. 2010) and *Cycas revoluta* (Thajuddin et al. 2010).

All accumulated data suggest that *Nostoc* is commonly occurring in the roots and a wide range of *Nostoc* strains appear to associate with the coralloid roots of cycads. A single species can be isolated from multiple cycad hosts (Lindblad and Costa 2002; Gehringer et al. 2010), but also one cycad plant may house multiple cyanoprokaryote strains (e.g., Lindblad et al. 1989; Zheng et al. 2002; Thajuddin et al. 2010) among which one can predominate (Caiola 1980). One morphotype was observed in each single nodule of *Zamia integrifolia* and *Dioon* sp. from Florida and confirmed biochemically on the bases of protein fingerprints and six enzymes, used routinely for distinguishing between cyanoprokaryote strains (Zimmerman and Rosen 1992). Studies of cyanobiont genetic diversity by PCR fingerprinting and 16sRNA based on natural and greenhouse species from *Bowenia*, *Cycas*, *Encephalartos*, *Lepidozamia*, *Macrozamia*, and *Zamia* revealed that although different strains of *Nostoc* exhibit symbiotic competence, in different roots from the same plant, different cyanobionts may occur, but in each coralloid root, only a single cyanobiont strain was detected (Costa et al. 1999, 2004; Rai et al. 2002; Gehringer et al. 2010; Cuddy et al. 2012). Plants growing in a close vicinity (ca. 1 m from each other) were inhabited by one and the same strain (Costa et al. 2004). A single explanation has not been adopted and it has to be investigated if this result is based simply on the strain availability in the surrounding environments, or host is supplied by a specific selective mechanism. Occurrence of more than one *Nostoc* strain in a single root was first reported by Zheng et al. (2002). Moreover, the same authors, in a study of cultivated *Macrozamia* from China and Sweden greenhouse specimens of this genus, pointed on different strains in apical, middle, and basal parts of a single root. These finding of Zheng et al. (2002) was debated by Costa et al. (2004), who explained it with the restrictions of the applied fingerprinting method and lack of controls. Later, Yamada et al. (2012) in an investigation of natural *Cycas revoluta* from Japan confirmed the possibility of developing more *Nostoc* strains in the same coralloid root (maximum three cyanobionts) and also in different roots of a single cycad plant. Yet, it remains uncertain if such co-occurrence is really unusual or it is caused by conditions of natural habitats with rich diversity of soil cyanoprokaryotes. Therefore, summarizing data and opinions of previous investigators, Yamada et al. (2012) supposed that lack of host specificity and of geographical structure of cyanobiont genomes can be explained by the possibility cycads to take the *Nostoc* strains from soils at random. In the same time, the genetic sequencing revealed differences between the soil algal flora and cycad cyanobionts. For example, different *Nostoc* strains were identified in the rhizospheres and inside *Dioon* sp. and *Zamia integrifolia* from Florida, and in addition, *Calothrix* and *Oscillatoria* were determined in the nearest soil, but not in cycads, whereas *Anabaena* was found to develop epiphytically on the root nodules (Zimmerman and Rosen 1992). These results suggest that successful infection requires specific compatibility with the host (Zimmerman and Rosen 1992). Successful recruitment of *Nostoc* by *Zamia* was achieved in the lab in two different experiments: in the first, both partners were separated and then re-united, and in the second a free-living soil *Nostoc* strain was used (Ow et al. 1999).

It is widely recognized that *Nostoc* is important partner due to its role of a N fixer, but it is also responsible for the production of neurotoxins, such as BMAA and its isomers, in these important relic plants (for details see Brenner et al. 2003; Stoyneva-Gärtner et al. 2025, this volume). Interestingly, this production is higher in cyanobionts (2–37 $\mu g\ g^{-1}$) of *Cycas micronesica* than in free-living algae (0.3 $\mu g\ g^{-1}$), and toxin concentrations are subsequently biomagnified through the food chain including flying foxes, reaching 6.6 $\mu g\ g^{-1}$ in the brain tissues of Chamorro people of Guam, who die from the amyotrophic lateral sclerosis/parkinsonism-dementia complex (Cox et al. 2003). However, the role of cyanobionts and of coralloid roots, as well as the biomagnification hypothesis, was strongly debated after the same toxins were found in *C. micronesica*, grown without cyanobionts (Marler et al. 2010; Snyder and Marler 2011; Chang et al. 2019). Cyanobionts of the epiphytic tropical *Zamia pseudoparasitica* contained gene, responsible for mycosporine biosynthesis (Bell-Doyon et al. 2020). MAAs long ago have been recognized as potent primary sunscreens, UV-protectors, typical for aeroterrestrial and extremophilic algae, and soil *Nostoc commune* was one of the first species in which they were recorded (for details see Stoyneva-Gärtner et al. 2020). Their finding in this epiphytic cycad was explained by the life in a high exposed habitat (Bell-Doyon et al. 2020).

2.1.8.2 Cyanosymbioses with Araucarians and Conifers

According to the fossils discovered in Early Cretaceous period, it is possible to suppose symbiotic associations between pine-like araucarians and cyanoprokaryotes (Harsh et al. 2018). The endophytic blue-green algae were discovered in a fossil root of the *Arauameylon pakurense* from Rajmahal Hills, India, which looked similar to coralloid cycadean roots (Sharma and Suthar 1989). In the same region, symbiotic nodules with N fixing blue-greens were discovered in other Early Createaceous fossil roots of a gymnosperm, which, most probably, was a coniferous species (Banerji and Ghosh 2002).

2.1.9 Cyanosymbioses with Angiosperms

To date the only widely recognized natural symbiosis between prokaryotic algae and angiosperms is this with *Gunnera*, but successful colonization of different crops by cyanoprokaryotes with establishing of novel endophytic associations was achieved experimentally (Gantar et al. 1991a, b; Álvarez et al. 2020, etc.) (Fig. 1).

2.1.9.1 Cyanosymbiosis with *Gunnera*

Gunnera, named in honor of the Norwegian bishop and botanist Johan Ernst Gunnerus (1718–1773), is the single genus of the family Gunneraceae, believed to have evolved in Cretaceous (90–100 MA) (Pawlowski 2009). It contains about 50 species, spread in tropical and subtropical regions of the southern hemisphere, occupying poor soils of moist areas commonly at higher elevations, but currently spread in other habitats due to their cultivation as ornamentals (Zimmerman and Bergman 1990; Bergman et al. 1992). The association with N-fixing algae, most probably, was advantageous for *Gunnera* plants in these relic conditions, and one possible explanation for the establishment of this cyanosymbiosis could be the assumed larger availability of *Nostoc* in habitats, occupied by *Gunnera* in the past (Pawlowski 2009). The origin of *Gunnera* dates about 95 MA back, and the partnership with cyanoprokaryotes seems to be nearly of the same age, thus being the latest symbiosis from evolutionary point of view (Bergman et al. 1992).

Experiments with *Gunnera tinctoria* demonstrated that the host is entirely dependent for its growth on N supply from the cyanobiont (Osborne et al. 1992) indicating the obtained benefits by the host. Successful reconstitution of *Gunnera-Nostoc* symbiosis was acquired in lab conditions, which allowed more detailed investigation of the process of invasion and symbiosis establishment (Bergman et al. 1992). It starts by attaching of *Nostoc* hormogonia to the mucilage at the stem apex of *Gunnera*, which then migrates to specific glands on the stem, formed on the basis of each petiole and filled with polysaccharide-rich acidic mucous (Zimmerman and Bergman 1990; Bergman et al. 1992; Elmerich and Newton 2007). According to Rasmussen et al. (1994), this acidic mucilage is the major component in a signaling system between the plant and the potential cyanobiont, functioning as a HIF, but also shown to stimulate its growth and to induce synthesis of 40-kDa and specific 65-kDa polypeptides in compatible *Nostoc* strains (Rasmussen et al. 1994). Attraction of potential symbiont and its further maintaining in vegetative stage may happen due to excretion of some soluble sugars from *Gunnera* stem glands (Khamar et al. 2010). Hormogonia of *Nostoc* enter the glands, glide through long channels to their bottom, and then develop intracellularly but still remain separated from the host cell protoplasm being surrounded by plasmalemma (Bergman et al. 1992; Usher et al. 2007). The ability of cyanobiont to penetrate the cell walls clearly distinguishes this symbiosis from the other plant cyanosymbioses and ensures the more efficient assimilation of N by the host due to its direct release, although the exact form of N transferred to the host is not known (Bergman et al. 1992). In addition, each gland cell forms protrusions by which the contact surface between the cyanobiont and the host is enlarged (Paracer and Ahmadjian 2000), obviously facilitating exchange of metabolites between the partners.

The cyanobionts of *Gunnera* were generally ascribed to *Nostoc*, but the symbiosis is of low species specificity in both controlled, experimental and natural, outdoor conditions (e.g., Zimmerman and Bergman 1990; Bergman et al. 1992; Osborne et al. 1992; Nilsson et al. 2000; de Cássia Mariotti et al. 2011). The first described *Nostoc gunnerae* was synonymized as *Trichormus gunnerae* (Komárek 2013). Genetic studies, including PCR fingerprinting, revealed that different strains can infect one plant, but commonly in one plant one species develops and it is possible that the same plant species from different localities is inhabited by different cyanobionts (Bergman et al. 1992; Guevara et al. 2002). The same result of revealing of a single strain in one host plant, and of diverse strains in different individuals, but this time along with differences in geographically distant regions of Chile, was obtained by Fernández-Martínez et al. (2013). In addition, in the same three-levelled genetic study of cyanobionts of *G. magellanica*, diversity of strains was detected in plants from the same sampling site (Fernández-Martínez et al. 2013). One and the same species, *G. mannicata*, grown outdoor or in a greenhouse, can be infected by different strains (Zimmerman and Bergman 1990). However, examination of 11 plants from different geographical regions by PCR fingerprinting demonstrated that a particular plant, and even the same gland, can contain more than one cyanobiont strain (Nilsson et al. 2000). The infection of *Gunnera* by cyanobionts from different strains was easily achieved in lab under controlled conditions (Zimmerman and Bergman 1990).

An important biologically active compound, neurotoxic BMAA, was produced by the *Gunnera-Nostoc* symbiosis (Cox et al. 2005). However, the toxin was not found in studies of aqueous root extracts of *G. mannicata*, widespread in Southern Brazil (de Cássia Mariotti et al. 2011). Since the presence of *Nostoc* was seen in the root tissues, the authors suggest that some other factors may be involved in the production and release of the neurotoxin. Here is important to note that earlier DNA studies of Zimmerman and Bergman (1990) showed that the *Nostoc* strain, isolated from *G. manicata*, differed from all other *Gunnera* isolates. Other cyanotoxins, nostopeptolides, and anabaenopeptins (for their characteristics see Gärtner et al. 2021;

Stoyneva-Gärtner et al. 2025, this volume) were found according to the genes that encode their transporters in the study of genomes of *Gunnera* cyanobionts (Warshan et al. 2018). However, it was supposed that in this case they, and nostopeptolides in particular, had the primary function to regulate the hormogonia motility (Liaimer et al. 2011, 2015), thus being important for establishment of the *Gunnera* cyanosymbiosis (Warshan et al. 2018).

2.1.9.2 Artificial Cyanosymbioses with Crops

The beneficial role of N fixers for rice (*Oryza sativa*) crops has long been recognized and currently is undisputable. Adherence of cyanoprokaryotes to rice roots was described by different authors (e.g., Svirčev et al. 1997; Nilsson et al. 2002; Whitton and Potts 2002), but little is known about the structures and mechanisms of formation of these associations (Álvarez et al. 2020). The first screening of 57 *Nostoc* cyanobionts demonstrated that 21 of them successfully colonized surfaces and intercellular spaces of rice roots exhibiting higher N fixation rates in comparison with free-living strains (Nilsson et al. 2002). Epiphytic development of *Nostoc* on the roots of rice was detected also by Álvarez et al. (2020), who conducted an experiment for inoculation of hydroponic rice cultures with symbiotically competent *Nostoc punctiforme* PCC 73102 and by non-symbiotic *Nostoc* sp. 7120. Successful results were obtained after application of the symbiotic strain 73102. Using hormogonia-rich inoculate, so strong adherence to the roots was achieved, that *Nostoc* could not be removed even after washing with running water. Moreover, regular observations showed that after this initial epiphytic growth, in 45-day-old plants *Nostoc* was found to develop inside the cells of the plant epidermis. This intracellular colonization indicated that more stable endophytic relationship had been established, which promoted the growth of the host plant in N-limited conditions. The N supply from the alga was proved by measuring of nitrogenase activity and by observations of *Nostoc* filaments with mature heterocytes in all rice tissues (Álvarez et al. 2020). Microscopic observations proved that algae reached epi- and exodermis but did not transverse the sclerenchyma and at longer cultivation, also colonize the root hairs (Álvarez et al. 2020). Earlier experiments with wheat (*Triticum vulgare*) in liquid cultures inoculated by soil *Nostoc* sp. 2S9B showed that it was able to colonize the wheat roots, penetrating the root epidermis and cortex, and occasionally to develop also intracellularly (Gantar et al. 1991a, b).

Different examples of successive co-culturing of crop and cyanoprokaryotes after inoculation of soil or amended compost by algae, which can create associations with plant roots, are available in the literature (e.g., Rai 2018; Galanakis 2021; Alvarez et al. 2021). In all these cases, algae act as biofertilizers, demonstrating the increase of plants growth and carbon content. Although these cyanosymbioses are out of the scope of this review, we mention the obtained artificial root associations to note their potential for future artificial symbioses and rhizosphere microbiome engineering with agricultural importance to get better crops from rice and wheat using symbiotic *Nostoc*, and from carrot, corn, cotton plants, ginseng, lucerne, mung bean and pea, papaver, potato, rice, sugar beet, tobacco, tomato, wheat, and *Chrysanthemum* through involving the free-living strains from the following genera: *Anabaena*, *Anacystis*, *Calothrix*, *Chroococcidiopsis*, *Chlorogloeopsis*, *Cylindrospermum*, *Gloeothece*, *Leptolyngbya*, *Mastigocladus*, *Nostoc*, *Oscillatoria*, and *Plectonema* (Rai 2018; Alvarez et al. 2021; Lee and Ryu 2021).

2.1.10 Cyanosymbioses with Animals ("Mobile Cyanosymbioses")

The investigations of polar bears in the Zoo of San Diego, California, revealed that inside their guard hairs, occupying more than half of the width, grow coccal blue-green algae with mucilage sheaths (Lewin and Robinson 1979) (Fig. 1). They were identified as *Aphanocapsa muscicola* (Syn. *Aphanocapsa montana*), exhibiting a slight difference with algae, found in the hairs of "green" polar bears from the Zoo of Fresno, California. Lewin and Robinson (1979) pointed that higher species diversity of endobionts may be expected due to the widespread greening of polar bears in zoo conditions but questioned occurrence of such event in the nature.

On complicated structure of the fissured hairs of the sloth fur, blue-green algae *Fischerella* sp., *Oscillatoria pilicola* and two unidentified species of the same genus, as well as *Nostoc* sp. (Thompson 1972; Wujek and Lincoln 1988; Kaup et al. 2021) were found along with 15 chlorophytes and one red alga (Kaup et al. 2021). Since cyanoprokaryotes represent minority of sloth symbionts, detailed discussion on sloth-algal symbiosis is provided in the end of the next paragraph.

2.2 *Symbioses with Chlorophyta and Other Eukaryotic Soil Algae*

Search of literature revealed that terrestrial algal symbioses involve mostly green algae and, rarely, species from yellow-green, red, euglenophyte and pyrrhophyte algae in partnership with bacteria, fungi, prokaryotic blue-green algae, lycophytes, gymnosperms, angiosperms, invertebrates, and vertebrates, but with restricted number of hosts inside each of these lineages (e.g., Reisser 1992). Considering that, as it was described before, the symbiont has to survive in new environment inside the host, it is possible to suggest that among the important traits of green algae that allow them to establish stable symbiotic associations, it is the resilience character of their multilayered cell walls, which usually contain the stable cellulose but also are known to contain different protective compounds such as algaenans, chitin, sporopollenin, etc. (Good and Chapman 1978; Stoyneva-Gärtner and Uzunov 2017). In addition, we have to mention the frequent presence of pyrenoid(s) in plastids of eukaryotic algae (Fig. 3), considered as significant CCM (Larkum et al. 2003; Zhan et al. 2018; Atkinson et al. 2020). Sometimes pyrenoids are embedded in the plastids and are difficult to be observed, as can be exemplified by the green filamentous genus *Stichococcus*, in which the presence of this specific structure was late proved by TEM (Gärtner et al. 2007) (Fig. 3). Another important feature is the possibility to reproduce by specific autospores. These mini-copies of the mother cell, formed in a vegetative way and completely developed inside it (Shubert and Gärtner 2015), seem to be extremely important in closed endosymbiotic environments with less chances for sexual reproduction. Moreover, autospores are the most effective reproduction units of eukaryotic algae in terms of time and energy spent (e.g., Stoyneva-Gärtner and Uzunov 2017). As it will be shown below, most of the identified symbiotic green algae, and the endosymbionts in particular, are coccal and capable of such autosporic reproduction (Shubert and Gärtner 2015). In the same time, most of these algae can produce motile reproductive cells, depending on the environmental conditions with increasing evidence on their production by more genera, traditionally known to lack them (e.g., Trainor 1963; Shubert and Gärtner 2015). Monadal (flagellate) algae and filamentous symbiotic algae can produce vegetatively way resting cells (cysts, akinetes, aplanospores), with a reproductive and dispersal functions (e.g., Ettl and Gärtner 2014; Stoyneva-Gärtner and Uzunov 2017). The local environmental factors may be supposed among the driving forces for establishment of symbioses with eukaryotic algae, with strong influence of the available nutrients. It is not to suppose, for example, finding of diatoms or silica-scaled synurophyceaen algae as endosymbionts due to lack of silicium in such environments.

2.2.1 Symbioses of Eukaryotic Algae with Bacteria

During the last few years, increasing evidence is accumulated on natural co-occurrence of algae and heterotrophic bacteria (Fig. 1), which perform concerted activities that would be impossible when partners live separately, some of which are indicative of symbiotic relationships (Donohue and Cogdell 2006; Yao et al. 2019). In nature, or in bioreactors, algae supplied bacteria with products released during photosynthesis, which, in turn, are promoting algal growth through remineralization of N, sulphur, and phosphorous or in more specific cases can provide vitamins or siderophores, important for iron binding (Yao et al. 2019). Therefore, modern biotechnology considers symbiotic microalgae–bacteria consortia as potential successful tool to improve microalgal biomass production and to enrich it with valuable chemical and energy compounds (Yao et al. 2019). Long bacteria were taken only as contaminants and different methods were proposed to obtain axenic algal monocultures for industrial needs, but today the new views on the complex relations between these organisms have driven the attention to possibilities for better utilization of these consortia, or initial symbioses, and thus, the use of mixed cultures becomes a new attractive trend (Lutzu and Dunford 2018; Yao et al. 2019). Since the majority of cultivated algae are obtained from aquatic habitats, the great potential of terrestrial algae yet is neither untapped nor realized (Stoyneva-Gärtner et al. 2020, 2022).

Results from experiments with co-cultures of green coccal *Auxenochlorella protothecoides* with *Escherichia coli* indicated commensal relations, in which the bacteria provided thiamine and its derivative metabolites (Higgins and Gheynst 2014; Higgins et al. 2016; Lutzu and Dunford 2018). This led to a significant increase (two to six times) of algal growth, algal lipid content (with twofold increase of neutral lipids), and glucose uptake, allowing to suggest use of commensalism as unexpensive biotechnological tool to increase the yield of valuable products (Higgins and Gheynst 2014; Higgins et al. 2016).

2.2.2 Symbioses of Eukaryotic Algae with Fungi

2.2.2.1 Natural Symbioses of Eukaryotic Algae with Fungi (Lichens)

The oldest fossil record of a lichen-like structure with unidentified coccal symbionts was already described in the former paragraph. The possibility these algae to be from the green lineage was supposed by Usher et al. (2007). Fossils of morphologically advanced lichens are relatively young, coming from Baltic amber of the Old World (ca. 50 MA) and Tertiary deposits (ca. 60 MA) with increasing number in Miocene (Nash III 2008).

Most of extant lichens (ca. 85%) contain green algae as sole photobionts (bipartite lichens), and a small number of tripartite lichens (ca. 4%) include them as primary photobionts along with cyanoprokaryotes as secondary photobionts (described in the former paragraph of this chapter) (Fig. 1). The anatomy, morphology, physiology, and ecology of these bipartite lichens, formed in a partnership with dikaryomycetous fungi as mycobionts, are quite well-known and their detailed description is beyond the scope of this study. Here we shall accent only on some paramount "evergreen" facts, needed for completeness of the description (based almost exclusively on Nash III 2008; Wirth et al. 2013; Ettl and Gärtner 2014):

1. Symbiotic green algae participate in lichens, in which mycobiont has leading role in shaping the thallus morphology. It is notable that lichen thalli significantly differ from free-living fungal mycelium and fruiting bodies and are recognizable as crustose, foliose (Fig. 5), or fruticose (Fig. 7) thalli (with some transitional forms known). Also the so-called leprose (Fig. 7) lichens are known, in which a real thallus is not formed, but algal cells and fungal hyphae are intermingled in shapeless greenish powder masses.
2. Up-to-now, photobionts (except those from green order Trentepohliales) are known to reproduce only strictly vegetatively.
3. The primary function of symbiotic green algae in this symbiosis is to supply the heterotrophic mycobiont with fixed carbon. In turn, they get a stable environment, in which they are protected from strong insolation and UV-radiation, and other unfavorable environmental factors by the hyphae of the mycobiont, which form the upper cortex of the lichen thallus. In this regard, lichen thalli have been

Fig. 7 Example of two terrestrial lichen habitats with different thalli types: (**a**, **b**) Fruticose lichen thalli of *Cladonia convoluta* on sand dunes near Arkutino, Bulgaria; (**c**, **d**) Leprose lichen thalli of *Lepraria* sp. on vertical soil slope along the mountain pathway in the region of Bozhenitsa, Stara Planina Mts (Bulgaria)

considered as self-contained miniature ecosystems (Honegger 1991), with the algal layer as the single primary producer.
4. Green photobionts always occupy a single layer in the thallus, named "algal layer," "photobiont layer," located between the hyphae of the upper cortex and of the medulla (the last usually laying above the lower cortex).
5. Due to slow growth of the lichen thalli, some details of lichen physiology and metabolite exchange have to be uncovered, but it is well-known that concerted efforts of both partners lead to formation of new compounds, secondary lichen metabolites. Most of these substances are useful for humans and find application in medicine, pharmacy, cosmetics, perfumery, textile dying, etc. with novel aspects increasingly discovered (e.g., Vaez and Javad Davarpanah 2021).
6. In this chapter, only terrestrial lichens are considered, but it has to be noted that lichens occur in freshwater and marine habitats, and as epiphytes and epilithes as well.
7. The transmission of cyanobionts occurs in both known ways, horizontal and vertical, separately or in combination, depending on the reproduction mode of the lichen.
8. The algae, which are involved as photobionts, can live also aposymbiotically, as it was demonstrated for the most widespread genus, *Trebouxia* s.l., long considered to be obligate symbiont (Famintzin and Baranietzky 1867; Tschermak-Woess 1978, 1988; Slocum et al. 1980; Bubrick et al. 1984; Gärtner 1985; Ahmadjian 1988; Garty and Delarea 1988; Mukhtar et al. 1994; Palmqvist et al. 1997; Kroken and Taylor 2000; Mikhailyuk et al. 2001; Gärtner and Stoyneva 2003; Lutzoni and Miadlikowska 2009; Beck et al. 2015; Ismail et al. 2017).
9. Lichen algae, and especially the widespread ones, usually possess a large, massive, central or cup-shaped and often lobed plastid with a big, visible pyrenoid.

As photobionts have been reported the following green algal genera: *Asterochloris*, *Cephaleuros*, *Chlorella* s.l., *Chlorosarcinopsis*, *Coccobotrys*, *Coccomyxa*, *Elliptochloris*, *Desmococcus*, *Dictyochloropsis*, *Dilabifilum*, *Diplosphaera*, *Gloeocystis*, *Leptosira*, *Myrmecia*, *Nannochloris*, *Phycopeltis*, *Pleurastrum*, "*Protococcus*", *Pseudochlorella*, *Stichococcus*, *Trentepohlia*, *Trebouxia* (incl. *Pseudotrebouxia*) and *Trochiscia* (Chlorophyta), as well as the yellow-green *Heterococcus* (Ochrophyta). Mostly coccal species are involved in lichen symbioses, and often the filamentous algae are reduced to short filaments or are even disintegrated to unicells.

The level of lichen specificity and plasticity of photobionts has been always disputable, with different opinions shared, ranging from the extremes of high specificity, resulted from long co-evolution to low specificity, being generally explained with the complicated taxonomy of the photobionts, which in the past were identified mainly according to their morphology. The modern development of algal taxonomy with larger application of molecular-genetic methods revealed a higher photobiont diversity than expected on the basis morphology, with an increasing number of publications on the topic.

It has been commonly accepted that each lichen species contains sole photobiont. Most probably, such paradigm reflects the constraints of the applied methods, because with development of research tools and intensifying of the studies, evidence is accumulating that more than one algal genotype and physiologically different type can be found in one and the same species, but also in an individual thallus (e.g., Nash III 1996; Helms et al. 2001; Romeike et al. 2002; Piercey-Normore 2006; Guzow-Krzemińska 2006; Casano et al. 2011; Muggia et al. 2010, 2013; Guzow-Krzemińska and Stocker-Wörgötter 2013; Moya et al. 2017; Medeiros et al. 2021). Diversity of fungi (e.g., Arnold et al. 2009; Muggia et al. 2016; Spribille et al. 2016; Medeiros et al. 2021) was also demonstrated but is out of the scope of our review. The diversity of genotypes can be interpreted as extending of ecological amplitude of the partners and providing adaptive advantages to stand changing environmental conditions (Romeike et al. 2002; Piercey-Normore 2006; Casano et al. 2011) or as a possibility for fungus to obtain maximum profit from concerted utilization of different algae that can use varying light of changing solar radiation and so to reach maximum photosynthesis effectivity (Piercey-Normore 2006). Another possible explanation looked in by Piercey-Normore (2006) was related with earlier suggestion that the co-occurrence of multiple algal genotypes in the same thallus may be related with the age (Fahselt et al. 1995). Some of these genotypes may be "founder events" (Piercey-Normore 2006), which are kept in small amounts, but can provide great flexibility to the population in case of environmental changes (Piercey-Normore 2006). Muggia et al. (2013), interestingly, compared the peculiar areolated (separated by cracks) thalli of *Protoparmeliopsis muralis* to "symbiotic playgrounds" in which different partnerships can be tested at a low risk. The higher diversity of such micro-scaled symbiotic associations was suggested to be useful in eventual case of changed habitat conditions with newly imposed selective constraints, when some of these combinations may outcompete the others (Muggia et al. 2013). Exactly for *P. muralis*, commonly recognized as one of the most successful and widespread species in the world, such low level of selectivity, reflected by the photobiont diversity, was proved (Guzow-Krzemińska 2006; Guzow-Krzemińska and Stocker-Wörgötter 2013). Moreover, in vitro experiments to re-synthesize this ubiquitous lichen demonstrated that in addition to its known diverse photobionts from the genus *Trebouxia*, another presumably incompatible green alga, *Asterochloris* sp., can be involved in the symbiotic interactions (Guzow-Krzemińska and Stocker-Wörgötter 2013). All these results presumed that the low specificity to photobionts, expressed in mycobionts capacity to form associations with different algae, is a beneficial trait of the lichen, which not only ensures its survival, but also facilitates successful colonization of new habitats (Guzow-Krzemińska and Stocker-Wörgötter 2013). In this colonization process, more strategies can be used, alone or mixed (for details see Sun and Friedman 2005; Blaha et al. 2006; Casano et al. 2011; Fernandez-Mendoza et al. 2011, 2017; Grube and Spribille 2012; Muggia et al. 2010, 2013; Medeiros et al. 2021).

In regard to maintained photobiont diversity, discovered and differently interpreted by the cited above authors, we would like to remind the brilliant paper of Padisák (1992), which adverts rare algae in phytoplankton of natural lakes as extremely important "deposit" of the "ecological memory" of the community. The latter term was defined as the "capacity of past states or experiences to influence present or future responses of the community" (Padisák 1992, p. 225). By making this link, we suppose similar role of different genotypes, maintained in lichen thalli, kept as genetic pools at microscopic scale, a hidden reserve, which may be useful in unpredictable environmental shifts. In the same time, likewise the rare algae, they function normally, contributing to the community production or, in case of lichens, to host life.

2.2.2.2 Artificial Symbioses of Eukaryotic Algae with Fungi

A co-culturing experiment, based on the different metabolism traits of the unicellular flagellate alga *Chlamydomonas reinhardtii* and the yeast *Saccharomyces cerevisiae*, demonstrated spontaneous formation of mutualistic relationship. This highlights the role of environmental changes as a driving force for free-living organisms to start symbiotic interactions (Hom and Murray 2014) (Fig. 1). The same engineered obligate mutualism was achieved between other yeasts (some isolated from soil) and four species of *Chlamydomonas*, but with different gain in terms of cell numbers, indicating that despite the phylogenetic broadness, the final success depends on trait peculiarities of involved species (Hom and Murray 2014). This was further shown by including of different fungal and algal species, such as unicellular coccal *Chlorella vulgaris*, yeasts (*Hanzenula polymorpha* and *Kluveromyces polysporus*), and other ascomycetes (*Aspergillus nidulans* and *Neurospora crassa*), in which synthetical partnerships were easily obtained, but lead only to physical complexes, without differences at cellular or organismal level (Hom and Murray 2014). The latter probably is explainable also with the shorter time of experience in comparison with millions of years of co-evolution.

Muggia et al. (2018) successfully embedded fungi and algae in alginate spheres to accomplish stable culture. Despite that these experiments include marine alga *Nanochloropsis*, and not soil algae, here we mention in this chapter to stress on this novel approach in study of symbiosis.

2.2.3 Symbioses with Bryophytes

Although rarely recorded, endophytic intra- and extracellular, as well as epiphytic symbioses of algae with liverworts and mosses (Fig. 1), occur in nature and need more attention in future research.

The green coccal alga *Chlorochytrium* (*C. lemnae*) is well-known as endobiont of small free-floating aquatic plant *Lemna* (duckweed), but it is almost neglected that it has been recorded as intercellular endophyte in subepidermal tissues of *Bryum pseudotriquetrum* (Bristol 1920), as intracellular endophyte in mosses *Bryum capillare*, *Funaria flavicans*, and *Tortella flavovirens* (Reese 1981), and by

Taylor (1951), as endophyte in the liverwort *Riella spiculata* (Reisser 1992; Trémouillaux-Guiller et al. 2002).

The green coccal unicellular *Chlorococcum* was found as epiphytic on the bonfire moss *Funaria hygrometrica* and common hair moss *Polytrichum commune* (Scheirer and Dolan 1983), but without describing morphological peculiarities of their surface in relation to retaining epiphytes (Solheim et al. 2004). SEM and TEM studies, along with lab culturing, revealed the presence of *Chlorococcum* on both leaf surfaces, often accompanied by rod-shaped bacterium and deuteromycete *Trichoderma viride* (Scheirer and Dolan 1983). Another green unicellular coccal alga, tentatively identified as *Scotinosphaera* sp. (Syn. *Kentrosphaera* sp.; Guiry and Guiry 2022), was reported as epiphyte on the moss *Bryum pseudotriquetrum* (Solheim et al. 2004).

2.2.4 Symbioses of Eukaryotic Algae with Lycophytes

A single example of fossil association between land plants and eukaryotic algae is the endophytic development of the colonial alga *Lageniastrum macrosporae*, which strongly resembles recent green flagellate, coenobial *Volvox*, inside megaspores of the lycophyte of *Sublagenicula nuda*-type, discovered by Renault (1894) (Fig. 1). The description of this association was based on the multiple algal occurrences within the thick megaspore walls, where they were always positioned just beneath the suture on the proximal surface, and it was speculated that since alga had to overcome the harsh conditions in the late Paleozoic environment, but did not produce resting cells, it adopted a strategy to colonize the resistant spores of the lycophyte (for details see Taylor and Krings 2005). *Volvox* is a freshwater inhabitant, but since some flagellate algae occur in soils (Ettl and Gärtner 2014) and exact identification in fossil state is uncertain, we include this ancient symbiosis to achieve completeness of the review.

2.2.5 Symbioses of Eukaryotic Algae with Gymnosperms

The symbioses of gymnosperms with green algae, although extremely interesting, are less known. Before their description provided below, we would like to note that they occur in the groups of ginkgoids and cycads (Fig. 1), which are the most plausible evolutionary relatives, sharing the motile spermatozoid, and are known to evolve in similar extremely moistened but disturbed riparian habitats (e.g., Royer et al. 2003).

2.2.5.1 Symbiosis with *Ginkgo biloba*

Intracellular symbioses of eukaryotic algae and higher plants are exceptional. Such an example is the green coccal alga *Coccomyxa* discovered from in vitro cell cultures of the "living fossil" *Ginkgo biloba* (maidenhair tree) with its proved universal worldwide distribution and vertical transmission (Trémouillaux-Guiller and Huss 2007; Trémouillaux-Guiller et al. 2002). The genetic university of the symbiont is explainable by the ancient origin of *Ginkgo*, in Middle Jurassic about 170 MA, and its intentional planting on all continents out of its native Eastern China sites. The most interesting in this unique symbiosis is that alga is photosynthetically inactive and resides intracellularly in an immature "precursor" form with a non-functional plastid (Trémouillaux-Guiller et al. 2002). These precursors evolve into mature algae and proliferate only in necrotizing host cells (Trémouillaux-Guiller and Huss 2007; Trémouillaux-Guiller et al. 2002). Storage bodies, probably of lipid nature, were found in the cytoplasm of both partners, presuming a possible involvement of the alga in metabolic pathways of its host (Trémouillaux-Guiller and Huss 2007). Occasionally, mature algae together with their precursor forms were detected by TEM in intact host cells of a green callus (Trémouillaux-Guiller et al. 2002). DNA barcoding placed the *Ginkgo* isolate among the green coccal genus *Coccomyxa*, a monophyletic group within the class Trebouxiophyceae, which contained seven species (Darienko et al. 2015). The symbiotic algae were nested in *Coccomyxa viridis*, together with other lichen photobionts and some free-living strains of the same species, thus confirming the first genus assignment by Trémouillaux-Guiller and Huss (2007). These authors noted the close features of algae from *Ginkgo* with lichen photobiont *Coccomyxa*, but later study of Darienko et al. (2015) revealed that lichen photobionts are not monophyletic within *the genus* and are related to free-living isolates of *Coccomyxa simplex*, *C. subellipsoidea*, and *C. viridis* (*photobionts in ascomycetous lichens*) *and C. dispar* (from a basidiomycetous lichen).

2.2.5.2 Symbiosis with Cycads

Green alga *Chlorococcum humicolum* was reported once as a symbiont of *Cycas revoluta* (Fernandes and Bhat 1945). Although considering that cycad symbionts are commonly involved de novo and possibility for such occasional capturing has not to be excluded, more further observations are needed to confirm eukaryotic green algae as stable symbionts in cycad roots. We also have to note that the taxonomy of *Chlorococcum* is quite complicated, and currently *C. humicolum* has been synonymized with *Chlorococcum infusionum*, considered as a freshwater alga (Guiry and Guiry 2022).

2.2.5.3 Symbiosis with Conifers

The epiphytic development of green filamentous *Trentepohlia* on the needles of *Abies alba* is discussed together with angiosperm symbioses.

2.2.6 Symbioses with Angiosperms

There is a widespread opinion that eukaryotic algae do not form intracellular or intercellular associations with flowering plants, but according to the profound review by Bristol (1920), *Chlorochytrium* (*C. lemnae*) is hosted by the following land plants: *Limnanthemum indicum* (found in yellowish spots raised on leaves), *Lychnis flos-cuculi* (in leaves), *Rumex obtusifolius* (in leaves, causing reddish color of the epidermis), and in both *Mentha aquatica* and *Peplis portula* in the intercellular spaces of leaves (Reisser 1992) (Fig. 1). Despite such spread of these associations, too less is known on the details of partners interactions and future investigations are needed, including more detailed taxonomy of the algal symbiont (including research on the closely related *Scotinosphaera*). The sole yellow-green alga, known as terrestrial endophyte, is the coccal *Chlorocloster solani* in potato (George 1962; Reisser 1992; Guiry and Guiry 2022).

Non-photosynthetic *Synchytrium borreriae*, thought to be closely related to *Chlorochytrium*, was found in galls on the leaves of *Borreira hispida*, and it is supposed that together with non-colored cells in the intracellular leaf spaces of *Salix repens*, these two inhabitants, long believed to be fungal parasites belonging to Chytridiomycota (the only eumycetes that reproduce by zoospores (Alexopolous et al. 2004)), in fact, are green algae (for details see Reisser 1992). Yet the taxonomy of all species (>200) of "*Synchytrium*," recognized as parasites of mosses, ferns, and angiosperms, is not completely resolved (James et al. 2006). Similar is the situation with the bizarre green alga *Rhodochytrium spilanthidis*, considered as parasite of tracheophytes (Reisser 1992). Since parasitism is not commonly accepted among symbiotic relationships, here we shall briefly mention the disputable relations between green algae and their hosts in cases of intercellular growth of green *Phyllosiphon* in stems and leaves of vascular plants, and the obligate endophytic growth of *Stomatochroon* (*S. reniformis*, with lost photosynthetic capacity) (Reisser 1992; Thompson and Wujek 1997; Brooks et al. 2015). Not long ago, a new variety, viz. *Stomatochroon reniformis* var. *chinensis*, was described from the leaf stomata of the fruit tree *Syzygium samarangense* (Zhu et al. 2014). Since no negative effects on the host have not been recorded regarding its physiology and structure, it was concluded that this variety was an endophyte, not a parasite. By 18s RNA, its close relation to green filamentous *Cephaleuros* was proved (Zhu et al. 2014). *Cephaleuros* contains 17 species that have been proved as endoparasites, plant pathogens with different level of damage in many tropical crops (Thompson and Wujek 1997; Brooks et al. 2015). Disputable are effects of the epiphytic growth of green filamentous *Phycopeltis* (containing sporopollenin in the cell walls; Good and Chapman 1978) and *Trentepohlia* seen as orange spots on a very broad range of vascular plants, among which are *Abies alba*, *Buxus*, *Hedera*, *Leskea*, and *Rubus*, but also tropical crops and ornamentals such as *Camelia* spp. and *Magnolia* spp. (for details see Thompson and Wujek 1997; Ettl and Gärtner 2014; Reisser 1992), all of which definitely need future detailed investigations.

2.2.7 Symbioses of Eukaryotic Algae with Animals ("Mobile" Algal Symbioses and Epibiotic Life)

Eukaryotic algae enter into symbioses with different interactions invertebrates and vertebrates (Fig. 1), some of which are yet disputable. The focus on symbioses of sloths in the beginning of the chapter is based on their recognition as models for mobile symbiotic systems.

2.2.7.1 Symbioses of Eukaryotic Algae with Sloths (Model "Mobile Symbioses")

Sloths are increasingly recognized as unusual "mobile ecosystems," bearing a high diversity of epibionts, or "dynamic, moving islands of biodiversity," considered as model systems for studies of biodiversity and partner interactions (Heaney et al. 2013; Borregaard et al. 2017; Kaup et al. 2021). They are well-known due to their peculiar "arboreal" mode of life, which they spend mostly hanging on trees in tropical rainforests of South America, but also for their extremely slow metabolism since their food comprises mainly, or exclusively, of tree leaves (Gilmore et al. 2001; Kaup et al. 2021). The common names of sloths in several different languages reflect this "lazy" life, but sloths move slowly through the forest and are able to swim. These placental mammals have evolved around 60 MA (O'Leary et al. 2013) and most species are extinct, but there are six extant species of two genera, *Bradypus* and *Choloepus*, the common ancestor of which is thought to have appeared 28 MA (Presslee et al. 2019). Both genera host different organisms, mostly related with their peculiar fur, which comprises of long hair of specific, groovy structure with crevices suitable for growth of green algae, "responsible" for its green coloration (Aiello 1985; Suutari et al. 2010; Kaup et al. 2021). The comparative phylogenetic and phylogeographic analysis revealed data that support the previously hypothesized symbiotic relationship between animals and fur algae, indicating coevolution and cospeciation (for *Bradypus* lineage in particular) (Fountain et al. 2017).

It has not been answered whether algae identified from sloths are capable of living on texturally smooth hair and the way of fur colonization is not fully known. Moreover, yet, it is even not clear whether algae assist in hair grooving, or they settle on already cracked hair threats, and it is also questioned whether they enter into the fissures or inhabit the outer parts of the ribs, with hypotheses on the topic and possible ways of colonization summarized in the current comprehensive paper by Kaup et al. (2021). The physical environment of the fur is characterized by gradients in light and temperature: darker (shady) and warmer, but more stable conditions of the hair basis near to skin, versus to more exposed and thus more varying conditions at hair tips that are more exposed (Kaup et al. 2021). Therefore, the sloth fur can be taken as a micro-ecosystem, the organization of which has been compared to stratified soil crusts and microbial mats, or to canopy-like, "layered" community organization of rich grasslands (Kaup et al. 2021). These fur ecosystems, from one side, receive different epibionts from different trees and from the ground communities, each of which hosts great diversity of organisms from different trophic levels, but also serves as dispersal vector for versatile organisms, which sloth "meet" during their slow movements, with ability to cover horizontal distances up to 38 m/day, but also moving up and down (Gilmore et al. 2001; Kaup et al. 2021). This complex "green" ecosystem is significant for sloth's well-being and security, considering that majority of their predators, such as spectacled owls, harpy eagles, ocelots, and tayra, can detect and attack them from the tree canopy (for details see Woollaston 2014; Kaup et al. 2021). Despite the lack of direct empirical tests, the idea of camouflage became a paradigm (Aiello 1985; Suutari et al. 2010; Pauli et al. 2014), but whether or how algae benefit sloths have been long discussed and it was even supposed that sloths simply are not able to clean themselves (Kaup et al. 2021). Some authors proposed nutritional role of fur algae finding support in the fact that remnants of green algal cells were found in stomach content (Pauli et al. 2014) or suggested contribution of algae to thermal isolation (Britton and Atkinson 1938; Goffart 1971; Montgomery and Sunquist 1978; Aiello 1985). As algae are known to be producers of versatile chemical compounds, the excretion of such substances was also suspected to play a role in promoting the health of sloths or as sunscreens, as shown by the discovery of MAAs in *Trichophilus welckeri* (Aiello 1985; Karsten et al. 2005; Suutari et al. 2010). It was also hypothesized that algal exopolysaccharides enhance bacterial growth, also beneficial for sloths (Suutari et al. 2010). Hypothesis that algal metabolites may protect sloths from aggressive Aztec ants that live mutualistically with trees of genus *Cecropia*, used by sloth as refuge sites and possibly for food, was proposed by Kaup et al. (2021). They found reason for this proposal in numerous documentations of undisturbed sloths hanging to ant or termite nests. Earlier, some ideas on connection between sloth moths on the fur and algae also emerged because adult moths from the genus *Cryptoses* may consume sloth/algal secretions on sloth hair (Kaup et al. 2021). Further ideas enlarge the sloth-algal symbiosis to a tripartite one because algae nourish the moths, then nutrients consumed by moths are released through defecation, increasing the amount of N on the fur, which, in turn, promote algal growth (Pauli et al. 2014). However, in addition to the algae, many other organisms inhabit sloth fur, the relations between which have to be clarified: arthropods, fungi, amoebozoans, cercozoans, apicomplexans, ciliates, and viruses (arboviruses) (Kaup et al. 2021).

Besides "green" sloths, with visible algae on the fur, there are "brown" sloths with invisible for naked eye algae (e.g., Suutari et al. 2010). However, according to Kaup et al. (2021), this should not be interpreted as real lack of fur algae but can be explained by different circumstances (including bathing of animals in zoos) or by living in artificial drier conditions, when algae can turn to a dormant or non-green state, because supplying of moisture (wetting) has shown to turn rapidly, in seconds to minutes, the brown sloths to green ones caused by visible greening, similar to changes that have been observed many times in soil algal crusts after raining or flooding (Goffart 1971; Kaup et al. 2021). Such drying-wetting changes of the fur coloration have been regularly observed during seasonal shifts of dry and rainy tropical seasons (for details see Kaup et al. 2021).

The algae on sloths have intrigued algologists since more than a century, with the best known and thought to be common sloth species, *Trichophilus welckeri* (Kaup et al. 2021), first identified in 1887 by Dutch phycologist Anna Antoinette Weber-van Bosse (1852–1942) (Weber-van Bosse 1887). However, this alga has been recorded on species from zoos but has not been found in nature (Suutari et al. 2010). In total, on the fur, algae from 19 mostly terrestrial genera have been discovered, many of which from surrounding soils: three blue-green (*Fischerella*, *Oscillatoria*, *Nostoc*), one red (*Rufusia*), and 15 green from division Chlorophyta (coccal unicellular or colonial *Asterochloris*, *Chlorella*, *Collinsiella*, *Dictyococcus*, *Myrmecia*, *Nannochloris*, *Planophila*, *Trebouxia* and filamentous *Printzina*, *Pseudendoclonium*, *Stichococcus*, *Trentepohlia*, *Trichophilus*, *Trichosarcina*, and *Ulothrix*) (Thompson 1972; Suutari et al. 2010; Wujek and Timpano 1986). By 18S amplicon sequencing also euglenophytes (*Petalomonas cantuscygni*) and pyrrhophytes from family Heterocapsaceae were revealed (Suutari et al. 2010).

However, it has to be stressed that much further work is necessary on identifying the algal symbionts and their relations with certain hosts species or genera from natural habitats and zoos, and greater algal biodiversity has to be expected, especially in regard the species composition (e.g., Pauli et al. 2014). A range of algae across and within different species of sloths was detected by amplicon metagenomic studies (Suutari et al. 2010). Along with some common algae (e.g., green *Trichophilus welckeri* and red *Rufusia pilicola*, formerly *Cyanoderma pilicola*), there are differences in the algal assemblages in terms of composition and numbers in both sloth genera, thought to be caused by the morphological peculiarities of their hair: (1) transverse cracks, which deepen and increase in quantity by the age, are typical for the three-fingered sloths (Wujek and Cocuzza 1986). When wet, these hairs swell considerably and retained moisture inside the cracks that was hypothesized to have a double function—to sustain algal growth (Aiello 1985), but also may contribute to buffer temperature changes in sloth body (Pauli et al. 2014, 2016); (2) vertical grooves of the hair are characteristic of the two-fingered sloth that does not absorb as much water and algae seemingly are growing inside them (Aiello 1985; Wujek and Cocuzza 1986). The algal diversity and specificity can also be dependent on the transmission mode, which yet needs to be clarified along with obligate of facultative character of symbionts, because all possible ways have been presumed (Kaup et al. 2021).

2.2.7.2 Symbioses of Eukaryotic Algae with Amphibians

The possibility for vertical transmission, currently suggested for salamander symbiotic algae (Kerney 2011; Lin and Bishop 2015), has some relation with land life of the amphibians and is the reason to include the amphibian-green algal symbioses in this review. The mutualistic symbiosis between the spotted salamander *Ambystoma maculatum* and green alga *Oophila amblystomatis* was discovered by Orr (1888) and further studied and described in detail by Gilbert (1942, 1944). Algae in this and other amphibians inhabit the transparent gelatinous egg capsules that surround their embryos, providing oxygen to the embryos, which excrete ammonia that is supplied to the algae, which benefit also from the protected environment (Hutchinson and Hammen 1958; Reisser 1992; Marco and Blaustein 2000; Small et al. 2014; Kerney 2011; Kim et al. 2014; Muto et al. 2017; Small and Bishop 2020). The oxygen supply is extremely important for the embryos because usually the eggs develop in conditions with low oxygen content because they are laid in small, shallow temporary ponds in woodlands, rich in detritus and decaying plant debris (Hutchinson and Hammen 1958). In addition, algae can excrete some antimicrobic substances to remove the pathogens from the eggs (Kim et al. 2014), and long ago it was described that masses of eggs inhabited by algae had lower mortality (Gilbert 1942, 1944; Hutchinson and Hammen 1958). Moreover, these eggs had earlier hatching time and faster growth than the algae-free clutches (Hutchinson and Hammen 1958).

Kerney et al. (2011) first demonstrated that the salamander-green algal symbiosis in case of *Ambystoma maculatum* is more intimate than previously reported, with algae able to invade the tissues and cells of the host, indicating that the interactions between both partners may be more complex than previously realized. Comparison of host cells with symbiotic algae with cells without algae, and of algae inside the animal cells with those in the egg capsule, was conducted by using de novo dual-RNA sequencing (Burns et al. 2017). This study demonstrated that intracellular algae are stressed and turn from oxidative metabolism to fermentation, whereas salamander cells did not exhibit stress hallmarks, which indicates their neutral or beneficial experience. Later, some competition for available inorganic carbon between the embryos and algae within the egg capsule environment was detected (for details see Burns et al. 2020).

Although the symbiotic alga *Oophila amblystomatis* was described about a century ago (Printz 1928), the information on its phylogeny and ecology is surprisingly low. Other taxonomic identifications include *Palmodactylon* sp. nov. and *Chlorochytrium* sp. (Gilbert Smith in Gilbert 1942), *Chlamydomonas gloeophila* (Ettl 1961), and *Chlamydomonas* sp. (Biebel 1969; Reisser 1992). Genetic studies of strains from North America and England particularly verified the chlamydomonad assignment of amphibia-associated algae (Kerney et al. 2011; Kim et al. 2014). They confirmed the discrete character of the genus *Oophila* and demonstrated that within this lineage, the algae from amphibians were nested into four possible subclades, without a clear relation with their hosts (Kim et al. 2014). Later study revealed two well-supported genetically *Oophila* clades (Nema et al. 2019).

Genetic studies of Kim et al. (2014) included investigation of two old strains, isolated from the amphibian eggs more than 60 years ago from two states of USA (New York and Connecticut) and identified morphologically as *Chlamydomonas gloeophila*. 18S DNA sequencing revealed their high similarity to the same, but free-living species, which is distant from the *Oophila* clade. As a result, it was supposed that *Chlamydomonas gloeophila* might occur in low amounts in amphibian egg clutches (Kim et al. 2014), thus confirming the earlier finding by Ettl (1961) but also showing that taxonomic identity and specificity of the amphibia-associated algae is still to be clarified.

Firstly, horizontal acquirement of algae was presumed, but such process has not been confirmed, thus provoking a set of studies regarding possibilities for vertical transfer of algae, supposed by Kerney et al. (2011). A subsequent genetically based investigation provided additional evidence for such vertical transfer, not only for the yellow spotted salamander but also for the wood frog deposit egg masses, with all algae nested among *Oophila* clade (Lin and Bishop 2015). This suggests the presence of more than known before obligate associations between amphibians and green algae.

The dwarf American toad, *Bufo americanus charlesmithii*, inhabits sites of strong solar insolation and high temperatures (25–40 °C), and this explains the greenish color of the late-stage tadpoles, some of which had already well-developed legs (Tumlison and Trauth 2006). The coloration follows the scattered pattern of the cutaneous blood vessels on the body and was caused by clutches of green flagellate alga *Chlorogonium*. The most interesting thing is the active swimming of these algae in the shallow pools to the skin of the tadpoles combined with the physiological and morphological changes of the algal cells (photosynthesis rates, dimensions, etc.). Tumlison and Trauth (2006) reasonably hypothesized the mutualistic relationship, in which gases are exchanged and reciprocally used along with the sun protection of the young frogs and warmth stress protection of both partners. Similar greening was observed, although not investigated microscopically, by the same authors in cases of other frogs, such as cricket frogs (*Acris crepitans*) and grey tree frogs (*Hyla versicolor*).

More symbioses between green algae and embryos of several species of tailed and tailless (anuran) amphibians, presumed as important for their survival and dispersal, are known (for details see Kerney 2011; Kim et al. 2014; Desnitskiy 2019) and need further, more detailed investigations.

Likewise in other symbioses, it is necessary to continue the studies and to investigate eventual interrelation with other organisms, which have been recorded from the egg envelopes, such as protozoa (including, in particular *Haptophyra* sp. which feeds on the algae), diatoms (usually on the surface of the egg masses), acanthocephalans, rotifers, nematodes, and an occasional annelid (Hutchinson and Hammen 1958).

2.2.7.3 Symbioses of Eukaryotic Algae with Other Animals

Many green algae can use terrestrial animals as their substrates, and epi- and endozoic algae have been found in the larvae of aquatic insects (including euglenophytes in their gut), but studies on their interactions with the hosts have not been conducted (for details see Hoffmann 1989; Reisser 1992). However, on *prima vista*, it seems that at least some of these algae use the rough surface as basic substrates for their aerophytic life but do not form stable associations, as it was shown for epimycotic algae on *Fomes* and *Trametes* (Stoyneva et al. 2015; Videv et al. 2017). The possibility that such epibiontic algae serve as camouflage and as a food source was hypothesised for a caterpillar collected in Australia that was completely covered by *Trebouxia* (Gribb 1965; Reisser 1992). Cases of achloric *Protoheca/Chlorella* pathogenous cutaneous infections of a broad range of hosts (snakes, beavers, cattle, lambs, deer, dogs, cats, and humans) are also known (Reisser 1992).

However, it is worth to mention the typical soil *Chlorosarcina* (Ettl and Gärtner 2014), found on the surfaces of the caudal lamellae of the naiads of damselflies (Odonata), but also inside them, with some respiratory relationships supposed (for details see Reisser 1992). In the damselfly larvae, was discovered the new species of euglenophytes, *Colacium libellae*, visible as a "conspicuous green plug" covering the rectum (Rosowski and Willey 1975; Reisser 1992).

A peculiar case of growth of the yellow-green coccal alga *Trypanochloris clausiliae* (Guiry and Guiry 2022) beneath the organic periostracum of the small terrestrial snail *Clausillia* was discovered by Geitler (1935). Considering the importance of this specific layer ("skin") for shell and snail development, special attention should be paid to this association in future studies.

3 Discussion

This review highlights the unprecedented capacity of the evolutionary antique phototrophs—cyanoprokaryotes and eukaryotic algae—to enter in efficient symbioses with a broad palette of organisms. Altogether 89 genera, most of which are green algae from Chlorophyta (44), followed by Cyanoprokaryota (38), Ochrophyta (Xanthophyceae, 3), Euglenophyta (2), Pyrrhophyta (1), and Rhodophyta (1) (Fig. 8), have been reported as partners in symbiotic relationships with heterotrophic and autotrophic hosts from more than 13 major groups of organisms (Figs. 1 and 9). However, inside these major lineages, the number of hosts, which are in tight coupling with soil algae, is more restricted. At the same time, considering the enormous algal biodiversity, including those of the specific group of soil algae and the number of species and genera with symbiotic competence, is quite restricted and the reason for this is far away from being understood. Although different traits of algae that ensure their survival in changing environments have been discussed in numerous papers in order to clarify their selective advantages, the analysis of existing literature does not allow us to determine definitely the features, responsible for

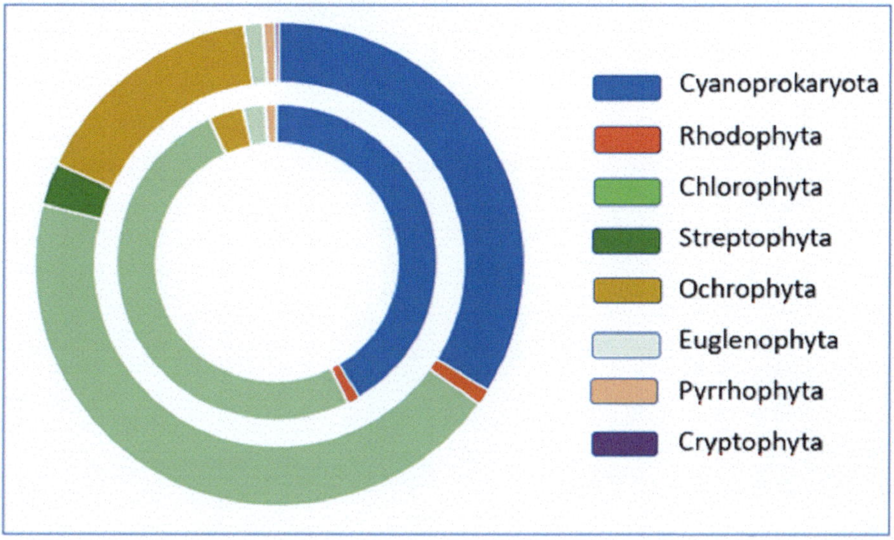

Fig. 8 Comparison of biodiversity of symbiotic soil algae (inner ring) with total biodiversity of soil algae (outer ring)

symbiotic effectiveness. Moreover, many strains or species of the same genus have abilities required for a wide variety of associations, as it was exemplified in case of *Nostoc* (Meeks and Elhai 2002). This can explain the species flexibility with the relative lack of discrimination between strains by hosts on the background of genus-specific symbioses (e.g., Zimmerman and Bergman 1990).

Most of the peculiarities, used to explain the success of symbionts, are shared with many other free-living genera of the same taxonomic group. As such, for cyanoprokaryotes can be enlisted the presence of heterocytes and mucous sheaths (Fig. 2), formation of gliding hormogonia (Fig. 4) and dormant akinetes, as well as diazotrophic abilities. Indeed, long ago Enderlin and Meeks (1983) underlined that hormogonia and N fixation are not a prerequisite for infection. For eukaryotic algae, such peculiarities appear to be the bearing of massive plastid(s) with one or more pyrenoids as CCM (Fig. 3), specific reproduction by autospores of coccal forms, by aplanospores and akinetes in most of the filamentous forms and formation of resting cysts in flagellate algae, together with well-developed multilayered cell wall. However, all these features can be found in many other algae inhabiting soil habitats. It has to be underlined that some of the enlisted features can be facultative, such as the formation of mucilage sheaths by some filamentous cyanoprokaryotes (e.g., Komárek 2013), and, therefore, shall not be discussed in detail below.

The analysis of the morphological and reproductive features of symbiotic algae revealed something which is obvious but has been generally ignored in discussions: "size matters" and larger algae, such as the terrestrial siphonal yellow-green *Vaucheria* and *Botrydium*, have not been involved in such partnerships. Regarding thallus morphology, both coccal and filamentous algae prevail among symbionts

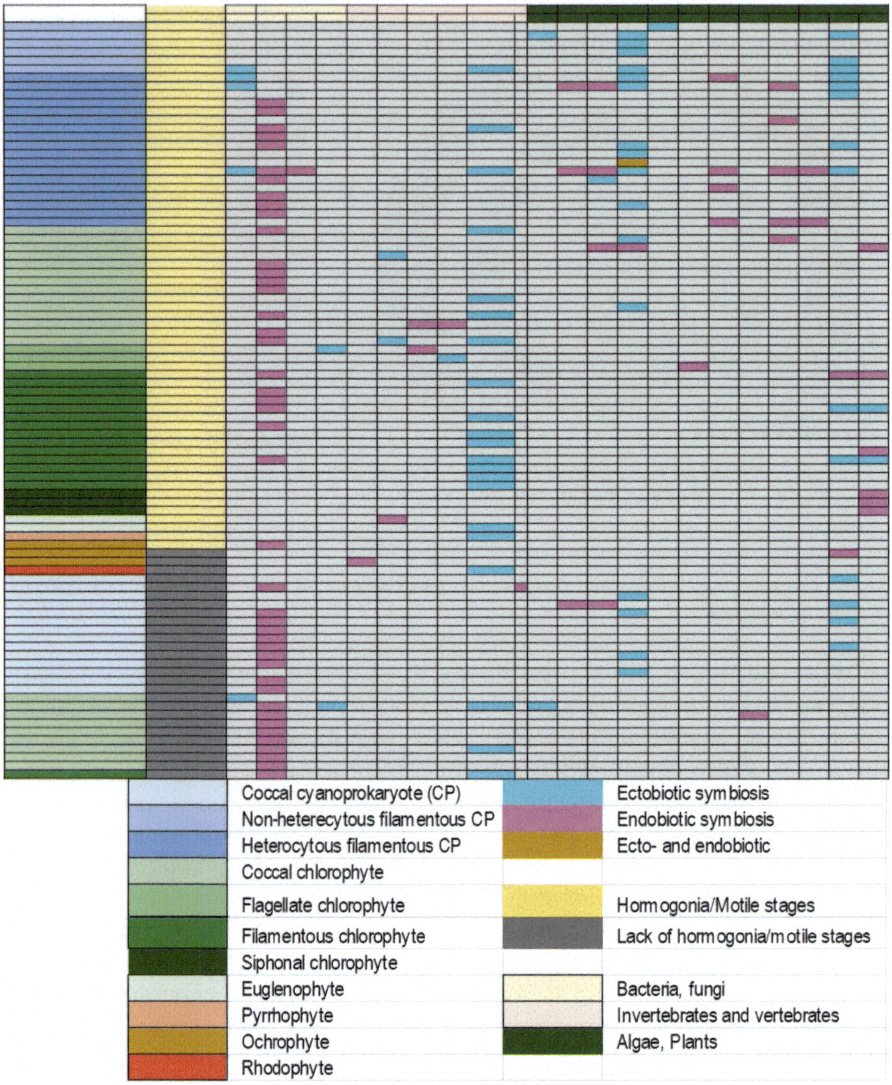

Fig. 9 Matrix of symbioses (ecto- and endo-) of soil algae and their main heterotrophic and autotrophic host types, distributed by genera, with indicated morphological type of the thallus (for blue-green and green algae) and presence/absence of gliding and motile stages. For details, see the text of the paper

(94%, Fig. 9). This is especially valid for the long-lasting and highly developed symbioses (e.g., lichens) and is logically explainable with the restricted size of the microenvironments, where energetically expensive flagellate forms are not necessary.

Undoubtedly, there is not yet a single accepted answer to the question of who initiated the relationships. In lichens, there are many examples of mycobiont hyphae actively searching for a new photobiont (e.g., Nash III 2008; Wirth et al. 2013). But we can presume that millions of years ago, "ladies' choice" could also happen on the ancient Earth near the shores of the waterbodies, near to temporary pools or after strong rains. In support of this is the finding of the earliest lichen fossil from the Doushantuo Formation in aquatic shallow habitat (Yuan et al. 2005). In such cases, motile stages seem to have been advantageous for entering in the host and establishment of the symbioses. The role of algal activity was many times stressed as a key trait of hormogonia-producing blue-greens, as was described in the relevant part of the paper.

The analysis of data, collected in this review, allows to state that hormogonia and other motile reproductive and dispersal stages can be produced by majority of symbiotic algae (62/89, or 70%) (Fig. 9). Moreover, these genera, capable to be "active invaders," obviously have more hosts in comparison with other algae (Fig. 9). Most eukaryotic algae, involved in symbiotic interactions (41/51, or 80%), include production of flagellate, motile spores (planospores, zoospores), and/or motile flagellate gametes (planogametes) among their proliferation modes, which are widely recognized as primordial in the evolution of reproduction, and this is especially valid for phylum Chlorophyta, which embraces all symbionts from the green lineage. In this regard, the recorded active movement of *Chlorogonium* to the skin of the sun-exposed tadpoles, currently happening in the shallow warm ephemeral pools (Tumlison and Trauth 2006), is providing support to our idea that ancient symbionts invaded actively their hosts. Earlier, discussing the fossils of *Lageniastrum* in lycophytes, Krings et al. (2005) also assumed that entrance into the intact spores might happen through their central sutures by the motile cells (swarmers), developed in both sexual and asexual reproduction of this volvocalean-like alga. For example, *Chlorochytrium*, a long-neglected terrestrial symbiont, is coccal in its vegetative state but reproduces by bi-flagellate gametes and additionally is able to form akinete-like thick-walled cells (Smith 1950). Likewise, the most widespread in lichens *Trebouxia* has not been observed to produce zoospores as a photobiont inside the lichens but readily makes them in cultures (Nash III 2008; Ettl and Gärtner 2014), suggesting that this feature is not necessary during the endobiotic life and for the vertical transfer. Exceptions from producers of motile spores or gametes are ten eukaryotic genera (20%), for which formation of zoospores is "not known" (e.g., Ettl and Gärtner 2014) but cannot be completely excluded. For example, the single exception from planospore (zoospore) and planogamete producers among filamentous symbiotic algae is the widely spread in soils *Stichococcus* (Ettl and Gärtner 2014). However, zoospores have been recorded in one species of its close relative *Pseudostichococcus monallantoides* (Moewus 1951; Pröschold and Darienko 2020; Guiry and Guiry 2022). On the other hand, the ten symbiotically competent genera with non-motile stages may be accepted as "the exception that confirms the rule," but it is to stress that taxonomy of most of them is not completely resolved and their identification is quite difficult even for experts. Such an example is *Chlorella* s.l., the misidentification of the "green balls" of which is not to be underestimated, especially in cases of direct morphological observations. Analyzing the results obtained from this review, it is possible to suppose that the lack of motile reproductive cells prevents most of microscopic species from the second green phylum, Streptophyta, to enter in symbioses. The soil algae from this taxonomic group that reproduce vegetatively through simple division or sexually through conjugation up-to-now have never been recorded as symbionts. In the same time, we want boldly to underline that the presence of flagella should not be taken as a single prerequisite for symbiotic competence. This can be exemplified with a few cases of euglenophytes (e.g., *Colacium*, *Petalomonas*) and pyrrhopytes (Heterocapsales), known as symbionts. This example confirms the complexity of selectivity factors, because euglenophytes and pyrrhophytes are very rarely found as soil inhabitants (Ettl and Gärtner 2014). In this regard, it seems that number of terrestrial symbionts from different taxonomic groups reflects the total algal biodiversity in soils and their relevant availability, outlining the environmental ability as important factor for symbiosis establishment (Fig. 8). Then, it is strange that one of the most widespread soil genera, the streptophyte genus *Klebsormidium*, equipped with well-developed resilient cell wall, having the ability to reproduce asexually and sexually with flagellate cells, but also vegetatively by fragmentation and by production of akinetes as resting stages (Lokhorst 1996; Uzunov et al. 2012; Ettl and Gärtner 2014; Stoyneva-Gärtner et al. 2019), according to our best knowledge, has not been recorded as symbiotic. Therefore, considering that *Klebsormidium* is often taken as one of the progenitors of the land plants (e.g., Sluiman et al. 2008; Turmel et al. 2013), it is possible to hypothesize that origins of symbiosis can be searched in the evolutionary history of participants and earlier evolved algae had some traits which gave them better chances to be accepted in complex symbiotic systems, increasing their capacity. In this regard, it has to be noted that land plants had more symbionts from ancient blue-green algae than from later evolved eukaryotic algae (Fig. 1). On the other side, this clearly confirms the greater significance of N fixation, because it is more reasonable to believe that land plants needed additional N supply more than single photosynthetic capacity.

The taxonomy of symbiotic algae is far away from being resolved, despite that most of them have been the subject of intense research considering morphology, ultrastructure, functionality, and genetic characteristics, and much more polyphasic studies are necessary to be accomplished. It is well understood that it is always to expect some hindering caused by horizontal transmission in cases when new symbionts are acquired de novo from the environment. By time, application of more modern microscopical, biochemical, physiological, ecological, paleontological, and molecular-genetic tools, according to novel approaches for cultivation and co-culturing of symbiotic partners, may bring more light on the symbiont diversity and specificity, as well as on genomic control and complex genomic interactions between partners.

Newly emerging data on the coexistence of more evolutionarily distant partners in symbioses long thought to be only bipartite but revealed to be more complex, involving at least three partners and larger metacommunities, on much greater biodiversity in both photobionts and hosts, as well as the modern views on micro- and macroecosystems, etc. will undoubtedly lead to new ideas for explaining the establishment, maintenance and spread of symbioses, whether fossil or extant, and the evolutionary forces behind them. It is necessary to stress the need for better investigation of many endo- and epibionts, for better understanding of symbiotic interactions, and for enlarging studies on different hosts, especially of "mobile substrata." Last but not least, the novel specific metabolites, produced by symbiotic systems, or their higher yield in comparison with free-living organisms, have to be noted with regard to the development of modern agriculture, biotechnology, and many other fields of human life. In this regard, it is necessary to go deeper in benefits for all partners and to uncover the novel organization level, features, or products which their partnerships create.

In conclusion, we would like to stress that symbiotic algae represent a fascinating group of organisms that has not been sufficiently explored and many interesting aspects of symbioses remain to be elucidated, so, we hope that this review will not only encourage, but also facilitate future researches in their efforts to unveil the enigma of symbiosis.

Acknowledgments This study is financed by the European Union-NextGenerationEU, through the National Recovery and Resilience Plan of the Republic of Bulgaria, project No. BG-RRP-2.004-0008. The authors are thankful to Dr K. Kosev, Director of the University Botanical Gardens of Sofia University "St. Klimnet Ohridski", for providing access and sampling of the exotic plant hosts grown in the Sofia Botanical Garden.

References

Adams DG, Duggan PS (2008) Cyanobacteria–bryophyte symbioses. J Exp Bot 59(5):1047–1058. https://doi.org/10.1093/jxb/ern005

Adams DG, Bergman B, Nierzwicki-Bauer SA et al (2013) Cyanobacterial-plant symbioses. In: Rosenberg E, DeLong EF, Lory S et al (eds) The prokaryotes – prokaryotic biology and symbiotic associations. Springer-Verlag, Berlin, pp 359–400. https://doi.org/10.1007/978-3-642-30194-0_17

Aguilera A, Klemenčič M, Sueldo DJ et al (2021) Cell death in cyanobacteria: current understanding and recommendations for a consensus on its nomenclature. Front Microbiol 12:631–654. https://doi.org/10.3389/fmicb.2021.631654

Ahern CP, Staff IA (1994) Symbiosis in cycads: the origin and development of coralloid roots in *Macrozamia communis* (Cycadeceae). Am J Bot 81:1559–1570. https://doi.org/10.1002/j.1537-2197.1994.tb11467.x

Ahmadjian V (1988) The lichen alga *Trebouxia*: does it occur free – living? Plant Syst Evol 158:243–247

Aiello A (1985) Sloth hair unanswered questions. In: Montgomery GG (ed) The evolution and ecology of armadillos, sloths and vermilinguas. Smithsonian Institution Press, Washington, DC, pp 213–218

Alexopolous CJ, Mims CW, Blackwell M (2004) Introductory mycology, 4th edn. Wiley, Hoboken

Álvarez C, Navarro JA, Molina-Heredia FP et al (2020) Endophytic colonization of rice (*Oryza sativa* L.) by the symbiotic strain *Nostoc punctiforme* PCC 73102. Mol Plant-Microbe Interact 33(8):1040–1045. https://doi.org/10.1094/MPMI-01-20-0015-SC

Alvarez AL, Weyers SL, Goemann HM et al (2021) Microalgae, soil and plants: a critical review of microalgae as renewable resources for agriculture. Algal Res 54:102–200. https://doi.org/10.1016/j.algal.2021.102200

Arnold AE, Miadlikowska J, Higgins KL et al (2009) Hyperdiverse fungal endophytes and endolichenic fungi elucidate the evolution of major ecological modes in the Ascomycota. Syst Biol 58:283–297

Arróniz-Crespo M, Pérez-Ortega S, De Los Ríos A et al (2014) Bryophyte-cyanobacteria associations during primary succession in recently deglaciated areas of Tierra del Fuego (Chile). PLoS One 9:e96081. https://doi.org/10.1371/journal.pone.0096081

Atkinson N, Mao Y, Chan KX et al (2020) Condensation of Rubisco into a proto-pyrenoid in higher plant chloroplasts. Nat Commun 11:6303. https://doi.org/10.1038/s41467-020-20132-0

Banerji J, Ghosh AK (2002) Mutualism/symbiosis from the Early Cretaceous (intertrappeans) of Rajmahal Basin Jharkhand, India. Curr Sci 83(9):1073–1074

Baulina OI, Lobakova ES (2003) Atypical cell forms overproducing extracellular substances in populations of cycad cyanobionts. Microbiology 72(6):701–712. https://doi.org/10.1023/B:MICI.0000008373.16642.e7

Bay G, Nahar N, Oubre M et al (2013) Boreal feather mosses secrete chemical signals to gain nitrogen. New Phytol 200(1):54–60. https://doi.org/10.1111/nph.12403

Beck A, Divakar PK, Zhang N et al (2015) Evidence of ancient horizontal gene transfer between fungi and the terrestrial alga *Trebouxia*. Org Divers Evol 15:235–248. https://doi.org/10.1007/s13127-014-0199-x

Becking JH (1978) Ecology and physiological adaptations of *Anabaena* in the *Azolla-Anabaena azollae* symbiosis. Ecol Bull 26:266–281

Bell-Doyon P, Laroche J, Saltonstall K et al (2020) Specialized bacteriome uncovered in the coralloid roots of the epiphytic gymnosperm, *Zamia pseudoparasitica*. Environ DNA 2:418–428. https://doi.org/10.1002/edn3.66

Bergman B, Johansson C, Soderback E (1992) The *Nostoc-Gunnera* symbiosis. New Phytol 122:379–400. https://doi.org/10.1111/j.1469-8137.1992.tb00067.x

Bergman B, Matveyev A, Rasmussen U (1996) Chemical signaling in cyanobacterial-plant symbioses. Trends Plant Sci 1:191–197. https://doi.org/10.1016/1360-1385(96)10021-2

Bergman B, Gallon JR, Rai AN et al (1997) N_2 fixation by non-heterocystous cyanobacteria. FEMS Microbiol Rev 19(3):139–185. https://doi.org/10.1111/j.1574-6976.1997.tb00296.x

Berrendero E, Valiente EF, Perona E et al (2016) Nitrogen fixation in a non-heterocystous cyanobacterial mat from a mountain river. Sci Rep 6:30920. https://doi.org/10.1038/srep30920

Biebel P (1969) Use of physiological and biochemical characteristics in distinguishing chlamydonomad algae associated with amphibian egg membranes. Int Bot Cong Abstr 11:15

Blaha J, Baloch E, Grube M (2006) High photobiont diversity in symbioses of the euryoecious lichen *Lecanora rupicola* (Lecanoraceae, Ascomycota). Biol J Linn Soc 88:283–293

Bond G, Scott GD (1955) An examination of some symbiotic systems for fixation of nitrogen. Ann Bot 19(1):67–77. https://doi.org/10.1093/oxfordjournals.aob.a083419

Borregaard MK, Amorim IR, Borges PAV et al (2017) Oceanic island biogeography through the lens of the general dynamic model: assessment and prospect. Biol Rev 92:830–853

Bouchard R, Peñaloza-Bojacá G, Toupin S et al (2020) Contrasting bacteriome of the hornwort *Leiosporoceros dussii* in two nearby sites with emphasis on the hornwort-cyanobacterial symbiosis. Symbiosis 81:39. https://doi.org/10.1007/s13199-020-00680-1

Boucher DH, James S, Keeler KH (1982) The ecology of mutualism. Annu Rev Ecol Syst 13(1):315–347

Brenner ED, Stevenson DW, Twigg RW (2003) Cycads: evolutionary innovations and the role of plant-derived neurotoxins. Trends Plant Sci 8(9):446–452. https://doi.org/10.1016/S1360-1385(03)00190-0

Brinkhuis H, Stefan S, Margaret EC et al (2006) Episodic fresh surface waters in the Eocene Arctic Ocean. Nature 441:606–609

Bristol BM (1920) A review of the genus *Chlorochytrium* Cohn. J Linn Soc Bot 45:1–28

Britton SW, Atkinson WE (1938) Poikilothermism in the sloth. J Mammal 19:94–99
Brooks F, Rindi F, Suto Y et al (2015) The Trentepohliales (Ulvophyceae, Chlorophyta): an unusual algal order and its novel plant pathogen—*Cephaleuros*. Plant Dis 99(6):740–753. https://doi.org/10.1094/PDIS-01-15-0029-FE
Bubrick P, Galun M, Frensdorff A (1984) Observations on free-living *Trebouxia* De Puymaly and *Pseudotrebouxia* Archibald, and evidence that both symbionts from *Xanthoria parientina* (L.) can be found free-living in nature. New Phytol 97(3):455–462
Bujak J (2007) The *Azolla* story. Climate change and Arctic hydrocarbons. GEO ExPro 4(4):66–72
Burns JA, Zhang H, Hill E et al (2017) Transcriptome analysis illuminates the nature of the intracellular interaction in a vertebrate-algal symbiosis. elife 6:e22054. https://doi.org/10.7554/eLife.22054
Burns JA, Kerney R, Duhamel S (2020) Heterotrophic carbon fixation in a salamander-alga symbiosis. Front Microbiol 11:1815. https://doi.org/10.3389/fmicb.2020.01815
Caiola MG (1980) On the phycobionts of the cycad coralloid roots. New Phytol 85:537–544. https://doi.org/10.1111/j.1469-8137.1980.tb00769.x
Campbell EL, Meeks JC (1989) Characteristics of hormogonia formation by symbiotic *Nostoc* spp. in response to the presence of *Anthoceros punctatus* or its extracellular products. Appl Environ Microbiol 55:125–131
Campbell EL, Wong FCY, Meeks JC (2003) DNA binding properties of the HrmR protein of *Nostoc punctiforme* responsible for transcriptional regulation of genes involved in the differentiation of hormogonia. Mol Microbiol 47:573–582
Campbell EL, Christman H, Meeks JC (2008) DNA microarray comparisons of plant factor- and nitrogen deprivation-induced hormogonia reveal decision-making transcriptional regulation patterns in *Nostoc punctiforme*. J Bacteriol 190(22):7382–7391. https://doi.org/10.1128/JB.00990-08
Car NG, Whitton BA (eds) (1982) The biology of cyanobacteria. Blackwell Scientific Publications, Oxford
Carrapiço F (1991) Are bacteria the third partner of the *Azolla-Anabaena* symbiosis? Plant Soil 137:157–160
Carrapiço F (2002) The *Azolla-Anabaena*-bacteria system as a natural microcosm. Proc SPIE 4495:261–265
Carrapiço F (2010) *Azolla* as a superorganism. Its implication in symbiotic studies. In: Seckbach J, Grube M (eds) Symbioses and stress: joint ventures in biology, Cellular origin, life in extreme habitats and astrobiology, vol 17. Springer, Dordrecht, pp 225–241. https://doi.org/10.1007/978-90-481-9449-0_11
Casano LM, del Campo EM, Garcia-Breijo FJ et al (2011) Two *Trebouxia* algae with different physiological performances are everpresent in lichen thalli of *Ramalina farinacea*. Coexistence versus competition? Environ Microbiol 13:806–818. https://doi.org/10.1111/j.1462-2920.2010.02386.x
Chang DCN, Grobelaar N, Coetzee J (1988) SEM observations on cyanobacteria-infected cycad coralloid roots. S Afr J Bot 54:491–495. https://doi.org/10.1016/s0254-6299(16)31284-4
Chang ACG, Tao C, Nan L et al (2019) Perspectives on endosymbiosis in coralloid roots: association of cycads and cyanobacteria. Front Microbiol 10:1888. https://doi.org/10.3389/fmicb.2019.01888
Christman HD, Campbell EL, Meeks JC (2011) Global transcription profiles of the nitrogen stress response resulting in heterocyst or hormogonium development in *Nostoc punctiforme*. J Bacteriol 193(24):6874–6886. https://doi.org/10.1128/JB.05999-11
Cohen MF, Meeks JC (1997) A hormogonium regulating locus, hrmUA, of the cyanobacterium *Nostoc punctiforme* strain ATCC 29133 and its response to an extract of a symbiotic plant partner *Anthoceros punctatus*. Mol Plant-Microbe Interact 10:280–289
Cohen MF, Wallis JG, Campbell EL et al (1994) Transposon mutagenesis of *Nostoc* sp. strain ATCC 29133, a filamentous cyanobacterium with multiple cellular differentiation alternatives. Microbiology 140:3233–3240

Collinson ME (2001) Cainozoic ferns and their distribution. Britonia 53:173–235. https://doi.org/10.1007/BF02812700

Collinson ME, Barke J, van der Burgh J et al (2009) A new species of the freshwater fern *Azolla* (Azollaceae) from the Eocene Arctic Ocean. Rev Palaeobot Palynol 155:1–14

Cornejo C, Scheidegger C (2016) Cyanobacterial gardens: the liverwort *Frullania asagrayana* acts as a reservoir of lichen photobionts. Environ Microbiol Rep 8(3):352–357. https://doi.org/10.1111/1758-2229.12386

Cornejo C, Nelson P, Stepanchikova I et al (2016) Contrasting pattern of photobiont diversity in the Atlantic and Pacific populations of *Erioderma pedicellatum* (Pannariaceae). Lichenologist 48(4):275–291. https://doi.org/10.1017/S0024282916000311

Costa J-L, Paulsrud P, Lindblad P (1999) Cyanobiont diversity within coralloid roots of selected cycad species. FEMS Microbiol Ecol 28:85–91

Costa JL, Paulsrud P, Rikkinen J et al (2001) Genetic diversity of *Nostoc* symbionts endophytically associated with two bryophyte species. Appl Environ Microbiol 67(9):4393–4396. https://doi.org/10.1128/AEM.67.9.4393-4396.2001

Costa JL, Romero EM, Lindblad P (2004) Sequence based data supports a single *Nostoc* strain in individual coralloid roots of cycads. FEMS Microbiol Ecol 49:481–487

Cox PA, Banack SA, Murch SJ (2003) Biomagnification of cyanobacterial neurotoxins and neurodegenerative disease among the Chamorro people of Guam. PNAS 100:13380–13383. https://doi.org/10.1073/pnas.2235808100

Cox PA, Banack SA, Murch SJ et al (2005) Diverse taxa of cyanobacteria produce β-N-methylamino-l-alanine, a neurotoxic amino acid. PNAS 102:5074–5078. https://doi.org/10.1073/pnas.0501526102

Cruz-Morales P, Corona-Gomez A, Selem-Mojica N et al (2017) The cycad coralloid root contains a diverse endophytic bacterial community with novel biosynthetic gene clusters unique to its microbiome. bioRxiv:1–53

Cuddy WS, Neilan BA, Gehringer MM (2012) Comparative analysis of cyanobacteria in the rhizosphere and as endosymbionts of cycads in drought-affected soils. FEMS Microbiol Ecol 80(1):204–215. https://doi.org/10.1111/j.1574-6941.2011.01288.x

Dal Grande F, Widmer I, Wagner HH et al (2012) Vertical and horizontal photobiont transmission within populations of a lichen symbiosis. Mol Ecol 21:3159–3172. https://doi.org/10.1111/j.1365-294X.2012.05482.x

Dalton DA, Chatfield JM (1985) A new nitrogen-fixing cyanophyte-hepatic association: *Nostoc* and *Porella*. Am J Bot 72(5):781–784. https://doi.org/10.2307/2443693

Darienko T, Gustavs L, Eggert A et al (2015) Evaluating the species boundaries of green microalgae (*Coccomyxa*, Trebouxiophyceae, Chlorophyta) using integrative taxonomy and DNA barcoding with further implications for the species identification in environmental samples. PLoS One 10(6):e0127838. https://doi.org/10.1371/journal.pone.0127838

De Bary HA (1878) Über Symbiose. Tageblatt für die Versammlung Deutscher Naturforscher und Aerzte (in Cassel) 51:121–126

De Bary A (1879) Die Erscheinung der Symbiose. In: Vortrag auf der Versammlung der Deutschen Naturforscher und Aertzezu Cassel. Verlag von Karl J. Trubner, Strasburg, pp 1–30

De Cássia Mariotti K, Barreto F, Schmitt GC et al (2011) Study of acute toxicity and investigation of the presence of β-N-methylamino-L-alanine in the *Gunnera manicata* L. a species native to Southern Brazil. Braz J Pharm Sci 47(3):623–628. https://doi.org/10.1590/S1984-82502011000300021

Del-Prado R, Divakar PK, Lumbsch HT et al (2016) Hidden genetic diversity in an asexually reproducing lichen forming fungal group. PLoS One 11(8):e0161031. https://doi.org/10.1371/journal.pone.0161031

DeLuca TH, Zackrisson O, Nilsson MC et al (2002) Quantifying nitrogen-fixation in feather moss carpets of boreal forests. Nature 419(6910):917–920. https://doi.org/10.1038/nature01051

Desirò A, Duckett JG, Pressel S et al (2013) Fungal symbioses in hornworts: a chequered history. Proc R Soc B 280:20130207. https://doi.org/10.1098/rspb.2013.020

Desnitskiy AG (2019) Facultative symbiosis of *Oophila amblystomatis* (Chlorophyceae) with amphibian eggs and embryos. Int J Plant Reprod Biol 11(2):103–106

Dick H, Stewart WDP (1980) The occurrence of fimbriae on a N2-fixing cyanobacterium which occurs in lichen symbioses. Arch Microbiol 124:107–109

Donohue TJ, Cogdell RJ (2006) Microorganisms and clean energy. Nat Rev Microbiol 4:800

Draganov S, Stoyneva M (1992) Algal flora of the Danube river (Bulgarian sector) and adjoined water basins. III. Algae from some adjoined water basins. Ann Univ Sofia Fac Biol 82(2):63–78

Duckett JG, Prasad AKSK (1974) Ultrastructural studies on the *Nostoc* colonies associated with *Blasia pusilla* and *Anthoceros* spp. J Bryol 8:181

Duckett JG, Toth R, Soni SL (1975) An ultrastructural study of the *Azolla–Anabaena azollae* relationship. New Phytol 75:111–118

Duckett JG, Prasad AKSK, Davies DA et al (1977) A cytological analysis of the *Nostoc*–bryophyte relationship. New Phytol 79:349–362

Duggan PS, Gottardello P, Adams DG (2007) Molecular analysis of genes in *Nostoc punctiforme* involved in pilus biogenesis and plant infection. J Bacteriol 189(12):4547–4551. https://doi.org/10.1128/JB.01927-06

Düll R, Düll-Wunder B (2008) Moose einfach und sicher bestimmen. Quelle & Meyer Verlag, Wiebelsheim

Dworkin M, Falkow S, Rosenberg E et al (eds) (2006) The prokaryotes. A handbook on the biology of bacteria: symbiotic associations, biotechnology, applied microbioogy, 3rd edn. Springer, Singapore

Elmerich C, Newton WE (eds) (2007) Associative and endophytic nitrogen-fixing bacteria and cyanobacterial associations. Springer Kluwer Academic Publishers, Dordrecht

Elvebakk A, Papaefthimiou D, Robertsen EH et al (2008) Phylogenetic patterns among *Nostoc* cyanobionts within bi-and tripartite lichens of the genus *Pannaria*. J Phycol 44:1049–1059. https://doi.org/10.1111/j.15298817.2008.00556.x

Enderlin CS, Meeks JC (1983) Pure culture and reconstitution of the *Anthoceros-Nostoc* symbiotic association. Planta 158:157–165

Ettl H (1961) Zwei neue *Chlamydomonaden*. Arch Protistenk Bd 105:273–284

Ettl H, Gärtner G (2014) Syllabus der Boden-, Luft- und Flechtenalgen, 2 ergänzte Auflage. Springer Spektrum, Berlin Heidelberg

Fahselt D, Alstrup V, Tavares S (1995) Enzyme polymorphism in *Umbilicaria cylindrica* in Northwest Greenland. Bryologist 98:118–122

Famintzin A, Baranietzky J (1867) Beitrag zur Entwickelungsgeschichte und der Gonidien und Zoosporenbildung bei *Physcia parietina*. Vorläufige Mitteilung. Botanische Zeitung 25:189–190

Fedrowitz K, Kaasalainen U, Rikkinen J (2011) Genotype variability of *Nostoc* symbionts associated with three epiphytic *Nephroma* species in a boreal forest landscape. Bryologist 114:220–230. https://doi.org/10.1639/0007-2745-114.1.220

Fedrowitz K, Kaasalainen U, Rikkinen J (2012) Geographic mosaic of symbiont selectivity in a genus of epiphytic cyanolichens. Ecol Evol 2:2291–2303. https://doi.org/10.1002/ece3.343

Fernandes F, Bhat JV (1945) A note on the association of *Chlorococcum humicolum* in the roots of *Cycas revoluta*. Curr Sci Assoc 14(9):235

Fernández-Martínez MA, de Los Ríos A, Sancho LG et al (2013) Diversity of endosymbiotic *Nostoc* in *Gunnera magellanica* from Tierra del Fuego, Chile. Microb Ecol 66(2):335–350. https://doi.org/10.1007/s00248-013-0223-2

Fernandez-Mendoza F, Domaschke S, Garcia MA et al (2011) Population structure of mycobionts and photobionts of the widespread lichen *Cetraria aculeata*. Mol Ecol 20:1208–1232

Fernandez-Mendoza F, Fleischhacker A, Kopun T et al (2017) ITS1 metabarcoding highlights low specificity of lichen mycobiomes at a local scale. Mol Ecol 26:4811–4830

Feyisa T, Amare T, Adgo E et al (2013) Symbiotic blue green algae (*Azolla*): a potential bio fertilizer for paddy rice production in Fogera Plain, Northwestern Ethiopia. Ethiop J Sci Technol 6(1):1–11

Fjerdingstad E (1976) *Anabaena* variabilis status *azollae*. Arch Hydrobiol Suppl 49:377–381
Fountain ED, Pauli JN, Mendoza JE et al (2017) Cophylogenetics and biogeography reveal a coevolved relationship between sloths and their symbiont algae. Mol Phylogenet Evol 110:73–80. https://doi.org/10.1016/j.ympev.2017.03.003
Frangedakis E, Shimamura M, Villarreal JC et al (2021) The hornworts: morphology, evolution and development. New Phytol 229(2):735–754. https://doi.org/10.1111/nph.16874
Frank AB (1877) Über die biologischen Verkältnisse des Thallus einiger Krustflechten. Beitr Biol Pflanz 2:123–200
Galanakis CM (ed) (2021) Microalgae. Cultivation, recovery of compounds and applications. Elsevier Academic Press, London
Gallon JR, Hashem MA, Chaplin AE (1991) Nitrogen fixation by *Oscillatoria* spp. under autotrophic and photoheterotrophic conditions. Microbiology 137(1):31–30. https://doi.org/10.1099/00221287-137-1-31
Gantar M, Kerby NW, Rowell P (1991a) Colonization of wheat (*Triticum vulgare* L.) by N_2-fixing cyanobacteria: II. An ultrastructural study. New Phytol 118(3):485–492. https://doi.org/10.1111/j.1469-8137.1991.tb00031.x
Gantar M, Kerby NW, Rowell P et al (1991b) Colonization of wheat (*Triticum vulgare* L.) by N_2-fixing cyanobacteria: I. A survey of soil cyanobacterial isolates forming associations with roots. New Phytol 118(3):477–483. https://doi.org/10.1111/j.1469-8137.1991.tb00030.x
Gärtner G (1985) Die Gattung *Trebouxia* Puymaly (Chlorellales, Chlorophyceae). Algol Stud 41:495–548
Gärtner G, Stoyneva MP (2003) First study of aerophytic cryptogams on monuments in Bulgaria. Ber Naturwiss-med Ver Innsb 90:73–82
Gärtner G, Fahrhofer K, Ingolić E et al (2007) Contribution to the taxonomy of terrestrial Stichococcus species (Chlorophyta, Trebouxiophyceae). In: Hindák F (ed) Book of abstracts. Symposium biology and taxonomy of green algae V, Smolenice, 25–29 June 2007, p 32
Gärtner G, Stoyneva-Gärtner M, Uzunov B (2021) Algal toxic compounds and their aeroterrestrial, airborne and other extremophilic producers with attention to soil and plant contamination: a review. Toxins 13:322. https://doi.org/10.3390/toxins13050322
Garty J, Delarea J (1988) Evidence of liberation of lichen ascospores in clusters and reports on contact between freeliving algal cells and germinating lichen ascospores under natural conditions. Can J Bot 66:2171–2177
Gehrig H, Schüssler A, Kluge M (1996) *Geosiphon pyriforme*, a fungus forming endocytobiosis with *Nostoc* (cyanobacteria), is an ancestral member of the *Glomales*: evidence by SSU rRNA analysis. J Mol Evol 43(1):71–81. https://doi.org/10.1007/BF02352301
Gehringer MM, Pengelly JJL, Cuddy WS et al (2010) Host selection of symbiotic cyanobacteria in 31 species of the Australian cycad genus: *Macrozamia* (Zamiaceae). Mol Plant-Microbe Interact 23:811–822
Geitler L (1930–1932) Cyanophyceae von Evropa. (Reprint 1985). Koelz Scientific Books, Konigstein
Geitler L (1935) *Trypanochloris*, eine neue grüne Alge in der schalen von Landschneken und ihre Begleitflora. Biol Gen 11:135–148
Geitler L (1959) Syncyanosen. In: Ruhland W (ed) Encyclopedia of plant physiology, vol 11. Heterotrophy. Springer Verlag, Berlin Göttingen Heidelberg, pp 530–545
Gentili F, Nilsson MC, Zackrisson O et al (2005) Physiological and molecular diversity of feather moss associative N2-fixing cyanobacteria. J Exp Bot 56(422):3121–3127. https://doi.org/10.1093/jxb/eri309
George EA (1962) *Chlorocloster solani* nov. spec., an algal endophyte in potato. New Phytol 62(2):170–172
Gilbert PW (1942) Observations on the eggs of *Ambystoma maculatum* with especial reference to the green algae found within the egg envelopes. Ecology 23:215–227
Gilbert PW (1944) The alga-egg relationship in *Ambystoma maculatum*, a case of symbiosis. Ecology 25:366–369

Gilmore DP, da Da Costa CP, Duarte DPF (2001) Sloth biology: an update on their physiological ecology, behavior and role as vectors of arthropods and arboviruses. Braz J Med Biol Res 34:9–25

Goffart M (1971) Function and form in the sloth. Pergamon Press, Oxford

Goffinet B, Shaw AJ (eds) (2009) Bryophyte biology, 2nd edn. Cambridge University Press, Cambridge. https://doi.org/10.1017/CBO9780511754807.004

Good BH, Chapman RL (1978) The ultrastructure of *Phycopeltis* (Chroolepidaceae: Chlorophyta). I. Sporopollenin in the cell walls. Am J Bot 65(1):27–33. https://doi.org/10.2307/2442549

Graham LE, Graham JM, Wilcox LW et al (2018) Evolutionary roots of plant microbiomes and biogeochemical impacts of nonvascular autotroph-microbiome systems over deep time. Int J Plant Sci 179(7):505–522. https://doi.org/10.1086/698709

Gribb AB (1965) A green alga growing on a caterpillar. Queensland Nat 17:5–6

Grilli M (1964) Infrastrutture di *Anabaena azollae* vivente nelle foglioline di *Azolla caroliniana*. Ann Microb Enzim 14:69–90

Grobbelaar N, Hattingh W, Marshall J (1986) The occurrence of coralloid roots on the South African species of the Cycadales and their ability to fix nitrogen symbiotically. S Afr J Bot 52:467–471

Grobbelaar N, Scott WE, Hattingh W et al (1987) The identification of the coralloid root endophytes of the Southern African cycads and the ability of the isolates to fix dinitrogen. S Afr J Bot 53:111–118. https://doi.org/10.1016/s0254-6299(16)31444-2

Grube M, Berg G (2009) Microbial consortia of bacteria and fungi with focus on the lichen symbiosis. Fungal Biol Rev 23:72–85

Grube M, Spribille T (2012) Exploring symbiont management in lichens. Mol Ecol 21:3098–3099

Guevara R, Armesto JJ, Caru M (2002) Genetic diversity of *Nostoc* microsymbionts from *Gunnera tinctoria* revealed by PCR-STRR fingerprinting. Microb Ecol 44(2):127–136. https://doi.org/10.1007/s00248-002-1019-y

Guiry MD, Guiry GM (2022) AlgaeBase. Available online http://www.algaebase.org/

Guo B, Zhang X-D, Wei Y-H (2012) Effects of light intensity on the growth of cyanobacteria isolated from *Huperzia serrata* (Thunb.) Trevis in liquid medium. Acta Photonica Sin 41(1):102–106

Gutiérrez-García K, Bustos-Díaz ED, Corona-Gómez JA et al (2018) Cycad coralloid roots contain bacterial communities including cyanobacteria and *Caulobacter* spp. that encode niche-specific biosynthetic gene clusters. Genome Biol Evol 11(1):319–334. https://doi.org/10.1093/gbe/evy266

Guzow-Krzemińska B (2006) Photobiont flexibility in the lichen *Protoparmeliopsis muralis* as revealed by ITS rDNA analyses. Lichenologist 38(5):469–476. https://doi.org/10.1017/S0024282906005068

Guzow-Krzemińska B, Stocker-Wörgötter E (2013) In vitro culturing and resynthesis of the mycobiont *Photoparmeliopsis muralis* with algal bionts. Lichenologist 45(1):65–76. https://doi.org/10.1017/S002428291200059X

Hamdi YA (1982) Application of nitrogen-fixing systems in soil management. FAO Soil Bull 49:1–193

Harsh R, Sharma BD, Bohra DR et al (2018) Mesozoic conifers of India – megafossils – an overview. Nelumbo 60(1):51–68. https://doi.org/10.20324/nelumbo/v60/2018/130178

Hashidoko Y, Nishizuka H, Tanaka M et al (2019) Isolation and characterization of 1-palmitoyl-2-linoleoyl-*sn*-glycerol as a hormogonium-inducing factor (HIF) from the coralloid roots of *Cycas revoluta* (Cycadaceae). Sci Rep 9:4751. https://doi.org/10.1038/s41598-019-39647-8

Heaney LR, Balete DS, Rickart EA (2013) Models of oceanic island biogeography: changing perspectives on biodiversity dynamics in archipelagoes. Front Biogeogr 5:249–257

Heckman DS, Geiser DM, Eidell BR et al (2001) Molecular evidence for the early colonization of land by fungi and plants. Science 293(5532):1129–1133. https://doi.org/10.1126/science.1061457

Helms G, Friedl T, Rambold G et al (2001) Identification of photobionts from the lichen family Physciaceae using algal-specific ITS rDNA sequencing. Lichenologist 33:73–86

Higgins BT, Gheynst JSV (2014) Effects of *Escherichia coli* on mixotrophic growth of *Chlorella minutissima* and production of biofuel precursors. PLoS One 9(5):e96807. https://doi.org/10.1371/journal.pone.0096807

Higgins BT, Gennity I, Samra S et al (2016) Cofactor symbiosis for enhanced algal growth, biofuel production, and wastewater treatment. Algal Res 17:308–315. https://doi.org/10.1016/j.algal.2016.05.024

Hirsh AM (2004) Plant-microbe symbioses: a review. Symbiosis 37(1–3):1–20

Ho Y-N, Mathew DC, Huang C-C (2017) Plant-microbe ecology: interactions of plants and symbiotic microbial communities. In: Yousaf Z (ed) Plant ecology. IntechOpen. https://doi.org/10.5772/intechopen.69088

Hobara S, McCalley C, Koba K et al (2006) Nitrogen fixation in surface soils and vegetation in an Arctic tundra watershed: a key source of atmospheric nitrogen. Arct Antarct Alp Res 38(3):363–372. https://doi.org/10.1657/1523-0430(2006)38[363:NFISSA]2.0.CO;2

Hoffmann L (1989) Algae of terrestrial habitats. Bot Rev 55:77–105

Hofmeister W (1862) On the germination, development, and fructification of the higher Cryptogamia. Ray Society, London. https://doi.org/10.5962/bhl.title.60930

Hom EF, Murray AW (2014) Plant-fungal ecology. Niche engineering demonstrates a latent capacity for fungal-algal mutualism. Science 345(6192):94–98. https://doi.org/10.1126/science.1253320

Honegger R (1991) Functional aspects of the lichen symbioses. Annu Revu Plant Physiol Plant Mol Biol 42:553–578

Honegger R, Edwards D, Axe L (2013) The earliest records of internally stratified cyanobacterial and algal lichens from the Lower Devonian of the Welsh Borderland. New Phytol 197:264–275

Hrouzek P, Tomek P, Lukešová A et al (2011) Cytotoxicity and secondary metabolites production in terrestrial *Nostoc* strains, originating from different climatic/geographic regions and habitats: is their cytotoxicity environmentally dependent? Environ Toxicol 26(4):345–358. https://doi.org/10.1002/tox.20561

Hutchinson VH, Hammen CS (1958) Oxygen utilization in the symbiosis of embryos of the salamander, *Ambystoma maculatum* and the alga, *Oophila amblystomatis*. Biol Bull 115(3):483–389

Ininbergs K, Bay G, Rasmussen U et al (2011) Composition and diversity of nifH genes of nitrogen-fixing cyanobacteria associated with boreal forest feather mosses. New Phytol 192:507–517

Ismail A, Marzuki SD, Yusof NBM et al (2017) Epiphytic terrestrial algae (*Trebouxia* sp.) as a biomarker using the free-air-carbon dioxide-enrichment (FACE) system. Biology 6(1):19. https://doi.org/10.3390/biology6010019

Issa A, Abd-Alla MH, Ohyama T (2014) Nitrogen fixing cyanobacteria: future prospect. In: Ohyama T (ed) Advances in biology and ecology of nitrogen fixation. InTech Open. https://doi.org/10.5772/56995

Ivanov D, Yaneva G, Potoroko I et al (2021) Contribution of cyanotoxins to the ecotoxicological role of lichens. Toxins 13:321. https://doi.org/10.3390/toxins13050321

James TY, Letcher PM, Longcore JE et al (2006) A molecular phylogeny of the flagellated fungi (Chytridiomycota) and description of a new phylum (Blastocladiomycota). Mycologia 98(6):860–871. https://doi.org/10.1080/15572536.2006.11832616

Johansen JR, Shubert E (2001) Algae in soils. Nova Hedwigia 123:297–306

Johansson C, Bergman B (1994) Reconstitution of the symbiosis of *Gunnera manicata* Linden: cyanobacterial specificity. New Phytol 126:643–652

Jones DL (ed) (2002) Cycads of the world, ancient plants in today's landscape, 2nd edn. Smithsonian Institution Press, Washington, DC

Joneson S, O'Brien H (2017) A molecular investigation of free-living and lichenized *Nostoc* sp. and symbiotic lifestyle determination. Bryologist 120(4):371–381. https://doi.org/10.1639/0007-2745-120.4.371

Jordan DC, McNicol PJ, Marshall MR (1978) Biological nitrogen fixation in the terrestrial environment of a high Arctic ecosystem (Truelove Lowland, Devon Island, N.W.T.). Can J Microbiol 24(6):643–649. https://doi.org/10.1139/m78-108

Kaasalainen U, Fewer DP, Jokela J et al (2012) Cyanobacteria produce a high variety of hepatotoxic peptides in lichen symbiosis. PNAS 109:5886–5891

Kanesaki Y, Hirose M, Hirose Y et al (2018) Draft genome sequence of the nitrogen-fixing and hormogonia-inducing cyanobacterium *Nostoc cycadae* strain WK-1, isolated from the coralloid roots of *Cycas revoluta*. Genome Announc 6(7):e00021-18. https://doi.org/10.1128/genomeA.00021-18

Karsten U, Friedl T, Schumann R et al (2005) Mycosporine-like amino acids and phylogenies in green algae: *Prasiola* and its relatives from the Trebouxiophyceae (Chlorophyta). J Phycol 41:557–566

Kaup M, Trull S, Hom EFY (2021) On the move: sloths and their epibionts as model mobile ecosystems. Biol Rev Camb Philos Soc 96(6):2638–2660. https://doi.org/10.1111/brv.12773

Kerney R (2011) Symbioses between salamander embryos and green algae. Symbiosis 54:107–117. https://doi.org/10.1007/s13199-011-0134-2

Kerney R, Kimb E, Hangarterc RP et al (2011) Intracellular invasion of green algae in a salamander host. PNAS 108(16):6497–6502. https://doi.org/10.1073/pnas.1018259108

Khamar HJ, Breathwaite EK, Prasse CE et al (2010) Multiple roles of soluble sugars in the establishment of *Gunnera-Nostoc* endosymbiosis. Plant Physiol 154:1381–1389

Khayatan B, Bains DK, Cheng MH et al (2017) A putative O-linked β-N-acetylglucosamine transferase is essential for hormogonium development and motility in the filamentous cyanobacterium *Nostoc punctiforme*. J Bacteriol 199:e00075-17

Kim E, Lin Y, Kerney R et al (2014) Phylogenetic analysis of algal symbionts associated with four North American amphibian egg masses. PLoS One 9(11):e108915. https://doi.org/10.1371/journal.pone.0108915

Kipp MA, Stüeken EE, Gehringer MM et al (2020) Exploring cycad foliage as an archive of the isotopic composition of atmospheric nitrogen. Geobiology 18(2):155–166. https://doi.org/10.1111/gbi.12374

Kluge M, Mollenhauer D, Mollenhauer R (1991) Photosynthetic carbon assimilation in *Geosiphon pyriforme* (Kützing) F. v. Wettstein, an endosymbiotic association of fungus and cyanobacterium. Planta 185(3):311–315

Kluge M, Mollenhauer D, Mollenhauer R et al (1992) *Geosiphon pyriforme*, an endosymbiotic consortium of a fungus and a cyanobacterium (*Nostoc*), fixes nitrogen. Bot Acta 105:343–344

Kluge M, Mollenhauer D, Mollenhauer R (1994) *Geosiphon pyriforme* (Kützing) von Wettstein, a promising system for studying endocyanoses. Prog Bot 55:130–141

Knapp E (1933) Über *Geosiphon pyriforme* Fr. Wettst., eine intrazelluläre Pilz-Algen-Symbiose. Ber Dtsch Bot Ges 51:210–217

Knight CD, Adams DG (1996) A method for studying chemotaxis in nitrogen fixing cyanobacterium-plant symbioses. Physiol Mol Plant Pathol 49:73–77

Komárek J (2013) Cyanoprokaryota: 3rd part: heterocystous genera. In: Süsswasserflora von Mitteleuropa. Springer Spectrum, Berlin/Heidelberg

Krings M, Grewing A, Taylor TN et al (2005) *Lageniastrum macrosporae* (fossil Volvocales, Lageniastraceae nov. fam.), an endophyte in megaspores from the Carboniferous of the French Massif Central. Geobios 38:451–465

Krings M, Hass H, Kerp H et al (2009) Endophytic cyanobacteria in a 400-million-year-old land plant: a scenario for the origin of a symbiosis? Rev Palaeobot Palynol 153(1–2):62–69. https://doi.org/10.1016/j.revpalbo.2008.06.006

Kroken S, Taylor JW (2000) Phylogenetic species, reproductive mode, and specificity of the green alga *Trebouxia* forming lichens with the fungal genus *Letharia*. Bryologist 103(4):645–660

Larkum AWD, Douglas SE, Raven JA (eds) (2003) Photosynthesis in algae. Kluwer Academic Publishers, Springer-Science + Business Media, Dordrecht

Lee SM, Ryu CM (2021) Algae as new kids in the beneficial plant microbiome. Front Plant Sci 12:599742. https://doi.org/10.3389/fpls.2021.599742

Leigh CJ (ed) (2002) Nitrogen fixation at the millennium. Elsevier, Amsterdam

Lewin R (1995) Symbiotic algae: definittions, quantification and evolution. Symbiosis 19:31–37

Lewin RA, Robinson PT (1979) The greening of polar bears in zoos. Nature 278(5703):445–447. https://doi.org/10.1038/278445a0

Li FW, Nishiyama T, Waller M et al (2020) *Anthoceros* genomes illuminate the origin of land plants and the unique biology of hornworts. Nat Plants 6(3):259–272. https://doi.org/10.1038/s41477-020-0618-2

Liaimer A, Jenke-Kodama H, Ishida K et al (2011) A polyketide interferes with cellular differentiation in the symbiotic cyanobacterium *Nostoc punctiforme*. Environ Microbiol Rep 3(5):550–558

Liaimer A, Helfrich EJ, Hinrichs K et al (2015) Nostopeptolide plays a governing role during cellular differentiation of the symbiotic cyanobacterium *Nostoc punctiforme*. PNAS 112(6):1862–1867. https://doi.org/10.1073/pnas.1419543112

Liaimer A, Jensen JB, Dittmann E (2016) A genetic and chemical perspective on symbiotic recruitment of cyanobacteria of the genus *Nostoc* into the host plant *Blasia pusilla* L. Front Microbiol 7:1693. https://doi.org/10.3389/fmicb.2016.01693

Liang F, Lindberg P, Lindblad P (2018) Engineering photoautotrophic carbon fixation for enhanced growth and productivity. Sustain Energ Fuels 2:2583–2600. https://doi.org/10.1039/C8SE00281A

Lin Y, Bishop CD (2015) Identification of free-living *Oophila* sp. (Chlorophyceae), an amphibian *Oophila amblystomatis* (Chlorophyceae) from yellow spotted salamander and wood frog breeding habitat. Phycologia 54(2):183–191. https://doi.org/10.2216/14-076.1

Lin C, Liu Z-Z, Zheng D-Y et al (1989) Re-establishment of symbiosis to *Anabaena*-free *Azolla*. In: Bothe H, de Bruijn FJ, Newton WE (eds) Nitrogen fixation: hundred years after: proceedings of the 7th international congress on N ≡ nitrogen fixation, Koln (Cologne), F.R.G., 13–20 March 1980, pp 551–559

Lindblad P (2009) Cyanobacteria in symbiosis with cycads. Microbiol Monogr 8:225–233. https://doi.org/10.1007/7171_2008_118

Lindblad P, Bergman B (1986) Glutamine synthetase: activity and localization in cyanobacteria of the cycads *Cycas revoluta* and *Zamia skinneri*. Planta 169(1):1–7

Lindblad P, Costa J-L (2002) The cyanobacterial-cycad symbiosis. Proc R Ir Acad 102B(1):31–33

Lindblad P, Bergman B, Hofsten AV et al (1985) The cyanobacterium-*Zamia* symbiosis: an ultrastructural study. New Phytol 101(4):707–716

Lindblad P, Rai AN, Bergman B (1987) The *Cycas revoluta-Nostoc* symbiosis: enzyme activities of nitrogen and carbon metabolism in the cyanobiont. J Gen Microbiol 1133(7):1695–1699. https://doi.org/10.1099/00221287-133-7-1695

Lindblad P, Haselkorn R, Bergman B et al (1989) Comparison of DNA restriction fragment length polymorphisms of *Nostoc* strains in and from cycads. Arch Microbiol 152(1):20–24. https://doi.org/10.1007/BF00447006

Liu X, Rousk K (2021) The moss traits that rule cyanobacterial colonization. Ann Bot 129(2):147–160. https://doi.org/10.1093/aob/mcab127

Lobakova ES, Dubravina GA, Zagoskina NV (2004) Formation of phenolic compounds in apogeotrophic roots of cycad plants. Russ J Plant Physiol 51:486–493. https://doi.org/10.1023/b:rupp.0000035741.31754.e6

Lokhorst GM (1996) Comparative taxonomic studies on the genus *Klebsormidium* (Charophyceae) in Europe. Cryptogam Stud 5:1–132

Lücking R, Barrie FR, Genney D (2014) *Dictyonema coppinsii*, a new name for the European species known as *Dictyonema interruptum* (Basidiomycota: Agaricales: Hygrophoraceae), with a validation of its photobiont *Rhizonema* (Cyanoprokariota: Nostocales: Rhizonemataceae). Lichenologist 46(3):261–267

Lumpkin TA (1993) Azollaceae Wettstein. Azolla Family. Flora N Am 2:338–342

Lutzoni F, Miadlikowska J (2009) Lichens. Curr Biol 19(13):R502–R503. https://doi.org/10.1016/j.cub.2009.04.034
Lutzoni F, Pagel M, Reeb V (2001) Major fungal lineages are derived from lichen symbiotic ancestors. Nature 411:937–940
Lutzoni F, Kauff F, Cox C et al (2004) Assembling the fungal tree of life: progress, classification, and evolution of subcellular traits. Am J Bot 91:1446–1480
Lutzu GA, Dunford NT (2018) Interactions of microalgae and other microorganisms for enhanced production of high-value compounds. Front Biosci 23:1487–1504
Maas W, Yetman D (2002) COSEWIC assessment and status report on the boreal felt lichen Erioderma pedicellatum in Canada. Committee on the Status of Endangered Wildlife of Canada, Ottawa
Madhusoodanan PV, Dominic TK (1996) Epiphytic cyanobacteria on mosses from the Western Ghats of Kerala. J Econ Tax Bot 20:355–359
Magain N, Sérusiaux E (2014) Do photobiont switch and cephalodia emancipation act as evolutionary drivers in the lichen symbiosis? A case study in the Pannariaceae (Peltigerales). PLoS One 9:e89876. https://doi.org/10.1371/journal.pone.0089876
Magain N, Sérusiaux E, Zhurbenko MP et al (2016) Disentangling the *Peltigera polydactylon* species complex by recognizing two new taxa, *P. polydactylon* subsp. *udeghe* and *P. seneca*. Herzogia 29:514–528. https://doi.org/10.13158/heia.29.2.2016.514
Magain N, Miadlikowska J, Goffinet B et al (2017) Macroevolution of specificity in cyanolichens of the genus *Peltigera* section *Polydactylon* (Lecanoromycetes, Ascomycota). Syst Biol 66:74–99. https://doi.org/10.1093/sysbio/syw065
Malar MC, Krüger M, Krüger C et al (2021) The genome of *Geosiphon pyriformis* reveals ancestral traits linked to the emergence of the arbuscular mycorrhizal symbiosis. Curr Biol 31(7):1570–1577.e4. https://doi.org/10.1016/j.cub.2021.01.058
Marco A, Blaustein AR (2000) Symbiosis with green algae affects survival and growth of northwestern salamander embryos. J Herpetol 34(4):617–621. https://doi.org/10.2307/1565283
Marler TE, Snyder LR, Shaw CA (2010) *Cycas micronesia* (Cycadales) plants devoid of endophytic cyanobacteria increase in beta-methylamino L alanine. Toxicon 56:563–568. https://doi.org/10.1016/j.toxicon.2010.05.015
Martin B, Schwab E (2012) Symbiosis: "living together" in chaos. Stud Hist Biol 4(4):7–25
McCurdy DW, Patrick JW, Offler CE (2008) Wall ingrowth formation in transfer cells: novel examples of localized wall deposition in plant cells. Curr Opin Plant Biol 11(6):653–661. https://doi.org/10.1016/j.pbi.2008.08.005
Medeiros ID, Mazur E, Miadlikowska J et al (2021) Turnover of lecanoroid mycobionts and their *Trebouxia* photobionts along an elevation gradient in Bolivia highlights the role of environment in structuring the lichen symbiosis. Front Microbiol 12:774839. https://doi.org/10.3389/fmicb.2021.774839
Meeks JC (1998) Symbiosis between nitrogen-fixing cyanobacteria and plants. Bioscience 48(4):266–276. https://doi.org/10.2307/1313353
Meeks JC (2003) Symbiotic interactions between *Nostoc punctiforme*, a multicellular cyanobacterium, and the hornwort *Anthoceros punctatus*. In: The 9th international symposium on nitrogen fixation with non-legumes, Leuuen, Belgium, 1–5 September 2002
Meeks JC, Elhai J (2002) Regulation of cellular differentiation in filamentous cyanobacteria in free-living and plant-associated symbiotic growth states. Microbiol Mol Biol Rev 66(1):94–121. https://doi.org/10.1128/MMBR.66.1.94-121.2002
Meeks JC, Campbell E, Hagen K et al (1999) Developmental alternatives of symbiotic *Nostoc punctiforme* in response to its symbiotic partner *Anthoceros punctatus*. In: Peschek GA, Loffelhardt W, Schmetterer G (eds) The phototrophic prokaryotes. Kluwer/Plenum, New York, pp 665–678. https://doi.org/10.1007/978-1-4615-4827-0_77
Meeks JC, Elhai J, Thiel T et al (2001) An overview of the genome of *Nostoc punctiforme*, a multicellular, symbiotic cyanobacterium. Photosynth Res 70:85–106

Metting B (1981) The systematics and ecology of soil algae. Bot Rev 47:195–312. https://doi.org/10.1007/BF02868854
Miadlikowska J, Richardson D, Magain N et al (2014) Phylogenetic placement, species delimitation, and cyanobiont identity of endangered aquatic *Peltigera* species (lichen-forming Ascomycota, Lecanoromycetes). Am J Bot 101:1141–1156. https://doi.org/10.3732/ajb.1400267
Mikhailyuk TI, Tsarenko PM, Nevo E et al (2001) Additions to the study of aerophytic eukaryotic algae of Israel. Int J Algae 11:371–390. https://doi.org/10.1615/InterJAlgae.v3.i3.30
Mishra AK, Tiwari DN, Rai AN (eds) (2018) Cyanobacteria: from basic science to applications. Elsevier, Amsterdam
Mitchell MC, Metodieva G, Metodiev MV et al (2017) Pyrenoid loss impairs carbon-concentrating mechanism induction and alters primary metabolism in *Chlamydomonas reinhardtii*. J Exp Bot 68(14):3891–3902. https://doi.org/10.1093/jxb/erx121
Moewus L (1951) Systematische Bestimmung einzelliger grüner Algen auf Grund von Kulturversuchen (*Sphaerosorus composita*, *Oocystis marina* und *Pseudostichococcus monallantoides*). Bot Notiser 1951:287–318
Mollenhauer D (1988) Weitere Untersuchungen an *Geosiphon pyriforme* – einer Lebensgemeinschaft von Pilz und Blaualge. Natur Mus 118:289–309
Mollenhauer D, Mollenhauer R (1988) *Geosiphon* cultures ahead. Endocyt Cell Res 5:69–73
Montgomery GG, Sunquist ME (1978) Habitat selection and use by two-toed and three-toed sloths. In: Montgomery GG (ed) The ecology of arboreal folivores. Smithsonian Institution Press, Washington, DC, pp 329–359
Moya P, Molins AA, Anez-Alberola FM et al (2017) Unexpected associated microalgal diversity in the lichen *Ramalina farinacea* is uncovered by pyrosequencing analyses. PLoS One 12:e0175091. https://doi.org/10.1371/journal.pone.0175091
Muggia L, Baloch E, Stabentheiner E et al (2010) Photobiont association and genetic diversity of the optionally lichenized fungus *Schizoxylon albescens*. FEMS Microbiol Ecol 75:255–272
Muggia L, Vancurova L, Škaloud P et al (2013) The symbiotic playground of lichen thalli – a highly flexible photobiont association in rock-inhabiting lichens. FEMS Microbiol Ecol 85(2):313–323. https://doi.org/10.1111/1574-6941.12120
Muggia L, Fleischhacker A, Kopun T et al (2016) Extremotolerant fungi from alpine rock lichens and their phylogenetic relationships. Fungal Divers 76:119–142. https://doi.org/10.1007/s13225-015-0343-8
Muggia L, Kraker S, Gößler T et al (2018) Enforced fungal-algal symbioses in alginate spheres. FEMS Microbiol Lett 365(14):fny115. https://doi.org/10.1093/femsle/fny115
Mukhtar A, Garty J, Galun M (1994) Does the lichen alga *Trebouxia* occur free-living in nature: further immunological evidence. Symbiosis 17:247–253
Muto K, Nishikawa K, Kamikawa R et al (2017) The roles of symbiotic green algae in eggs of *Hynobius nigrescens*, an oxygen and ammonia in the symbiotic relationship amphibian endemic to Japan. Phycol Res 65(2):171–174. https://doi.org/10.1111/pre.12173
Nash TH III (1996) Lichen biology. Cambridge University Press, Cambridge
Nash TH III (ed) (2008) Lichen biology, 2nd edn. Cambridge University Press, Cambridge
Nathanielsz CP, Staff IA (1975a) A mode of entry of blue-green algae into the apogeotropic roots of *Macrozamia comunis*. Am J Bot 62:232–235. https://doi.org/10.2307/2441856
Nathanielsz CP, Staff IA (1975b) On the occurrence of intracellular blue-green algae in cortical cells of the apogeotropic roots of *Macrozamia communis* L. Johnson. Ann Bot 39(3):363–368. https://doi.org/10.1093/oxfordjournals.aob.a084949
Nelsen MP, Lücking R, Grube M et al (2009) Unravelling the phylogenetic relationships of lichenised fungi in Dothideomyceta. Stud Mycol 64:135–144. https://doi.org/10.3114/sim.2009.64.07
Nelson JM, Hauser DA, Gudiño JA et al (2019) Complete genomes of symbiotic cyanobacteria clarify the evolution of vanadium-nitrogenase. Genome Biol Evol 11(7):1959–1964. https://doi.org/10.1093/gbe/evz137

Nema M, Hanson ML, Müller KM (2019) Phylogeny of the egg-loving green alga *Oophila amblystomatis* (Chlamydomonadales) and its response to the herbicides atrazine and 2,4-D. Symbiosis 77(1):23–39. https://doi.org/10.1007/s13199-018-0564-1

Nilsson M, Bergman B, Rasmussen U (2000) Cyanobacterial diversity in geographically related and distant host plants of the genus *Gunnera*. Arch Microbiol 173(2):97–102. https://doi.org/10.1007/s002039900113

Nilsson M, Bhattacharya J, Rai AN et al (2002) Colonization of roots of rice (*Oryza sativa*) by symbiotic *Nostoc* strains. New Phytol 156(3):517–525. https://doi.org/10.1046/j.1469-8137.2002.00534.x

Nilsson M, Rasmussen U, Bergman B (2006) Cyanobacterial chemotaxis to extracts of host and nonhost plants. FEMS Microbiol Ecol 55:382–390. https://doi.org/10.1111/j.1574-6941.2005.00043.x

O'Brien HE, Miadlikowska J, Lutzoni F (2005) Assessing host specialization in symbiotic cyanobacteria associated with four closely related species of the lichen fungus *Peltigera*. Eur J Phycol 40:363–378. https://doi.org/10.1080/09670260500342647

O'Brien HE, Miadlikowska J, Lutzoni F (2009) Assessing reproductive isolation in highly diverse communities of the lichen-forming fungal genus *Peltigera*. Evolution 63:2076–2086. https://doi.org/10.1111/j.1558-5646.2009.00685.x

O'Brien HE, Miadlikowska J, Lutzoni F (2013) Assessing population structure and host specialization in lichenized cyanobacteria. New Phytol 198:557–566. https://doi.org/10.1111/nph.12165

O'Leary MA, Bloch JI, Flynn JJ et al (2013) The placental mammal ancestor and the post-K-Pg radiation of placentals. Science 339(6120):662–667. https://doi.org/10.1126/science.1229237

Obukowicz M, Schaller M, Kennedy GS (1981) Ultrastructure and phenolic histochemistry of the *Cycas revoluta-Anabaena* symbiosis. New Phytol 87(4):751–775. https://doi.org/10.1111/j.1469-8137.1981.tb01711.x

Ogle H, Brown J (eds) (1997) Plant pathogens and plant diseases. Rockvalle Publications, Armidale

Orr H (1888) Note on the development of amphibians, chiefly concerning the central nervous system; with additional observations on the hypophysis, mouth, and the appendages and skeleton of the head. Q J Microsc Sci 29:295–324

Osborne B, Cullen A, Jones P et al (1992) Use of nitrogen by the *Nostoc-Gunnera tinctoria* (Molina) Mirbel symbiosis. New Phytol 120(4):481–487

Otálora MA, Salvador C, Martínez Iet al (2013) Does the reproductive strategy affect the transmission and genetic diversity of bionts in Cyanolichens? A case study using two closely related species. Microb Ecol 65:517–530. https://doi.org/10.1007/s00248-012-0136-5

Ow MC, Gantar M, Elhai J (1999) Reconstitution of a cycad-cyanobacterial association. Symbiosis 27:125–134

Padisák J (1992) Seasonal succession of phytoplankton in a large shallow lake (Balaton, Hungary). A dynamic approach to ecological memory, its possible role and mechanisms. J Ecol 80(2):217–230. https://doi.org/10.2307/2261008

Palmqvist K, de los Rios A, Ascaso C et al (1997) Photosynthetic carbon acquisition in the lichen photobionts *Coccomyxa* and *Trebouxia* (Chlorophyta). Physiol Plant 101:67–76

Papaefthimiou D, van Hove C, Lejeune A et al (2008) Diversity and host specificity of *Azolla* cyanobionts. J Phycol 44:60–70. https://doi.org/10.1111/j.1529-8817.2007.00448.x

Paracer S, Ahmadjian V (eds) (2000) Symbiosis. An introduction to biological associations, 2nd edn. Oxford University Press, New York

Pardo-De la Hoz CJ, Magain N, Lutzoni F et al (2018) Contrasting symbiotic patterns in two closely related lineages of trimembered lichens of the genus *Peltigera*. Front Microbiol 16(9):27–70. https://doi.org/10.3389/fmicb.2018.02770

Pathak J, Ahmed H, Rajneesh et al (2019) Genetic regulation of scytonemin and mycosporine-like amino acids (MAAs) biosynthesis in cyanobacteria. Plant Gene 17:100172. https://doi.org/10.1016/j.plgene.2019.100172

Pauli JN, Mendoza JE, Steffan SA et al (2014) A syndrome of mutualism reinforces the lifestyle of a sloth. Proc R Soc B Biol Sci 281:20133006

Pauli JN, Peery MZ, Fountain ED et al (2016) Arboreal folivores limit their energetic output, all the way to slothfulness. Am Nat 188:196–204

Paulsrud P, Rikkinen J, Lindblad P (1998) Cyanobiont specificity in some *Nostoc*-containing lichens and in a *Peltigera aphthosa* photosymbiodeme. New Phytol 139:517–524. https://doi.org/10.1046/j.1469-8137.1998.00220.x

Pawlowski K (ed) (2009) Prokaryotic symbionts in plants. Elsevier

Pecundo MH, Chang ACG, Chen T et al (2021) Full-length 16S rRNA and ITS gene sequencing revealed rich microbial flora in roots of *Cycas* spp. in China. Evol Bioinformatics Online 17:1176934321989713. https://doi.org/10.1177/1176934321989713

Pereira AL (2017) The unique symbiotic system between a fern and a cyanobacterium, *Azolla-Anabaena azollae*: their potential as biofertilizer, feed, and remediation. IntechOpen. https://doi.org/10.5772/intechopen.70466

Pereira AL, Carrapiço F (2007) Histochemistry of simple hairs from the foliar cavities of *Azolla filiculoides*. Plant Biosyst 141:323–328. https://doi.org/10.1080/11263500701627588

Pereira AL, Carrapiço F (2016) An extra sheath around the heterocysts of Anabaena azollae from the aquatic macrophyte *Azolla filiculoides* Lamarck. Bot Lett 163:449. https://doi.org/10.1080/23818107.2016.1234975

Pereira AL, Vasconcelos V (2014) Classiication and phylogeny of the endosymbiotic cyanobacterium *Anabaena azollae* Strasburger: an answered question? Int J Syst Evol 64:1830–1840. https://doi.org/10.1099/IJS.0.059238-0

Peters GA (1991) *Azolla* and other plant-cyanobacteria symbioses: aspects of form and function. Plant Soil 137:25–36. https://doi.org/10.1007/BF02187428

Peters GA, Meeks JC (1989) The Azolla-Anabaena Symbiosis: Basic Biology. Ann Rev Plant Biol 40:193–210

Peters GA, Perkins SK (1993) The *Azolla-Anabaena* symbiosis: endophyte continuity in the *Azolla* life-cycle is facilitated by epidermal trichomes. II. Re-establishment of the symbiosis following gametogenesis and embryogenesis. New Phytol 123(1):65–75

Peters JW, Szilagyi RK (2006) Exploring new frontiers of nitrogenase structure and mechanism. Curr Opin Chem Biol 10(2):101–108. https://doi.org/10.1016/j.cbpa.2006.02.01

Pi C-Y, Liu X, Wang Z et al (2018) Bryophyte-cyanobacteria symbioses and their nitrogen fixation capacity—a review. Chin J Plant Ecol 42(4):407–418. https://doi.org/10.17521/cjpe.2017.0191

Piercey-Normore MD (2006) The lichen-forming ascomycete *Evernia mesomorpha* associates with multiple genotypes of *Trebouxia jamesii*. New Phytol 169(2):331–344. https://doi.org/10.1111/j.1469-8137.2005.01576.x

Poelt J, Mayrhofer H (1988) Über Cyanotrophie bei Flechten. Plant Syst Evol 158:265–281

Pratte BS, Thiel T (2021) Comparative genomic insights into culturable symbiotic cyanobacteria from the water fern *Azolla*. Microb Genom 7(6):000595. https://doi.org/10.1099/mgen.0.000595

Presslee S, Slater GJ, Pujos F et al (2019) Palaeoproteomics resolves sloth relationships. Nat Ecol Evol 3(7):1121–1130. https://doi.org/10.1038/s41559-019-0909-z

Printz H (1928) Chlorophyceae. In: Engler A, Prantl K (eds) Die natürlichen Pflanzenfamilien, vol 3. W. Engelmannl, Leipzig, pp 1–463

Pröschold T, Darienko T (2020) The green puzzle *Stichococcus* (Trebouxiophyceae, Chlorophyta): new generic and species concept among this widely distributed genus. Phytotaxa 441(2):113–142. https://doi.org/10.11646/phytotaxa.441.2.2

Rai AN (ed) (1990) CRC handbook of symbiotic cyanobacteria. CRC Press, Boca Raton

Rai AN (ed) (2018) Handbook of symbiotic cyanobacteria in symbiosis, CRC handbook of symbiotic cyanobacteria. CRC Press, Boca Raton

Rai AN, Lindblad P, Bergman B (1986) Absence of the glutamine-synthetase-linked methylammonium (ammonium)-transport system in the cyanobiont of *Cycas*-cyanobacterial symbiosis. Planta 169(3):379–381. https://doi.org/10.1007/BF00392134

Rai AN, Soderback E, Bergman B (2000) Cyanobacterium-plant symbioses. New Phytol 147(3):449–481. https://doi.org/10.1046/j.1469-8137.2000.00720.x

Rai AN, Bergman B, Rasmussen U (eds) (2002) Cyanobacteria in symbiosis. Kluwer Academic Publishers, Dordrecht
Ran L, Larsson J, Vigil-Stenman T et al (2010) Genome erosion in a nitrogen-fixing vertically transmitted endosymbiotic multicellular cyanobacterium. PLoS One 5(7):e11486. https://doi.org/10.1371/journal.pone.0011486
Rasmussen U, Johansson C (2002) Diversity and specificity in cyanobacterial symbioses. Proc R Ir Acad 102B:53–56
Rasmussen U, Johansson C, Bergman B (1994) Early communication in the *Gunnera-Nostoc* symbiosis: plant induced cell differentiation and protein synthesis in the cyanobacterium. Mol Plant-Microbe Interact 7:696–702. https://doi.org/10.1094/MPMI-7-0696
Raven JA, Allen JF (2003) Genomics and chloroplast evolution: what did cyanobacteria do for plants? Genome Biol 4(3):209. https://doi.org/10.1186/gb-2003-4-3-209
Raven PH, Evert RF, Eichhorn SE (2005) Biology of plants, 7th edn. W.H. Freeman and Co, New York
Reese WD (1981) "Chlorochytrium", a green alga endophytic in Musci. Bryologist 84:75–78
Reinke J (1872) Parasitische *Anabaena* in Wurzeln der Cycadeen. Göttingen Nachrichten 57:107
Reisser W (ed) (1992) Algae and symbioses. Plants, animals, fungi, viruses, interactions explred. Biopress Limited, Bristol
Renault B (1894) Sur quelques nouveaux parasites des Lépidodendrons. Société d'Histoire Naturelle d'Autun. Procès-Verbaux des Séances 1893:168–178
Renzaglia KS (1978) A comparative morphology and developmental anatomy of the Anthocerotophyta. J Hattori Bot Lab 44:31–90
Renzaglia KS (1982) A comparative developmental investigation of the gametophyte generation in the Metzgeriales (Hepatophyta), Bryophytorum bibliotheca, vol 24. J. Cramer, Vaduz
Renzaglia KS, Duff RJT, Nickrent DL et al (2000) Vegetative and reproductive innovations of early land plants: implications for a unified phylogeny. Philos Trans R Soc Lond B Biol Sci 355(1398):769–93
Rikkinen J (2013) Molecular studies on cyanobacterial diversity in lichen symbioses. MycoKeys 6:3–32. https://doi.org/10.3897/mycokeys.6.3869
Rikkinen J, Virtanen V (2008) Genetic diversity in cyanobacterial symbionts of thalloid bryophytes. J Exp Bot 59(5):1013–1021. https://doi.org/10.1093/jxb/ern003
Rikkinen J, Oksanen I, Lohtander K (2002) Lichen guilds share related cyanobacterial symbionts. Science 297(5580):357. https://doi.org/10.1126/science.1072961
Rodgers GA, Henriksson E (1976) Associations between the blue green algae Anabaena variabilis and Nostoc muscorum and the moss Funaria hygrometrica with reference to the colonization of Surtsey, Iceland. Acta Bot Isl 4:10–15
Rodgers GA, Stewart WDP (1977) The cyanophyte-hepatic symbiosis. I. Morphology and physiology. New Phytol 78:441–458
Romeike J, Friedl T, Helms G et al (2002) Genetic diversity of algal and fungal partners in four species of *Umbilicaria* (lichenized ascomycetes) along a transect of the Antarctic peninsula. Mol Biol Evol 19:1209–1217
Rosowski JR, Willey RL (1975) *Colacium libellee* sp. nov. (Euglenophyceae), a photosynthetic inhabitant of the larval damselfly rectum. J Phycol 11(3):310–315; 975. https://doi.org/10.1111/j.1529-8817.1975.tb02785.x
Royer DL, Hickey LJ, Wing SL (2003) Ecological conservatism in the "living fossil" *Ginkgo*. Paleobiology 29(1):84–104
Scheirer DC, Brasell HM (1984) Epifluorescence microscopy for the study of nitrogen fixing blue green algae associated with *Funaria hygrometrica*, Bryophyta. Am J Bot 71(4):461–465
Scheirer DC, Dolan HA (1983) Bryophyte leaf epiflora: an SEM and TEM study of *Polytrichum Commune* Hedw. Am J Bot 70(5):712–718. https://doi.org/10.2307/2443125
Schnepf E (1964) Zur Feinstruktur von *Geosiphon pyriforme*. Arch Micobiol 49:112–131
Schuergers N, Wilde A (2015) Appendages of the cyanobacterial cell. Life (Basel) 5(1):700–715. https://doi.org/10.3390/life5010700

Schüßler A (2002) Molecular phylogeny, taxonomy, and evolution of *Geosiphon pyriformis* and arbuscular mycorrhizal fungi. Plant Soil 244(1/2):75–83. https://doi.org/10.1023/A:1020238728910
Schüßler A (2012) The *Geosiphon–Nostoc* endosymbiosis and its role as a model for arbuscular mycorrhiza research. In: Hock B (ed) Fungal associations, The mycota: a comprehensive treatise on fungi as experimental systems for basic and applied research, vol 9. Springer, Berlin, Heidelberg, pp 77–79. https://doi.org/10.1007/978-3-642-30826-0_5
Schüßler A, Kluge M (2001) *Geosiphon pyriforme*, an endocytosymbiosis between fungus and cyanobacteria, and its meaning as a model system for arbuscular mycorrhizal research. In: Hock B (ed) Fungal associations, The mycota, vol IX. Springer-Verlag, Berlin, pp 151–161. https://doi.org/10.1007/978-3-662-07334-6_9
Schüßler A, Wolf E (2005) *Geosiphon pyriformis*—a glomeromycotan soil fungus forming endosymbiosis with cyanobacteria. In: Declerck S, Fortin JA, Strullu DG (eds) In vitro culture of mycorrhizas, Soil biology, vol 4. Springer, Berlin. https://doi.org/10.1007/3-540-27331-X_15
Schüßler A, Mollenhauer D, Schnepf E et al (1994) *Geosiphon pyriforme*, an endosymbiotic association of fungus and cyanobacteria: the spore structure resembles that of arbuscular mycorrhizal fungi. Bot Acta 107(1):36–45. https://doi.org/10.1111/j.1438-8677.1994.tb00406.x
Sharma BD, Suthar OP (1989) Algal symbiotic gymnospermous roots from the Mesozoic of Rajmahal Hills, India. Phytomorphology 39:161–163
Shi DJ, Hall DO (1988) *Azolla* and immobilized cyanobacteria (blue-green algae): from traditional agriculture to biotechnology. Plants Today 1:5–12
Shubert E, Gärtner G (2015) Chapter 7 – Nonmotile coccoid and colonial green algae. In: Wehr JD, Sheath RG, Kociolek JP (eds) Freshwater algae of North America, 2nd edn. Academic Press, London, pp 315–373
Silaban A, Bai R, Gutierrez-Wing MT et al (2014) Effect of organic carbon, C:N ratio and light on the growth and lipid productivity of microalgae/cyanobacteria coculture. Eng Life Sci 14(1):47–56. https://doi.org/10.1002/elsc.201200219
Singh PK (1998) Water fern Azolla: a proven biofertilizer for wetland rice. In: Kaushik B.D. (ed) Soil plant microbe interaction in relation to integrated nutrient management, Venus Printers and Publishers, New Delhi, pp 64–75
Singh G, Dal Grande F, Divakar PK et al (2017) Fungal–algal association patterns in lichen symbiosis linked to macroclimate. New Phytol 214:317–329. https://doi.org/10.1111/nph.14366
Singh P, Kumar A, Singh VK, Shrivistava A (eds) (2020) Advances in cyanobacterial biology, 1st edn. Elsevier Academic Press, London
Slocum RD, Ahmadjian V, Hildreth KC (1980) Zoosporogenesis in *Trebouxia gelatinosa*: ultrastructure potential for zoospore release and implications for the lichen association. Lichenologist 12(2):173–187
Sluiman H, Guihal C, Mudimu O (2008) Assessing phylogenetic affinities and species delimitations in Klebsormidiales (Streptophyta): nuclear-encoded rDNA phylogenies and its secondary structure models in Klebsormidium, Hormidiella, and Entransia. J Phycol 44:183–195
Small DP, Bishop CD (2020) Physiological benefits and latent effects of an algal-salamander symbiosis. Comp Biochem Physiol A Mol Integr Physiol 246:110715. https://doi.org/10.1016/j.cbpa.2020.110715
Small DP, Bennett RS, Bishop CD (2014) The roles of oxygen and ammonia in the symbiotic relationship between the spotted salamander *Ambystoma maculatum* and the green alga *Oophila amblystomatis* during embryonic development. Symbiosis 64:1–10. https://doi.org/10.1007/s13199-014-0297-8
Smith GM (1950) The fresh-water algae of the United States, 2nd edn. McGraw-Hill Book Company Inc., New York
Snyder LR, Marler TE (2011) Rethinking cycad metabolite research. Commun Integr Biol 4:86–88. https://doi.org/10.4161/cib.4.1.14084
Solheim B, Johanson U, Callaghan TV et al (2002) The nitrogen fixation potential of arctic cryptogram species is influenced by enhanced UV-B radiation. Oecologia 133(1):90–93. https://doi.org/10.1007/s00442-002-0963-z

Solheim B, Wiggen H, Roberg S et al (2004) Associations between Arctic cyanobacteria and mosses. Symbiosis 37:169–187
Sprent J, Raven J (1985) Evolution of nitrogen-fixing symbioses. Proc R Soc Edinb B Biol Sci 85(3–4):215–237. https://doi.org/10.1017/S0269727000004036
Spribille T, Tuovinen V, Resl P et al (2016) Basidiomycete yeasts in the cortex of ascomycete macrolichens. Science 353:488–492. https://doi.org/10.1126/science.aaf8287
Stewart WDP (1977a) A botanical ramble among the blue-green algae. Br Phycol J 12(2):89–115. https://doi.org/10.1080/00071617700650111
Stewart WDP (1977b) Blue-green algae. In: Hardy RWF, Silver WS (eds) A treatise of dinitrogen fixation, section III biology. Wiley/Interscience, New York/London, pp 63–123
Stewart WDP, Rowell P, Rai AN (1983) Cyanobacteria-eukaryotic plant symbioses. Ann Microbiol 134B:205–228. https://doi.org/10.1016/s0769-2609(83)80106-9
Stewart KJ, Coxson D, Grogan P (2011) Nitrogen inputs by associative cyanobacteria across a low arctic tundra landscape. Arct Antarct Alp Res 43(2):267–278. https://doi.org/10.1657/1938-4246-43.2.267
Stoyneva MP (1991) Algal flora of the Danube River (Bulgarian sector) and adjoined water basins. Dissertation, Sofia University
Stoyneva M (1995) Algal flora of the Danube river (Bulgarian sector) and adjoined water basins. V. Algal flora of the water bodies adjacent to the Lake of Srebarna. Ann Univ Sofia Fac Biol 88(2):5–19
Stoyneva MP, Uzunov BA, Gärtner G (2015) Aerophytic green algae, epimycotic on *Fomes fomentarius* (L. ex Fr.) Kickx. Ann Sofia Univ Fac Biol 99(2):19–25
Stoyneva-Gärtner MP, Uzunov BA (2017) Bases of algology and mycology. J & MJ Books, Sofia
Stoyneva-Gärtner M, Uzunov B, Gärtner G et al (2019) Review on the biotechnological and nanotechnological potential of the streptophyte genus *Klebsormidium* with pilot data on its phycoprospecting and polyphasic identification in Bulgaria. Biotechnol Biotechnol Equip 33(1):559–578. https://doi.org/10.1080/13102818.2019.1593887
Stoyneva-Gärtner M, Uzunov B, Gärtner G (2020) Enigmatic microalgae from aeroterrestrial and extreme habitats in cosmetics: the potential of the untapped natural sources. Cosmetics 7:27. https://doi.org/10.3390/cosmetics7020027
Stoyneva-Gärtner M, Uzunov B, Gärtner G (2022) Aeroterrestrial and extremophilic microalgae as promising sources for lipids and lipid nanoparticles in dermal cosmetics. Cosmetics 9:11. https://doi.org/10.3390/cosmetics9010011
Stoyneva-Gärtner M, Gärtner G, Uzunov B (2025) Toxic compounds of free-living and symbiotic soil algae. In: Soil algae: morphology, ecology and biotechnological applications. Springer, Singapore
Strasburger E (1873) Über *Azolla*. Hermann Dabis Verlag, Jena
Suárez-Moo PJ, Vovides AP, Griffith MP et al (2019) Unlocking a high bacterial diversity in the coralloid root microbiome from the cycad genus *Dioon*. PLoS One 14(2):e0211271. https://doi.org/10.1371/journal.pone.0211271
Sun HJ, Friedman EI (2005) Communities adjust their temperature optima by shifting producer-to-consumer ratio, shown in lichens as models: II. Verification. Microb Ecol 49:528–535
Suutari M, Majaneva M, Fewer DP, Voirin B, Aiello A, Friedl T, Chiarello AG, Blomster J (2010) Molecular evidence for a diverse green algal community growing in the hair of sloths and a specific association with *Trichophilus welckeri* (Chlorophyta, Ulvophyceae). BMC Evol Biol 10:86
Svenning MM, Eriksson T, Rasmussen U (2005) Phylogeny of symbiotic cyanobacteria within the genus *Nostoc* based on 16S rDNA sequence analyses. Arch Microbiol 183(1):19–26. https://doi.org/10.1007/s00203-004-0740-y
Svirčev Z, Tamas I, Nenin P et al (1997) Co-cultivation of N2-fixing cyanobacteria and some agriculturally important plants in liquid and sand cultures. Appl Soil Ecol 6(3):301–308. https://doi.org/10.1016/s0929-1393(97)00022-x
Taylor J (1951) A new Australian hepatic. Kew Bull 1:45–47
Taylor TN, Krings M (2005) Fossil microorganisms and land plants: associations and interactions. Symbiosis 40:119–135

Thajuddin N, Muralitharan G, Sundaramoorthy M et al (2010) Morphological and genetic diversity of symbiotic cyanobacteria from cycads. J Basic Microbiol 50(3):254–265. https://doi.org/10.1002/jobm.200900343

Thompson RH (1972) Algae from the hair of the sloth *Bradypus*. J Phycol 8(S2):8

Thompson RH, Wujek DE (1997) Trentepohliales: *Cephaleuros*, *Phycopeltis* and *Stomatochroon*. Morphology, taxonomy and ecology. Science Publishers, Enfield

Trainor FR (1963) Zoospores in *Scenedesmus obliquus*. Science 142(3600):1673–1674. https://doi.org/10.1126/science.142.3600.1673

Trémouillaux-Guiller J, Huss VA (2007) A cryptic intracellular green alga in *Ginkgo biloba*: ribosomal DNA markers reveal worldwide distribution. Planta 226(2):553–557. https://doi.org/10.1007/s00425-007-0526-y

Trémouillaux-Guiller J, Rohr T, Rohr R, Huss VA (2002) Discovery of an endophytic alga in *Ginkgo biloba*. Am J Bot 89(5):727–733. https://doi.org/10.3732/ajb.89.5.727

Tschermak-Woess E (1978) *Myrmecia reticulata* as a phycobiont and free-living *Trebouxia* – the problem of *Stenocybe septata*. Lichenologist 10:69–79

Tschermak-Woess E (1988) The algal partner. In: Galun M (ed) CRC handbook of lichenology, vol I. CRC Press, Boca Raton, pp 39–44

Tuan DT, Thuyet TQ (1979) Use of *Azolla* in rice production in Vietnam. In: Nitrogen and rice symposium proceedings. International Rice Research Institute, Manila, pp 395–405

Tumlison P, Trauth CE (2006) A novel facultative mutualistic relationship between bufonid tadpoles and flagellated green algae. Herpetol Conserv Biol 1(1):51–55

Turmel M, Otis C, Lemieux C (2013) Tracing the evolution of streptophyte algae and their mitochondrial genome. Genome Biol Evol 5:1817–1835

Usher KM, Bergman B, Raven JA (2007) Exploring cyanobacterial mutualisms. Annu Rev Ecol Evol Syst 38:255–273. https://doi.org/10.1146/annurev

Uzunov BA, Gärtner G, Stoyneva MP (2012) Notes on the akinete-forming strain of the green alga *Klebsormidium dissectum* (Streptophyta) from Pirin Mts., Bulgaria. Phyton Ann Rei Bot 52(1):139–144

Vaez M, Javad Davarpanah S (2021) New insights into the biological activity of lichens: bioavailable secondary metabolites of *Umbilicaria decussata* as potential anticoagulants. Chem Biodivers 18(5):e2100080. https://doi.org/10.1002/cbdv.202100080

Videv P, Gärtner G, Uzunov BA et al (2017) Epimycotic algae on the medicinal fungus *Trametes versicolor* (L.) Lloyd. Int J Adv Res Bot 3(2):18–26. https://doi.org/10.20431/2455-4316.0302004

Villarreal AJC, Renzaglia KS (2006) Structure and development of *Nostoc* strands in *Leiosporoceros dussii* (Anthocerotophyta): a novel symbiosis in land plants. Am J Bot 93(5):693–705. https://doi.org/10.3732/ajb.93.5.693

Villarreal AJC, Renzaglia KS (2015) The hornworts: important advancements in early land plant evolution. J Bryol 37(3):157–170. https://doi.org/10.1179/1743282015Y.0000000016

von Wettstein F (1915) *Geosiphon* Fr Wettst, eine neue, interessante Siphonee. Österr Bot Z 65:145–156

Wagner GM (1997) *Azolla*: a review of its biology and utilization. Bot Rev 63(1):1–26. https://doi.org/10.1007/BF02857915

Warming E (1895) Plantesamfund – Grundtræk af den økologiske Plantegeografi. P.G. Philipsens Forlag, Kjøbenhavn

Warshan D, Espinoza JL, Stuart RK et al (2017) Feathermoss and epiphytic *Nostoc* cooperate differently: expanding the spectrum of plant-cyanobacteria symbiosis. ISME J 11(12):2821–2833. https://doi.org/10.1038/ismej.2017.134

Warshan D, Liaimer A, Pederson E et al (2018) Genomic changes associated with the evolutionary transitions of *Nostoc* to a plant symbiont. Mol Biol Evol 35(5):1160–1175. https://doi.org/10.1093/molbev/msy029

Weber-van Bosse A (1887) Étude sur les algues parasites des Paresseux. Natuurk Verh Holl Maatsch Wetensch Haarlem III 5:10

West NJ, Adams DG (1997) Phenotypic and genotypic comparison of symbiotic and free-living cyanobacteria from a single field site. Appl Environ Microbiol 63:4479–4484. https://doi.org/10.1128/aem.63.11.4479-4484.1997

Whitton BA (ed) (2012) Ecology of cyanobacteria II: their diversity in space and time. Springer, Dordrecht

Whitton B, Potts M (2002) Ecology of cyanobacteria: their diversity in time and space. Springer, Dordrecht

Whitton BA, Roger PA (1989) Use of blue-green algae and *Azolla* in rice culture. In: Campbell R, Macdonald RM (eds) Microbial inoculation of crop plants, vol 25. IRL Press at Oxford University, Oxford, pp 89–100

Willis KG, McElwain JC (2002) The evolution of plants. Oxford University Press, Oxford

Wirth V, Hauck M, Schultz M (2013) Die Flechten Deutschlands. Ulmer, Stuttgart

Wittmann W, Bergersen FR, Kennedy GS (1965) The coralloid roots of *Macrozamia communis* L. Johnson. Aust J Biol Sci 18:1129–1134

Woollaston V (2014) Sloths move so slowly that algae grows on their fur – and it can be used as camouflage and food. https://www.dailymail.co.uk/sciencetech/article-2532635/Sloths-slowly-ALGAE-grows-fur-used-camouflage-food.html

Wujek DE, Cocuzza JM (1986) Morphology of hair of two- and three-toed sloths (*Edentata: Bradypodidae*). Rev Biol Trop 34:243–246

Wujek DE, Lincoln TA (1988) Ultrastructure and taxonomy of *Oscillatoria pilicola*, a new species of bluegreen alga from sloth hair. Brenesia 29:1–6

Wujek DE, Timpano P (1986) *Rufusia* new genus (Porphyridiales, Phragmonemataceae), a new red alga from sloth hair. Brenesia 25–26:163–168

Yamada S, Ohkubo S, Miyashita H et al (2012) Genetic diversity of symbiotic cyanobacteria in *Cycas revoluta* (Cycadaceae). FEMS Microbiol Ecol 81(3):696–706. https://doi.org/10.1111/j.1574-6941.2012.01403.x

Yao S, Lyu S, An Y et al (2019) Microalgae–bacteria symbiosis in microalgal growth and biofuel production: a review. J Appl Microbiol 126:359–368. https://doi.org/10.1111/jam.14095

Yeung T, Kwan M, Adler L et al (2017) Increased methane production in cyanobacteria and methanogenic microbe co-cultures. Bioresour Technol 243:686–692. https://doi.org/10.1016/j.biortech.2017.06.126

Yuan X, Xiao S, Taylor TN (2005) Lichen-like symbiosis 600 million years ago. Science 308:1017–1020. https://doi.org/10.1126/SCIENCE.1111347

Zackrisson O, DeLuca TH, Nilsson M-C et al (2004) Nitrogen fixation increases with successional age in boreal forests. Ecology 85:3327–3334. https://doi.org/10.1890/04-0461

Zhan Y, Marchand CH, Maes A et al (2018) Pyrenoid functions revealed by proteomics in *Chlamydomonas reinhardtii*. PloS One 13(2):e0185039. https://doi.org/10.1371/journal.pone.0185039

Zhang XJ, Feng J, Wang G et al (2014) A morphological and phylogenetic study of a filamentous cyanobacterium *Microcoleus vaginatus*, associated with the moss *Mnium cuspidatum*. Symbiosis 64:43–51. https://doi.org/10.1007/s13199-014-0301-3

Zheng Y, Chiang TY, Huang CL et al (2018) Highly diverse endophytes in roots of *Cycas bifida* (Cycadaceae), an ancient but endangered gymnosperm. J Microbiol 56:337–345. https://doi.org/10.1007/s12275-018-7438-3

Zheng WW, Song TY, Bao XD et al (2002) High cyanobacterial diversity in coralloid roots of cycads revealed by PCR fingerprinting. FEMS Microbiol Ecol 40:215–222

Zhu C (1982) Fine structure of blue-green algae and the cells lined along the endophyte cavity in the coralloid root of *Cycas revoluta*. Acta Bot Sin 24(2):109–114

Zhu H, Zhao Z-J, Xia S et al (2014) Morphology and phylogenetic position of *Stomatochroon reniformis var. chinensis var. nov*. (Trentepohliales, Ulvophyceae), a rare endobiotic alga from China. Phycologia 53(5):493–501. https://doi.org/10.2216/14-010.1

Zilliges Y, Kehr JC, Meissner S et al (2011) The cyanobacterial hepatotoxin microcystin binds to proteins and increases the fitness of microcystis under oxidative stress conditions. PLoS One 6(3):e17615. https://doi.org/10.1371/journal.pone.0017615

Zimmerman WJ, Bergman B (1990) The Gunnera symbiosis: DNA restriction fragment length polymorphism and protein comparisons of *Nostoc* symbionts. Microb Ecol 19:291–302. https://doi.org/10.1007/BF02017173

Zimmerman WJ, Rosen BH (1992) Cyanobiont diversity within and among cycads of one field site. Can J Microbiol 38:1324–1328. https://doi.org/10.1139/m92-218

Physiological Ecology of Soil Algae

V. Sivamurugan, S. Murugesan, and N. Shanthi

Abstract Algae were discovered to be edible when man began to search the natural world for food. Protein levels in these algae were higher than in animals and plants, and they required far less water to generate than other species. Subsequently, food security is deemed vital for the existing population and global climate change, and algae are an alternative meal for individuals suffering from nutritional shortages. Since algae is abundant in antioxidant components, its bioactive chemicals can treat a variety of conditions, including high blood pressure, heart disease, diabetes, and cancer. These algae effectively utilize carbon dioxide emissions, increasing oxygen production and preserving the balance of the planet. Additionally, these algae store carbon dioxide and use it to create a variety of useful and profitable goods. When man looked for algae not only as food but also as an alternative energy source, they discovered that biofuel from algae is eco-system-friendly and reduces global warming to some extent. Algae also has several uses in agriculture, including improving soil fertility, restoring soil, controlling pests, forming microbial crusts, treating agricultural effluent, and recycling fresh water. Algae not only as food but also as fertilizer is an important contribution to plant growth by improving its physiological antagonistic reactions. Because this bio-fertilizer is rich in micro- and macronutrients, fiber ash is rich in carbohydrates and plant growth-promoting factors. It is environmentally friendly and inexpensive, making it a suitable biofertilizer for alternative agriculture. Technical training in kelp production and quality management for producers can lead to more sustainable kelp development. Thus, it is an inevitable fact that the quality and utility of these multipurpose algae should be further enhanced through research-validated techniques.

The micro- and macronutrients in this biofertilizer make fiber ash rich in carbohydrates and plant growth-promoting elements. It is an appropriate biofertilizer for alternative agriculture because it is affordable and environmentally benign. A more sustainable kelp development can be achieved by providing farmers with technical

V. Sivamurugan
PG and Research Department of Chemistry, Pachaiyappa's College, Chennai, India

S. Murugesan · N. Shanthi (✉)
PG and Research Department of Botany, Pachaiyappa's College, Chennai, India

A. Mishra, A. Varma (eds.), *Soil Algae*, https://doi.org/10.1007/978-981-97-9873-5_3

training in kelp production and quality control. Therefore, it is given that through procedures supported by research, the quality and usefulness of these multipurpose algae should be further improved.

Keywords Food security · Biofertilizer · Soil conditioner · Soil crust

1 Introduction

Humanity has grasped the value of the earth's crop for billions of years and continues to use it in food, medicine, and many other necessities. Man has recorded his history in each era with many changes and developments in agriculture according to the changing needs of the people. To fight malnutrition and hunger in the twenty-first century, a sufficient amount of quality food is needed, and the transformation of food production systems into sustainable, efficient, and environmentally friendly processes (Von Braun 2007). We need to develop new technologies to meet our needs through them. Mankind has learned that algae can be very useful when one is looking for other safe food using these new methods.

There is no good awareness among people about the benefits of algae. People are accustomed to seeing the effect of algae on a daily basis and are unable to understand the potential benefits. Macroalgae and microalgae are active protein sources, and some species have similar protein levels, if not higher than normal plant and animal protein sources (Bleakley and Hayes 2017). The production of algae protein has more benefits than conventional methods of protein production. For example, in terms of production and the amount of healthy food, and in terms of water consumption, it is much lower than animal and other protein proteins (Bleakley and Hayes 2017). As a result, algae appear to be an alternative source of protein in the future. This is because food security is of paramount importance due to population growth and global climate change.

Algae are proposed to be a source of active chemicals in the food and pharmaceutical industries. They are also considered to be the most important economic activity in modern conditions. With the growing popularity of active food to improve or maintaining health, researchers are focusing on their efforts to discover new chemicals that have significant health benefits. These include the potential use of algae as a source of active nutrients such as lipids, proteins, polysaccharides, phenolic, carotenoids, and other compounds and various options to enhance the production of an important metabolite, or using the tools and information provided by marine biotechnology or by developing the various factors involved in mass production (Ibáñez and Cifuentes 2012). Because dried seaweed is high in dietary fiber and a variety of other bioactive compounds, it has the potential to improve the nutritional quality of the product (Kim and Li 2011). The pharmaceutical industry is showing great interest in seaweed phytochemicals for therapeutic purposes. Seaweed has been shown to have antioxidant properties that protect the human body from free radicals and slow the progression of many chronic diseases such as high blood pressure, heart disease, diabetes, and cancer (Kolanjinathan et al. 2014; Collins et al. 2016).

Seaweed plays an important role in soil fertility, soil restitution, agricultural pest control, microbiological crust formation, agricultural wastewater treatment, and recycling of fresh water. Agriculture is essential for human civilization. Agricultural success is largely determined by the level of soil fertility. Algae, like other species of animals, can help improve soil qualities such as carbon content, texture, air permeability, and nitrogen fixation (Hamed 2007). Cyanobacterial-based biofertilizers can bring low cost to farmers who wish to increase soil fertility because they are naturally safe and high in nitrogen. Emerging farmers can use biofertilizers to increase and enhance yields while keeping the soil healthy for themselves and their communities. When the algae are exposed to a deficiency of P, they release foreign phosphatase rapidly (Healy 1973). Algae can also emit other substances and affect the pH of the surrounding environment, allowing absorbent P to be reached (Grobbelaar 1983). Algae play an economic role in soil regeneration by increasing soil fertility and improving plant conditions under certain environmental conditions (Pandey et al. 2005; Prabu and Udayasoorian 2007).

Not only did these algae change the planet's atmosphere by creating oxygen at the beginning of life's history, but they also set the stage for the emergence of eukaryotic species. Algae have always been more important than most people know from ancient times until now. Another reason to say that "our lives depend on the fight" today is the production of oxygen by algae. Algae are a major producer when the aquatic environment depends on them, so those who love seafood should thank them. Algae have become a possible option in many of the most expensive climate change modes (Chapman 2013). Algae can be used in many ways to reduce carbon emissions in the atmosphere. Apart from being a very efficient carbon dioxide storage solution, they can be easily used in a variety of other products, or commercial and commercial products (Daniel Cressey 2017).

Algae have a high potential for converting CO_2 from the atmosphere into beneficial products, including carbohydrates, lipids, and other bioactive compounds. Biofuels made from this biomass have renewable benefits and contribute very little to environmental damage and global warming (Khan et al. 2019). Despite the fact that algae are effective source of bioenergy and biopharmaceuticals in general, there are major limitations and obstacles that removed technology from the pilot to the industrial sector. Algae are now being promoted as a viable solution for humanity. It is being explored in a variety of fields, eradicating algae culture for biomass production, softening biomass first, and improving the fermentation process. Algal bioethanol production is a highly critical concern (Khan et al. 2019). However, the long list of barriers listed above hindered the full use of algae. Obstacles are gradually overcome, and opportunities arise. This is something that next-generation firms understand and invest in now, but it is also something that ordinary people can think of as they try to reduce their impact on the environment.

Although many researchers have published numerous articles on algae in the past, in this chapter, we have provided very detailed information on algae. This will be very important in a society that is facing many problems right now. Eventually, there are various creatures on earth and we can get various benefits from them. But there are many benefits to this alga, from the air we breathe to the food we eat and the fuel we use in cars, the biofertilizer of plants and the source of bioactive compounds that can be used for various human diseases. Researchers thus view algae as a promising future on earth.

2 Food Security

Food production with a growing population is very important and necessary in today's world. The most urgent thing is to establish new technologies at the same time and thereby producing sustainable food to meet human needs. A person faces many global challenges on how to meet the need for food security and biodiversity conservation in order to produce potential quality and nutritious food products that are kept environmentally safe (Greene Gerber et al. 2016). The food we have destroys a lot of water, and deforestation is harmful to the environment. So, man needs another source of food and a diet for both animals and humans. Another possible source of food is algae.

The ocean is called the earth's lungs because algae provide 80% oxygen in the atmosphere (Cikoš et al. 2021). Algae are important for human health and for the world. Algae are considered an open-source food; one important factor to consider is its composition and nutrient content. The composition of the nutrients varies greatly between algal species, and indeed among the same species, the nutrient content can vary greatly depending on the growth area, the composition of the media, and the temperature and light prevailing. At the same time, algae are not good food, but cultivating them properly can turn them into good food. Their curiosity, investment, and consumption increase, and the market value goes up.

Many studies record the potential for nutritional or active content of various algae, but several limited studies measure the availability of vitamins and phytochemicals from algal foods (Wells et al. 2017). Algae cultivation is a common agricultural extension; we already do not replace any agricultural crops. Microalgae photosynthetic organisms can be used to improve processes that can accumulate micro- and macronutrients, such as iron, vitamins, proteins, PUFAs, and produce nutritious biomass to be used as food additives. The essential nutrients that should be considered are protein and lipid content, as well as vitamins and minerals, all known to have a positive effect on the health of mortals. Microalgae has emerged as a promising stable source of protein. By the middle of this century, algae may make up 18% of the protein sources in various markets. However, the elements associated with food safety of algae are now less well known, especially the presence of contaminants, allergies, or harmful substances produced during the processing of small algae. Therefore, the timeframe for marketing microalgae and different protein sources varies (Cofrades et al. 2008). The most common microalgae grown for human consumption are *Chlorella, Arthrospira, Dunaliella, Nannochloris, Nitzschia, Crypthecodinium, Schizochytrium, Tetraselmins,* and *Skeletonema* spp.

Many microalgae foods (i) are composed of a small amount of dried algae, (ii) added to microalgae, or (iii) added to a specific compound extracted from microalgae, such as PUFAs or pigments (Lafarga et al. 2019).

Nostoc, Arthrospira, and *Aphanizomenon* protein-rich microalgae have been a feature of the human weight loss program based on the fact centuries ago (Spolaore et al. 2006). Spanish historians find the Aztecs eating a blue cake made from *Arthrospira* (Cornish and Garbary 2010). The use of microbial algae for food and biochemical purposes was once recommended in 1952 at the Algae Mass-Culture Symposium, although some growth was made in the early 1940s. The first *Chlorella* industrial facilities were developed in Japan, and Mexico pioneered the planting of *Arthrospira* in the 1970s (Spolaore et al. 2006). Although the diversity of microalgae species in the environment is estimated at between 200,000 and 800,000, only a few are used for food purposes (Wolkers et al. 2011). In the USA, legal recognition of the sale of algae and components is under the auspices of the Food and Drug Administration (FDA), which may grant Generally Recognized As Safe (GRAS) status in the product. In Europe, a member state official conducts an initial evaluation of a new product, which is later approved through the European Commission (EC) in the absence of objections by member states. In the event of a dispute, the European Food Safety Authority (EFSA) is obliged to conduct a safety assessment of novel foods (Cofrades et al. 2008).

Implementing various conservation and sustainability strategies for algae foods and using whole microalgal biomass in food, rather than extracting only algal bioactive compounds, is recommended. Adding fewer algae to our regular diet is a good idea to improve our daily diet to get nutritious and healthy portions. Compared to conventional protein sources, microalgae provide important natural benefits. Microalgae needs only 2.5 square meters of land per kilogram of protein produced, while beef requires 144,258 square meters. Microalgae can also be grown on uncultivated soil by adding very little fresh water. Unfortunately, the technology for microalgae is still in its infancy, making it costly, especially in terms of harvesting and depleting water. Before microalgae can become a widely used protein source, further study and technological advances are needed (Martín et al. 2018).

In addition to providing protein, microalgae provide its high-value nutrients, omega-three, and six-course polyunsaturated fatty acids where the content can compete with that of marine fish and reduce chemical infection and high purity. In addition, microalgae pigment can act as antioxidants in combination with many habitats that enhance durability and natural color.

Humans eat not only small algae but also large algae. Small algae have been consumed for hundreds of years in many countries, especially in Asia, including Japan and other parts of the Orient (Wijesinghe and Jeon 2012), they have received attention in recent years. Its bioactive properties have attracted large interest and earned the moniker of "new superfoods" in 2017 (Cofrades et al. 2008). Brown algae are the most commonly eaten (66.5%), followed by red (33%), and green (33%) (Lorenzo et al. 2017). The macroalgal species *Fucus* are considered safe for human consumption. *Himanthalia elongata, Fucus serratus, Fucus vesiculosus, Undaria pinnatifida, Ascophyllum nodosum,* and *Laminaria pinnatifida* are examples of *Undaria pinnatifida, Alaria esculenta, Laminaria saccharina, Laminaria japonica,* and *Laminaria digitata* (CEVA 2014). Studies have shown that macroalgae (*Grateloupia filicina, Chondrus crispus, Ulva lactuca*) have a higher fiber content than plants (Yuan et al. 2009), which may benefit human health. Dietary fiber is known to promote the growth of healthy intestinal flora, as well as reduce the risk of diabetes, obesity, and hypercholesterolemia, due to its ability to absorb organic molecules such as glucose and cholesterol (Bocanegra et al. 2009). Despite the fact that macroalgae contain a small amount of total lipids, their quality profile has a certain interest from the diet due to the presence of 3 and 6 polyunsaturated fatty acids (PUFAs), which have functions such as cardiovascular and coronary protection and reducing arteriosclerosis (Khan et al. 2007; Plaza et al. 2008). Ash contents of edible macroalgae contain higher levels of macro- and microelements than those found in edible plants, with levels of Na, K, Ca, and Mg ranging from 8.083 to 17.875 mg/100 g and Fe levels, Zn, Mn, and Cu ranging from 5.1 to 15.2 mg/100 g. According to this fact, brown and red macroalgae can be used as food ingredients to achieve the recommended daily use of minerals and trace elements. For example, the consumption of 10 g of *Ulva lactuca* provides 70% of daily Mg and more than half the requirement of Fe (Yuan et al. 2009). *Heanthalia elongata* is the most exploited because of its mineral composition and is a good substitute for salt, as it reduces salt consumption and health-related issues (Circuncisão et al. 2018).

Seaweed and microalgae are considered protein sources. Other types of seaweed and microalgae are recognized to include degrees of protein similar to those of other common sources of protein, such as meat, egg, soybeans, and milk (Gouveia et al. 2008). Using algae for protein production is more beneficial than relying on high-protein plants in production cages, both in terms of efficiency and food output. Marine plants and microalgae have a high protein yield in each region (2.5–7.5 tons/Ha/year and 4–15 tons/Ha/year, respectively) unlike terrestrial crops, such as soybeans, pulse legumes, and wheat (0.6–1.2) tons/Ha/year, 1–2 tons/Ha/year, and 1.1 tons/Ha/year, respectively) (Van Krimpen et al. 2013). Organic farming already requires about 75% of the world's total protein-rich fresh water that requires 100 times more water than the equivalent amount of protein ever produced from plant sources (Wallace 2000). Marine algae now no longer need clean water or arable land to grow, expanding resources that can be used to produce more food or different cash crops (Van Krimpen et al. 2013). In addition, due to the harsh environment and phototropic environment, algae are often exposed to excessive oxidative stress and free-radical stress (Sampath-Wiley et al. 2008). For example, carotene, chlorophyll, phycobiliprotein and polyphenols (e.g., catechins, flavonols, and phlorotannins) can convey the benefits of firmness to the consumer when consumed (Wang et al. 2009).

The current food system has wiped out our entire area with erosion, deforestation. Pollution of water resources, depletion of groundwater, and depletion of soil fertility have made food production questionable. It is therefore necessary to produce food with a growing population. So, people are looking for another way to do more research, especially how to eat algae as food because algae are a permanent source of food rich in vitamins, minerals, proteins, and polysaccharides. It eliminates the daily need of nutrition. Algae in the field of biotechnology are a blessing in disguise, offering the potential to boost the bioeconomy through their efficient use in the food industry.

3 Significance of Algae for Value-Added Products

Most people do not have a good idea about algae. There is no proper awareness among our people about how algae can be used in so many ways. Algae can be used as food, medicine, and biofuels and have many economic values. Algae are important components in the marine environment and provide a wide range of important ecosystem functions. Algae have a significant impact on global production by playing key roles in nutrient degradation, protecting coasts from harmful waves, and aiding in carbon sequestration (Teagle et al. 2017). Microalgae, like phytoplankton, form the basis of the seafood chain and are a major source of omega 3 fatty acids (Reisewitz et al. 2006; Bertocci et al. 2015), while macroalgae are an important habitat in coastal ecosystems, supporting various types of planning and management (Adarme-Vega et al. 2012; Bertocci et al. 2015). Algae are increasingly being used in commercial systems.

Algae biomass, already well established as food in Asia, is gradually being introduced into Western foods, which can be used by humans directly or in a variety of applications (Peteiro 2018). Based on archaeological evidence from 14,000 years ago in Chile (Dillehay et al. 2008) and early textual interpretations, algae have been part of the human diet for thousands of years (e.g., in China, 300 AD; in Ireland, 600 AD; Newton 1951; Aaronson 1986; Turner 2003; Gantar and Svircev 2008; Craigie 2010). Microalgae grow food and food additives over macroalgae (Gantar and Svircev 2008; Chacón-Lee and González-Mario 2010; FAO 2016). FAO (2014) estimates that humans consume 38% of the 23.8 million seaweeds harvested in 2012 in ways that appear to them as seaweed (e.g., alginates, carrageenans) and used as dietary and beverage agents. Human consumption of algal food varies from country to country, with annual consumption in Japan from 9.6 (2014) to 9.8 (2014) in recent years (2010–2014).

In addition, the genus *Arthrospira* has been used in Western diets since the 1970s and is now called "high food" because of its nutritional properties (Vonshak and Tomaselli 2002). Microalgae are also frequently used as a food and nutrition supplement, and some species provide an important supply of antioxidants, pigment, fats, and vitamins (Hemantkumar and Rahimbhai 2020; Kinley et al. 2020). Overall, a strong focus on health and the widespread use of dietary supplements have resulted in an increasing demand for algal products worldwide. Algae are increasingly marketed as "active foods" or "nutritious foods" in addition to their nutritional value; these terms, which have no legal status in many countries, describe foods containing bioactive compounds, or phytochemicals, which may be more beneficial to health than basic foods (e.g., anti-inflammatories and disease prevention (Hafting et al. 2012). Industrial, regulatory, and nutritional factors all play a role in the journey from algal research to the introduction of new food products or dietary supplements (Borowitzka 2013; Finley et al. 2014).

Many recent studies show a broader interest in algal diet and/or its potential dietary potential (Pangestuti and Kim 2011; Hafting et al. 2015). Not only this, these algae have brought about changes in the agricultural sector. Both macro- and microalgae contain a variety of chemicals that promote germination, leaf or stem growth, flowering, and can be used as a biological protection against plant diseases (Olaizola 2003) and are found in various coastal areas. Many of the nutrients are still in the discarded biomass after oil and carbohydrates have been found in macroalgae and microalgae. The remaining biomass will be used as fertilizer. As a natural biofertilizer, cyanobacteria play an important role in maintaining and enhancing soil fertility to grow rice that grows and thrives (Song et al. 2005).

Many studies report on the nutritional or bioactive content of various algae, but few test the bioavailability of nutrients and phytochemicals in algal foods. From the "research" of algal to the definition of healthy food, the achievement of bioaccessibility, and the subsequent performance of bioactivity, the design and construction of medium-sized farming systems to produce high-quality commercial products, phycologists have many opportunities to collaborate with other scientists and doctors in emerging areas. This is because algae are a rich source of bioactive chemicals that can be used for commercial purposes. Algae provide a combination of various drugs, proteins, vaccines, and nutrients that are not available or are too expensive to create in animals and plants (Olaizola 2003).

In addition, other species of marine plants collected along the European coast, such as *Laminaria digitata, L. hyperborea, Ascophyllum nodosum,* and *Gelidium corneum*, are used as feedstock to extract food, medicinal, biomedical, and biotechnological markers of hydrocolloids alginate and agar (Peteiro 2018). Algae biomass has also been investigated as a supplement to cattle feed to improve weight while reducing the release of enteric methane (Makkar et al. 2016; Wan et al. 2019). Bioremediation and biomonitoring (Reis et al. 2016), biofuel generation (Darda et al. 2019; Ravanal et al. 2019), and biopolymers (e.g., bioplastics) are among the new algal biomass applications currently under investigation (Abdul Khalil et al. 2017; Zhang et al. 2019). Recent research, one in the early stages of development, has highlighted the potential for these marine resources to be used in new health solutions (Bogie et al. 2019; Kwon et al. 2020).

Marine biomass can also be used for a variety of fatty foods, including co-fired bioelectricity, bioethanol, biodiesel, bio-oil, and biomethane for fermentation. Thanks to their continued technology for converting fossil fuels, algae biofuels have great commercial potential. Algae contain more than 50% starch, which can be converted to ethanol. The biorefinery is currently being investigated as a way to increase environmental sustainability and economic viability of existing industrial processes. It explores the different modes that can be used to process, extract, and certify natural algae products (Zhang and Thomsen 2019). Bioactive compounds with anticancer properties are one of the target areas for microalgae.

Although algae is very valuable, there are many obstacles on growing and selling algae. Producing algae-based food products and extracts requires high costs. A major obstacle to success in food-based food production is the lack of stimulus-based food production and low awareness of its health benefits. Also, the development of modern science and research publications has provided a clear definition of the importance of algae in the biotechnology industry, such as biofuel, biofertilizer, and food additives. The various valuable products found in algae have a high value, but the sale is still in its early stages, and could be considered a gateway to the multi-billion-dollar industry in the near future.

4 Role of Algae as Soil Conditioner

Soil conditioner refers to a chemical compound used to nourish the soil which helps in the growth of various plants. It increases air permeability, water-holding capacity, and nutrient performance in the soil due to soil texture and its level improves supporting vegetation. It lowers essential nutrients and minerals with a concentrated amount of soil that improves soil quality and makes it more fertile.

4.1 Different Soil Conditioners

Fertilizer is mainly available as inorganic, organic, and biofertilizer. Inorganic nitrogenous fertilizer comes as sodium nitrate, also called chilates, and ammonium nitrate, which is used to quickly supply nitrogen to the soil. Ammonium chloride, ammonium nitrate, and ammonium sulfate are also used as standard nitrogenous fertilizers.

Urea is another method that contains much of the nitrogen used as a fertilizer. Potassium is a key component of fertilizers. Its organic outsource is kelp, which is used to increase potassium content in the soil, and should only be used if its deficiency is noticeable. Di-ammonium hydrogen phosphate is used as a fertilizer for better plant growth, but when used regularly it slightly increases the pH of the soil.

Organic nitrogen comes naturally from plant and animal products. Supporting nitrogenous component in soil oil groups, cottonseed meal, corn gluten meal, skin, and fish diet is also used to increase soil nitrogen content.

Organic fertilizers are prepared from organic sources including Farmyard Manure (FYM), vermicompost, etc. FYM is one of the oldest methods used to grow crops made from a mixture of rotten manure and urine of farm animals and their waste and left over of the fodder supplied to animals. It is rich in nutrients and increases soil, chemical, and biological properties. Vermicompost is one of the most effective biofertilizers that causes no damage to the soil or plant. It is a product of a variety of rotten vegetables, food waste, bedding material, and vermicast.

4.2 Algae as a Soil Conditioner

Algae as a condition of the soil are biological, natural, and a source of renewable energy. It retains the water and essential nutrients needed for proper plant growth. Blue green algae bind and release auxin and amino compounds that promote plant growth, used mainly in rice plants (Jayant kr sahu 2008). Blue green algae have also been very effective in treating seeds instead of chemicals, which are used similarly to increase yield, weight, and oil content and to promote resistance to Tikka disease in the groundnut plant (Sujanya and Chandra 2012).

Dry algae also act as humus in the soil, which helps to bind the soil and makes it beneficial to use as a biofertilizer. So, algal fertilizer increases the water-holding capacity of the soil and increases the cellular space between the soil molecules because of the proper supply of air to the root system.

4.3 Soil Fertility

Green manure has also been shown to promote root growth and produce good yields (Mandimba et al. 1998). Algal biomass contains a high percentage of macronutrients, an astonishing amount of micronutrients and amino acids, other growth regulators, polyamines, natural enzymes carbohydrates, proteins, and vitamins used to improve plant growth and yield (Abd El Moniem and Abd-Allah 2008).

The addition of organic matter to the soil affects the chemical properties of the fertilized soil, as well as the biological activity of the microflora, which determines the biochemical nature of soil fertility. Algae in the soil nourish living organisms, provide nitrogen and stabilize the soil, helping to improve its fertility. It promotes seed germination and the growth of other microorganisms. Algae, found in different soils, can help the soil improve its features, such as carbon content, texture, and water maturity (Ibraheem 2007). Cyanobacteria fix nitrogen up to 30% and this will increase soil fertility and increase plant production (Dey et al. 2010). Prasanna et al. (2012) reported that inoculation cyanobacteria in the soil increased rice yields by 19% and increased nitrogen fixation. In addition, these microorganisms can reduce the use of inorganic fertilizers, alleviate stress in agroecosystems, and help reclaim wastelands (Pathak et al. 2018).

Biostimulants and biofertilizers are considered to be environmentally friendly and effective in synthetic products such as fertilizers, plant protection products, and improving plant growth, yield, quality, nutrient uptake, tolerance of abiotic stress, and shelf life of harvested products (Yakhin et al. 2017; Rouphael and Colla 2018). The production of microalgal biomass in polluted water can transform pollutants into sustainable biofertilizer and benefit young people with commercial opportunities in crop production (Coppens et al. 2016). Using biostimulants in agriculture, therefore, has the following benefits (Ronga et al. 2019):

(i) Increase in crop production

(ii) Increase in the efficiency of nutrient utilization
(iii) Improved crop quality

Algae-based wastewater treatment, biofertilizer production, soil ecosystem protection, plant or plant growth, and food safety are key areas of focus (Singh et al. 2019; Deepika and Mubarak Ali 2020). Khan et al. (2019) suggested the idea of integrating algae culture based on polluted water with biomass exploitation as a biofertilizer. Microalgae-based plant biostimulation can also be caused by mutations in microbial communities living in both the phyllosphere and the rhizosphere (Renuka et al. 2018).

Several studies have been conducted using microalgal treatments as growth promoters for various important crops (Hajnal-Jafari et al. 2016). Despite the potential of green microalgae for improved soil fertility, there is no known research conducted on the utilization of *C. sorokiniana* as a natural growth promoter for *Brassica rapa chinensis*. Besides enhancing stress tolerance, nutrient uptake, growth, and yield, seaweed-based biostimulants have also been shown to help reduce seed dormancy and enhance root systems, flowering (Ali et al. 2019), fruit quality, and taste (Kapur et al. 2018), and even the quality of produce (Li and Mattson 2015). These all-round effects lead to improved crop productivity.

Marine algae are widely used in agriculture because of their biostimulant activity (Rouphael et al. 2017). Rosalba Mireya Hernández-Herrera et al. (2014) examined the effect of different concentrations (0.2%, 0.4%, and 1.0%) of liquid seaweed extracts (LSEs) made from two green seaweeds namely, *Ulva lactuca and Caulerpa sertularioides,* as well as two brown seaweeds, namely *Padina gymnospora,* and *Sargassum liebmannii* as biostimulants in the growth and development of tomatoes (*Solanum lycopersicum*) under the greenhouse and laboratory conditions, using two foliar sprays. Sutharsan et al. (2014) evaluated the effect of foliar use of *Sargassum crassifolium* extracted from different concentrations (concentrations (10%, 20%, 50%, and 100%) for use on tomato plants five times 3 weeks after planting and the results were recorded 2 weeks later. Jebasingh et al. (2015) determined the effect of seaweed liquid extract (SLE) *of Laurencia pinnatifida, Surgassum duplicatum*, and *Caulerpa scalpelliformis* on seed germination and growth of the *Vigna mungo* legume plant. Sivasangari Ramya et al. (2015) studied the impact on growth, biochemistry, and brinjal yields using liquid extracts from the brown sea alga *Stoechospermum marginatum*. Nerissa Ali et al. (2016) observed an effect planted under tropical conditions with the release of alkaline seaweed made from *Ascophyllum nodosum* (ASWE) in tomato plants (*Lycopersicum esculentum*). Beneficial effects of marine vegetation extraction applications include improved seedling development, flowering and fruit set, crop yields, and tolerance for various abiotic stressors (De Pascale et al. 2018).

Algal biomass has been widely used in agriculture, but by the twentieth century, products derived from algal extracts have attracted the attention of farmers around the world (Dmytryk and Chojnacka 2018). In fact, a variety of organic compounds extracted from algae and cyanobacteria (e.g., phenols, terpenoids, free fatty acids (FFA), polysaccharides, and carotenoids) have been shown to have beneficial effects on plant production (Pan et al. 2019). According to the literature, algal metabolites can play an important role in (Ronga et al. 2019):

(i) Soil pollution and fertilizer application
(ii) Plant protection from biotic and abiotic stress factors
(iii) Crop development

In addition, microalgae and cyanobacteria also release phytohormones, known for their role as plant growth promoters (Han et al. 2018; Pan et al. 2019). Considering their potential benefits to sustainable agricultural development, both biomass and extracts from microalgae and cyanobacteria are commercially available (Górka et al. 2018).

Green manure was also found to promote root growth and produce good yields (Mandimba et al. 1998). Dry green algae contain a high percentage of macronutrients, numerous micronutrients, and amino acids (Mahmoud 2001) as a soil condition and to successfully grow grain crops due to improved nutrient uptake and pest resistance (Prakash and Nikhil 2014).

4.4 Improving Its Carbon, Nitrogen, and Phosphorous Content

Soil C: N: P stoichiometry plays an important role in the circulation of elements in the ecosystem, but its importance in the conversion of P to paddy soil remains unclear. Some algal and cyanobacterial species and their extracts are known for their positive effect on soil level. Well-known functions include their ability to

(i) Convert atmospheric nitrogen into nitrogen compounds that plants can easily absorb
(ii) Provide essential nutrients for plants
(iii) Improve the physical and chemical environment

Microalgae and cyanobacteria can act as a source of organic matter in high-yielding plants, which must be provided with sufficient concentration to grow suitable plants (Renuka et al. 2018). In addition, these organisms often release polymeric substances out of polymeric, increase the levels of organic carbon in the soil, and promote the growth of other microorganisms. Karthikeyan et al. (2007) have shown that the use of cyanobacterial suspensions (*Calothrix ghosei*, *Hapalosiphon intricatus*, and *Nostoc* sp.) in wheat production increased the concentration of organic carbon in the soil, while promoting higher grain yields. Recently, Renuka et al. (2018) examined the impact of unicellular and filamentous microalgal cyanobacterial consortia on soil composition and grain quality of the wheat crop. Besides increasing the availability of micronutrients in the soil (especially zinc, iron, copper, and manganese), the consortium studied also improved its organic carbon content.

Spirulina platensis biomass has been shown to enhance macronutrients in soil (nitrogen, phosphorus, and potassium) (Aly and Esawy 2008), acting as a biofortification agent, enhancing plant protein content (Kalpana et al. 2014) and increasing plant growth. Organic amendments can increase the availability of P in the soil by reducing the formation of irreversible adsorption structures in the soil (Siddique and Robinson 2003) or by increasing the flow of P in the soil (Mohanty et al. 2006). Ahmed et al. (2019) found that the combined use of compost and inorganic fertilizers increased the fraction of labile P, and different soils (pH, SOM, P, and Olsen-P) showed different relationships with P fractions.

Microalgae and cyanobacteria absorb nitrogen and phosphorus from various sources and store them in their biomass. As these nutrients are important for the development of high-quality plants, microalgal/cyanobacterial biomass can be considered an important source of these nutrients, avoiding the use of traditional chemical fertilizers (Win et al. 2018; Ronga et al. 2019). Using *Nostoc* sp. plant cells in the rice field, Innok et al. (2009) found an increase in grain yield compared to those obtained with chemical-based nitrogen fertilizers. Swarnalakshmi et al. (2013) examined the use of biofilm composed of cyanobacterium *Anabaena torulosa* in wheat plant which significantly increased nitrogen availability in the soil. Considering the observed improvements in nitrogen availability, we may conclude that cyanobacteria can be used instead of chemical nitrogen fertilizers, which can reduce crop production costs by 25–40%. Finally, cyanobacteria are known for their ability to form soil crusts that prevent nitrogen leaching/runoff and, consequently, the risks of nitrogen contamination (Renuka et al. 2018).

Other microelements important for plant growth and development (e.g., potassium, magnesium, sulfur and iron) can also be found in microalgal/cyanobacterial biomass. Therefore, the use of microalgae and cyanobacteria in the soil can improve the availability of both macro- and micronutrients in plant growth. These improvements in nutrient availability have already been reported by Renuka et al. (2017) and Marks et al. (2019).

4.5 Supporting Microorganism Growth

Soil organisms make up a small part of the whole soil, and their presence and function are essential for building and modifying soil and chemical properties. Soil biota can be synthesized in size in microbiota, mesobiota, and macrobiota. Microbiota contains viruses, actinomycetes, fungi, algae, and protozoa.

The soil catches many fungal pathogens of plants, which attack the roots mainly. Some are very special and can be suppressed by avoiding endangered plants for years. White rot fungus (Basidiomycetes) can eliminate the oxidation of lignin by producing an oxidizing environment with groups of enzymes such as laccases and lignin and manganese peroxidase. Mycorrhizal–plant symbiosis is less specific and more common than other interactions between microorganisms, for example, *Rhizobium* repair. It is estimated that 80% of all plants can be associated with mycorrhiza.

Cyanobacteria and algae are also often able to fix nitrogen. Algae and cyanobacteria combine and grow on the surface of very dry or barren soils and can contribute significantly to the flow of system nutrients.

Protozoa are blocked in water-filled holes so that they can move and be eaten, and they can form encapsulations or cyst, enabling them to live longer under adverse conditions. Mesobiota contains nematodes, enchytraeids, collembola or springtails, larvae, rotifers, and small insects. Nematodes cover different trophic levels in the soil and cannot be classified as food groups. Roots or nematodes for feeding plants are often called parasitic nematodes, as this can greatly affect crop yields if they occur in large numbers. Collembola as a group is believed to be omnivorous, eating detritus, germs, fungi, or other parts of the mesofauna. Oribatid larvae feed mainly on detritus and fungi in the soil, while carnivorous caterpillars feed on nematodes and other mesofauna groups. The mesobiota plays an important role in the circulation of nutrients in the soil. In low-quality substrates, this also increases nutrient uptake through mineral digestion, leading to a better response to decay of the remaining substrate (Clarholm 2005).

Macrobiota contains earthworms, mollusks, and large arthropods. Earthworms have also been shown to significantly increase the dispersal of microorganisms in the soil through their movement. They also increase the speed of transport and, therefore, the maturation of cellular nutrients such as nitrate. Macropores also increase air infiltration into the soil and water infiltration.

Many soil processes, especially those related to genetic modification and the digestion of organic minerals, are controlled by soil biota. These processes are essential for the soil to function and perform essential ecosystem services such as storage and filtering of nutrients and water, and providing a basis for crop production. The level of soil and biota living within it are highly correlated. Soil properties such as structure, penetration rate, and water-holding capacity are influenced by macrobiota activity mainly. Although many biological processes in the soil are resistant and can withstand severe stress, important processes can be reduced or stopped if exposed to adverse conditions.

Earth microorganisms, such as viruses, algae, fungi, actinomycetes, and protozoa and infective agents such as microorganisms are within the major resources of tiny diversity activities (Andreote et al. 2014). These tiny soil insects perform many important functions as well as other potentially dangerous effects (Andreote et al. 1996).

Soil fertility and its composition from mineral bedrock involve a variety of chemical interactions, physical, and biological processes. Soil growth rate is controlled by other factors such as location, climate, time, bedrock type, plants, and bacteria (Egli et al. 2011) which is why the nutritional status is determined by the quality and identification of bacteria in the soil (Lombard et al. 2011). Three types of equipment are commonly studied and developed to explain how microscopic activity can lead to improved plant growth:

(i) By deceptive plant hormonal signaling (Verbon and Liberman 2016)
(ii) By pathogenic microscopic type competition or expulsion (Mendes et al. 2013)

(iii) By increasing the bioavailability of nutrients found in the soil (Van-der et al. 2008)

Soil microbes serve as a link between the soil and the plant roots, facilitating nutrient recycling, organic matter decomposition, and quickly responding to changes in the soil environment, acting as key indicators of environmental activity (Jacoby et al. 2017). To increase crop yields, a range of non-symbiotic bacteria (*Azotobacter*, *Azospirillum*, *Bacillus*, and *Klebsiella* sp.) and symbiotic bacteria (*Rhizobium* sp.) are now used worldwide (Cocking 2003).

Rhizobacteria that promote plant growth (PGPR) are considered being beneficial microorganisms that should be used as a biofertilizer to promote plant growth and crop yield (Ojuederie et al. 2019). Various species of rhizobacteria such as *Pseudomonas*, *Azospirillum*, *Azotobacter*, *Klebsiella*, *Enterobacter*, *Alcaligenes*, *Arthrobacter*, *Burkholderia*, *Bacillus*, and *Serratia* have been investigated to improve growth and productivity of various plants (Zafar-Ul-Hye et al. 2019).

Beneficial soil microbiota such as plant growth-promoting microorganisms (PGPMs), comprising specific groups of bacteria and fungi, provide essential agroecosystem services that support plant growth (Rouphael and Colla 2020) and ameliorates soil productivity (Santos et al. 2019). They maintain key agroecological cycles fundamental for soil nutrient enrichment, crop nutrient improvement, plant tolerance to biotic and abiotic stresses, biocontrol of pests and diseases, and water uptake enhancement (Lobo et al. 2019; Goswami and Deka 2020). They are actively involved in healthy plant development and growth through secretion of hormonal growth regulators and resistance induction against phytopathogens (Dakora et al. 2015). Besides, versatile plant growth-promoting microorganisms could be used to bioremediate polluted fields and increase the land available for production, as with heavy metal-polluted soils (Gouda et al. 2018). These agroecosystem services are primarily important in supporting crop production in smallholder agroecosystems, which are characteristically defined by limited resource inputs.

Plant growth-promoting microorganisms enhance the availability of P, K, Zn, Se, and Fe in the soil through biochemical processes such as solubilization, chelation, mineralization, and oxidation and reduction reactions (Rouphael and Colla 2020). Plant growth-promoting microorganisms are also known to secrete phytohormones such as auxins (Azizoglu 2019), cytokinin, abscisic acid, ethylene, brassinosteroids, jasmonic acid, salicylic acid, strigolactones, and gibberellins (Saad et al. 2020) that act as plant growth stimulators and stress controllers. Additional methods efficient in augmenting indigenous soil PGPMs can be integrated to enhance microbial biofunctionality. These include the use of organic amendments such as farmyard manure and vermicompost (Koskey et al. 2017), biotechnological approaches such as plant breeding (Bakker et al. 2012), and crop management practices such as agroforestry, rotation, intercropping, cover cropping, and practicing reduced soil disturbance (Hontoria et al. 2019; Elagib and Al-Saidi 2020). Commercial microbial inoculants contain single species or multiple strains of rhizobia, *Pseudomonas* spp., *Azotobacter* spp., *Bacillus* spp., *Trichoderma* spp., *Aspergillus* spp., and *Glomus* spp. which have been largely used in smallholder agroecosystems for crop production (Adeyemi et al. 2020).

5 Building Soil Crusts

Biological soil crusts (BSCs) are unique soils called cryptogamic, cryptobiotic, microbiotic, microfloral, microphytic or organogenic crusts, and just biocrusts (Belnap et al. 2016). They contain a variety of prokaryotic and eukaryotic organisms that can be autotrophic, heterotrophic, or saprotrophic (Belnap et al. 2016). They colonize the soil and form a living crust cover (Belnap et al. 2016). As a result, biocrusts have a major impact on poor or dehydrated habitats that are less susceptible to high vegetation, often dominating regions with harsh environmental conditions, especially in arid regions around the world (Weber et al. 2016; Felde et al. 2017).

Formal soil layers are thin, dense, continuous layers of uncoated soil particles over milled and open soil. The plaque builds up when the closed soil dries up after rain or irrigation. It can also show that the soil has high sodium, which increases soil dispersal when wet with rain or irrigation. Soil microorganisms usually bind to soil particles in areas where the vegetation is scarred and water is limited to form a biological crust. Crusts alter soil characteristics such as water availability, nutrient content, air permeability, and the risk of erosion, and thus affect plant life directly or indirectly.

Algae stabilize the soil by combining small particles into larger particles. This binding can be achieved by binding the soil particles to the grip of algae fibers, or by attaching them to mucilaginous sacs. This binding increases the organic content in the crust, improving soil resistance to both wind and water erosions. Erosion of air and water is a major problem facing deserts, and if we introduce algae in those arid areas, after the formation of the crust, it will not be easy to erode and can be easily restored to agriculture, and later fertile enough to support a large carbon sink. Many researchers have studied and reported land coverage from arid and desert areas (Maestre et al. 2011), but such studies, which are very limited to the Asian and Indian subcontinent, have been studied by many researchers (Vinoth et al. 2017).

Phototrophic microbial communities, especially algae, are important components of the global ecosystem, and are on top of the soil layer and remain in a stable state during melting. However, these insects in tropical countries like India break the dormancy within a few hours of receiving water, especially during the monsoon (Kumar and Adhikary 2015).

Cyanobacteria and eukaryotic microalgae in soil crusts are important components of soil ecosystems that improve soil structure, stabilize erosion, and increase rainwater infiltration in areas with vegetation growth (Johansen 1993). Cyanobacterial genera commonly occur in the biological soils of *Chroococcus, Gloeocapsa, Leptolyngbya, Microcoleus, Nostoc, Phormidium,* and *Scytonema* (Büdel et al. 2016). The most common algal taxa found in these communities are members of the Klebsormidiophyceae, Zygnematophyceae (Streptophyta), Chlorophyceae, Trebouxiophyceae, Ulvophyceae (Chlorophyta), Diatomea, Eustigmatophyceae, and Xanthophyceae (Ochrophyta) (Büdel et al. 2016; Pushkareva et al. 2016).

Microalgae and cyanobacteria in biocrusts are well adapted to their most common habitats. An important feature of the phototrophic earth microorganisms is the availability of water, which is strongly influenced by salt. Although the effects of desiccation are closely related to those of salt, both pressures show two different types of dehydration. While under conditions of salt stress, algal cells are often in contact with liquids of reduced water intensity, desiccation release leads to severe cellular dehydration. However, both drought and salt stress prevention procedures work equally well. Tolerance to desiccation is attributed to the formation of water-retaining structures, such as mucilage layers thickened by the extraction of exopolymeric substances (EPS), as well as the formation of biofilms, cell colonies, and aggregates (Karsten and Holzinger 2012). Next, exopolymeric substances have been proposed to have a salt retention effect (Rossi et al. 2018). Gustavs et al. (2010) have shown that some of the earth's algae can produce organic osmolytes, which play another important role in desiccation tolerance and are the most common form of halotolerant algae to tolerate high salinity (Karsten et al. 2014). These salt-tolerant microalgae may be the ones that will replenish large amounts of potash tailings through the production of synthetic biocrust.

Recent research has reported the effective use of biocrust formation by injecting cyanobacteria in different areas to improve the growth and restoration of damaged areas (Lan et al. 2014). However, the selection of suitable algal isolate for the formation of a synthetic crust is essential for the success of recovery (Román et al. 2018; Giraldo-Silva et al. 2019).

Cyanobacteria and green algae are important factors and are the only major producers during (a) premature sequencing; (b) residential rehabilitation procedures under constant anthropogenic pressure; and (c) soil biological soils of high-demand habitats, for example tropical (Büdel 1999) and cold deserts (Büdel et al. 2016).

Green algae are a major part of soil biological crusts in mountainous habitats (Remias 2012). In their high habitats, these algae are exposed to harsh and highly volatile natural conditions, especially strong radiation, including ultraviolet radiation (Pichrtová et al. 2013) and dehydration leading to depletion. Therefore, green algae that still live in these areas should already have prevention or protection strategies, as well as remedial measures.

Green algae will serve as role models for learning desiccation tolerance or image protection techniques because of their natural ability to tolerate adverse conditions. The presence of biological crust plays an important role in the environment; for the real character, it can improve the formation of soil pores, reduce soil compaction (Meng et al. 2011), reduce rainfall (Belnap 2010), and effectively reduce rainfall. In extreme cases, such as water scarcity, malnutrition, high temperatures, biological crusts have the potential to survive. They can gradually improve soil quality and the environment, playing a key role in preventing and controlling soil erosion and desertification (Gao et al. 2017). With its chemical character it can alter soil pH content, essential plant nutrients, and efficiency, and increase organic matter, total nitrogen, total phosphorus, and total potassium (Gao et al. 2017). With its biological character, it can increase the activity of enzymes such as soil urease, invertase, catalase, and dehydrogenase.

6 Soil Reclamation

Salt has been an important historical factor that has influenced agricultural life. About 10% of the excavated area is covered by a variety of salt-affected soils, and the Asian continent is the largest salt-affected area.

The soil holds millions of microbes that contribute to improving soil fertility and increasing plant growth. The physical and chemical properties of the soil depend on the number and quality of soil organisms, the pH, and the energy of the redox. These influence the formation and strength of microscopic community and soil functions (Lombard et al. 2011).

Improvement of soil quality, plant nutrition, and the preservation of plant health are important roles for soil microorganisms in agriculture (Lugtenberg 2015). Organic matter degradation will be done with the help of these microorganisms in the soil (Schulz et al. 2013). In addition, it is noteworthy that many species of bacteria have been used in the digestion of organic matter in the soil. i.e. bioremediation of soil contamination (Zaidi et al. 2008).

These tiny soil insects perform many important functions as well as dangerous hazards. In the soil profile, the impact of soil biota is as varied and complex as the same action may be harmful or positive, depending on its nature (Parkin et al. 1996). Plants, on the other hand, show a different range of interactions with these microorganisms that live in the soil and extend a wide range of biological opportunities (competitive, exploitative, neutral, communal, and mutualistic). However, in the present context, the effects of pathogenic reduction, such as infection and herbivory, are widely studied (Zhang et al. 2013).

As interdisciplinary interactions with microorganisms are considered at the forefront of agricultural management and biological factors especially in global change, the impact of environmental stress factors must be considered, as they affect proper intercrop management interactions (Zolla et al. 2013). Soil formation with high soil fertility is the result of more than a hundred years of "evolution" of soil.

Increasing soil salts, high soluble salts, sodicity, and highly volatile sodium are all major global degradation problems (Vanessa et al. 2009). The remodeling of such soils basically requires the replacement of calcium-replaced sodium with subsequent sodium replacement. It is usually done by adding gypsum ($CaSO_4$, $2H_2O$) or pyrite ($FeSO_4$) followed by excessive salt immersion by flooding or heavy irrigation. Singh (1961) suggested that cyanobacteria can regenerate alkaline soils because they grow successfully in saline/alkaline soils when most plants fail to grow. Pandey et al. (2005), Jaiswal et al. (2010), Murtaza et al. (2011), Venus Singh and Singh (2015), and Jan et al. (2017) also highlight the role of cyanobacteria in restoring alkaline saline soils.

Besides their importance in soil fertility with essential nutrients and in the restoration of soil structures, microalgae and cyanobacteria can also improve certain soil characteristics, such as concentration, porosity, penetration, air permeability, and humidity (Baweja et al. 2019; Pan et al. 2019). Typically, the development of these traits is associated with polysaccharides released by several microalgal/cyanobacterial species (Sharma et al. 2012; Baweja et al. 2019). These polysaccharides often form an adhesive sheath and gelatinous that contributes to soil compaction and prevents erosion (Baweja et al. 2019). These polysaccharides play an important role in the soil's air permeability and water retention capacity, which is important for maintaining soil temperature, pH, and salinity (Baweja et al. 2019; Pan et al. 2019). The development of these soil structures greatly improves plant production, as they promote root growth and increase microbial activity. Therefore, microalgae and cyanobacteria have been widely used in restoring soil structures and in improving soil fertility. Hu et al. (2002) have shown that resistance to erosion of uncultivated sand is increased when there are algae in desert soils, such as *Microcoleus vaginatus*, *Phormidium tenue, Scytonema javanicum, Nostoc* sp., and *Desmococcus olivaceus*, concluding that this increase was associated with higher concentrations of crusts in identified algae-treated samples.

Cyanobacteria can not only survive, but thrive in conditions that are considered uninhabitable, tolerating exhaustion, high temperatures, excessive pH, and high salinity, showing their ability to adapt to extreme conditions. Cyanobacteria can restore alkaline soils and fertility can be improved to grow cereal, sugarcane, and horticultural crops. The physiological characteristics of cyanobacteria in large amounts of salt include (a) the accumulation of osmoprotective compounds, (b) the maintenance of low intracellular ion concentrations of inorganic ions, and (c) the expression of salt-stress proteins. Exposure to cyanobacterial cells in various abiotic pressures has resulted in the rapid expression of several stress-regulated proteins and the modification of the protein synthesis system. Protection in the alkaline environment is provided by the combination of certain fatty acids and sucrose- and osmotic-stress-induced proteins. In cyanobacteria, the accumulation of internal osmoticum in the form of inorganic ions and the inhibition of intracellular Na^+ accumulation by the reduction of Na influx and by efficient mechanisms of efflux or metabolic modification have been extensively investigated. Na^+ extrusion in cyanobacteria is driven by the Na^+/H^+ antiporter, which is enhanced by the enhanced activity of cytochrome oxidase. Inhibition of sodium ion penetration appears to be a major mechanism of cyanobacteria survival against salt stress and a combination of salt-stress proteins found in cyanobacteria.

Cyanobacterial infusion causes soil enrichment with stable nitrogen, improved soil structure, and decreased pH, electrical conductivity (EC), and Na^+. Extracellular polysaccharides released by cyanobacteria have been reported to be responsible for binding soil particles, thus leading to the formation of a stronger and more durable structure that improves soil stability and protects against erosion. The potential impact of these pests on agriculture through their use, such as soil heaters, crop growth controllers, and soil health enhancements, has been well documented. In addition to improving rice yields, cyanobacteria produce beneficial and indirect beneficial effects on the physical, chemical, and biological properties of soil and water in the soil and water in rice fields, which are important for agriculture. Field cleaning may not be effective in salting soil restitution, and the addition of cyanobacterium inoculum and gypsum supplementation is required prior to irrigation to improve saline soils (Singh and Dhar 2009).

The difficulty of land reclamation in arid and semi-arid areas is often because of the salty conditions of extensive areas of soil. Several studies have been conducted on the effect of salt on growth, metabolism, and yield of plants and algae (Tang et al. 2007). Other growth regulators such as gibberellic acid (GA3) were used to improve plant salt tolerance. From an economic point of view, growth controls are expensive and ineffective when used in large quantities. Algae play an economic role in regenerating soils that increase soil fertility and improve plant conditions under certain environmental conditions (Nisha et al. 2007).

A major problem in land reclamation in arid and semi-arid areas is the salinity of large areas of soil. Ordinary plants do not tolerate this high salt content, and although the addition of growth regulators such as gibberellic acid (GA3) improves plant salt tolerance, they are very expensive and cannot be used on a large scale. However, marine algae are more tolerant of salt and can easily grow in these conditions. Over time, algae can not only increase the fertility of the soil in these regions, but it can also reduce salt levels and thus prepare the soil for plant growth.

7 Climate Change

The problem of climate change due to the increase in the greenhouse gases in the atmosphere is a problem which threatens our way of life; 33% of the world's land surface area is covered by deserts. This land area covered by deserts consists of nutrient-devoid soil with very little agricultural potential. However, this land can be regenerated in such a manner so that it can temporarily serve as agricultural land, and ultimately become a carbon sink like the Amazon Rainforests to help combat the problem of climate change. Both these problems can be combated together by using eco-friendly methods for large-scale soil regeneration.

There are many anthropogenic and industrial activities such as coal and other fossil fuel combustion, deforestation, urbanization, and industrialization that cause emission of greenhouse gases and change in climate. Although there are some preventive actions taken, still natural calamities such as flood, snow storm, and change in seasons are occurring globally. However, the nature itself has its own healing capacity to recover from damage. Microbial communities play a vital role in the global carbon cycle (Fig. 1), as they fix atmospheric carbon, aid plant development, and break down or modify organic matter in the environment.

Microorganisms aid in the reduction of global warming, which has implications for critical ecological processes such as nitrogen cycle (Fig. 2), which rely on

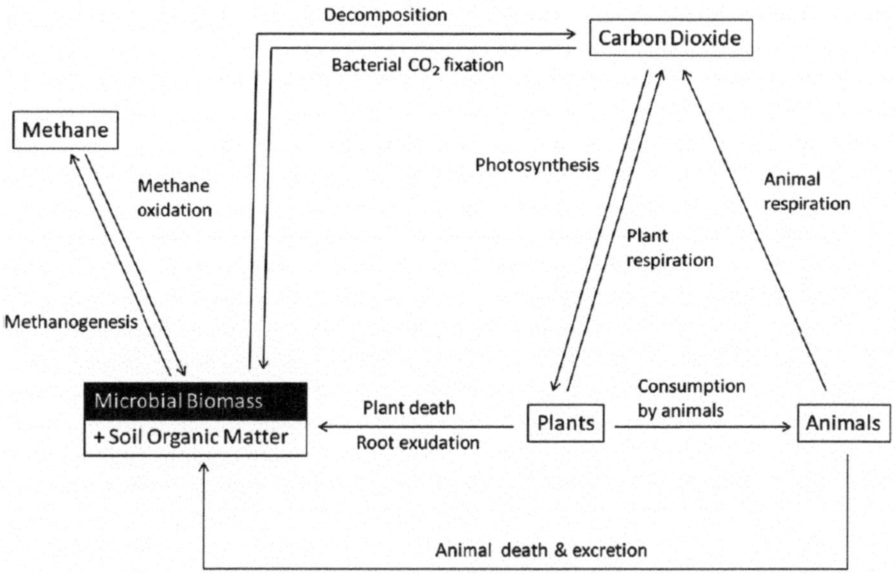

Fig. 1 Microbial communities and carbon cycle. (Credit: Abatenh et al. 2018)

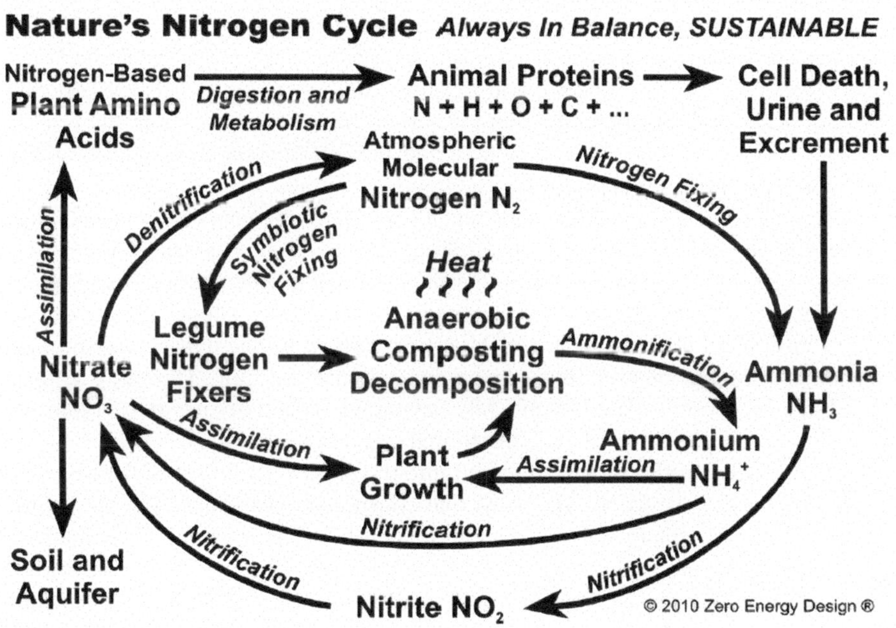

Fig. 2 Microbial communities and nitrogen cycle. (Credit: Abatenh et al. 2018)

microbial activity. Microorganisms are critical in the breakdown and transformation of dead organic matter into forms that other organisms can use. In order to achieve their primary goal of survival through reproduction, soil microorganisms require the transportation of carbon between environmental compartments. Bacteria utilize both organic and inorganic forms of carbon as carbon and energy sources. The terrestrial carbon cycle is dominated by the balance between photosynthesis and respiration.

Abatenh et al. (2018) reported on the role of microbial community for the treatment of greenhouse gases and reduction of pollutants. The biogeochemical reactions such as nitrogen fixation, humus production, and carbon cycle controlled by microorganisms have led to significant contribution to the reduction of GHGs emission. Algae present on soil increase organic matter, provide nitrogen and stabilize the soil, which helps in improving of its fertility. This promotes germination of seeds and growth of other microorganisms. Algae which are found in different soil types may help the soil to improve its characteristics such as carbon content, texture, and aeration (Ibraheem 2007). Cyanobacteria fix nitrogen up to 30%, and this will increase soil fertility and increase crop production (Dey et al. 2010). Prasanna et al. (2012) reported that inoculation of cyanobacteria to the soil increased the rice yield by 19% and an increase in nitrogen fixation. They are environmentally friendly biofertilizers in increasing crop productivity, especially in lowland rice ecosystems. It improves soil texture and quality by increasing aeration, water-holding capacity, and nutrient viability, all of which help to support growth.

A great response to climate warming is being produced by microbial degradation of soil carbon. Microbial biomass and enzymes are effective methods for boosting warming because they efficiently degrade carbon-based organic matter and release hazardous emissions into the atmosphere and prevent climate change at the same time. The effects of temperature on enzyme activity and microbial physiological properties are direct. The carbon response of soil microorganisms to climate change is determined by their efficiency in utilizing carbon. It also increases the water-holding capacity of soil and increase the intermolecular space between soil molecules due to which proper aeration is supplied to the root system. Samolov et al. (2020) observed that the presence of biocrusts formed by the eukaryotic and prokaryotic microorganisms stabilizes the carbon and nitrogen ratio. The study demonstrated that biocrusts isolated from Chile have a wide diversity of microorganisms mainly composed of chlorophyta, streptophyta, and cyanobacteria.

Methane is a greenhouse gas, released into the atmosphere most of the time due to microbial activities. Microorganisms that consume methane are essential for the Earth's climate to remain stable. Methane is used by bacteria to fuel their metabolism. Methanotrophic bacteria use methane as their only energy source and convert it to carbon dioxide during digestion. These bacteria have the ability to consume large amounts of methane, which aids in the reduction of methane emissions from factories and landfills. Microorganisms require a large number of methane molecules. In anaerobic environments, such as deep-packed muck, carbon dioxide is easily transformed to methane. This process is carried out by methanogenic bacteria. Although it is projected that ocean warming and acidification will affect coral reef biodiversity, the combined impacts of these stressors on overall biodiversity have been studied. Ocean warming boosted species richness considerably compared to current control conditions, whereas acidification decreased richness dramatically. The structure of the cryptobenthic population on coral reefs, which includes a diverse range of species, is influenced by various environmental factors (Timmers et al. 2021).

8 Environmental Management

Environmental evaluation of soil and water resources is aided by research into the nature of the many algal species. Pollutants have an effect on microorganisms, as seen by soil algae's response. The nature and level of contaminants present in bodies can be evaluated as an environmental pollution goal based on an examination of algal colonies isolated from soil and water. Dirborne and Ramanujam (2017) explored 124 different types of soil algae species distributed among two types of forests, namely pine forests and broadleaf forests. The study showed 21% of cyanobacteria present in subtropical broadleaf forests, whereas pine forest soils have 20% higher amount of diatoms and 13% cyanobacteria. The study has demonstrated that the diversity of soil algae has a strong correlation with soil characteristics. The growth of soil algae indicates the fertile nature of the soil, and they act as soil environmental health indicators. Algal growth is beneficial to the soil by absorbing heavy metals and xenobiotic organic compounds such as pesticides, insecticides, dyes, and amines. Eco-toxicological assessment of soil or water samples can be done by evaluating types and level of algal growth for environmental pollution management (Meeranayak et al. 2020).

Understanding the ecology of soil algae can aid in environmental conservation by preserving the natural character of the soil as well as critical nutrients for plant growth (Ray and Thomas 2012). For instance, the growth of higher saline-tolerant halophilic algae has been isolated from soil surrounded in potash mining areas (Sommer et al. 2020). The role of environmental conditions on soil diatom produc tivity as well as the seasonal and geographical patterns of diatom assemblages were examined by Foets et al. (2020). For a period of 12 months, we collected soil algal samples every 4 weeks in 16 locations along the Attert River watershed (Luxembourg). Pigment analysis with high-performance liquid chromatography was then used to calculate algal abundances. Our findings show that diatom productivity is primarily influenced by factors linked to soil moisture availability, resulting in seasonal variations, whereas green algae concentrations remained constant throughout the study period.

Microalgae have also gained increasing interest as a source of renewable energy (e.g., biofuel production) and agriculture (Alvarez et al. 2021) and for industrial and domestic wastewater bioremediation (Renuka et al. 2017), algae-based wastewater remediation, biofertilizer processing, soil ecosystem protection, plants or crops growth, and food safety chain (Singh et al. 2019; Deepika and Mubarak Ali 2020).

Silvarrey Barruffa et al. (2021) explored the harmful effects of cyanobacteria on eutrophication and discussed the preventive measures/adaptation strategies needed to be taken against climate change.

9 Conclusion and Future Perspectives

As soil additives, dry microalgae boost plant nutrition, which improves all physiological reactions that contribute to healthy growth. Dry algae per kilogram of soil can enhance soil fertility, plant nutritional requirements for high harvests, and reduce pollution. Composting is a straightforward and practical approach for managing waste algal biomass taken from such habitats, and it is most probable that algae composting will be used for this purpose in the future, with the compost product being acceptable for use as a fertilizer. As a result, algae biomass has been combined with other organic wastes to create a compostable blend.

Microalgal extracts had a number of positive impacts, including better crop yield and quality, higher nutrient uptake, and promoted resistance to abiotic stresses. Microalgae can be used in organic or excellent agricultural practices. Microalgae could help to replace traditional chemical fertilizers in alternative farming in the future. Farmlands can benefit from cyanobacterial biofertilizers because they are both environmentally friendly and financially effective. Seaweeds are a good source of macro- and micronutrients, fibers, ash, phenol, carbs, and plant growth hormones, among other things. Plants' physiological and biochemical characteristics are improved by growth-promoting chemicals generated by biofertilizers. Biofertilizers are also financially promising in addition to these benefits. In comparison to chemical fertilizers, they are also less expensive. Seaweed has a lot of promise as growth stimulants or even agricultural soil conditioners.

To better understand the changes generated by microalgal biostimulants and biofertilizers in crop productions, agronomic, physiological, chemical, biochemical, and molecular investigations are needed. One feasible strategy to employ microalgal and cyanobacterial biomass in agriculture while avoiding excessive process costs is to promote the biorefinery idea, in which the acquired biomass should be fully used in order to generate many products for various applications at the same time. Furthermore, more research investigations concentrating on the influence of each metabolite/biomass on crops are required for a better definition of the most appropriate compounds for agricultural activities.

Future study will focus on designing and developing effective microalgae-based products with biostimulation properties. Technical training in algal fertilizer production and quality management for producers can result in more sustainable algal fertilizer growth. To build a thorough, varied, and relevant decision system for restoring arid lands based on the induction of biocrust formation, more research is needed into the applicability of cyanobacteria inoculation in different ecosystems with different soil types and climate conditions.

References

Aaronson, S (1986) A role for algae as human food in antiquity. Food and Foodways 1(3), 311–315.

Abatenh E, Gizaw B, Tsegaye Z, Tefera G (2018) Microbial function on climate change – review. Environ Pollut Climate Change 2:147

Abd El Moniem EA, Abd-Allah ASE (2008) Effect of green algae cells extract as foliar spray on vegetative growth, yield and berries quality of superior grapevines. J Amer Eur Agric and Environ Sci 4(4):427–433

Abdul Khalil HPS, Saurabh CK, Tye YY, Lai TK, Easa AM, Rosamah E (2017) Seaweed based sustainable films and composites for food andpharmaceutical applications: a review. Renew Sust Energ Rev 77:353–362

Adarme-Vega TC, Lim DKY, Timmins M, Vernen F, Li Y, Schenk P (2012) Microalgal biofactories: a promising approach towards sustainableomega-3 fatty acid production. Microb Cell Factories 11:96

Adeyemi NO, Atayese MO, Olubode AA, Akan ME (2020) Effect of commercial arbuscular mycorrhizal fungi inoculant on growth and yield of soybean under controlled and natural field conditions. J Plant Nutr 43:487–499. https://doi.org/10.1080/01904167.2019.1685101

Ahmed W, Jing H, Kaillou L, Qaswar M, Khan MN, Jin C (2019) Changes in phosphorus fractions associated with soil chemical properties under long-term organic and inorganic fertilization in paddy soils of southern China. PLoS One 14:e0216881

Ali N, Farrell A, Ramsubhag A, Jayaraman J (2016) The effect of *Ascophyllum nodosum* extract on the growth, yield and fruit quality of tomato grown under tropical conditions. J App Phys 28:1353–1362

Ali O, Ramsubhag A, Jayaraman J (2019) Biostimulatory activities of Ascophyllum nodosum extract in tomato and sweet pepper crops in a tropical environment. PLoS One 14:e0216710

Alvarez AL, Weyers SL, Goemann HM, Peyton BM, Gardner RD (2021) Microalgae, soil and plants: a critical review of microalgae as renewable resources for agriculture. Algal Res 54:102200

Aly MS, Esawy MA (2008) Evaluation of *Spirulina platensis* as biostimulatory for organic farming systems genet. Eng Biotechnol J 6:1–7

Andreote FD, Gumiere T, Durrer A (1996) Exploring interactions of plant microbiomes. Trends Biotechnol 29(12):569–572.

Andreote FD, Gumiere T, Durrer A (2014) A exploring interactions of plant microbiomes. Sci Agric 71:528–539

Azizoglu U (2019) Bacillus thuringiensis as a biofertilizer and biostimulator: a mini-review of the little-known plant growth-promoting properties of Bt. Curr Microbiol 76:1379–1385

Bakker MG, Manter DK, Sheflin AM, Weir TL, Vivanco JM (2012) Harnessing the rhizosphere microbiome through plant breeding and agricultural management. Plant Soil 360:1–13

Baweja P, Kumar S, Kumar G (2019) Organic fertilizer from algae: a novel approach towards sustainable agriculture. In: Giri B, Prasad R, Wu Q-S, Varma A (eds) Biofertilizers for sustainable agriculture and environment. Springer, Cham, pp 353–370

Belnap J (2010) The potential roles of biological soil crusts in dryland hydrologic cycles[J]. Hydrol Process 20(15):3159–3178

Belnap J, Büdel B, Lange OL (2016) Biological soil crusts as an integral component of desert environments. In Biological Soil Crusts: Structure, Function, and Management (pp. 3–30). Springer, Cham.

Bertocci I, Araújo R, Oliveira P, Sousa-Pinto I (2015) Potential effects ofkelp species on local fisheries. J Appl Ecol 52:1216–1226

Bleakley S, Hayes M (2017) Algal proteins: extraction, application, and challenges concerning production. Food Secur 6:33

Bocanegra A, Bastida S, Benedi J, Rodenas S, Sanchez-Muniz F (2009) Characteristics and nutritional and cardiovascular-health properties of seaweeds. J Med Food 12:236–258

Bogie J, Hoeks C, Schepers M, Tiane A, Cuypers A, Leijten F (2019) Dietary Sargassum fusiforme improves memory and reduces amyloid plaque load in an Alzheimer's disease mouse model. Sci Rep 9:4908
Borowitzka MA (2013) Dunaliella: biology, production, and markets. In: Richmond A, Hu Q (eds) Handbook of microalgal culture. Wiley-Blackwell, Chichester, pp 359–368
Büdel B (1999) Ecophysiology of photosynthetic lichens. Cryptogamic Botany 10(3):170–176.
Büdel B, Dulic T, Darienko T (2016) Cyanobacteria and algae of biological soil crusts. In: Weber B, del Burkhard B, Belnap J (eds) Biological soil crusts: an organizing principle in drylands. Springer, Cham, pp 55–80
CEVA. (2014). Edible seaweed and French regulation-Synthesis made by CEVA (31/03/2014).
Chapman RL (2013) Algae: the world's most important "plants"—an introduction. Mitig Adapt Strateg Glob Chang 18(1):5–12
Chacón-Lee T, González-Mariño GE (2010) Microalgae for "healthy" foods—possibilities and challenges. Compr Rev Food Sci Food Saf 9(6):655–675.
Cikoš A-M, Jerković I, Molnar M, Šubarić D, Jokić S (2021) New trends for macroalgal natural products applications. Nat Prod Res 35:7
Circuncisão AR, Catarino MD, Cardoso SM, Silva AMS (2018) Minerals from macroalgae origin: health benefits and risks for consumers. Mar Drugs 16:400
Clarholm M (2005) Soil protozoa: an under-researched microbial group gaining momentum. Soil Biol Biochem 37:811–817. [Review of the role of protozoa in nutrient cycling in soil ecosystems]
Cocking EC (2003) Endophytic colonization of plant roots by nitrogen-fixing bacteria. Plant Soil 252:169–175
Cofrades L-L, Solas B, Jiménez-Colmenero. (2008) Influence of different types and proportions of added edible seaweeds on characteristic of low-salt gel/emulsion meat systems. Meat Sci 79:767–776
Collins KG, Fitzgerald GF, Stanton C, Ross RP (2016) Looking beyond the terrestrial: the potential of seaweed derived bioactives to treat non-communicable diseases. Mar Drugs 14(60):1–31
Coppens J, Grunert O, Van Den Hende S, Vanhoutte I, Boon N, Haesaert G, De Gelder L (2016) The use ofmicroalgae as a high-value organic slow-release fertiliser results in tomatoes with increased carotenoid and sugar levels. J Appl Phycol 28:2367–2377
Cornish ML, Garbary DJ (2010) Antioxidants from macroalgae: potential applications to human health and nutrition. Algae 25:155–171
Craigie JS (2010) Seaweed extract stimuli in plant science and agriculture. J Appl Phycol 23:371–393
Cressey D (2017) Climate change is making algal blooms worse. Nature 2:84
Dakora FD, Matiru VN, Kanu AS (2015) Rhizosphere ecology of lumichrome and riboflavin, two bacterial signal molecules eliciting developmental changes in plants. Front Plant Sci 6:700
Darda S, Papalas T, Zabaniotou A (2019) Biofuels journey in Europe: currently the way to low carbon economy sustainability is still a challenge. J Clean Prod 208:575–588
De Pascale S, Rouphael Y, Colla G (2018) Plant biostimulants: innovative tool for enhancing plant nutrition in organic farming. Eur J Hortic Sci 82:277–285
Deepika P, Mubarak Ali DJB (2020) Production and assessment of microalgal liquid fertilizer for the enhanced growth of four crop plants. Biocatal Agric Biotechnol 28:101701
Dey HS, Tayung K, Bastia AK (2010) Occurrence of nitrogen–fixing cyanobacteria in local rice fields of Orissa, India. Ecoprint 17:77–85
Dillehay TD, Ramirez C, Pino M, Collins MB, Rossen J, Pinot-Navarro JD (2008) Monte Verde: seaweed, food, medicine and the peopling of South America. Science 320:784–789
Dirborne CM, Ramanujam P (2017) Diversity and ecology of soil algae in broadleaf sacred grove and pine forest in East Khasi Hills, Meghalaya. Nelumbo 59:195
Dmytryk A, Chojnacka K (2018) Algae as fertilizers, biostimulants, and regulators of plant growth. In: Chojnacka K, Wieczorek PP, Schroeder G, Michalak I (eds) Algae biomass: characteristics and applications. Springer, Cham, pp 115–122

Egli M, Wernli M, Burga C, Kneisel C, Mavris C (2011) Fast but spatially scattered smectite-formation in the proglacial area Morteratsch: an evaluation using GIS. Geoderma 164:11–21
Elagib NA, Al-Saidi M (2020) Balancing the benefits from the water– energy–land–food nexus through agroforestry in the Sahel. Sci Total Environ 742:140509
FAO (2014) The state of the world fisheries and aquaculture 2014. FAO, Rome, p 223
FAO (2016) The state of the world fisheries and aquaculture 2016. Contribution to Food Security and Nutrition for All, Rome, p 200
Felde VJMNL, Chamizo S, Felix-Henningsen P, Drahorad SL (2017) What stabilizes biological soil crusts in the negev desert? Plant Soil 429:9–18
Finley JW, Ellwood K, Hoadley J (2014) Launching a new food product or dietary supplement in the United States: industrial, regulatory, and nutritional considerations. Annu Rev Nutr 34:421–447
Foets J, Boon N, Schaumann GE (2020) Microbial hotspots on clay mineral surfaces in soils: An emerging understanding of interactions with soil biogeochemistry. Front Microbiol 11:1682.
Gantar M, Svircev Z (2008) Microalgae and cyanobacteria: food for thought. J Phycol 44:260–268
Gao LQ, Bowker MA, Xu MX (2017) Biological soil crusts decrease erodibility by modifying inherent soil properties on the Loess Plateau, China. Soil Biol Biochem 2017(105):49–58
Giraldo-Silva A, Nelson C, Bargter NN, Garcia-Pichel F (2019) Nursing biocrusts: isolation, cultivation, and fitness test of indigenous cyanobacteria. Restor Ecol 27:793–803
Górka B, Korzeniowska K, Lipok J, Wieczorek PP (2018) The biomass of algae and algal extracts in agricultural production. In: Algae biomass: characteristics and applications. Springer Science and Business Media LLC, Berlin, pp 103–114
Goswami M, Deka S (2020) Plant growth-promoting rhizobacteria— alleviators of abiotic stresses in soil: a review. Pedosphere 30:40–61
Gouda S, Kerry RG, Das G, Paramithiotis S, Shin HS, Patra JK (2018) Revitalization of plant growth promoting rhizobacteria for sustainable development in agriculture. Microbiol. Res 206:131–140
Gouveia L, Batista AP, Sousa I, Raymundo A, Bandarra N (2008) Microalgae in novel food products. In: Konstantinos N, Papadopoulos PP (eds) Food chemistry research development. Nova Science Publishers, New York, pp 75–112
Grobbelaar JU (1983) Availability of phosphorus and other nutrients to algae. In T.R.G. Gray & J.R. Postgate (Eds.), Microbial Ecology. Cambridge University Press pp. 219–225.
Greene Gerber LN, Tester JW, Beal CM, Huntley ME, Sills DL (2016) Target cultivation and financ-ing parameters for sustainable production of fuel and feed from microalgae. Environ Sci Technol 50(7):3333–3341
Gustavs L, Eggert A, Michalik D, Karsten U (2010) Physiological and biochemical responses of green microalgae from different habitats to osmotic and matric stress. Protoplasma 243:3–14
Hafting JT, Critchley AT, Cornish ML, Hubley SA, Archibald AF (2012) On-land cultivation of functional seaweed products for human usage. J Appl Phycol 24:385–392
Hafting JT, Craigie JS, Stengel DB, Loureiro RR, Buschmann AH, Yarish C, Edwards MD (2015) Prospects and challenges for industrial production of seaweed bioactives. J Appl Phycol 27(5): 2261–2277.
Hajnal-Jafari TI, Duric SS, Stamenov DR (2016) Influence of green algae Chlorella vulgaris on initial growth of different agricultural crops. Matica Srpska J Nat Sci 130(5):29–33
Hamed SMM (2007) Studies on nitrogen fixing cyanobacteria. M. Sc. Thesis, Botany Department, Faculty of Science, Beni-Suef University, Beni-Suef
Han X, Zeng H, Bartocci P, Fantozzi F, Yan Y (2018) Phytohormones and effects on growth and metabolites of microalgae: a review. Fermentation 4:25
Healy FP (1973) Characteristics of phosphorus deficiency in anabaena. J Phycol 9:383–394
Hemantkumar JN, Rahimbhai MI (2020) Microalgae and its use in nutraceuticals and food supplements. In: Vítová M (ed) Microalgae: from physiology to application. IntechOpen, London

Hernández-Herrera RM, Santacruz-Ruvalcaba F, Ruiz-López MA, Norrie J, Hernández-Carmona G (2014) Effect of liquid seaweed extracts on growth of tomato seedlings (*Solanum lycopersicum*L.). J App Phys 26(1):619–628

Hontoria C, García-González I, Quemada M, Roldán A, Alguacil MM (2019) The cover crop determines the AMF community composition in soil and in roots of maize after a ten-year continuous crop rotation. Sci Total Environ 660:913–922

Hu C, Liu Y, Song L, Zhang D (2002) Effect of desert soil algae on the stabilization of fine sands. Environ Boil Fishes 14:281–292

Ibáñez E, Cifuentes A (2012) Benefits of using algae as natural sources of functional ingredients. J Sci Food Agric 93(4):703–709

Ibraheem IBM (2007) Cyanobacteria as alternative biological conditioners for bioremediation of barren soil. Egyptian J Phycol 8:99–116

Innok S, Chunleuchanon S, Boonkerd N, Teaumroong N (2009) Cyanobacterial akinete induction and its application as biofertilizer for rice cultivation. Environ Boil Fishes 21:737–744

Jacoby R, Peukert M, Succurro A, Koprivova A, Kopriva S (2017) The role of soil microorganisms in plant mineral nutrition-current knowledge and future directions. Front Plant Sci 8:1617–1622

Jaiswal P, Kashyap AK, Prasanna R, Singh PK (2010) Evaluating the potential of *N. calcicola* and its biocarbonate resistant mutant as bioameleorating agents for 'usar' soil. Indian J Microbiol 50:12–18

Jan ZS, Sultan AT, Wasiullah and Ahmad W. (2017) The role of cyanobacteria on availability of major plant nutrients and soil organic matter to rice crop under saline soil condition. Sarhad J Agric 33(4):566–572

Jebasingh SE, Lakshmikandan M, Vasanthakumar P, Sivaraman K (2015) Improved seedling growth and seed germination in legume crop *Vigna mungo* (L.) Hepper utilizing marine macro algal extracts. Proc Natl Acad Sci India B 85(2):643–651

Johansen JR (1993) Cryptogamic crusts of semiarid and arid lands of North America. J Phycol 29:140–147

Kalpana P, Sai G, Bramari L, Anitha (2014) Biofortification of *Amaranthus gangeticus* using *Spirulina platensis* as microbial inoculant to enhance iron levels. Int J Res Appl Nat Soc Sci 2:103–110

Kapur B, Sarıda MA, Çeliktopuz E, Kafkas E, Payda S, Kargı S (2018) Health and taste related compounds in strawberries under various irrigation regimes and bio-stimulant application. Food Chem 263:67–73

Karsten U, Holzinger A (2012) Light, temperature, and desiccation effects on photosynthetic activity, and drought-induced ultrastructural changes in the green alga Klebsormidium dissectum (Streptophyta) from a high alpine soil crust. Microb Ecol 63:51–63

Karsten U, Herburger K, Holzinger A (2014) Dehydration, temperature, and light tolerance in members of the aeroterrestrial green algal genus Interfilum (Streptophyta) from biogeographically different temperate soils. J Phycol 50:804–816

Karthikeyan N, Prasanna R, Nain L, Kaushik BD (2007) Evaluating the potential of plant growth promoting cyanobacteria as inoculants for wheat. Eur J Soil Biol 43:23–30

Khan MN, Cho JY, Lee MC, Kang JY, Park NG, Fujii H, Hong YK (2007) Isolation of two anti-inflammatory and one pro-inflammatory polyunsaturated fatty acids from the brown seaweed Undaria pinnatifida. J Agric Food Chem 55:6984–6988

Khan SA, Sharma GK, Malla FA, Kumar A, Rashmi, Gupta N (2019) Microalgae based biofertilizers: a biorefinery approach to phycoremediate wastewater and harvest biodiesel and manure. J Clean Prod 211:1412–1419

Kim SK, Li YX (2011) Medicinal benefits of sulfated polysaccharides from sea vegetables. Adv Food Nutr Res 64:391–402

Kinley RD, Martinez-Fernandez G, Matthews MK, de Nys R, Magnusson M (2020) Mitigating the carbon footprint and improving productivity of ruminant livestock agriculture using a red seaweed. J Cleans Prod 259:120836

Kolanjinathan K, Ganesh P, Saranraj P (2014) Pharmacological importance of seaweeds: a review. World J Fish Mar Sci 6(1):1–15
Koskey G, Mburu SW, Njeru EM, Kimiti JM, Ombori O, Maingi JM (2017) Potential of native rhizobia in enhancing nitrogen fixation and yields of climbing beans (Phaseolus vulgaris L.) in contrasting environments of eastern Kenya. Front Plant Sci 8:443
Kumar D, Adhikary SP (2015) Diversity, moleculary phylogeny and metabolic activity of cyanobacteria in biological soil crusts from Santiniketan (India). J Appl Phycol 27:339–349
Kwon PS, Oh H, Kwon S, Jin W, Zhang F, Fraser K (2020) Sulfated polysaccharides effectively inhibit SARS-CoV-2 in vitro. Cell Discov 6:50
Lafarga T, Acién-Fernández FG, Castellari M, Villaró S, Bobo G, Aguiló-Aguayo I (2019) Effect of microalgae incorporation on the physicochemical, nutritional, and sensorial properties of an innovative broccoli soup. Food Sci Technol 111:167–174
Lan S, Zhang Q, Wu L, Liu Y, Zhang D, Hu C (2014) Artificially accelerating the reversal of desertification: cyanobacterial inoculation facilitates the succession of vegetation communities. Environ Sci Technol 48:307–315
Li Y, Mattson NS (2015) Effects of seaweed extract application rate and method on post-production life of petunia and tomato transplants. Horttechnology 25:505–510
Lobo CB, Juárez Tomás MS, Viruel E, Ferrero MA, Lucca ME (2019) Development of low-cost formulations of plant growth-promoting bacteria to be used as inoculants in beneficial agricultural technologies. Microbiol Res 219:12–25
Lombard N, Prestat E, van Elsas JD, Simonet P (2011) Soilspecific limitations for access and analysis of soil microbial communities by metagenomics. FEMS Microbiol Ecol 78:31–49
Lorenzo JM, Agregán R, Munekata PES, Franco D, Carballo J, Sahin S, Lacomba R, Barba FJ (2017) Proximate composition and nutritional value of three macroalgae: Ascophyllum nodosum, Fucus vesiculosus and Bifurcaria bifurcata. Mar Drugs 15:360
Lugtenberg B (2015) Life of microbes in the rhizosphere. In: Lugtenberg B (ed) Principles of plant-microbe interactions. Springer International Publishing Switzerland, Heidelberg, pp 7–15
Maestre FT, Bowker MA, Cantón Y (2011) Ecology and functional roles of biological soil crusts in semi-arid ecosystems of Spain. J Arid Environ 75:1282–1291
Mahmoud MS (2001) Nutritional status and growth of maize plants as affected by Green microalgae as soil additives. J Biol Sci 1:475–479
Makkar HPS, Tran G, Heuze V, Giger-Reverdin S, Lessire M, Lebas F, Ankers P (2016) Seaweeds for livestock diets: a review. Anim Feed Sci Technol 212:1–17
Mandimba GR, Okomba GD, Pandzou. (1998) Nodulated legumes as green manure: an alternative source of nitrogen for non-fixing and poor fixing crops. Int J Trop Agric 16:131–145
Marks EA, Montero O, Rad JC (2019) The biostimulating effects of viable microalgal cells applied to a calcareous soil: increases in bacterial biomass, phosphorus scavenging, and precipitation of carbonates. Sci Total Environ 692:784–790
Martín P, Caporgno, Mathys A (2018) Trends in microalgae incorporation into innovative food products with potential health benefits. Front Nutr 5:58
Meeranayak UFJ, Nadaf RD, Toragall MM, Nadaf U (2020) The role of algae in sustainable environment: a review. J Algal Biomass Util 11:28–34
Mendes R, Garbeva P, Raaijmakers JM (2013) The rhizosphere microbiome: significance of plant beneficial, plant pathogenic, and human pathogenic microorganisms. FEMS Microbiol Rev 37:634–663
Meng J, Bu CF, Zhang XC (2011) Developmental characteristics of biological soil crust under different vegetation types in wind-water erosion crisscross region, Northern Shaanxi Province, China[J]. J Northwest For Univ 26(4):41–46
Mohanty S, Kumar N, Rajan AR (2006) Availability and uptake of phosphorus from organic manures in groundnut (*Arachis hypogea* L.)–corn (Zea mays L.) sequence using radio tracer technique. Geoderma 133:225–230
Murtaza B, Murtaza G, Zia-ar-Rehman M, Ghafoor A, Abubakar S, Sabir M (2011) Reclamation of salt affected soils using amendments and growing wheat crop. Soil Environ 30(2):130–136

Newton L (1951) Seaweed utilization. Sampson Low, London
Nisha R, Kaushik A, Kaushik CP (2007) Effect of indigenous cyanobacterial application on structural stability and productivity of an organically poor semiarid soil. Geoderma 138:49–56
Ojuederie OB, Olanrewaju OS, Babalola OO (2019) Plant growth promoting rhizobacterial mitigation of drought stress in crop plants: implications for sustainable agriculture. Agronomy 9:712
Olaizola M (2003) Commercial development of microalgal biotechnology: from the test tube to the marketplace. Biomol Eng 20:459–466
Pan S, Jeevanandam J, Danquah MK (2019) Benefits of algal extracts in sustainable agriculture. In: Grand challenges in marine Biotechnology. Springer Science and Business Media LLC, Berlin, pp 501–534
Pandey KD, Shukla PN, Giri DD, Kashyap AK (2005) Cyanobacteria in alkaline soil and the effect of cyanobacteria inoculation with pyrite amendments on their reclamation. Biol Fertil Soils 41:451–457
Pangestuti R, Kim SK (2011) Biological activities and health benefit effects of natural pigments derived from marine algae. J Funct Foods 3:255–266
Parkin TB, Doran JW, Vizcaino EF (1996) Field and laboratory tests of soil respiration. In: Doran JW, Jones AJ (eds) Methods for assessing soil quality. SSSA Spec. Publ. 49. SSSA, Madiso, pp 231–246
Pathak J, Rajneesh MPK, Singh SP, Hader DP, Sinha RP (2018) Cyanobacterial farming for environment friendly sustainable agriculture practices: innovations and perspectives. Front Environ Sci 6:13
Peteiro C (2018) Alginate production from marine macroalgae, with emphasis on kelp farming. In: Rehm BHA, Moradali F (eds) Alginates and their biomedical applications. Springer, pp 27–66
Pichrtová M, Remias D, Lewis LA, Holzinger A (2013) Changes in phenolic compounds and cellular ultrastructure of Arctic and Antarctic strains of Zygnema (Zygnematales, Streptophyta) after exposure to experimentally enhanced UV to PAR ratio. Microb Ecol 65:68–83
Plaza M, Cifuentes A, Ibáñez E (2008) In the search of new functional food ingredients from algae. Trends in Food Science & Technology 19(1):31–39.
Prabu PC, Udayasoorian C (2007) Native cyanobacteria W estiellopsis (TL-2) sp for reclaiming paper mill effluent polluted saline sodic soil habitat of India. EJEAFChe 6(2):1775–1786
Prakash S, Nikhil K (2014) Algae as soil conditioner. Int J Eng Tech Res. ISSN: 2321-0869 2(4):68–70
Prasanna R, Joshi M, Rana A, Shivay YS, Nain L (2012) Influence of co-inoculation of bacteria-cyanobacteria on crop yield and C–N sequestration in soil under rice crop. World J Microbiol Biotechnol 28:1223–1235
Pushkareva E, Johansen JR, Elster J (2016) A review of the ecology, ecophysiology and biodiversity of microalgae in Arctic soil crusts. Polar Biol 39:2227–2240
Ravanal MC, Camus C, Buschmann AH, Gimpel J, Olivera-Nappa A, Salazar O (2019) Production of bioethanol from brown algae. In: Hosseini M (ed) Advances in feedstock conversion technologies for alternative fuels and bioproducts. Woodhead Publishing Series in Energy, Cambridge, pp 69–88
Ray JG, Thomas TB (2012) Ecology and diversity of green-algae of tropical oxicdystrustepts soils in relation to different soil parameters and vegetation. Res J Soil Biol 4:42–68
Reis A, Mendes A, Lobo-Fernandes H, Empis J, Novais J (2016) Influence of photoperiod and light intensity on the growth and production of pigments and fatty acids by Spirulina maxima. Bioresour Technol 97(15):979–985
Reisewitz SE, Estes JA, Simenstad CA (2006) Indirect food web interactions: sea otters and kelp forest fishes in the Aleutian archipelago. Oecologia 146:623–631
Remias D (2012) Cell structure and physiology of alpine snow and freshwater algae. In Snow and Ice-Related Microorganisms (pp. 109–119). Springer.
Renuka N, Prasanna R, Sood A, Bansal R, Bidyarani N, Singh R, Shivay YS, Nain L, Ahluwalia AS (2017) Wastewater grown microalgal biomass as inoculants for improving micronutrient availability in wheat. Rhizosphere 3:150–159

Renuka N, Guldhe A, Prasanna R, Singh P, Bux F (2018) Microalgae as multi-functional options in modern agriculture: current trends, prospects and challenges. Biotechnol Adv 36:1255–1273
Román JR, Roncero-Ramos B, Chamizo S, Rodríguez-Caballero E, Cantón Y (2018) Restoring soil functions by means of cyanobacteria inoculation: importance of soil conditions and species selection. Land Degrad Dev 29:3184–3193
Ronga D, Biazzi E, Parati K, Carminati D, Carminati E, Tava A (2019) Microalgal biostimulants and biofertilisers in crop productions. Agronomy 9:192
Rossi F, Mugnai G, De Philippis R (2018) Complex role of the polymeric matrix in biological soil crusts. Plant Soil 429:19–34
Rouphael Y, Colla G (2018) Synergistic biostimulatory action: designing the next generation of plant biostimulants for sustainable agriculture. Front Plant Sci 9:1655
Rouphael Y, Colla G (2020) Editorial: biostimulants in agriculture. Front Plant Sci 11:40
Rouphael Y, Cardarelli M, Lucini L, Rea E, Colla G (2017) Nutrient solution concentration affects growth, mineral composition, phenolic acids, and flavonoids in leaves of artichoke and cardoon. Hortic Environ Biotechnol 58(5):439–447
Saad MM, Eida AA, Hirt H (2020) Tailoring plant-associated microbial inoculants in agriculture: a roadmap for successful application. J Exp Bot 71:3878–3901
Samolov E, Baumann K, Büdel B, Jung P, Leinweber P, Mikhailyuk T, Karsten U, Glaser K (2020) Biodiversity of algae and cyanobacteria in biological soil crusts collected along a climatic gradient in Chile using an integrative approach. Microorganisms 8:1047
Sampath-Wiley P, Neefus CD, Jahnke LS (2008) Seasonal effects of sun exposure and emersion on intertidal seaweed physiology: fluctuations in antioxidant contents, photosynthetic pigments and photosynthetic efficiency in the red algae Porphyra umbilicalis Kützing (Rhodophyta, Bangiales). J Exp Mar Biol Ecol 361:83–91
Santos MS, Nogueira MA, Hungria M (2019) Microbial inoculants: reviewing the past, discussing the present and previewing an outstanding future for the use of beneficial bacteria in agriculture. AMB Express 9:1–13
Schulz S, Brankatschk R, Dümig A, Kögel-Knabner, Schloter M (2013) The role of microorganisms at different stages of ecosystem development for soil formation. Biogeosciences 10:3983–3996
Sharma R, Khokhar M, Jat R, Khandelwal S (2012) Role of algae and cyanobacteria in sustainable agriculture system. Wud Pecker J Agric Res 1:381–388
Siddique MT, Robinson JS (2003) Phosphorus sorption and availability in soils amended with animal manures and sewage sludge. J Environ Qual 32:1114–1121
Silvarrey Barruffa S, Calvo Carrasco D, Serrano Vergara J (2021) New insights into the antioxidant capacity of marine algal compounds. J Appl Phycol 33(2):677–687
Singh RN (1961) Role of blue – green algae in nitrogen economy of Indian agriculture. Indian Council of Agricultural Research, New Delhi
Singh NK, Dhar DW (2009) Genetic engineering, biofertilisation, soil quality and organic farming, In: Cyanobacterial reclamation of salt-affected soil. Springer, pp 243–276
Singh V, Singh DV (2015) Cyanobacteria modulated changes and its impact on bioremediation of saline-alkaline soils. Bangladesh J Bot 44(4):653–658
Singh JS, Kumar A, Singh M (2019) Cyanobacteria: a sustainable and commercial bio-resource in production of bio-fertilizer and bio-fuel from waste waters. Environ Sustain Indic 3:100008
Sivasangari Ramya S, Vijayanand N, Rathinavel S (2015) Foliar application of liquid biofertilizer of brown alga *Stoehospermum marginatum* on growth, biochemical and yield of *Solanum melongena*. Int J Recycl Org Waste Agricul 4(3):167–173
Sommer V, Karsten U, Glaser K (2020) Halophilic algal communities in biological soil crusts isolated from potash tailings pile areas. Front Ecol Evol 8:46
Song T, Mårtensson LT, Eriksson W, Rasmussen ZU (2005) Biodiversity and seasonal variation of the cyanobacterial assemblage in a rice paddy field in Fujian, China. FEMS Microbiol Ecol 54(1):131–140
Spolaore P, Joannis-Cassan C, Duran E, Isambert A (2006) Commercial applications of microalgae- review. J Biosci Bioeng 101:87–96

Sujanya S, Chandra S (2012) Effect of biological agent as suitable for chemical seed treatment on groundnut cultivation. Indian J Innov Dev 1:47–53

Sutharsan S, Nishanthi S, Srikrishnah S (2014) Effects of foliar application of seaweed (*Sargassum crassifolium*) liquid extract on the performance of *Lycopersicon esculentum* mill. in Sandy Regosol of Batticaloa District Sri Lanka. Am Eurasian J Agric Environ Sci 14(12):1386–1396

Swarnalakshmi K, Prasanna R, Kumar A, Pattnaik S, Chakravarty K, Shivay YS, Singh R, Saxena AK (2013) Evaluating the influence of novel cyanobacterial biofilmed biofertilizers on soil fertility and plant nutrition in wheat. Eur J Soil Biol 55:107–116

Tang D, Shi S, Li D, Hu C, Liu Y (2007) Physiological and biochemical responses of Scytonema javanicum (cyanobacterium) to salt stress. J Arid Environ 71(3):312–320

Teagle H, Hawkins SJ, Moore PJ, Smale DA (2017) The role of kelpspecies as biogenic habitat formers in coastal marine ecosystems. J Exp Mar Biol Ecol 492:81–98

Timmers MA, Jury CP, Vicente J, Bahr KD, Webb MK, Toonen RJ (2021) Biodiversity of coral reef cryptobiota shuffles but does not decline under the combined stressors of ocean warming and acidification. Proc Natl Acad Sci U S A 118:e2103275118

Turner NJ (2003) The ethnobotany of edible seaweed (Porphyra abbottae and related species; Rhodophyta: Bangiales) and its use by First Nations on the Pacific Coast of Canada. Can J Bot 81:283–293

Van Krimpen M, Bikker P, Van der Meer I, Van der Peet-Schwering C, Vereijken J (2013) Cultivation, processing and nutritional aspects for pigs and poultry of European protein sources as alternatives for imported soybean products. Wageningen UR Livestock Research, Lelystad

Van-der H, Bardgett MGA, Van-Straalen NM (2008) The unseen majority: soil microbes as drivers of plant diversity and productivity in terrestrial ecosystems. Ecol Lett 11(3):296–310

Vanessa NL, Wong Ram CD, Richard SBG (2009) Carbon dynamics of sodic and saline soils following gypsum and organic material additions: a laboratory incubation. Appl Soil Ecol 41:29–40

Verbon EH, Liberman LM (2016) Beneficial microbes affect endogenous mechanisms controlling root development. Trends Plant Sci 21:218–229

Vinoth M, Muruganantham P, Jeevanantham G, Hussain MJ, Balagura B, Ahamed AK (2017) Distribution of Cyanbacteria in biological soil crusts in sacred groves forest of Ariyalur and Pudukottai district, Tamil Nadu, India. Res J Life Sci Bioinform Pharm Chem Sci 3(3):215–241

Von Braun J (2007) The world food situation: new driving forces and required actions. International Food Policy Research Institute, Washington

Vonshak A, Tomaselli L (2002) Arthrospira (Spirulina): systematics and ecophysioiogy. In: Whitton BA, Potts M (eds) The ecology of cyanobacteria: their diversity in time and space. Springer, Dordrecht, pp 505–522

Wallace J (2000) Increasing agricultural water use efficiency to meet future food production. Agric Ecosyst Environ 82:105–119

Wan AHL, Davies SJ, Soler-Vila A, Fitzgerald R, Johnson MP (2019) Macroalgae as a sustainable aquafeed ingredient. Rev Aquac 11:458–492

Wang B, Li Y, Wu N, Lan CQ (2009) CO2 Bio-mitigation using microalgae. Appl Microbiol Biotechnol, 79(5):707–718

Weber B, Büdel B, Belnap J (2016) Biological soil crusts: an organizing principle in drylands. Springer, Cham

Wells PP, Craigie JS, Raven JA, Merchant SS, Helliwell KE, Smith AG, Camire ME, Susan H (2017) Algae as nutritional and functional food sources: revisiting our understanding. J Appl Phycol 29(2):949–982

Wijesinghe W, Jeon YJ (2012) Enzyme-assistant extraction (EAE) of bioactive components: a useful approach for recovery of industrially important metabolites from seaweeds: a review. Fitoterapia 83:6–12

Win TT, Barone GD, Secundo F, Fu P (2018) Algal biofertilizers and plant growth stimulants for sustainable agriculture. Ind Biotechnol 14:203–211

Wolkers H, Barbosa M, Kleinegris DMM, Bosma R, Wijffels RH (2011) Microalgae: the green gold of the future? In: Harmsen P (ed) Large-scale sustainable cultivation of microalgae for the production of bulk commodities. Propress, Wageningen, pp 1–34

Yakhin OI, Lubyanov AA, Yakhin IA, Brown PH (2017) Biostimulants in plant science: global perspective. Front Plant Sci 7:2049

Yuan YV, Westcott ND, Hu C, Kitts DD (2009) Mycosporine-like amino acid composition of the edible red alga, Palmaria palmata (dulse) harvested from the west and east coasts of Grand Manan Island, New Brunswick. Food Chem 112:321–328

Zafar-Ul-Hye M, Danish S, Abbas M, Ahmad M, Munir TM (2019) ACC deaminase producing PGPR Bacillus amyloliquefaciens and Agrobacterium fabrum along with biochar improve wheat productivity under drought stress. Agronomy 9:343

Zaidi S, Usmani S, Singh BR, Musarrat J (2008) Significance of Bacillus subtilis strains SJ-101 as a bioinoculant for concurrent plant growth promotion and nickel accumulation in Brassica Juncea. Chemosphere 64:991–997

Zhang X, Thomsen M (2019) Biomolecular composition and revenue explained by interactions between extrinsic factors and endogenous rhythms of Saccharina latissima. Mar Drugs 17:107

Zhang X, Hu Q, Sommerfeld M, Chen Y (2019) Microalgal biotechnology: A romising tool for the greening of industry and environment. Algal Research 37:51–63

Zhang Y, Lubberstedt T, Xu ML (2013) The genetic and molecular basis of plant resistance to pathogens. J Genet Genomics 40:23–35

Zolla G, Bakker MG, Badri DV, Chaparro JM, Sheflin AM (2013) Understanding root-microbiome interactions. In: de Bruijn FJ (ed) Molecular microbial ecology of the rhizosphere, Wiley Blackwell, vol 2, pp 745–754

Enzymes From Soil Algae and Their Applications

Swati Mishra Shukla, Anubhav Pradeep, and Vijayata Singh

Abstract Microalgae have a great potential for enzyme synthesis for several applications in the industry; hence, they are referred to as biofactories. In comparison with other microbial cells, microalgae have the advantage of having a photoautotrophic attribute with minimal nutritional requirements. Enzymes are widely investigated in aquatic microalgae while some of them are studied to explore their abundance in soil and their role in the environment like laccases in green algae. The group of taxa that demonstrate Laccase enzyme activity includes taxa *Tetracystis* sp., *Trebouxia* sp., *Scenedesmus* sp., *Chlamydomonas* sp., *Chlorococcum* sp., *Nostoc muscorum, and Chlorella vulgaris. Nostoc ellipsosporum* are significantly responsible for phenolic/pesticide degradation. For example, enzymes for biodiesel production and for the production high-value-added lipids (acyl-CoA diacylglycerol acyltransferase in *Chlorella ellipsoidea*) and enzymes utilized in healthcare applications (polytone synthase, l-asparaginase, Lycopene β-cyclase, β-Carotene Oxygenase, found in *Chlamydomonas reinhardtii, Chlorella vulgaris*). Enzymes like glutathione reductase, dismutase, and laccases play significant role to removing toxic metals, pesticides, and phenolic pollutants. This environmental breakdown of phenolic pollutants contributes as a fundamental process of biodegradation. The source of these enzymes is the group of algal taxa includes *Scenedesmus* sp., *Chlorella* sp., *Trebouxia* sp. This chapter is an attempt to achieve the objective of "state-of-the-art" enzymes by illustrating the methods and enzymatic regulations and focusing on the microalgal enzymes in industrial, health, or environmental applications.

S. M. Shukla (✉)
Department of Botany, Banaras Hindu University, Varanasi, India

A. Pradeep
Department of Ophthalmology, University of Illinois, Chicago, IL, USA
e-mail: anubhavp@uic.edu

V. Singh
Department of Bioscience and Pest Management, Colorado State University, Fort Collins, CO, USA

A. Mishra, A. Varma (eds.), *Soil Algae*, https://doi.org/10.1007/978-981-97-9873-5_4

Keywords Triacyl glycerides (TAGs) · Ribulose 1,5 bis-phosphate carboxylase/oxygenase (RuBisCO) · ADP glucose phosphorylase (AGPase) · Phosphatase (PO_4^{-3})

1 Introduction

Enzymes are considered the most dexterous catalysts for industrial application. Only certain fungal or bacterial enzymes (proteases cellulases, and amylases) lead the enzymatic market globally; nevertheless, photoautotrophic microalgae have remarkable metabolic paths over N, P, and C biogeochemical cycles, which are utilized as enzyme sources in applications of biomonitoring, bioremediation, and industry. Moreover, their survival strategies and adaptations in the metabolism make them ideal sources for novel enzymes. Here is an attempt to review the available knowledge about the enzymes produced from microalgae and their application along with suggesting their possible accessibility in soil algae. The potential applications of enzymes are determined in various sectors ranging from water pollution monitoring, mining effluents treatment, human health, carbon sequestration, etc.

1.1 Algae as an Architect of Evolution

Soil algae have been on this planet since the Precambrian era, dominated by the cyanobacterial colony as a pioneering community further followed by eukaryotic coccoid and filamentous algal forms. Cyanobacteria, a prokaryotic group of microalgae, constitute the majority of biomass in almost all types of habitats on the earth surface, followed by desmids and diatoms. Algae, an integral constituent of the colonial synusia, perform a major role in primary and secondary plant community succession by acting. The terrestrial ecosystem relies on two factors, abiotic and biotic; abiotic factors include water, nutrients, organic matter (carbon), and minerals, while the biotic factors include microbes where microalgae play a vital role.

1.2 Role of Soil Algae in Maintaining Soil Health

Soil microalgae play a role as the food web base in soil and provide food for other non-photosynthetic organisms in the soil. They protect associated plant root systems from the attack of disease-causing pathogens and help to make nitrogen available to the plants by converting atmospheric free N_2 to NO_3^-, preventing soil erosion by binding soil particles and as a cementing agent. They also help in increasing the water retention capacity of soil for a longer time due to hygroscopic mucilage secretion. The capsular and sheath proteoglycans produced by green and blue-green algae play a significant role in soil neogenesis. Heger et al. (2012) observed an alteration in the terrestrial diatoms community in response to diverse agricultural applications involving organic and conventional farming systems. Stanek-Tarkowska and Noga (2012) have also defined a possible correlation between farming along-side fertilization systems and alteration in organic matter content, pH, and moisture availability of soils in the presence of microalgae.

1.3 *Unnoticed Soil Inhabitant Algal Community*

The occurrences of algae in almost all terrestrial environments are found both on and beneath the soil surfaces on earth. According to their occurrences, they can be either saphophytes or cryptophytes.

Saphophytes are surface algae, most of the species of Myxophyceae found upon the surface of the soil, for example, *Mesotaenium* sp., *Botrydium* sp., *Protosiphon* sp., *Oedocladium* sp., *Vaucheria* sp., *Fritschiella* sp. While Cryptophytes are subterranean in habitat and occur inside the soil. The algal species include *Nostoc, Anabaena, Aulosira, Scytonema, Tolypothrix, Nodularia, Hapalosiphon, Euglena, Chlorella*, and *Chlamydomonas* have been reported from the agricultural lands or other terrestrial forms, where few of them also have ability to fix the atmospheric nitrogen in the soil to enrich the fertility of the agricultural fields. The eukaryotic algal flora on mountains represents monoid and coccoid forms of division, Chlorophyta, Xanthophyta, and a filamentous genus *Klebsormidium* from Streptophyta (Tschaikner et al. 2007). The soil algal flora contains members of the Cyanophyta, Chlorophphyta, Euglenophyta, Chrysophyta, and Rhodophyta. Approximately more than 38 prokaryotic genera and 147 genera of eukaryotic algae comprise terrestrial species. Consequently, bryophytes and vascular plants' evolution made the soil less exposed and limited the light availability to the surface, resulting in the declination of the ecological importance of soil algae. Despite being ancient algal forms, their study is relatively recent and certainly less investigated in comparison with aquatic studies (Table 1).

1.4 *Microalgal Enzymes: Economic Importance*

What makes algal enzymes important and useful for a commercial applications? According to De Clerk et al. (2013), approximately 72,500 existing algal species represent different groups. But only the group of taxa capable of high biomass production and higher metabolic growth are considered as commercially and economically significant. The high biomass production is dependent on suitable environmental conditions, and biocatalysts catalyze all biochemical reactions needed to drive the cell metabolism. According to Faber (2011), enzymes are defined as "Biocatalysts," the organic macromolecules that catalyze biological reactions. Enzymes are

Table 1 The graph shows the presence and absence of enzymes in aquatic and soil algal forms

	Laccase	Carbonic anhydrase	Hydrogenases	Lipoxygenases	Nitrilase	Nitrogenase	Phosphatase	Thiolase
Aquatic- algae	–	+	+	+	+	+	+	+
Soil-algae	+	*	*	–	–, *	+	*	–, *

Here + represents “presence,” – is for “absence,” and * represents “may be found” but not explored and reported yet described to have a specific substrate-binding ability and are widely applied in the manufacturing of pharmaceuticals, food processing industry, biofuel production, detergent industry, and paper production industries to recuperate, balance, and regulate the production (Table 2). The target enzymes in microalgae for industrial purposes mostly include amylases, laccases, galactosidases, proteases, lipases, cellulases, phytases, and certain antioxidant enzymes like superoxide dismutases, peroxidases, and carbonic anhydrases. Along with other enzymes which linked with carbon concentration and carbohydrate accumulation Fig. 1.

For example, in the hydrolysis process to break chemical bonds, the enzymes that catalyze the reaction are hydrolases. Hydrolases that are significant for industrial purposes are cellulases exploited in the production of biofuel (Cao and Huimin 2002), amylases in syrup production (Nigam 2013), besides galactosidases,

Table 2 Soil-microalgal enzymes show potential application in industrial or biotechnological uses

Enzymes	Sources	Actual/potential uses	References
Laccases	*Tetracystis aeria*, *C. moewusii*, *Chlorococcum* sp.	Breakdown of environmental pollutants like triclosan and 17α-ethinylestradiol	Otto et al. (2015)
Carbonic anhydrase	*Chlamydomonas reindhardtii*	Used in the protection of transferred organs	Bejaoui et al. (2015)
		In the CO_2 sequestration process	
Hydrogenases	*C. reinhardtii, Spirulina* sp., *Anabaena* sp., *Chlorella vulgaris, C. angutae*	In the synthesis of bio-hydrogen	Contreras et al. (2008)
Lipoxygenases	Aquatic algae	In the production of oxylipins (for biological defense mechanism)	Chen et al. (2015) and Schnitzler et al. (2001)
Nitrilases	*Chlorella* sp., *Scenedesmus* sp., *Acidovorax facilis*	In nitriles degradation process, cyanide detoxification, bio-treatment process in gold mine effluents	Sohoni et al. (2015)
Nitrogenase	*Gloethece* sp., *Plectonema* sp., *Pseudoanabaena* sp., *Anabaena* sp., *Calothrix* sp., *Nodularia* sp., *Nostoc* sp., *Stigonema* sp., *Richelia intracellularis*	In biological nitrogen fixation process, which has significant impact on agricultural practices	Bergman et al. (2012)
Phosphatases	*Scenedesmus* sp., *Chlorella vulgaris*, *Anacystis* sp.	As a bioindicator of phosphorous, heavy metals and mannitol manufacturing industries	Litchman and Nguyen (2008) and Tekaya et al. (2013)
Thiolases	*Bumilleriopsis* sp., Euglena sp.	In sterols, biobutanol, isoprenoid synthesis	Winkler et al. (1988)

Adapted and modified from Hernández et al. (2017)

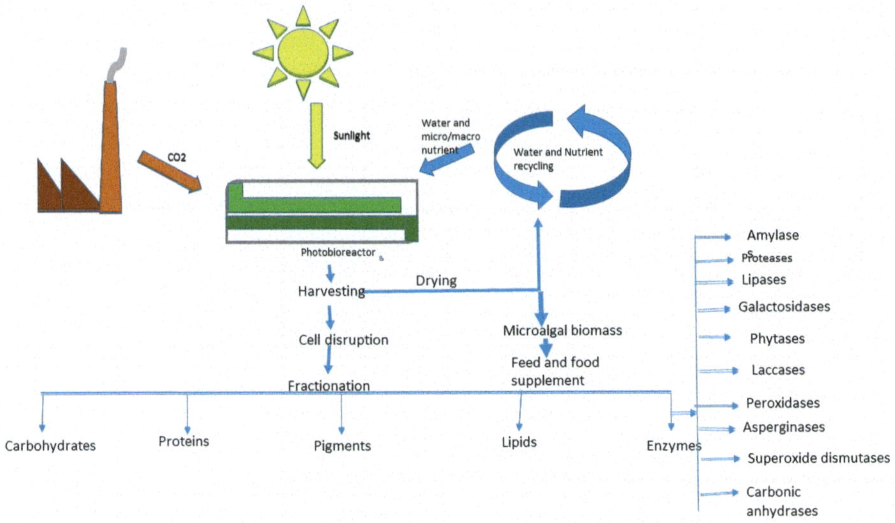

Fig. 1 Microalgae as Biofactory: Microalgae cultivation in photobioreactor, utilizing CO_2 emission from factories, sunlight and water/nutrient from recycling process. Cells processed through harvesting, cell disruption and fractionation steps and after separating biomass the other products including enzymes extracted at industrial level

phytases, and papain, widely applied in processing of food for industrial purposes (Fernandes 2010), and other hydrolases exploited in several pharmaceutical applications (Vellard 2003).

However, microbial enzymes dominate the global market for a while; still, the potential application of biocatalysts for many industrial purposes is immeasurable. Different microalgal enzymes that have a diverse range of catalytic properties are preferably considered proficient catalysts compared with chemical catalysts and are in high demand in many industrial applications (Choi et al. 2015).

2 Microalgal Enzymes Production Methods

2.1 *Production in Biorefineries*

At the commercial level, achieving the demand and profitable large-scale production of algae-derived products predominantly requires optimization in culture and processing technologies.

Estimation of algal biomass output found that only 20–40% of it is being used to generate biological fuels, while the rest of 60–80% comprises great-value coproducts such as enzymes, algal meal, nutraceuticals, organic pigments, and omega-3 fatty acids (Subhadra and Edwards 2011) (Fig. 1).

The hydrothermal technique of liquefaction induces the industrial production of oil but at the cost of a rise in temperature (250–350 °C) and also an increase in pressure (1500–3000 psi), which cause denaturation, minor destruction of biomass's nutrient value, and inhibition of its recycling process (Frank et al. 2013).

To address this problem, Samorì et al. (2013) used a switchable solvent (*N*, *N*-dimethyl cyclohexyl-amine) to extract lipids straight from the liquid cultures of algae, with encouraging results and low toxicity. Alternative solvents such as α-pinene, p-cymene, and d-limonene are presently being investigated (Dejoye et al. 2013).

In general, raw algal enzymes are extracted using buffers to provide protection from denaturation and maintain their enzymatic activity. Davison and Davison's (1987) buffers offer successful extraction of enzymes involved in photosynthesis and brown algae-originated kinase, while Gerard and Driscoll's (1996) buffer was found appropriate for isolation of RuBisCo from green microalgae (Bischof et al. 2002). The PO_4^{-3} buffer (pH 7) supplemented with non-ionic detergents (Brij 99 0.1%) was recommended for the extraction of lipoxygenase from cyanobacteria (Beneytout et al. 1989).

Additionally, the French pressure extrusion technique to disrupt unicellular algae cells without denaturation of enzymes was used by Duggan and Anderson (1975). This method has been frequently applied for the separation of several essential enzymes in cyanobacteria.

Ultrasonic disruption can break down algal cells while leaving enzyme activity unaffected; this method was exploited to isolate specific enzymes from cyanobacteria (Smith et al. 1967).

Another enzyme separation method called immunoprecipitation, for example, superoxide dismutase from blue-green algae, was isolated using antibodies and precipitated with ammonium sulfate at a 33% concentration (Asada et al. 1975). The cell disruption method for enzyme isolation is also effective for lipid extraction and thus is implemented effortlessly into a biorefinery setup.

2.2 *Technology of Genetic Overexpression and Transgenics*

With the emergence of genetic engineering techniques, several tools offer different possible ways to enhance the production of enzymes or metabolites. For microalgal cells, the genetic transformation technique which involves spinning with glass beads or silicon carbide whiskers, electroporation method, biolistic microparticle bombardment method (Purton et al. 2013), and *A. tumefaciens*-mediated method have been applied (Radakovits et al. 2010).

Transgenic or overexpression of acyl-ACP thioesterases induces fatty acids production (Radakovits et al. 2010) by determining fatty acids chain length in biodiesel production. Another crucial enzyme that needs to be manipulated is AGPase, which plays a main role in starch production reaction (Zabawinski et al. 2001).

The genetic manipulation of transcription factors for target enzymes, aimed at overexpression proved as a highly promising technique for avoiding the restrictions of biochemical stress that, needs to be thoroughly researched (Courchesne et al. 2009). The genome of algal viruses (phaeovirus) represents a source of strong promoters that might be used to yield heterologous forms of enzymes in algae (Walker et al. 2005). For example, the GmDof4 gene hetero-expressed from soybean considerably increases the amount of lipid without affecting the growth rate of *C. ellipsoidea* culture under mixotrophic conditions (Zhang et al. 2014). Thus, genetic manipulation and overexpression of genes for vital enzymes involving metabolic engineering in algae could be a quick and feasible approach for bioactive compounds production or as a source of biofuels.

3 Enzymatic Regulation in Algae

The components that affect the growth of algae directly influence the enzymatic regulation mechanisms. Alterations in pH, quality of light, temperature range, nutrient (nitrogen, phosphorus, sulfur) availability, salinity, and the presence of xenobiotics directly influence the metabolic pathway resulting in altered enzyme production. The factors that influence the production and accumulation of enzymes in the cell are mentioned below.

(a) *Stimulus of Light*:
Light quality affects the algal growth and development of metabolites for biomass and lipids production in *Chlorella vulgaris*. Blue light stimulus is comparatively stronger than white light (Atta et al. 2013) and the antenna complex LHCSR, a protective protein trigger under excess of light energy (reported in *Chlamydomonas reinhardtii*) by Peers et al. 2009. Proteins help in protection function in association with the thylakoid-associated Ser-Thr kinase Stt7 to balance the light absorption capability of PS II and PS I (Bellafiore et al. 2005). This results in the photosynthetic enzymes regulation and biomass development, alongside the induction of stress-resistant enzymes. As studies suggest in *Euglena* sp., *Chlamydomonas* sp., *Synechococcus* sp., and *Synechocystis* sp., enzymes like N-glyceraldehyde, 3-phosphate dehydrogenase, fructose 1,6 bisphosphatase, and ribulose 5-phosphate kinase are resistant to H_2O_2 and hydroxylamine (Takeda et al. 1995) offer role in biotechnological benefits.

(b) *Stimulus of CO_2 Availability*:
Variations in photosynthetic enzymes have been noticed as light quality or CO_2 concentration changes. The exposure of low concentrations of CO_2 to *C. reinhardtii* activates the formation of pyrenoid structures, causing high levels of 1,5-bisphosphate carboxylase/oxygenase to accumulate for CO_2 fixation (Kuchitsu et al. 1991).

(c) *Stimulus as Nutrient Availability/Starvation*:
According to a previous study, lipid buildup is triggered by a lack of N and P because algal enzymatic pools are influenced by nutrient availability/deficiency.
For example, phosphorus-rich ribosomes (Ågren 2004) synthesize enzymes; hence, phosphate restriction has a significant effect on enzyme manufacturing (Courchesne et al. 2009). In consequence, it was noticed in *Anabaena* sp. that phosphorus starvation negatively affects the production of enzymes, while nitrogen starvation might result in the stimulation of proteases (Berges and Falkowski 1998) and nitrogenases (heterocystous or non-heterocystous) for N_2 fixation and H_2 generation studied (Masukawa et al. 2014). Proteases enhance self-protein digestion as a source of carbon and energy for triacylglycerol anabolism, and they favor activation of diacylglycerol: acyl CoA acyltransferase gene transcript under nitrogen-deficient conditions (Msanne et al. 2012). During sulfur deficiency and anoxic circumstances, chlorophytes switch to fermentation and produce [Fe]-hydrogenases to serve to deliver energy for anaerobic reactions (Winkler et al. 2002).

(d) *Stimulus in the Form of Herbicides' Presence*:
Algae can sustain and degrade some herbicides with their ability to have certain specific enzymes; *Chlorella* sp. can resist and sustain itself in the presence of metflurazon, due to the N demethylase activity of cytochrome P450 monooxygenase (Thies et al. 1996).
Reports mentioned that herbicides as stimulants also induce the development of antioxidants in *Chlorella* sp. such as SOD, peroxidase, and catalase (Romero et al. 2011), while in *Scenedesmus* sp., ascorbate peroxidase, glutathione reductase, and glutathione S transferase (Geoffroy et al. 2002).

4 Target Biocatalysts for Industrial Production

4.1 Laccases

Laccases belong to the group of oxidoreductases that catalyze the oxidation reaction of a broad range of substrates. Laccases are naturally involved in biosynthesis as well as in degradation of the lignin. To date, laccases and laccase-like multi-Cu oxidases are reported in prokaryotes, lichens, fungi, insects, and higher plants (Lisov et al. 2007). However, a green alga *Tetracystis* that inhabits soils worldwide (Ettl and Gärtner 1995) is a perfect model to study phenoloxidases of algae from the perspective of soil environments. Laccases isolation from lichens species harboring green algae also put forward the opportunity for algal and fungal symbionts to have potential accountability for the production of the enzyme.

Applications

- The potential functions of *Tetracystis aeria* involve the breakdown of toxic phenolic substances in soil or in defense actions against pathogens (Claus 2004).

- Brown and Bold (1964) studied the metabolic action of *T. aeria* metabolizing starch extracellularly to grow heterotrophically in the presence of glucose or fructose as a carbon source.

4.2 Carbonic Anhydrase

CAs are prevalent metal enzymes (Zn^{2+} containing active site) discovered in prokaryotes as well as eukaryotes existing in various forms: α -CAs (algae, and green plant cytoplasm); β -CAs (primarily in algae, and chloroplasts of both mono and dicotyledons); γ-CAs (archaea), and δ- and ζ-CAs (marine diatoms). These isoforms play a dynamic role in the biological process of acid-base regulation, respiration, and CO_2 and HCO_3-transportation (Chegwidden et al. 2000).

Applications

- These enzymes were discovered in the periplasmic region of *Chlamydomonas reindhardtii* (Fukuzawa et al. 1990) and have been used in human health as a liver defender and to improve the preservation of transferred organs (Bejaoui et al. 2015).
- The study of this enzyme utilized in CO_2 sequestration and involvement in organs transfer could provide a new byproduct as well as an environmental advantage in the context of climate alteration.

4.3 Hydrogenases

Hydrogenases (H2ases) catalyze reversible molecular hydrogen oxygenation reactions. This is a significant enzyme in photobiological H2 generation and plays a key role in microbial metabolism (Vignais et al. 2001). Based on the catalytic core of the enzyme, hydrogenases are divided into three categories: [Fe]-H2ases (Fe-containing) [NiFe]-H2ases (Ni & Fe containing), and metal-free H2ases (Gutekunst et al. 2006).

Applications

- Compared to traditional physicochemical approaches, these enzymes used to assist in biological H_2 synthesis have environmental benefits. Biological--photolysis reaction involves the conversion of H_2O in the presence of sunlight into H_2 and O_2 in microalgae and/or cyanobacteria during photosynthetic processes.
- *Chlamydomonas* sp., studied as a model in bio-hydrogen generation research, yielded satisfactory results. There are a few taxa utilized for the biological photoproduction of hydrogen *Spirulina* sp. and *Anabaena* sp. (Dutta et al. 2005), *Chlorella vulgaris*, *Chlamydomonas angusta*, and *C. reinhardtii.*

4.4 Lipoxygenases

Lipoxygenase (LOXs) catalyzes deoxygenation reactions that involve conversion to cis, trans-conjugated monohydroperoxides from unsaturated fatty acids.

Applications

- LOX is the first and most important enzyme in the oxylipin pathway; it is present in different organisms including red algae *Pyropia haitanensis* known as the richest life form for oxylipin source, and might play their role in defense or immunity.
- In the diatoms, the toxicity of aldehydic oxylipins found against bacteria, fungi, and even invertebrates (Adolph et al. 2004) suggests that LOXs assist in the induction of oxylipins as a novel agent in the area of biocontrol.

4.5 Nitrilase

Nitrilases are adaptable enzymes that perform nitrile metabolization and may be utilized instead of strong acids or base catalysts for nitrile hydrolysis (Djambaski et al. 2009). Nitrilase is basically carboxylic acid where –OH has been replaced with cyanide and has the general structural formula R–CN.

Applications

- Large-scale synthesis of nitrile compounds, predominantly utilized as solvents, precursor material for pharmaceuticals manufacturing, herbicides production, and other industrially important chemicals (Gupta et al. 2010).
- Nitrilases convert cyanide to β-cyanoalanine and γ-glutamyl-β-cyanoalanine, which can be used as a C and N source in *Chlorella* sp. (Gurbuz et al. 2004). For example, *Scenedesmus* sp. can detoxify up to 400 mgL^{-1} cyanide in 68 hours (Gurbuz et al. 2004). *Scenedesmus obliquus* is suggested as an indicator in gold mine effluents for biotreatment because it is often employed in mining and is present in significant amounts in wastewater (Gurbuz et al. 2009). Nitrilase--producing algae proved to be a good choice for bioremediation; however, further investigation is required to improve nitrilase production in algae.

4.6 Nitrogenase

Nitrogenase is an enzyme complex made up of two parts: dinitrogenase (Mo-Fe protein) and dinitrogenase reductase (Fe protein); dinitrogenase is an α2β2 tetramer coded by the nifD and nifK genes, and dinitrogenase reductase is a dimer coded by the nifH gene (Bergman et al. 1997). The dinitrogenase reductase takes electrons from ferredoxin/flavodoxin and transfers them to dinitrogenase, which uses the reducing power to convert N_2 to NH_4^+ in an ATP-dependent process.

In *Gloethece* sp., *Plectonema* sp., and *Pseudanabaena* sp. nifH, nifD, and nifK are grouped in one operon (Soto-Urzúa and Baca 2001). When nitrogen supplies (NH^{4+}, NO^{3-}) are scarce, few cyanobacterial strains can develop particular cell forms, called heterocysts, in response to environmental indications (Whitton and Potts 2012). Nitrogenase synthesis and heterocyst generation are suppressed when nitrogen sources are accessible (Graham et al. 2009).

Heterocytes are specific structures reformed in cyanobacteria to perform nitrogen fixation that provides an anaerobic environment for nitrogenase to reduce N_2. Some cyanobacteria, on the other hand, lack a heterocyst and rely on different systems of oxygen protection to enable nitrogenase catalysis. In these systems, the photosynthetic, respiratory, and N_2 fixation processes may be separated spatially and temporally (Berman-Frank et al. 2003). Some N_2-fixing species, such as *Anabaena*, *Aphanothece*, *Gloeothece*, *Plectonema*, *Oscillatoria*, and *Nodularia*, have a distinct spatial separation between N_2 fixation and photosynthesis. To regulate the expression of photosynthetic and N_2 fixation genes for temporal differentiation, in a circadian rhythm, the first expression of (psaA, psbA genes) allows several hours of difference from the genes' nifHDK expression.

Applications

- In many natural and manmade environments, cyanobacterial nitrogen fixation has become quite important in terms of ecology. Globally, terrestrial nitrogen fixation is the estimation of 100–200 Tg per year primarily due to the great nitrogen fixation power of cyanobacteria.
- This nitrogen fixation is extremely beneficial to ecosystems and human activities, such as 5 g of nitrogen per m2 per year in rice crops, which can contribute 70–110 $kgNha^{-1}$ per culture cycle.

4.7 *Phosphatase*

Phosphatases are a family of enzymes that make available inorganic forms of phosphorous from organic sources, allowing organisms to use it. Alkaline phosphatases are reported in marine and green microalgae owing to the alkaline water (pH = 8.3); they are not free enzymes localized in the exterior part of the alga (cell wall or periplasmic space; Hernández et al. 1994). Due to their sensitivity to phosphorous, alkaline phosphatase was reported in the aquatic algal forms *Gelidium*, *Cladophora*, *Polysiphonia*, *Stypocaulon*, *Desmodesmus*, and *Astrionella* proved to be beneficial for environmental monitoring (Hernández et al. 2002).

Applications

- Alkaline phosphatase isolated from *Scenedesmus* sp. could be a good predictor of the internal P state inside the cell (Litchman and Nguyen 2008).

- Phosphatases can be used to monitor water quality studied in *Chlorella vulgaris* as a biosensor for heavy metals (Cd^{2+} and Zn^{2+}) which also plays a critical role in the continuous monitoring of pollution control in water bodies (Durrieu and Tran-Minh 2002).
- When *Scenedesmus* sp. and *Aphanothece* sp. were exposed to Ni^{2+}, Zn^{2+}, and Cd^{2+}, their alkaline phosphatases decreased (Awasthi 2012). For a rapid detection of heavy metal pollution in different water bodies, the sensitivity of AP to heavy metals in combination with the immobilized algal cells proved beneficial.
- Phosphatase converts mannitol to active forms such as mannitol 1-phosphate and mannitol. Mannitol has medical and industrial implications and in algae works as an osmoprotectant (Dittami et al. 2011). It is used as a neuroprotective substance in the treatment of Parkinson's disease (Shaltiel-Karyo et al. 2013), as well as to reduce elevated intracranial pressure and protect the kidneys during various procedures (Shawkat et al. 2012).

Due to their specific role in Aquatic phosphorous cycle, algal phosphatases are utilized as heavy metal biosensors also used to make mannitol from fructose in the future.

4.8 Thiolase

Thiolases are enzymes found in both prokaryotic and eukaryotic organisms, including algae. This enzyme family consists of two types: catabolic and metabolic thiolase. Both of these enzymes catalyze to produce acetoacetyl-CoA.

The substrate spectrum of catabolic thiolase is diverse, including larger substrates like 3-ketodecanoyl-CoA, which is involved in lipid metabolism. Metabolic thiolase, on the other hand, is essential for converting the universal metabolic precursor acetyl-CoA into other biotechnologically important carbon skeletons that exclusively need acetyl-CoA and acetoacetyl-CoA as substrates.

Applications

- The mevalonic acid pathway uses thiolase in the production of various sterols (Lopes et al. 2013). For example, ergosterol, manufactured from acetyl CoA, is the most common sterol found in fungi and algae (Miller et al. 2012). The significance of this enzyme as a repressive influence on malignant growth and tumors is evident (Kobori et al. 2007).
- Thiolase is also a critical enzyme in the manufacture of carotenoids studied in *Cyanophora*, *Galdieria* (extremophiles), *Cyanidium* (hot soil algae), and *Euglena*, among other organisms; other Chlorophyceae algae have lost this metabolic pathway (Lohr et al. 2012). Isoprenoid metabolic modification in algae is gaining attraction because of its potential industrial applications (Tippmann et al. 2013).

- PHB works as an essential precursor of biodegradable polymers and biobutanol (Reiße et al. 2014) can be produced using acetoacetyl-CoA by thiolase II and 3-ketothiolase.
- Algal thiolase is essential because it is an important enzyme in the manufacture of isoprenoids and sterols, utilized in the production of biofuels and bioplastics using an enzymatic cascade.

5 Applications of Enzymes in Different Industrial Sectors

5.1 Enzymes for High-Value-Added Lipids and Biodiesel Production

An enormous amount of lipids is stored in microalgal cells (Chisti 2007) in the form of polyunsaturated fatty acids (PUFA) and triglycerides (TAGs). These two forms are widely acknowledged for their biotechnological application perspective, predominantly for the production of nutraceuticals and biodiesel (Bellou et al. 2014). TAGs are deposited and are located in the cytosol as lipid droplets are promptly utilized to manufacture biodiesel by following transesterification reactions catalyzed by acid or base, Fig. 2. However, PUFAs have been utilized and proved

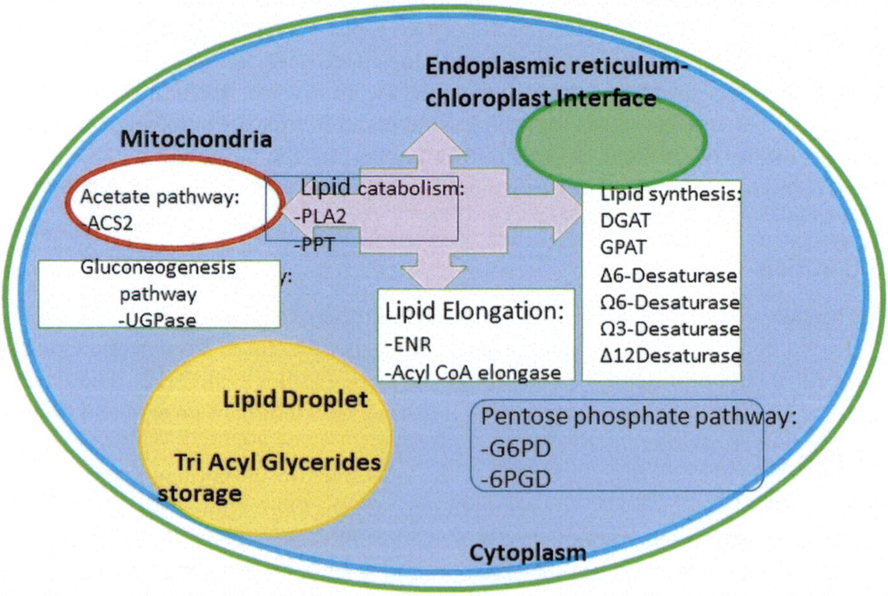

Fig. 2 Schematic representation of key enzymes engineered for TAGs and PUFAs in microalgal cell for the production of high value-added lipids. Enzymes are localized in subcellular compartments as single lipid droplet where TAGs are accumulated is shown here in the cell

beneficial in health sectors (Wells et al. 2017; Caporgno and Mathys 2018), for example, eicosapentaenoic acid (EPA) and Ω-3 fatty acids such as docosahexaenoic acid (DHA).

TAG biosynthetic pathway insights provided that acyl-CoA diacylglycerol acyltransferase (DGAT) is involved in the final reaction of TAGs production and has been an extensively studied enzyme (Xu et al. 2018). This enzyme exists in three different forms (DGATs 1–2–3) catalyzing the acyl-CoA-dependent formation of TAGs using n-1,2-diacylglycerol (DAG) as precursors (Lung and Weselake 2006). Those soil algae that have been studied and utilized in biodiesel production include *Chlorella ellipsoidea* by Guo et al. (2017), oleaginous microalgae *Neochloris oleoabundans* (Klaitong et al. 2017), and *Nostoc oculata.*

5.2 Enzymes for Healthcare Application

Other microalgal enzymes, which are involved in carotenoid synthesis, have been widely considered valuable in the manufacturing of nutraceutical and cosmeceutical compounds. The frequently studied enzymes are β-carotene oxygenase, lycopene--β-cyclase, phytoene synthase, phytoene desaturase, β-carotene hydroxylase, and zeaxanthin epoxidase (Saini et al. 2019) involved in carotenoid synthesis analyzed in *C. reinhardtii* (Table 2). Recently, CRISPR/Cas9 (clustered regularly interspaced short palindromic repeats/CRISPR-associated protein 9) has come to the limelight and is utilized for specific and highly proficient "knock-out" of vital genes (Cong et al. 2013), for example, CRISPR/Cas9 mediated knock-out of the zeaxanthin epoxidase (ZEP) gene in *C. reinhardtii* (Baek et al. 2018). Alternative studies have been performed focusing on the enzymes that participate in oxylipin synthesis (De los Reyes et al. 2014; Lauritano et al. 2016; Ávila-Román et al. 2018). Oxylipins are secondary metabolites and they have proved their application as antipredator and anti-cancerous substances (Table 3).

5.3 Enzymes for Bioremediation

Bioremediation is the degradation and transformation of toxic pollutants into less dangerous metabolites by the application of microorganisms and their enzymes. The potential of microalgae to bloom in environments that are rich in nutrients (e.g., eutrophic environments) has the ability to bio-sequestrate heavy metal ions, and represents an ideal candidate organism for bioremediation approaches (Mathimani and Pugazhendhi 2019). A novel approach to enzymatic bioremediation involves the direct use of purified or partially purified enzymes from microalgae to detoxify a specific pollutant or toxicant (Sharma et al. 2018). A recent study focused on Cr (VI) reduction in *C. vulgaris* (Yen et al. 2017) has been reported to involve both a biological route, through the putative enzyme chromium reductase, and a

Table 3 Enzymes from soil microalgae utilized in healthcare applications

S. no.	Enzyme	Microalgae	Habitat	Class	Main result	Reference
1.	β-Carotene Oxygenase	*Chlorella* sp.	Soil	Trebouxiophyceae	Enhanced astaxanthin production under combined high light stress	
2.	I-asparaginase	*Chlamydomonas* sp.	Soil	Chlorophyceae	An enzyme purified and tested	
3.	I-asparaginase	*Chlorella vulgaris*	Soil	Chlorophyceae	Searching for new I-asparaginase. Screening of 40 microalgal isolates	Mahdikhanlou and Ebrahimnezhad (2014)
4.	Lycopene β-cyclase	*Chlamydomonas reinhardtii*	Soil	Chlorophyceae	Enhanced gene expression under high light stress	Couso et al. (2012)
5.	Lipolytic acid I					
6.	Polytone synthase	*Chlamydomonas reinhardtii*	Soil	Chlorophyceae	Increase in violaxanthin and lutein production	Couso et al. (2012)

Adapted and modified from Vingiani et al. (2019)

non-biological route. Using the scavenger molecule glutathione (GSH), several strains of microalgae are reported to be capable of achieving Cr (IV) removal from water bodies like *Scenedesmus* sp. and *Chlorella* species (Pradhan et al. 2019). Also, lead (Pb) contamination removal has been reported in two lichenic microalgal strains from the *Trebouxia* genus (TR1 and TR9) (del Álvarez et al. 2012) and Cu, cd, and As contamination removal from *Anabaena* sp.

6 Future Perspectives

As an excellent source for biorefinery and high biomass producers, algae also can deliver bioenergetics and high-value byproducts like pigments, toxins, and enzymes.

This chapter summarizes the facts for different algal enzymes demonstrate invigorating properties for industry, biomonitoring, and bioremediation which claims a prospect for biotechnological development. Also suggesting a possibility for biocatalyst production is the bio-prospection of enzymes from algae either in photoreactors or through heterologous expression in bacteria.

The best thing about algal enzymes is the advantage of algae being produced without using C or N sources and utilizing small quantities of toxic compounds. Additionally, they inhabit countless environmental conditions with a wide diversity of forms and sizes (Singh and Singh 2015). Despite the potential part of algae as a source of novel enzymes, soil algae have not held much attention in the current literature. Also, due to the reliability of proper taxonomic treatment, standard methods of cultivation and discrete interpretation of physiological attributes, plant mass characteristics, and fluctuating morphological properties of axenic clones became more complex to explore eukaryotic soil algae.

To emphasize the importance of soil algae, their occurrences in abundance from free-living, epiphytic to symbiotic forms offer plenty of opportunities for future research. However, efforts are continuing to understand metabolic insights, and enzymes exert changes in the enzymatic pool to alter the metabolic pathways in metabolic modification, resulting in the accumulation or decomposition of metabolites to maintain cell vitality.

In algae, the metabolic reaction of the cell to disparity in the environmental factors is vital; any alteration in growth components can manifest the cell metabolism and lead to accelerated enzyme production. The known facts about enzymatic regulations in response to different stimuli are very important to an algal enzyme co--production in biorefinery.

Heavy metals are supposed to induce the production of antioxidant enzymes in microalgae (Arunakumara and Xuecheng 2008), which could offer an immense role in bioremediation and antioxidant production at the industrial level. Thus, the recommended application of algal enzymes in different industrial activities and environmental protection is an impending prospect.

Typically, enzyme characterization and extraction are widely reported in the majority of aquatic algal forms, while soil microalgae are inadequate for bioproduct extraction at the commercial level (Table 1). Including the above-mentioned roles, soil algae cultivation provides advantages and remedies over desertification, water pollution control, and reclamation of "Usar" land (Singh 1950). Reports suggest that extreme habitat microalgae enzymes would be valuable over the chemical enzymes, as these microalgal forms would possess the capability to biosynthesize special enzymes to synthesize complex polymers and perform detoxification events for endurance in an adversary environment. Enzymes with these exceptional features including tolerance to a wide range of pH, thermophilic nature, stability of enzyme's function over a range of temperatures, and other severe conditions of reaction (Nigam 2013) are specifically endorsed for commercial and industrial purposes.

Misra et al. (2016) developed the first comprehensive database named the "database of enzymes of microalgal biofuel feedstock." This database is focused on enzymes involved in the lipid accumulation in 15 different algal species whose whole genomes were sequenced and available in the database. This web domain provides integration for querying enzymes involved in lipid biosynthesis and analysis related to metabolic pathways and is freely accessible at the website http://bbprof.immt.res.in/embf. This available information ought to be endorsed as an assistance tool to develop genetically modified high lipid-yielding algal strains with higher growth rates in soil systems that have not been studied and reported yet.

Acknowledgment First of all, my heartfelt gratitude to the Almighty who gave me strength to write this chapter and also thankful to my kids Aaroh Shukla and Pradipti Shukla who helped me a lot in their own unique way.

References

Adolph S, Bach S, Blondel M, Cueff A, Moreau M, Pohnert G, Poulet SA, Wichard T, Zuccaro A (2004) Cytotoxicity of diatom-derived oxylipins in organisms longing to different phyla. J Exp Biol 207:2935–2946

Ågren GI (2004) The C: N: P stoichiometry of autotrophs – theory and observations. Ecol Lett 7:185–191

Arunakumara KKIU, Xuecheng Z (2008) Heavy metal bioaccumulation and toxicity with special reference to microalgae. J Ocean Univ China 7:60–64

Asada K, Yoshikawa K, Takahashi M, Maeda Y, Enmanji K (1975) Superoxide dismutases from a blue-green alga, Plectonema boryanum. J Biol Chem 250(8):2801–7. PMID: 164453

Atta M, Idris A, Bukhari A, Wahidin S (2013) The intensity of blue LED light: a potential stimulus for biomass and lipid content in freshwater microalgae *Chlorella vulgaris*. Bioresour Technol 148:373–378

Ávila-Román J, Talero E, de los Reyes C, García-Mauriño S, Motilva V (2018) Microalgae-derived oxylipins decrease inflammatory mediators by regulating the subcellular location of NFB and PPAR. Pharmacol Res 128:220–230

Awasthi M (2012) Relevance of alkaline phosphatase activity of immobilized green algae and cyanobacteria for heavy metals toxicity monitoring. J Mater Environ Sci 3:446–451

Baek K, Yu J, Jeong J, Sim SJ, Bae S, Jin E (2018) Photoautotrophic production of macular pigment in a Chlamydomonas reinhardtii strain generated by using DNA-free CRISPR-Cas9 RNP-mediated mutagenesis. Biotechnol Bioeng 115:719–728

Bejaoui M, Sokol H, Marteau P (2015) Targeting the Microbiome in Inflammatory Bowel Disease: Critical Evaluation of Current Concepts and Moving to New Horizons. Dig Dis 33 Suppl 1:105–112. https://doi.org/10.1159/000437104. Epub (2015). PMID: 26366577

Bellafiore S, Barneche F, Peltier G, Rochaix JD (2005) State transitions and light adaptations require chloroplast thylakoid protein kinase STN7. Nature 433:892–895

Bellou S, Baeshen MN, Elazzazy AM, Aggeli D, Sayegh F, Aggelis G (2014) Microalgal lipids biochemistry and biotechnological perspectives. Biotechnol Adv 32:1476–1493

Beneytout J, Andrianarison R, Rakotoarisoa Z, Tixier M (1989) Properties of a lipoxygenase in green algae (Oscillatoria sp.). Plant Physiol 91:367–372

Berges JA, Falkowski PG (1998) Physiological stress and cell death in marine phytoplankton: induction of proteases in response to nitrogen of light limitation. Limnol Oceanogr 43:129–135

Bergman B, Gallon JR, Rai AN, Stal LJ (1997) N_2 fixation by non-heterocystous cyanobacteria. FEMS Microbiol Rev 19:139–185

Bergman B, Sandh G, Lin S, Larsson J, Carpenter EJ (2012) *Trichodesmium* – a widespread marine cyanobacterium with unusual nitrogen fixation properties. FEMS Microbiol 37:286–302

Berman-Frank I, Lundgren P, Falkowski P (2003) Nitrogen fixation and photosynthetic oxygen evolution in cyanobacteria. Res Microbiol 154:157–164

Bischof K, Kräbs G, Wiencke C, Hanelt D (2002) Solar ultraviolet radiation affects the activity of ribulose-1,5-biphosphate carboxylase-oxygenase and the composition of photosynthetic and xanthophyll cycle pigments in the intertidal green alga Ulva lactuca L. Planta 215:502–509

Brown RM, Bold HC (1964) Comparative studies of the algal genera Tetracystis and Chlorococcum, vol 6417. University of Texas Publication, pp 1–213

Chen H, Zhu Z, Chen J, Yang R, Luo Q, Xu J, Shan H, Yan X (2015) A multifunctional lipoxygenase from Pyropia haitanensis—The cloned and functioned complex eukaryotic algae oxylipin pathway enzyme. Algal Res 12:316–327

Cao Y, Huimin T (2002) Effects of cellulase on the modification of cellulose. Carbohydr Res 337:1291–1296

Caporgno MP, Mathys A (2018) Trends in microalgae incorporation into innovative food products with potential health benefits. Front Nutr 5:1–10

Chegwidden WR, Dodgson SJ, Spencer IM (2000) The roles of carbonic anhydrase in metabolism, cell growth and cancer in animals. In: Chegwidden WR, Carter ND, Edwards YH (eds) The carbonic anhydrases: new horizons. Birkhauser Verlag, Basel, pp 343–363

Chisti Y (2007) Biodiesel from microalgae. Biotechnol Adv 25(3):294–306

Choi J-M, Han SS, Kim J-S (2015) Industrial applications of enzyme biocatalysis: current status and future aspects. Biotechnol Adv 33:1443–1454

Claus H (2004) Laccases: structure, reactions, distribution. Micron 35:93–96

Cong L, Ran FA, Cox D, Lin S, Barretto R, Habib N, Hsu PD, Wu X, Jiang W, Marrani LA (2013) Multiplex genome engineering using CRISPR/Cas systems. Science 339:819–823

Contreras JB, Scott JA, Mendoza CL, Espinal G, Zapata Z (2008) Potencial de las algas verdes para la producción fotobiólogica de hidrógeno. Ciencia y Sociedad 33:307–326

Courchesne NMD, Parisien A, Wang B, Lan CQ (2009) Enhancement of lipid production using biochemical, genetic and transcription factor engineering approaches. J Biotechnol 141:31–41

Couso I, Vila M, Vigara J, Cordero BF, Vargas MÁ, Rodríguez H, León R (2012) Synthesis of carotenoids and regulation of the carotenoid biosynthesis pathway in response to high light stress in the unicellular microalga *Chlamydomonas reinhardtii*. Eur J Phycol 47:223–232

Davison IR, Davison JO (1987) The effect of growth temperature on enzyme activities in the brown algae *Laminaria saccharina*. Br Phycol J 22:77–87

De Clerk O, Guiry M, Leliaert F, Samyn Y, Verbruggen H (2013) Algal taxonomy: a road to nowhere? J Phycol 49:215–225

De los Reyes C, Ávila-Román J, Ortega MJ, de la Jara A, García-Mauriño S, Motilva V, Zubía E (2014) Oxylipins from the microalgae *Chlamydomonas debaryana Nannochloropsis gaditana* and their activity as TNF- inhibitors. Phytochemistry 102:152–161

Dejoye C, Vian MA, Chemat F (2013) New procedure for extraction of algal lipids from wet biomass: a green clean and scalable process. Bioresour Technol 134:271–275 del Álvarez R, Hoyo A, García-Breijo F, del Reig-Armiñana J, Campo EM, Guéra A, Barreno E, Casano LM (2012) Different strategies to achieve Pb-tolerance by the twoTrebouxiaalgae coexisting in the lichen Ramalina farinacea. J Plant Physiol 169:1797–1806

Dittami SM, Gravot A, Renault D, Goulitquer S, Eggert A, Bochereau A, Boyen C, Tonon C (2011) Integrative analysis of metabolite and transcript abundance during the short-term response to saline and oxidative stress in the brown alga Ectocarpus silicolosus. Plant Cell Environ 34(4):629–642

Djambaski P, Aleksieva P, Emanuilova E, Chernev G, Spasova D, Nacheva L, Kabaivanova L, Miranda IM, Samuneva B (2009) Sol-gel nanomaterials with algal heteropolysaccharide for immobilization of microbial cells, producing A-galactosidase and nitrilase. Biotechnol Biotechnol Equip 23:1270–1274

Duggan JX, Anderson LE (1975) Light-regulation of enzyme activity in Anacystis nidulans (Richt.). Planta 122:293–297

Durrieu C, Tran-Minh C (2002) Optical algal biosensor using alkaline phosphatase for determination of heavy metals. Ecotoxicol Environ Saf 51:206–209

Dutta D, Debojyoti De, Surabhi C, Sanjoy KB (2005) Hydrogen production by cyanobacteria. Microb Cell Factories 4:1–11

Ettl H, Gärtner G (1995) Syllabus der Boden-, Luft- und Flechtenalgen. Gustav Fischer Verlag, Stuttgart

Faber K (2011) Biotransformations in organic chemistry. https://doi.org/10.1007/978-3-642-17393-6

Fernandes P (2010) Enzymes in food processing: a condensed overview on strategies for better biocatalysts. Enzyme Res:1–19

Frank ED, Elgowainy A, Han J, Wang Z (2013) Life cycle comparison of hydrothermal liquefaction and lipid extraction pathways to renewable diesel from algae. Mitig Adapt Strateg Glob Chang 18:137–158

Fukuzawa H, Fujiwara S, Yamamoto YD, Sese ML, Miyachi S (1990) cDNA cloning, sequence, and expression of carbonic anhydrase in Chlamydomonas reinhardti i: regulation by environmental CO2 concentration. Proc Natl Acad Sci USA 87:4383–4387

Geoffroy L, Teisseire H, Couderchet M, Vernet G (2002) Effect of oxyfluorfen and diuron alone and in mixture on antioxidative enzymes of Scenedesmus obliquus. Pestic Biochem Physiol 72:178–185

Gerard VA, Driscoll T (1996) A spectrophotometric assay for RuBisCO activity: applications to the kelp Laminaria saccharina and implications for radiometric assays. J Phycol 32:880–884

Graham L, Graham J, Wilcox L (2009) Algae. Pearson Benjamin Cummings, San Francisco

Guo X, Fan C, Chen Y, Wang J, Yin W, Wang RRC, Hu Z (2017) Identification and characterization of an efficient acyl-CoA: diacylglycerol acyl transferase1 (DGAT1) gene from the microalga *Chlorella ellipsoidea*. BMC Plant Biol 17:1–16

Gupta V, Gaind S, Kumar P, Sood N, Kumar A (2010) Purification and characterization of intracellular nitrilases from *Rhodococcus* sp potential role of periplasmic nitrilase. Afr J Microbiol Res 4:1148–1153

Gurbuz F, Ciftci H, Akcil A, Karahan AG (2004) Microbial detoxification of cyanide solutions: a new biotechnological approach using algae. Hydrometallurgy 72:167–176

Gurbuz F, Ciftci H, A. Akcil A. (2009) Biodegradation of cyanide containing effluents by Scenedesmus obliquus. J Hazard Mater 162:74–79

Gutekunst K, Hoffmann D, Lommer M, Eggert M, Susuki I, Schulz-Friedrich R, Appel J (2006) Metal dependence and intracellular regulation of the bidirectional Ni Fe-hydrogenase in *Synechocystis* sp. PCC 6803. Int J Hydrog Energy 31:1452–1459

Heger T, Straub F, Mitchell E (2012) Impact of farming practices on soil diatoms and testate amoebae: a pilot study in the DOK-trial at Therwil, Switzerland. Eur J Soil Biol 49:31–36

Hernández I, Niell FX, Fernández JA (1994) Alkaline phosphatase activity in marine macrophytes. Study of its localization in some widespread species in the southern Spain. Mar Biol 120:501–509

Hernández I, Niell FX, Whitton BA (2002) Phosphatase activity of benthic marine algae. An overview. J Appl Phycol 14:475–487

Hernández E, Benítez CL, Hernández C (2017) Algal enzymes, biotechnological potential uses: a review. Cymbella 3(1):1–15

Klaitong P, Fa-aroonsawat S, Chungjatupornchai W (2017) Accelerated triacylglycerol production and altered fatty acid composition in oleaginous microalga Neochloris oleoabundans by overexpression of diacylglycerol acyltransferase 2. Microb Cell Factories 16:1–10

Kobori M, Yoshida M, Ohnishi-Kameyama M, Shinmoto H (2007) Ergosterol peroxide from an edible mushroom suppresses inflammatory responses in RAW264.7 macrophages and growth of HT29 colon adenocarcinoma cells. Br J Pharmacol 150:209–219

Kuchitsu K, Tsuzuki M, Miyachi S (1991) Polypeptide composition and enzyme activities of the pyrenoid and its regulation by CO2 concentration in unicellular green algae. Can J Bot 69:1062–1069

Lauritano C, Romano G, Roncalli V, Amoresano A, Fontanarosa C, Bastianini M, Braga F, Carotenuto Y, Ianora A (2016) New oxylipins produced at the end of a diatom bloom and their effects on copepod reproductive success and gene expression levels. Harmful Algae 55:221–229

Lisov A, Zavarzina A, Zavarzin A, Leontievsky A (2007) Laccases produced by lichens of the order Peltigerales. FEMS Microbiol Lett 275:46–52

Litchman E, Nguyen BLV (2008) Alkaline phosphatase activity as a function of internal phosphorous concentration in freshwater phytoplankton. J Phycol 44:1379–1383

Lohr M, Schwender J, Polle JEW (2012) Isoprenoid biosynthesis in eukaryotic phototrophs: a spotlight on algae. Plant Sci 185–186:9–22

Lopes G, Sousa C, Valentão P, Andrade PB (2013) Sterols in algae and health. In: Hernández--Ledesma B, Herrero M (eds) Bioactive compounds from marine foods: plant and animal sources. Wiley, Chichester, pp 173–191

Lung SC, Weselake RJ (2006) Diacylglycerol acyltransferase: a key mediator of plant triacylglycerol synthesis. Lipids 41:1073–1088

Mahdikhanlou K, Ebrahimnezhad H (2014) Plant leaf classification using centroid distance and axis of least inertia method. In 2014 22nd Iranian conference on electrical engineering (ICEE), pp 1690–1694. IEEE

Masukawa H, Sakurai H, Hausinger RP, Inoue K (2014) Sustained photobiological hydrogen production in the presence N_2 by nitrogenase mutants of the heterocyst-forming cyanobacterium Anabaena. Int J Hydrog Energy 39:19444–19451

Mathimani T, Pugazhendhi A (2019) Utilization of algae for biofuel, bio-products and bioremediation. Biocatal Agric Biotechnol 17:326–330

Miller MB, Haubrich BA, Wang Q, Snell WJ, Nes WD (2012) Evolutionarily conserved Δ 25(27)-olefin ergosterol biosynthesis pathway in the alga *Chlamydomonas reinhardtii*. J Lipid Res 53:1636–1645

Misra N, Panda PK, Parida BK, Mishra BK (2016) dEMBF: a comprehensive database of enzymes of microalgal biofuel feedstock. PLoS One 11:146–158

Msanne J, Xu D, Konda AR, Casas-Mollano JA, Awada T, Cahoon EB, Cerutti H (2012) Metabolic and gene expression changes triggered by nitrogen deprivation in the photoautotrophically grown microalgae Chlamydomonas reinhardii and Coccomyxa sp. C-169. Phytochemistry 75:50–59

Nigam PS (2013) Microbial enzymes with special characteristics for biotechnological applications. Biomolecules 3:597–611

Otto B, Beuchel C, Liers C, Reisser W, Harms H, Schlosser D (2015) Laccase-like enzyme activities from chlorophycean green algae with potential for bioconversion of phenolic pollutants. FEMS Microbiol Lett 362:11

Peers G, Truong TB, Ostendorf E, Busch A, Elrad D, Grossman AR, Hippler M, Niyogi KK (2009) An ancient light-harvesting protein is critical for the regulation of algal photosynthesis. Nature 462:511–521

Pradhan D, Sukla LB, Mishra BB, Devi N (2019) Biosorption for removal of hexavalent chromium using microalgae Scenedesmus sp. J Clean Prod 209:617–629
Purton S, Szaub JB, Wannathong T, Young REB, Economou C (2013) Genetic engineering of algal chloroplasts: progress and prospects. Russ J Plant Physiol 60:491–499
Radakovits R, Robert EJ, Al Darzins, Matthew CP (2010) Genetic engineering of algae for enhanced biofuel production. Eukaryot Cell 9(4):486–501
Reiße S, Garbe D, Brück T (2014) Meiothermus ruber thiolase – a new process stable enzyme for improved butanol synthesis. Biochemie 103:16–22
Romero DM, Ríos de Molina MC, Juárez AB (2011) Oxidative stress induced by a commercial glyphosate formulation in a tolerant strain of Chlorella kessleri. Ecotoxicol Environ Saf 74:741–747
Saini DK, Chakdar H, Pabbi S, Shukla P (2019) Enhancing production of microalgal biopigments through metabolic and genetic engineering. Crit Rev Food Sci Nutr:1–15
Samorì C, López Barreiro D, Vet R, Pezzolesi L, Brilman W, Galletti P, Tagliavini E (2013) Effective lipid extraction from algae cultures using switchable solvents. Green Chem 15(2):353–356
Schnitzler I, Georg P, Mark H, Wilhelm B (2001) Chemical defense of brown algae (Dictyopteris spp.) against the herbivorous amphipod Ampithoe longimana. Oecologia 126:515–521
Shaltiel-Karyo R, Frenkel-Pinter M, Rockenstein E, Patrick C, Levy-Sakin M, Schiller A, Egoz--Matia N, Masliah E, Segal D, Gazit E (2013) A blood-brain barrier (BBB) disrupter is also a potent α-synuclein (α-syn) aggregation inhibitor: a novel dual mechanism of mannitol for the treatment of Parkinson disease (PD). J Biol Chem 288:17579–17588
Sharma B, Dangi AK, Shukla P (2018) Contemporary enzyme based technologies for bioremediation: a review. J Environ Manag 210:10–22
Shawkat W, Westwood H, Mortimer A (2012) Mannitol: a review of its clinical uses. Contin Edu Anesth Crit Care Pain 12:82–85
Singh R (1950) Reclamation of 'Usar' lands in India through Blue-green Algae. Nature 165:325–326
Singh SP, Singh P (2015) Effect of temperature and light on the growth of algae species: a review. Renew Sust Energ Rev 50:431–444
Smith AJ, London J, Stanier RY (1967) Biochemical basis of obligate autotrophy in blue-green algae and Thiobacilli. J Bacteriol 94:972–983
Sohoni SV, Nelapati D, Sathe S, Javadekar-Subhedar V, Gaikaiwari RP, Wangikar PP (2015) Optimization of high cell density fermentation process for recombinant nitrilase production in *E. coli*. Bioresour Technol 188:202–208
Soto-Urzúa L, Baca BE (2001) Mechanisms for protecting nitrogenase from inactivation by oxygen. Rev Latinoam Microbiol 43(1):37–49
Stanek-Tarkowska J, Noga T (2012) Diversity of diatoms (Bacillariophyceae) in the soil under traditional tillage and reduced tillage. Inzynieria Ekologiczna 30:287–296
Subhadra BG, Edwards M (2011) Coproduct market analysis and water footprint of simulated commercial algal biorefineries. Appl Energy 88:3515–3523
Takeda T, Yokota A, Shigeoka S (1995) Resistance of photosynthesis to hydrogen peroxide in algae. Plant Cell Physiol 36:1089–1095
Tekaya N, Saiapina O, Ben Ouada H, Lagarde F, Ben Ouada H, Jaffrezic-Renault N (2013) Ultra--sensitive conductometric detection of heavy metals based on inhibition of alkaline phosphatase activity from *Arthrospira platensis*. Bioelectrochemistry 90:24–29
Thies F, Backhous T, Bossmann B, Grimme LH (1996) Xenobiotic biotransformation in unicellular green algae. Involvement of cytochrome P450 in the activation and selectivity of the pyridazinone pro-herbicide metflurazon. Plant Physiol 112:361–370
Tippmann S, Chen Y, Siewers V, Nielsen J (2013) From flavors and pharmaceuticals to advance biofuels: production of isoprenoids in Saccharomyces cerevisiae. Biotechnol J 8:1435–1444
Tschaikner A, Ingolic E, Gärtner G (2007) Observations in a new isolate of Coelastrella terrestris (Reisigl) Hegewald & Haganata (Chlorophyceae, Seenedesmaceae) from alpine soil (Tyrol, Austria). Phyton 46:237–245

Vellard M (2003) The enzyme as drug: application of enzymes as pharmaceuticals. Curr Opin Biotechnol 14:444–450

Vignais PM, Billoud B, Meyer J (2001) Classification and phylogeny of hydrogenases. FEMS Microbiol Rev 25:455–501

Vingiani GM, De Pasquale L, Adrianna I, Dobson Alan DW, Chiara (2019) Microalgal enzymes with biotechnological applications. Mar Drugs 17(8):459

Walker TL, Collet C, Purton S (2005) Algal transgenics in the genomic era. J Phycol 41:1077–1093

Wells ML, Potin P, Craigie JS, Raven JA, Merchant SS, Helliwell KE, Smith AG, Camire ME, Brawley SH (2017) Algae as nutritional and functional food sources: revisiting our understanding. J Appl Phycol 29:949–982

Whitton BA, Potts M (2012) Introduction to cyanobacteria. In: Witton BA (ed) The ecology of Cyanobacteria II: their diversity in time and space. Springer, Dordrecht, pp 1–13

Winkler U, Säftel W, Stabenau H (1988) β-Oxidation of fatty acids in algae: localization of thiolase and acyl- CoA oxidizing enzymes in three different organisms. Planta 175:91–98

Winkler M, Hemschemeier A, Gotor C, Melis A, Happe T (2002) [Fe]-hydrogenases in green algae: photo-fermentation and hydrogen evolution under sulfur deprivation. Int J Hydrog Energy 27:1431–1439

Xu Y, Caldo KMP, Pal-Nath D, Ozga J, Lemieux MJ, Weselake RJ, Chen G (2018) Properties and biotechnological applications of Acyl-CoA: diacylglycerol acyltransferase an phospholipid: diacylglycerol acyltransferase from terrestrial plants and microalgae. Lipids 53:663–688

Yen HW, Chen PW, Hsu CY, Lee L (2017) The use of autotrophic *Chlorella vulgarisum* chromium (VI) reduction under different reduction conditions. J Taiwan Inst Chem Eng 74:1–6

Zabawinski C, Van Den Koornhuyse N, D'Hulst C, Schlichting R, Giersch C, Delrue B, Lacroix JM, Preiss J, Ball S (2001) Starchless mutants of Chlamydomonas reinhardtii lack the small subunit of a heterotetrameric ADP-glucose pyrophosphorylase. J Bacteriol 183(3):1069–1077

Zhang J, Hao Q, Bai L, Xu J, Yin W, Song L (2014) Overexpression of the soybean transcription factor GmDof4 significantly enhances the lipid content of Chlorella ellipsoidea. Biotechnol Biofuels 7(1):128

Role of Soil Algae in Bioremediation of Pollutants

Abhishek Saxena, Bharti Mishra, Rashi Tyagi, Raya Bhattacharjya, Pankaj Kumar Singh, and Archana Tiwari

Abstract The key global crisis is soil pollution and its related issues, which are affecting soil biodiversity and ecosystem of soil. The current study includes the challenges and limitations associated with the soil pollutants caused by uncontrolled anthropogenic activities. Soil-algae technology is the emerging biotechnological approach for soil pollution control. This chapter gives an insight into soil bioremediation, control of soil pollution, and factors affecting algae survival in soil. Major emphasis is given to soil-algae diversity, which elaborates biotechnological aspects of algae in control of soil pollution and imparts global sustainable development goals.

Keywords Algae · Biodiversity · Bioremediation · Soil pollution · Soil ecosystem

1 Introduction

Soil has been recognized as the fundamental source of food and productivity, which in turn caters to the fulfillment of several Sustainable Development Goals. However, increasing anthropogenic activities like unsustainable use of soil, deforestation, leaching of harmful chemicals have led to unprecedented contamination and pollution of the natural soil microbiome, thus revealing the earnestness in resolving this global issue (FAO and ITPS 2015). Therefore, in order to develop awareness and stimulate organizations around the world to contribute and manage sustainable use of soils, the FAO established the Global Soil Partnership (GSP) in 2012 (Erdogan et al. 2021). This was an initiative to bridge the existing gaps in the soil management strategies and strengthen the synergy of actions between nations.

A. Saxena · B. Mishra · R. Tyagi · R. Bhattacharjya · P. K. Singh · A. Tiwari (✉)
Diatom Research Laboratory, Amity Institute of Biotechnology, Amity University, Noida, Uttar Pradesh, India
e-mail: atiwari9@amity.edu

A. Mishra, A. Varma (eds.), *Soil Algae*, https://doi.org/10.1007/978-981-97-9873-5_5

Besides GSP, the United Nations (UN) Decade on Ecosystem Restoration is another global movement where almost 70 countries have participated for the revival and restoration of ecosystems, including farmlands, during this period between 2021 and 2030 (Valderrábano et al. 2021). Their primary objectives include reestablishing and remediating the soil health and nutrition. Few other prominent initiatives that have been included in the former umbrella movement are the Intergovernmental Science-Policy Platform on Biodiversity and Ecosystem Services (IPBES), aiming at restoration of land, fresh, and marine ecosystems; the Bonn Challenge, which is an initiative to restore almost 350 million hectare of degraded land; United Nations Convention to Combat Desertification (UNCCD), an agreement for sustainable land management; and the International Blue Carbon Initiative, which focuses on mitigating change in global climate by restoring coastal and marine ecosystems (Valderrábano et al. 2021).

With rapid increase in urbanization and industrialization, there has been an outburst of several chemically hazardous compounds, synthetic chemicals as well as other natural pollutants, which have raised the bar of concern both for the environment and for human health (Ben Chekroun et al. 2014). Since these chemicals in soil are persistent even at low critical levels, they have led to a plethora of human health problems as well as threats to wildlife (Pavlostathis et al. 2001).

2 Why Soil Remediation Is Required?

Although both geogenic and anthropogenic activities have contributed to the pollution of the soils, rising industrial activities, agriculture, lifestyle, and climate change are the key players (Fig. 1). Due to the unsustainable practices of using soil as a resource available in an infinite form, over the years, we have sensitized the

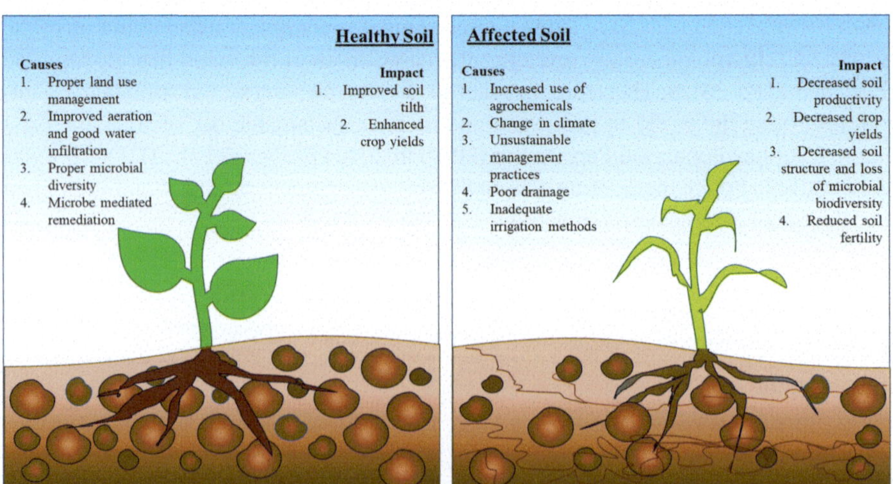

Fig. 1 Depiction of the contrasting properties of a healthy and affected soil

importance and urge to preserve and manage soil flora and fauna. Since different management regimes have different patterns of exposure, the degree of contaminants also differs variably. The increasing concern of arsenicosis in humans is due to the contamination of soils with almost 4.6–10.1% of arsenic, which also bioaccumulates in the crops grown on them (Rodrigo-Comino et al. 2020). Besides this, an overexposure of soils to selenium, lead, trace elements like cadmium, chromium, cobalt, and copper also augments the bioavailability of these hazardous pollutants to humans and other wildlife, leading to the chain of serious health conditions (Bihaqi and Zawia, 2013; Deng and Wilson, 2001). In addition to these, the presence of emerging contaminants like VOCs and PAHs has even threatened the risk of cancer in the Asia-Pacific regions (Ali et al. 2020).

As different pollutants have different physiochemical properties, their biodegradability also differs accordingly (Fu and Secundo 2016). Thus, to manage, undertake, and implement measures to prevent, reduce, and remediate the soil pollutant levels, the United Nations Environmental Assembly, in 2017, had implemented these measures (UNEA 3).

Besides physical remediation levels, biological- based remediation methods have been considered as the most appropriate and environmentally safe and sustainable mitigation methods (Rodrigues et al. 2017). Alongside phytoremediation, algae-based remediation and biomonitoring to evaluate the ecotoxicity levels and environmental risk of pollutants have been considered crucial in the recent years (Nam and An 2021). In addition to this, since diverse range of algal species have different degrees of sensitivity, they are screened and deployed accordingly. One such study had reported the varying efficiency of four different algal specie in biodegradation of PAHs by *Scenedesmus platydiscus*, *S. quadricauda*, *Chlorella vulgaris*, and *Raphidocolis captricornutum*, where the latter was the most effective able to degrade almost 78% of the hydrocarbon (Lei et al. 2007). Since algae, in combination with cyanobacteria and lichens are considered pioneers of the primary succession on land (Bhattacharjya et al. 2020), their ability to colonize and form coating in the form of biofilms on the soil particles serves as the most crucial step in retaining soil moisture content. Hence, they stabilize the soil- binding capacity and promote positive pedogenesis (Tiwari and Kaur 2014). Algae have been known to exist in the wild as a divergent organism in strong synergy with other microbial communities (Tiwari 2016). Thus, in order to enhance the degree of pollutant degradation efficiency, the consortia of algae with existing diazotrophs of the soil has been attributed as the key regulators of soil fertility, carbon, and nutrient composition (Subash Chandra Bose et al. 2011, 2013).

3 Soil-Algae Diversity

In nature, algae are of four types, i.e., yellow, green, red, and brown algae. Yellow algae (mustard algae) are a form of green algae, which adhere to the surface of walls, pool bottoms, and objects like ladders, filters, and pool lights. On the other hand, green algae (Chlorophyta) are widely present in freshwater and on the soil surface. Red algae are known as Rhodophyta commonly found in seawater. Brown algae (Phaeophyta) are the largest among them, which are commonly observed as kelps or seaweed (FAO-UN 2021). Soil is the principal non-aqueous habitat for algae (Zenova et al. 1995). They are the leading microbes found in the terrestrial ecosystem due to their wide morphological and physiological adaptability (Metting 1981; Hoffmann 1989). In the agriculture field, algae are the chief source of nitrogen and carbon to other soil microbes and also participate in increasing soil fertility due to their rich organic content (Mishra and Pabbi 2004; Tiwari 2016). Soil- algae interaction is very significant as the excretory products release growth-promoting substances, for example, amino acids, hormones, organic acids, and vitamins (Roger and Reynaud 1982; Wilson 2006). The presence of algae in soil stabilizes the soil ecosystem and lessens the soil erosion (Evans and Johansen 1999; Hu et al. 2004).

In the soil ecosystem, two major groups of algae are found, i.e., green algae and diatoms. Basically, soil algae are classified into four types, i.e., Cyanophyta (blue-green algae), Chlorophyta (grass-green algae), Xanthophyta (yellow-green algae), and Bacillariophyta (diatoms). Among them, Cyanophyta (blue-green algae) and Chlorophyta (grass-green algae) are the most abundant species in soil. Chlorophyta and Bacillariophyta are majorly found in the temperate region, whereas blue-green algae are found in tropical soils. Common species from green algae (Chlorophyta) are *Chlorella*, *Chlamydomonas*, and *Chlorococcum*, while *Navicula* and *Pinnularia* are the most common species from Bacillariophyta that are most dominant in soil. Cyanobacteria, the blue-green algae, strengthen the soil particles and prevent soil erosion (Khan and Rao 2019).

They are motile species that can be found beneath the earth's surface. However, earthworms and rainwater may transport them to deeper soil surfaces. In the soil ecosystem, algae populations range from 10,000 to 100,000 cells per gram of soil (Khan and Rao 2019). Owing to their photosynthetic nature, algae introduce organic matter to the soil, release polysaccharide content, and ultimately enhance soil aggregation. The polysaccharide content from algae increases soil porosity, aggregation, and water-holding capacity (Roger and Reynaud 1982; Choudhary et al. 2007). Algae improve the recovery of saline and metals in the soil, thus increasing soil quality (Rai et al. 1998; Whitton and Potts 2000). In previous studies, the addition of farm soil with algae boosts the grain yields up to 25% (Yanni 1992; Gurung and Prasad 2005; Song et al. 2005). Cyanobacteria, the blue-green algae, are the most favorite biofertilizer because of their oxygenic photosynthesis and nitrogen fixation capacity (Relwani 1963; Ernst et al. 1992). Examples of nitrogen-fixing species are *Anabaena*, *Calothrix*, *Nostoc*, *Schizothrix*, and *Scytonema* (Kannaiyan 1990; Rai et al. 2000). The green algae survive well on moist soil and acidic soil which is non-flooded, whereas diatoms choose to survive well in land rich in organic matter. Dead and decaying algal cells together with other soil microbes combine to form the crust of organic matter which in turn prevents erosion of soil nutrients (Bhatnagar et al. 2008). Advantageously, soil-algae diversity maintains soil fertility, increases the organic carbon content, reduces soil erosion, increases the water- holding capacity of the soil, and liberates a large amount of oxygen to the soil (Tiwari 2014). The emphasis on soil-algae diversity in the future recognizes the dynamic region of interacting processes, which is important for the sustainability of human

Table 1 Tabulated representation of different class of algae present in the soil ecosystem

Algal species found in soil				
Cyanobacteria	Chlorophytes	Diatoms	Euglenoids	Reference
Anabaena viguieri, *Anabaena* sp., *Aphanothece stagnina*, *Gloeothece palea*, *Gloeocapsa* sp., *L. putealis*, *Lyngbya major*, *Lyngbya* sp., *Microcystis* sp., *Microcoleus paludosus*, *M. subtorulosus*, *Nostoc commune*, *Nostoc* sp., *Oscillatoria chlorine*, *Synechococcus aeruginosus*, *O. curviceps*, *O. princeps*, *O. pseudogeminata*, *O. tenuis*, *Oscillatoria* sp., *Phormidium mucicola*, *Phormidium* sp., *Oscillatoria limosa*, *Phormidium rubescens*, *Nostoc azollae*, *Symploca borealis*, *Aphanocapsa parietina*, *Spirulina agilissima*, *Calothrix confervicola*, *Schizothrix fuscescens*, *Oscillatoria limosa*, *Phormidium rubescens*, *SpirulIna agilissima*, *Schizothrix fuscescens*	*Actinotaenium cucurbita*, *Characium guttula*, *Apiocystis* sp., *Chlorella vulgaris*, *Chlamydomonas reinhardtii*, *Cladophora* sp., *Closterium* sp., *Cosmarium laeve*, *C. quadrum*, *C. regnellii*, *Klebsormidium flaccidum*, *Scenedesmus quadricauda*, *Ulothrix variabilis*, *Haematococus lacustris*, *Mesotaerium mirificum*, *Closterium monilliferum*, *Chlamydomonas reinhardtii*, *Microspora* sp., *Chlorella vulgaris*, *Closterium monilliferum*, *Penium spinospermum*, *Chlamydomonas reinhardtii*, *Leptosira terrestris*	*Achnanthes exigua*, *Amphora montana*, *Encyonema leei*, *Eunotia soleirolii*, *Hantzschia amphioxys*, *Na. gregaria*, *Na. mutica*, *Na. pelliculosa*, *Na. gregaria*, *Na. mutica*, *Na. pelliculosa*, *Na. gregaria*, *Na. mutica*, *Na. pelliculosa*, *Na. seminulum*, *Na. rostellata*, *Na. subminuscula*, *Navicula* sp., *Nitzschia amphibian*, *Ni. amphibioides*, *Ni. levidensis*, *Ni. obtus*, *P. intermedia*, *Ni. palea*, *Ni. pusilla*, *Pinnularia borealis*, *P. gibba*, *P. obscura*, *P. subcapitata*, *P. viridis*, *Pinularia* sp., *Surirella robusta*, *Pinnularia perirrorata*, *Navicula cincta*, *Navicula cryptotenella*, *Adlafia bryophila*, *Adlafia brockmannii*, *Sellaphora atomoides*, *Mayamaea excelsa*, *Stauroneis parathermicola*, *Stauroneis thermicola*, *Mayamaea arida*, *Placoneis hambergii*, *Placoneis ignorata*, *Placoneis paraelginensis*, *Geissleria paludosa*, *Luticola cf. micra*, *Humidophila contenta*, *Halamphora montana*, *Planothidium lanceolatum*, *Eunotia botuliformis*, *Fragilariforma virescens*, *Navicula cryptonella*, *Pinnularia viridis*, *Surirella acuminata*, *Frustulia rhomboides*, *Navicula cryptotenella*, *Pinnularia viridis*, *Surirella acuminata*, *Frustulia rhomboides*	*Euglena* sp., *Lepocinclis ovum*, *Phacus acuminatus*, *Lepocinclis globules*, *Ochromonas danica*, *Phacus acuminatus*, *Lepocinclis globules*, *Ochromonas danica*, *Euglena acus*	Lin et al. (2013), Adesalu and Olugbemi (2015) and Antonelli et al. (2017)

civilization. The abundance of different types of algae found in the soil ecosystem is presented in Table 1.

4 Factors Affecting Algae Survival in Soil

There are many physical factors that affect the activity of the soil biota (Killham 1994). The main ones are temperature, pH, moisture, soil mineralogy, and light.

4.1 Temperature

Temperature impacts the soil biota activity directly by controlling the ratio of physiological action like activity of the enzyme and secondarily by changing physicochemical qualities like nutrient transport and dispersion, mineral weathering, and soil moisture, among others. Biological activity rises with rising temperature within certain limitations. Organisms have a temperature range in which their biological activity takes place. The temperature range in which typical soil organisms can be active is around 0–60 °C, while no one group is possible to be effective over the full variety. A few organisms that are known to exist beyond the optimum temperature range of the soil often develop existence mechanisms to stay alive. Glade (1914) discovered that when moist, brief durations of 60–90 °C would not kill some Cyanophyceae. When dry, they could withstand temperatures of up to 95 °C. According to Booth (1946), all soil algae appear to die at 110 °C. Glade (1914) discovered that *Cylindrospermum* spores were more robust than vegetative cells, demonstrating the relevance of past soil conditions in spore development (Demeter 1956). Extreme climates, such as volcano hot springs and eruptions, may reach temperatures of hundreds of degrees. Microorganisms were originally considered to be able to withstand temperatures of over 400 °C, but then again this has since existed decreased to roughly 121 °C. There is a temperature optimum inside of an organism's range of temperature where the biological function operates optimally. Further than this point, cellular functions become less effective, and as the temperature approaches its upper limit, permanent changes in cell characteristics occur, leading to cell death. In comparison to the organisms better suited to the lesser limit, organisms with a high temperature limit are less active at the lower extreme.

4.2 pH

The microorganisms come up with a range of pH in which they are effective, analogous to temperature. Microorganisms may withstand extremities within specified limitations, although this typically necessitates the cell using energy to maintain the proper inner cellular pH (pH 7.0). Only a rare species (bacteria and archaea) can endure pH levels as low as 1.0 or as high as 11, but then these are not situations encountered in farming soils. Bigger animals of the soil, rather than microbes, may be more precise to pH. Because earthworms are vulnerable to low pH, their purpose is switched by another species (enchytraeids in forests) or the soil develops in a diverse way in forest and peatland environments (pH 5) (peatlands). The solubility of elements is directly affected by pH. Aluminum becomes more soluble and so more accessible to organisms at an acidic pH, resulting in higher toxicity. At pH extremes, essential minerals might become unavailable. Phosphorus and manganese, for example, come to be progressively scarce at elevated pH levels.

4.3 Soil Moisture

Moisture of the soil has two effects on the soil biota. Water is required for living, enzyme function, and metabolic activity, as well as serving as a solution for biological substances as well as other components. Soil moisture has a physical effect on the soil temperature (water is an excellent source of energy) as well as oxygenation. The mobility along with the predation of microorganisms into the soil is influenced by the degree to which soil apertures are packed. Plants may not be capable of extracting enough water through their roots in extremely dry soil due to the energy required to pull water from narrow pores. The permanent wilting point is the moment at which the plant can no longer recover. In moist environments, however, oxygen does not travel as rapidly all through the soil, and the quantities accessible to organisms may be decreased, leading to anoxic environments. Fungi are much more impervious to water distress than bacteria, while certain bacteria can develop resilient endospores at extreme circumstances, allowing the species to live unless more favorable conditions exist. The cell may create chemicals to help it survive in a dearth. Actinomyces, which can withstand arid circumstances better than fungi, may defend them during droughts by producing proline like amino acid.

4.4 Soil Mineralogy

Bacteria and fungi utilize soil minerals as a natural resource of power and for soil microbes (chemoautotrophs) or because of the formation of acids like sulfuric acid (Uroz et al. 2009). Bacteria have been demonstrated to create minerals in addition to weathering minerals (Párraga et al. 1998). Several nutrients, notably ammonium, may get tied up with the clay and inaccessible depending on the pH. Clays contain many places on its surfaces that can attach molecules and entire microorganisms, impacting cell migration throughout soil and organic matter degradation.

4.5 Light

Light intensity and quality directly affects organisms found near the soil surface. Plants, algae, and microorganisms utilize the energies following the sunshine to synthesize the carbohydrates. Much of this constituent (photosynthate) works its aspect further into soil microorganisms through with the roots as well as leaf fall in plants. Some of the organic matter is absolutely inputted into the soil by means of microalgae and cyanobacteria, also by way of the discharge of molecules after the cells and whenever the cell goes dead. Some wavelengths of daylight are further harmful to organisms than others. The radiation of the ultraviolet damages the DNA, causing metamorphoses in animals that can lead to death or biochemical alterations such as changes in enzymes and metabolic pathways. Organic molecules may undergo structural changes as a result of photochemical depreciation, making them more easily destroyed.

5 Bioremediation of Soil Pollutants Using Algae

Soil contamination has increased around the world via anthropogenic activities (e.g., use of agrochemical, smelting, mining, wars, and training for military, different industries, consumption of fossil fuels, improper waste disposal, and irrigation) that become the topic of concern. There are few studies by Fu and Secundo (2016) of different countries that are facing the problem of soil contamination; for example, the agricultural waste, mineral mining, and manufacturing industries creating different chemical waste, etc. are causing soil pollution in Australia. In China, a survey was done in 2014 where approximately twenty million hectares of farmland were found contaminated with heavy metals that affect the food supply by around ten million tons every year. Similarly, in Europe, there were found 340,000 contaminated sites, in which only 15% remediation was successfully achieved (Fu and Secundo 2016). Increasing use of chemical fertilizers, oil spills, irrigation and sewage run offs containing heavy metal and trace elements are known as the chief causes of soil pollution affecting both human and environment. Unfortunately, the soil has been contaminated with heavy metals that need to be prevented via inexpensive methods to control soil pollution and must use advanced tools and technology to reduce the negative or dangerous impact on the environment (Liu et al. 2018; Guo et al. 2020).

To reduce the chances of soil contamination, there are in-situ (at the site of pollution or polluted place) and ex-situ (outside the polluted site) techniques used. The in-situ process does not create the hinderance in soil and related microorganisms to it, and the exposure of the contaminants to people also reduces as it does not require digging and transportation of polluted soil to the treatment site, it is a simple and cost-effective process in comparison to ex-situ. However, few criteria of field required to observe carefully, such as soil permeability, weather, pH, nutrient availability, chemical property in terms of potential of deep leaching, and contamination depth. Examples of in-situ processes are encapsulation, vitrification, surface capping, soil flushing, immobilization, soil flushing, electrokinetic, phytoremediation, and bioremediation. Since Ex-situ process requires excavation of soil, followed by transportation, dumping, and refilling of the sites, it includes an extra expenditure as compared to the in situ approach. However, implementing ex-situ methods involves rapid and regulated study of the affected site combining physical, chemical as well as biological remediation techniques (Liu et al. 2018; Guo et al. 2020).

Chemical methods for soil remediation can be achieved by adsorption methods using specific media, immobilization, chemical oxidation and extraction, precipitation, chemical leaching, chemical fixation, nano-remediation, and ion exchange. For example, chemical oxidation can be used in both ex-situ and in-situ techniques. This method can be successfully applied to remediate soil contaminated with hydrocarbon. In this method, ex-situ is achieved by mixing the oxidizing agents (potassium permanganate, hydrogen peroxide, hypochlorite, chlorine dioxide, etc.) with polluted soil. However, the ex-situ process is costly than the in-situ process, due to the extra expense of excavation. Chemical treatment has some disadvantages as it forms by-products that further co-contaminate the soil and create hindrance to the soil system and related microorganisms (Sharma et al. 2018; Rada et al. 2019).

Physical methods for soil remediation (such as heavy metals) in the polluted site depend on types of waste. This method removes approx. all the contaminants from the soil. Physical treatment varies from waste to waste and their particle size. The contaminated soil with heavy metal can be treated by heat treatment, electro-remediation, vitrification process, and soil replacement methods. But there is a drawback with the physical method i.e., in this technique, the further processing needed, required a very expensive instrument and it's a very time-consuming process (Sharma et al. 2018).

However, to find or utilize the cost-effective, non-toxic, and eco-friendly method for soil remediation, the use of the biological method is a known more applicable option than others, in which several biological agents (microorganisms), such as algae, bacteria, and fungi, can participate to degrade, remove, and immobilize contaminants from soil. The microorganisms utilize their metabolic machinery to reduce the toxicity of pollutants for soil remediation and they convert the harmful contaminants into less or non-toxic compounds such as the conversion of organic pollutants into water and CO_2. The microorganism can modify their metabolic pathway according to their environment and can synthesize desirable enzymes for degrading pollutants.

The contamination strategy by using microorganisms depends on environmental conditions, pollutant types, sources of phosphorous and nitrogen, and their concentration. Among all, the phycoremediation strategy is preferable technology where the algae (macroalgae or microalgae) are used for removing harmful contaminants from soil. Algae are used for remediation purposes due to their effective role and low cost. They are very sensitive toward the toxic chemicals; thus, they are used in soil as bioindicators to evaluate the genotoxicity, ecotoxicity, and the risk of

Table 2 List of pollutants and their removal by different algae

S. no.	Remediated pollutants	Algae	Reference
1.	DDT, Dieldrin, and Photodieldrin	*Ankistrodesmus amalloide*	Koul and Taak (2018)
2.	Copper (Cu)	*Sargassum* sp. and *Ulva fasciata*	Koul and Taak (2018)
3.	Tetracycline	*Chlorella vulgaris*	Koul and Taak (2018)
4.	Polycyclic aromatic hydrocarbons (PAHs)	*Scenedesmus platydiscus*, *Scenedesmus quadricauda*, *Chlorella vulgaris*, and *Raphidocoliscap tricornutum*	Fu and Secundo (2016)
5.	Lead (Pb), cadmium (Cd)	*Nitzschia pelliculosa*	Marella et al. (2020a, b)
6.	Manganese (Mn), copper (Cu), zinc (Zn), cobalt (Co)	*Nitzschia closterium*	Marella et al. (2020a, b)
7.	Mercury (Hg)	*Chetoceros costatum*	Marella et al. (2020a, b)
8.	Mesotrione (herbicide)	*Ankistrodesmus fusiformis*, *Amphora coffeaeformis*, and *Pediastrum tetras*	Ben Chekroun et al. (2014)
9.	Phyrene, phenanthrene, and naphthalene	*Chlorella*, *Scenedesmus*, and *Selenastrum*	Filippo Dell' Anno et al. (2021)

contaminants for environment. Algal species are known to develop several cope up mechanisms for combating the oxidative stress caused due to heavy metal pollution. These include synthesizing binding proteins like PCs (poly(-y-glutamyl-cysteinyl) glycines, compartmentalization, and adsorption of elements into the vacuole. It was given that in *Thalassiosira weissflogii* and *T. peudonana*, phytochelatins found at higher concentration under the metal stress, whereas *Chlorella vulgaris* under the copper and chromium stress conditions produced a higher concentration of proline (Fu and Secundo, 2016; Koul and Taak 2018) (Table 2).

6 Biotechnological Aspects of Algae in Bioremediation

In soil, presence of hazardous and toxic hydrocarbons is the major source of contaminants that result from improper treatment and directly dump into soil, leading to infected surfaces and sub-surfaces which cause forest fires. Chemicals such as trichloroethylene, biphenyls, polycyclic aromatic hydrocarbons, and many others are xenobiotics that are recalcitrant to biodegradation due to their toxic nature and high molecular weight (Shang et al. 2001). Due to intensified energy and industrial effluents, the soil pollution is quite persistent at nearby industrial sites (Germaine et al. 2009). Coupled with carcinogenicity, accumulations of toxic organic compounds in soil are a cause of severe concern. Industrial compounds containing nitrogen are generated in large quantities for usage in petroleum products, dyes, polymers, insecticides, explosives, and pharmaceuticals. Different nitroaromatics, nitrate esters, and nitrogen-containing heterocycles are among the chemical groups involved. The United States Environmental Protection Agency has classed many of these substances as hazardous because they are poisonous and pose a concern to human health.

PAHs are common contaminants found in high concentrations at industrial sites connected with the petroleum, coal tar, gas generation, and wood preservation industries. PAHs are classified as environmental priority pollutants because of their carcinogenic and mutagenic effects (Doty 2008). They are less easily degraded than many other organic molecules, making them stable and recalcitrant in soils. Though conceivable, bioremediation of PAH-contaminated soils is impeded by these chemicals' limited bioavailability, making their removal a lengthy and arduous procedure. However, some PAH-degrading bacteria appear to have evolved to cope with PAHs' unfavorable physicochemical features. Recent research has revealed that these strange organisms typically belong to a small number of bacterial or fungal genera. In bioremediation processes, optimizing the environmental conditions guarantees that physiological and biochemical activities are directed toward biodegradation of the target pollutants.

Combining physiological and genetic techniques allows us to gain a better knowledge of many aspects of xenobiotic biodegradation. However, biotechnological approaches for destroying hazardous wastes have a numerous advantage over traditional physicochemical methods. When properly operated, biotechnology techniques have the potential to completely destroy organic wastes. The gain of knowledge can help in designing new microorganisms for bioremediation. Understanding of current metabolic pathways and processes may be useful in creating new metabolic pathways by reducing the accumulation of inhibitory metabolites and increasing the substrate specificity (Head and Bailey 2003). It is possible to expand the spectrum of substrates that an organism can use by genetically modifying metabolic pathways. Aromatic hydrocarbon dioxygenases, which are members of the Rieske non-heme iron oxygenase family, have broad substrate specificity and catalyze enantiospecific reactions with a variety of substrates (Gibson and Parales 2000). Recently, hybrid dioxygenases with improved PCB, TCE, and other aromatic hydrocarbon degradation capabilities were examined (Furukawa 2000).

The projected research activities in the field of remediation would aid in the development of advanced bioprocess technology. Marine enzymes could be an innovative waste treatment option for reducing industrial waste. To treat contaminated wastes and places, a variety of bioremediation techniques have been developed. Consider three basic principles when choosing the best treatment strategy for a specific site: the pollutant's ability to undergo biological transformation into less toxic products (biochemistry), the contaminant's accessibility to microorganisms (bioavailability), and the possibility of optimizing biological activity (bioactivity). Biological remediation, which uses microbes and plants to remove dangerous toxins, is typically thought to be a safer and less expensive option (Dua et al. 2002). During soil bioremediation procedures, microorganisms play a key role in degrading or mineralizing different pollutants and converting non-toxic by-products.

Recent developments in the molecular genetics of degradation, as well as studies on enzyme-tailoring and DNA-shuffling, have been reviewed. Bioremediation has become a rapidly expanding field as a result of breakthroughs in biotechnology, and it has been commercially employed for the management of hazardous wastes and contaminated locations.

7 Mechanism Involved in Remediating Pollutants

There are two levels of technique for remediating pollutants: (i) biological or physical/chemical treatments and (ii) in-situ or ex-situ. The biological treatment of pollutants may be divided separately owing to their sustainability, both environmentally and economic feasibility. The criteria that are considered for the adoption of the bioremediation technique are nature of pollutants, amount of pollution, environment, site or place, cost dynamics, and environmental policies. The success of any bioremediation process is based upon the amount of oxygen available, pH of the habitat, temperature, abiotic factors, and nutrient concentrations. The mechanism behind treatment or bioremediation of contaminated sites comprising soil pollutants is categorized into biosorption and bioaccumulation or, in simple words, a passive adsorption mechanism.

7.1 Biosorption

Biosorption is the capacity of any biological material to collect or accumulate pollutants from the surroundings through physicochemical pathways or metabolically mediated uptake methods. Researchers are working on finding the most appropriate economical way to remove pollutants using biological species. The mechanism of biosorption is inexpensive over bioaccumulation because in the case of biosorption the biomass can be directly used from industrial effluents, processed, and utilized. However, low selectivity might be due to the interaction of biomass with the physiochemical process (Lee and Chang 2011). It is a low-energy-demand technique wherein heavy metals can be adsorbed by ion exchange, chelation, adsorption, diffusion, precipitation, complexation, and crystallization method (Marella et al. 2020a, b). Accumulation of heavy metals is very harmful to fauna and flora and causes several health hazards. In biosorption, biosorbents are added to the solution containing adsorbate; initially, the bioabsorbent must be immersed in the solution containing absorbate and incubated to attain equilibrium. After some time, the absorbent containing absorbate detached. The entire mechanism is reversible which requires no new addition of nutrients. It is a single-step process that is nontoxic (Mishra and Malik 2013). The absorbents can be industrial by-products, agricultural waste material, and microbial biosorbents (bacteria, yeast, algae, and fungi). However, algae are the most suitable and appropriate biosorbents because they effectively removed pollutants from wastewater and simultaneously utilized the available nutrients for their growth and development. Algae have strong capacity for sorption due to their large surface area to volume ratio, global availability, and metal regeneration capacity (Marella et al. 2020a, b).

7.2 *Bioaccumulation*

Bioaccumulation is a biological phenomenon, a metabolically active process that involves the absorption of pollutants by algae into their intracellular space, often using an import system using importer complexes that translocate through the lipid bilayer. Sequestration by proteins and peptide ligands within the storage system allows the entry into the intracellular space by the heavy metals (Mrvčić et al. 2012). In bioaccumulation, microbes use their proteins to consume heavy metals, followed by sequestration into intracellular space to accumulate them to their cellular process for catalysis, signaling, and stabilizing biomolecular charges (Lee and Chang 2011). Genetically engineered species are considered strongly as a recombinant expression that provides improved translation and transcription of the import and storage systems for enhanced bioaccumulation, i.e., absorption and sequestration of heavy metal ions (Mrvčić et al. 2012; Chojnacka 2009, 2010; Deng and Wilson 2001; de Silóniz et al. 2002). Once pollutants release to the water bodies, they join with the fresh water and other watery fluids, such as industrial wastewater, sewage, and run-off from mining and agricultural fields after rainfall. All these effluents are health and the environment (Diep et al. 2018; Koller and Saleh 2018).

Bioaccumulation is a cost effective and ecofriendly approach because of its higher capacity to remove pollutants. More investigation and innovation are needed to further improvise the process of bioaccumulation because with increasing demand for metal resources and clean freshwater, more amount of clean water is required, and this can be achieved by adopting sustainable strategies. For example, shellfish can be a good indicator of the bioaccumulation of heavy metals.

8 Challenges and Limitations

Several persistent hazardous and toxic pollutants easily seep into the soil ecosystem, which causes health problems and environmental damage. Anthropogenic activities such as agricultural, domestic, industrial, municipal, and military activities are major sources of soil pollutants. Soil biodiversity is an important aspect that can suitably be incorporated; however, keeping the record and its analysis is also a challenge. As indicated by FAO, soil pollution is a worldwide threat that is affecting one-third of soil across the globe. To recover 1 cm of the arable soil layer, 1000 years are required. It is extremely challenging to control the load of inorganic toxic compounds, organic wastes, sewage and sewage sludge, heavy metal pollutants, and organic pesticides that are causing damage to the soil (FAO and UNEP 2021).

Physicochemical and biological methods are very well implemented to control the issue of soil pollution. Among biological methods, bioremediation and phytoremediation restore soil health and preserve soil fertility. Algae, fungi, and bacteria are currently being studied and actively engaged by the researchers to control the soil pollution and make it a less toxic form (Verma 2021). However, alongside the challenges, the limitations of the biological approach cannot be hidden. Several environmental parameters of soil, such as moisture, temperature, acidity or alkalinity of the soil, type of soil and its verities, and other related soil features, pose a big problem to the biological species to control the soil pollution through either bioaccumulation or bioabsorption (Ashraf et al. 2014). The need of the hour is the implementation of new technologies and related multidisciplinary methods that must be incorporated to overcome the challenges or limitations in controlling soil pollution. In this context, omics, modeling, new species, or genetically engineered organisms will give a better understanding of the soil-related problems and their pollutants (Biswas and Sarkar 2018). A new emerging technology is nanobioremediation, which could overcome the related issues in a more economic, political, social, and industrial manner (Suriyaprakash et al. 2021). Apart from this, the application of sole algal bioremediation can be a sustainable, ecofriendly, cost-effective approach to control soil pollution and its contamination, which not only face future challenges but also protect the soil functionality and its biodiversity through its natural origin to achieve the status of global sustainable developmental goals (Sutherland et al. 2021).

9 Conclusion

Soil-algae bioremediation is highly desirable over other microbial treatments as it prevails over the necessity of carbon and other nutrients. Soil-algae interactions are more efficient in the treatment of hazardous pollutants, which also provide energy benefits such as renewable energy by fixing carbon dioxide. To quantify the interaction between soil and algae in bioremediation, more investigation is required both at the molecular and metabolic levels. However, it has been emphasized that the transformation of pollutants to biomass is highly challenging (almost negligible) as compared to wastewater treatment. More research must be carried out toward this direction, and researchers must be encouraged to study more about soil-algae diversity and bioremediation of pollutants. Hence, soil bioremediation is one of the significant biotechnological aspects that should not be overlooked.

Acknowledgments We are thankful to the Department of Biotechnology (DBT), New Delhi, India, for providing financial assistance under project Grant No. BT/PR/15650/AAQ/3/815/2016 for conducting research.

Declaration of Competing Interest The author declares no conflict of interest.

References

Adesalu TA, Olugbemi OM (2015) Soil Algae: a case study of two vegetable farmlands in Lagos and Ogun states, Southwest Nigeria. IFE J Sci 17(3):765–772

Anno FD, Rastelli E, Sansone C, Brunet C, Ianora A, Anno AD (2021) Bacteria, fungi and microalgae for the bioremediation of marine sediments contaminated by petroleum hydrocarbons in the Omics Era. Microorganisms 9:1695

Antonelli M, Wetzel CE, Ector L, Teuling AJ, Pfister L (2017) On the potential for terrestrial diatom communities and diatom indices to identify anthropic disturbance in soils. Ecol Indic 75:73–81

Ashraf MA, Maah MD, Yusoff I (2014) Soil contamination, risk assessment and remediation. In: Hernandez-Soriano MC (ed) Environmental risk assessment of soil contamination. IntechOpen. https://doi.org/10.5772/57287. https://www.intechopen.com/chapters/46032

Ali ME, Hoque ME, Safdar Hossain SK et al (2020) Nanoadsorbents for wastewater treatment: next generation biotechnological solution. Int J Environ Sci Technol 17:4095–4132. https://doi.org/10.1007/s13762-020-02755-4

Ben Chekroun K, Sánchez E, Baghour M (2014) The role of algae in bioremediation of organic pollutants. Int J Environ Res Public Health 1(2):19–32. http://www.journalissues.org/irjpeh

Bhatnagar A, Makandar MB, Garg MK, Bhatnagar M (2008) Community structure and diversity of cyanobacteria and green algae in the soils of Thar Desert (India). J Arid Environ 72(2):73–83. https://doi.org/10.1016/j.jaridenv.2007.05.007

Bhattacharjya R, Begum A, Tiwari A (2020) Role of algae–fungi relationship in sustainable agriculture:227–254. https://doi.org/10.1007/978-3-030-45971-0_10

Biswas R, Sarkar A (2018) Omics' tools in soil microbiology: the state-of-the-art 2018. In: Advances in soil microbiology: recent trends and future prospects. https://doi.org/10.1007/978-981-10-6178-3_3

Booth WE (1946) The thermal death point of certain soil inhabiting algae. Ptoc Montana Acad Sci 5(6):21–23

Bihaqi SW, Zawia NH (2013). Enhanced taupathy and AD-like pathology in aged primate brains decades after infantile exposure to lead (Pb). Neurotoxicology. 39:95–101 https://doi.org/10.1016/j.neuro.2013.07.010

Chojnacka K (2009) Biosorption and bioaccumulation in practice. Nova Science Publishers, New York

Chojnacka K (2010) Biosorption and bioaccumulation – the prospects for practical applications. Environ Int 36(3):299–307

Choudhary KK, Singh SS, Mishra AK (2007) Nitrogen fixing cyanobacteria and their potential applications. In: Gupta RK, Pandey VD (eds) Advances in applied phycology. Daya Publishing House, New Delhi, pp 142–154

de Silóniz MI, Balsalobre L, Alba C, Valderrama MJ, Peinado JM (2002) Feasibility of copper uptake by the yeast *Pichia guilliermondii* isolated from sewage sludge. Res Microbiol 153(3):173–180

Demeter O (1956) Uber modiyikationenbei Cyanophyceen. Arch Mikriobiol 24:105–133

Deng X, Wilson D (2001) Bioaccumulation of mercury from wastewater by genetically engineered *Escherichia coli*. Appl Microbiol Biotechnol 56(1):276–279

Diep P, Mahadevan R, Yakunin AF (2018) Heavy metal removal by bioaccumulation using genetically engineered microorganisms. Front Bioeng Biotechnol 6:157

Doty SL (2008) Tansley review: enhancing phytoremediation through the use of transgenics and endophytes. New Phytol 179:318–333

Dua M, Singh A, Sethunathan N, Johri A (2002) Biotechnology and bioremediation: successes and limitations. Appl Microbiol Biotechnol 59(2):143–152

Erdogan HE, Havlicek E, Dazzi C, Montanarella L, Van Liedekerke M, Vrščaj B et al (2021) Soil conservation and sustainable development goals(SDGs) achievement in Europe and central

Asia: which role for the European soil partnership? Int Soil Water Conserv Res 9(3):360–369. https://doi.org/10.1016/J.ISWCR.2021.02.003

Ernst A, Black T, Cai Y, Panoff JM, Tiwari DN, Wolk CP (1992) Synthesis of nitrogenase in mutants of the cyanobacterium *Anabaena* sp. PCC 7120 affected in heterocyst development. J Bacteriol 174:6025–6032

Evans RD, Johansen JR (1999) Microbiotic crusts and ecosystem processes. Crit Rev Plant Sci 18:183–225

FAO & ITPS (2015) Status of the world's soil resources (SWSR) e main report. Food and Agriculture Organization of the United Nations and Intergovernmental Technical Panel on Soils, Rome

FAO and UNEP (2021) Global assessment of soil pollution – summary for policy makers. FAO, Rome. https://doi.org/10.4060/cb4827en

FAO-UN (2021) Food and Agriculture Organization of the United Nations NPA-Algae. https://www.fao.org/agriculture/crops/thematic-sitemap/theme/spi/soil-biodiversity/soil-organisms/by-type/algae/en/

Fu P, Secundo F (2016) Algae and their bacterial consortia for soil bioremediation. Chem Eng Trans 49:427–432

Furukawa K (2000) Engineering dioxygenases for efficient degradation of environmental pollutants. Curr Opin Biotechnol 11:244–249

Germaine KJ, Keogh E, Ryan D, Dowling DN (2009) Bacterial endophyte-mediated naphthalene phytoprotection and phytoremediation. FEMS Microbiol Lettr 296:226–234

Gibson DT, Parales RE (2000) Aromatic hydrocarbons dioxygenases in environmental biotechnology. Curr Opin Biotechnol 11:236–243

Glade R (1914) Zur Kenntnis der Gattung *Cylindrospermam* Cohns. Beitr. Z Biol d Pflanzen 12:295–343

Guo M, Song W, Tian J (2020) Biochar-facilitated soil remediation: mechanisms and efficacy variations. Front Environ Sci 8:521512. https://doi.org/10.3389/fenvs.2020.521512

Gurung S, Prasad BN (2005) Azolla and cyanobacteria (BGA): potential biofertilizers for rice. Sci World 3:85–89

Head IM, Bailey MJ (2003) Environmental biotechnology. Methodological advances spawn new concepts in environmental biotechnology. Curr Opin Biotechnol 14:247–247

Hoffmann L (1989) Algae of terrestrial habitats. Bot Rev 55:77–105

Hu CX, Zhang DL, Liu YD (2004) Research progress on algae of the microbial crusts in arid and semiarid regions. Prog Nat Sci 14:289–295

Kannaiyan S (1990) Biotechnology of biofertilizer for rice. Tamil Nadu Agricultural University, Coimbatore, p 212

Khan A, Rao TS (2019) Molecular evolution of xenobiotic degrading genes and mobile DNA elements in soil bacteria. In: Das S, Dash HR (eds) Microbial diversity in the genomic era. Academic, pp 657–678. https://doi.org/10.1016/B978-0-12-814849-5.00036-8

Killham K (1994) Soil ecology. Cambridge University Press. Ann Bot 75(1):102–103, Cambridge, p 242. https://doi.org/10.1016/S0305-7364(05)80017-4

Koller M, Saleh HM (2018) Introductory chapter: introducing heavy metals. In: Heavy metals, pp 3–11. https://doi.org/10.5772/intechopen.74783

Koul B, Taak P (2018) Biotechnological strategies for effective remediation of polluted soils. In: Soil remediation through algae, plants and animals, pp 129–195. https://doi.org/10.1007/978-981-13-2420-8_7

Lee YC, Chang SP (2011) The biosorption of heavy metals from aqueous solution by *Spirogyra* and *Cladophora* filamentous macroalgae. Bioresour Technol 102(9):5297–5304

Lei AP, Hu ZL, Wong YS, Tam NFY (2007) Removal of fluoranthene and pyrene by different microalgal species. Bioresour Technol 98(2):273–280

Lin CS, Chou TL, Wu JT (2013) Biodiversity of soil algae in the farmlands of mid-Taiwan. Bot Stud 54(1):41. https://doi.org/10.1186/1999-3110-54-41

Liu L, Li W, Song W, Guo M (2018) Remediation techniques for heavy metal-contaminated soils: principles and applicability. Sci Total Environ 633:206–219. https://doi.org/10.1016/j.scitotenv.2018.03.161

Marella TK, Saxena A, Tiwari A (2020a) Diatom mediated heavy metal remediation: a review. Bioresour Technol 305:123068. https://doi.org/10.1016/j.biortech.2020.123068

Marella TK, López-Pacheco IY, Parra-Saldivar R, Dixit S, Tiwari A (2020b) Wealth from waste: diatoms as tools for phycoremediation of wastewater and for obtaining value from the biomass. Sci Total Environ 1(724):137960

Metting B (1981) The systematics and ecology of soil algae. Bot Rev 47:195–312

Mishra A, Malik A (2013) Recent advances in microbial metal bioaccumulation. Crit Rev Environ Sci Technol 43(11):1162–1222

Mishra U, Pabbi S (2004) Cyanobacteria: a potential biofertilizer for rice. Resonance 9:6–10

Mrvčić J, Stanzer D, Šolić E, Stehlik-Tomas V (2012) Interaction of lactic acid bacteria with metal ions: opportunities for improving food safety and quality. World J Microbiol Biotechnol 28(9):2771–2782

Nam SH, An YJ (2021) Perspectives on microalgae as model organisms toward the standardization of soil algal toxicity test methods. Comp Biochem Physiol Part C Toxicol Pharmacol 249:109144. https://doi.org/10.1016/j.cbpc.2021.109144

Párraga CA, Brelstaff G, Troscianko T, Moorehead IR (1998) Color and luminance information in natural scenes. J Opt Soc Am A Opt Image Sci Vis 15(3):563–569

Pavlostathis SG, Prytula MT, Yeh DH (2001) Potential and limitation of microbial reductive dichlorination for bioremediation application. In: Proceeding of the first European bioremediation conference, Chania, pp 101–104

Rada EC, Andreottola G, Istrate IA, Viotti P, Conti F, Magaril ER (2019) Remediation of soil polluted by organic compounds through chemical oxidation and phytoremediation combined with DCT. Int J Environ Res Public Health 16(17):3179. https://doi.org/10.3390/ijerph16173179

Rai LC, Tyagi B, Rai PK, Mallick N (1998) Inter-active effects of UV-B and heavy metal (Cu and Pb) on nitrogen and phosphorus metabolism of a N_2 fixing cyanobacterium *Anabaena doliolum*. Environ Exp Biol 39:221–223

Rai LC, Kumar HD, Mohn FH, Soeder CJ (2000) Services of algae to the environment. J Microbiol Biotechnol 10:119–136

Relwani LL (1963) Role of blue green algae on paddy field. Curr Sci 32:417–418

Rodrigo-Comino J, Lopez Vicente M, Kumar V, Rodríguez-Seijo A, Valkó O, Rojas C et al (2020) Soil science challenges in a new era: a transdisciplinary overview of relevant topics. Air Soil Water Res 13:1e17. https://doi.org/10.1177/1178622120977491

Rodrigues ET, Pardal MA, Gante C, Loureiro J, Lopes I (2017) Determination and validation of an aquatic maximum acceptable concentration-environmental quality standard (MAC-EQS) value for the agricultural fungicide azoxystrobin. Environ Pollut 221:150–158

Roger PA, Reynaud PA (1982) Free-living blue-green algae in tropical soils. Martinus Nijhoff Publisher, La Hague

Shang QT, Doty SL, Wilson AM, Howald WN, Gordon MP (2001) Trichloroethylene oxidative metabolism in plants: the trichloroethanol pathway. Phytochemistry 58:1055–1065

Sharma S, Tiwari S, Hasan A, Saxena V, Pandey LM (2018) Recent advances in conventional and contemporary methods for remediation of heavy metal-contaminated soils. 3 Biotech 8(4):216. https://doi.org/10.1007/s13205-018-1237-8

Song T, Martensson L, Eriksson T, Zheng W, Rasmessen U (2005) Biodiversity and seasonal variation of the cyanobacterial assemblage in rice paddy field in Fujian, China. FEMS Microbiol Ecol 54:131–140

Subash Chandra Bose SR, Balasubramanian R, Megharaj M, Venkateswarlu K, Naidu R (2011) Consortia of cyanobacteria/microalgae and bacteria: biotechnological potential. Biotechnol Adv 29(6):896–907

Subash Chandra Bose SR, Ramakrishnan B, Megharaj M, Venkateswarlu K, Naidu R (2013) Mixotrophic cyanobacteria and microalgae as distinctive biological agents for organic pollutant degradation. Environ Int 51:59–72
Suriyaprakash R, Dowlath MJH, Karuppannan SK, Arunachalam KD (2021) Chapter 12 – Soil contamination, risk assessment and nanobioremediation. In: Amrane A, Mohan D, Nguyen TA, Assadi AA, Yasin G (eds). Elsevier, Nanomaterials for soil remediation, pp 239–258., ISBN:9780128228913. https://doi.org/10.1016/B978-0-12-822891-3.00012-8
Sutherland DL, McCauley J, Labeeuw L, Ray P, Kuzhiumparambil U, Hall C, Doblin M, Nguyen LN, Ralph PJ (2021) How microalgal biotechnology can assist with the UN Sustainable Development Goals for natural resource management. Curr Res Environ Sustain 3:100050., ISSN:2666-0490. https://doi.org/10.1016/j.crsust.2021.100050
Tiwari A (2014) Cyanobacteria: nature, potentials and applications. Astral International Publishing House, New Delhi
Tiwari A (2016) Algal application in horticulture: novel approaches to wards sustainable agriculture. Ann Hortic 9(2):117–120. https://doi.org/10.5958/0976-4623.2016.00048.7
Tiwari A, Kaur A (2014) Allelopathic impact of cyanobacteria on pathogenic fungi. Int J Pure App Biosci 2(3):63–70
Uroz S, Calvaruso C, Turpault MP, Frey-Klett P (2009) Mineral weathering by bacteria: ecology, factors, and mechanisms. Trends Microbiol 17(8):378–387. https://doi.org/10.1016/j.tim.2009.05.004
Valderrábano M, Nelson C, Nicholson E, Etter A, Carwardine J, Hallett JG et al (2021) Using ecosystem risk assessment science in ecosystem restoration: a guide to applying the Red List of Ecosystems to ecosystem restoration. In: Using ecosystem risk assessment science in ecosystem restoration: a guide to applying the Red List of Ecosystems to ecosystem restoration. https://doi.org/10.2305/iucn.ch.2021.19.en
Verma A (2021) Bioremediation techniques for soil pollution: an introduction. Intech Open. https://doi.org/10.5772/intechopen.99028. https://www.intechopen.com/online-first/78227
Whitton AB, Potts M (2000) The ecology of cyanophyta. Kluwer Academic Publishers, Dordrecht, pp 233–255
Wilson LT (2006) Cyanobacteria: a potential nitrogen source in rice fields. Texas Rice 6:9–10
Yanni YG (1992) The effect of cyanobacteria and Azolla on the performance of rice under different levels of fertilizer nitrogen. World J Microbiol Biotechnol 8:132–136
Zenova GM, Shtina EA, Dedysh SN, Glagoleva OB, Likhacheva AA, Gracheva TA (1995) Ecological relations of algae in biocenoses. Mikrobiologiya 64:121–133

The Impact on Soil Ecology of the Algal Community

Smriti Shukla, Divya Upadhyay, Arti Mishra, Tanu Jindal, and Kartikeya Shukla

Abstract In soil ecosystems, the algal community is essential because it affects a range of biogeochemical processes and promotes general soil health. This chapter sheds light on the diverse effects that algae have on the soil environment by examining the complex interactions between the algal community and soil ecology. The interactions between soil-dwelling organisms and algal communities are dynamic and reciprocal, with both positive and negative effects on soil structure, overall soil fertility, nutrient cycling, and microbial diversity. Algae promote microbial activity and nutrient availability by releasing organic compounds and exudates, which enhance soil fertility. Furthermore, some algal species develop symbiotic relationships with soil fungi to form mycorrhizal networks, which improve plant uptake of nutrients. However, excessive algal growth can also pose challenges, such as nutrient imbalances and alterations in soil pH, which may negatively affect other soil organisms. This chapter examines the growth conditions and distribution of soil algae, focusing on environmental factors that affect their abundance and presence. This chapter discusses the physiological ecology of soil algae, offering insights into their responses to changing environmental conditions and adaptive mechanisms. It focuses on the soil ecology of the algal community, elucidating the intricate relationships between soil-dwelling organisms and the dynamic interactions that shape soil ecosystems. Agriculture and its intersection with soil algae are examined, emphasizing the potential benefits and challenges posed by these microorganisms in the context of sustainable agriculture. Soil algae culture is explored as a potential

---

S. Shukla
Amity Institute of Environmental Toxicology, Safety and Management, Amity University, Noida, India

D. Upadhyay · T. Jindal · K. Shukla (✉)
Amity Institute of Environmental Sciences, Amity University, Noida, India
e-mail: tjindal@amity.edu

A. Mishra
Umeå Plant Science Center, Department of Plant Physiology, Umeå University, Umeå, Sweden

Department of Botany, Hansraj College, University of Delhi, Delhi, India

A. Mishra, A. Varma (eds.), *Soil Algae*, https://doi.org/10.1007/978-981-97-9873-5_6

avenue for harnessing the beneficial aspects of these microorganisms for agricultural and environmental applications. This chapter concludes with a summary of key findings and insights, providing a foundation for future research directions and perspectives on harnessing soil algae for sustainable land management practices. This chapter also aims to contribute to the growing body of knowledge surrounding soil algae, offering a holistic perspective that spans their ecological roles, distribution patterns, physiological adaptations, and applications in agriculture.

Keywords Algal community · Soil health · Soil-dwelling organisms · Sustainable agriculture · Physiological adaptations

1 Introduction

The soil is a mineral medium in which organisms live. These organisms play a key part in soil development because of their influence on fertility levels. They are essential for plant growth. Mineral layers combined in varying texture, shape, consistency, color, chemical properties, biological properties, and other properties with certain organic substances that differentiate it from its parental compounds. Supporting soil algae growth and their contribution to agricultural production will encourage the presence of enough moisture and phosphorus, as well as sunlight. Several micro- and macroorganisms are present in the soil that play a significant role in the development and fertility of the ground. Algae that belong to the micro group are the most frequent organisms found in the soil (Hoffman 1998).

Algae are omnipresent in the ecosystem and fluctuate greatly. In most of the soils, in which moisture and sunshine are accessible, algae have chlorophylls, and they are found in most of the soils (Hoffman 1998). Soil algae are known to interact with plant roots and to produce hormones that stimulate root development and increase other beneficial root activities of microorganisms linked with them. In certain cases, agricultural operations and human activities impact the interaction of soil biota, such as microorganisms, animals, and plants. Plague, chemical fertilizers, and agricultural methods in soil habors impact the structure of plants, animal life, and the soil community. Although various studies have been carried out in other parts of the globe, it is still difficult to make general conclusions about flora variety and its effect on the functioning of ecosystems (Metting 1981; Johansen 1993).

Soil algae such as higher plants may synthesize their own foods. They may be split into four different categories, blue-green, green, yellow-green and diatoms, in almost every habitat. At the soil surface and in subsurface layers of wet land, significant quantities are present. As biofertilizers, cyanobacteria have considerable potential (Adesalu and Olugbemi 2015). Biodiversity was shown as a key indicator of landscape quality and as a fundamental instrument to evaluate the effect, success, and impact of remedial activities by studies of bioindicators. Research over the impact of algal components on soil nutrients (Eneje and Donatus 2009) has been similarly carried out in Nigeria. The results of their experiments reveal that the soil nutrient status and structural stability were affected by the composting amendment. The algae compost improved the stability and nutrients of the soil. More studies have focused on water systems in Nigeria (Adesalu and Nwankwo 2005; Adesalu et al. 2008).

2 Soil Algae

A large comprehensive documentation, by contrast, indicates the focus on aquatic creatures. Different studies have investigated the probable sig-to-soil fertility of a photosynthetic microflora, which synthesizes carbon and fixes the nitrogen independently as a group. Many connections between environmental factors and the occurrence and composition of soil algal populations have to be determined, however (Prescott 1956). Some phycologists regard the terrestrial algae as degraded forms that make aquatic environments more natural. However, there is a tendency for a particular algal soil flora to appear healthy, while a number of species thrive in water (Tiffany 1951). Only terrestrial forms are recognized as *Botrydium*, the prototypon form, *Bumilleria exilis*, *chlorococcum humicola*, and a number of species such as *Zygnema*, *Zygogonium*, *Oedogonium*, *Botrydiopsis*, *Vaucheria*, and *Microcoleus* (Fritsch 1907; Tiffany 1951). Soil algae have been investigated primarily in descriptive and taxonomic studies in a wide range of geographically divided areas, including in England, Canada, Denmark, Africa, Germany, Scandinavia, Brazil, the Belgian Congo, Algeria, India, Japan, New South Wales, Australia, Cuba, Jamaica, Russia, lower Central America, and northern South America. In the United States, efforts have been made to list the plants of algal flora of the diverse soil environments within the dry Southwest.

Several recent studies have focused on soil Chlorophyceae grown in pure farming. The study highlights, before taxonomic determination of a soil species, the need of extensive observation in pure culture. Soil algae have been reported as the earliest successive stages of plant nutrient-poor substrates such as recent volcanic deposits (Trueb 1888; Russell 1952). Surface shapes act as major colonizers of stony, low fertile soils (Osborne 1950). The interwoven growth of algal species such as *Porphyrosiphon notarisii*, *Zygogonium ericetorum*, *Palmogloea vaginata californica* (Droue 1943; Shxelds 1957; Shxelds et al. 1957; Shxelds and Francis 1962; Durrell and Shields 1961) consolidates the surface, leading to the formation of a soil crust, which is particularly obvious if vascular plants do not form a constant ground covering. This algae-stabilized surface layer increases penetration, reduces runoff (Booth 1941; Elwell et al. 1939), and provides a substratum for the germination of seeds of higher plants during moist weather. After death, algae generate surface humus and perform solvent action on certain soil minerals preserving the availability of reserve components for higher plants in half-available form (Chodat 1928;

Martin 1949; Allxsos et al. 1937). Algae can play a substantial role in raising soil nitrogen, particularly in semidesert settings. Desert rock blue-green algae fix nitrogen, causing the buildup of nitrate (Allxsos and Morris 1930; Lowe and Moyse 1934; Waksman 1952; Foot 1947).

Many species, which do not fix nitrogen, give both amino and inorganic nitrogen indirectly to the substrate after death and breakdown (Fletcher and Martin 1948; Shxelds et al. 1957). The *Pleurococcus vulgaris* and *Protococcus viridis* soil algae and many others seem to avoid losing soil ammonia and nitrate leaching by turning those forms into organic nitrogen (Waksman 1952). The release of oxygen from rice roots and increased soil phosphate are crucial for blue-green algae (Haraison and Aiyer 1913; Fuller and Rogers 1952). A significant number of soil algae are proposed for their probable relevance, the survival of subsoil darkness, vitality conservation for long periods of drought, and energy storage (Waksman 1952).

3 Soil Fertility

Soil fertility refers to a soil's inherent ability to provide essential nutrients in adequate amounts and appropriate proportions. Microbials can promote soil fertility, as they are "built-in" soil regulators and catalysts to recycle nutrients to accessible inorganic forms and offer early warning of land deterioration. They present an excellent opportunity to improve soil fertility. The reasons that restrict soil fertility by microorganisms are explored, including environmental pollutants and climate change. The final section of this chapter highlights techniques for the control of soil fertility by means of microorganisms, such as organic farming, zero laying, mixed crop production, nano-biofertilization, biopesticides, and soil carbon diagnostics (Bharti and Dotaniya 2017).

Soil fertility is an important factor in determining agricultural environment. Texture, organic matter, soil microbial biomass, air, water, and other factors all contribute to the complexity of soil. Soil functionality is a measure of microbial diversity in the soil as well as crop yield potential (Singh et al. 2011a, b, c). In general, soil fertility refers to a soil's ability to supply nutrients to a crop, facilitate tillage, promote seed penetration, and support plant development (Dotaniya and Kushwah 2013). Soil fertility boosts microbial variety and population, or, to put it another way, soil microbial population boosts fertility in agricultural systems (Singh 2013; Singh and Singh 2013; Singh and Pandey 2013; Singh 2014). Man has been practicing agriculture since ancient times. Most agricultural methods are now elementary in nature and yield primitive results. In ancient times, cutting and burning were used to cultivate food crops to fulfill the needs of hungry mouths. However, as the world's population grew at a rapid rate, and the need for food increased, a shift in agricultural technologies in the form of the green revolution occurred in the mid-1960s. With the passage of time, people began to recognize the importance of soil fertility in connection to agricultural yield all over the world. Fertilizers containing nitrogen (N) have changed agricultural output and food supply in both developed and developing countries (Meena et al. 2013). Many nations, including India, experienced food scarcity during this period. Crop stagnation and soil health decrease were seen as a result of extensive exploration of arable areas and excessive use of chemical fertilizers, which is a serious problem in both emerging and developed countries (Singh 2015). The intense farming depleted soil nutrients, causing nutrient deficiency symptoms in plants and lowering crop productivity and quality. Farmers' present practices in relation to food grain production have limited the recycling of organics and other agricultural waste, resulting in a decrease in soil organic matter (Shukla et al. 2013).

As a result, the scientific community throughout the world is reconsidering age-old methods, such as the utilization of natural resources, agricultural residues, microbes, and organic amendments (Singh and Strong 2016). In this period, the usage of microorganisms is rapidly expanding, and agricultural crop productivity has risen from a state of stagnation. The relevance of soil microbial variety and population in agricultural crop productivity rose as a result of this revolution. The dynamics of nutrient release from any source in soil are mostly controlled by soil microbes. It plays an important role in the plant nutrition cycle, soil characteristics, organic matter breakdown, and organic and inorganic degradation in soil. Heavy metals, insecticides, herbicides, and other pollutants are all affected by soil microorganisms. They break down the chemicals into smaller, less dangerous components.

4 Role of Soil Algae

The ecological variables that influence the dispersal of algae are comparable to those that influence seed plant distribution (Tiffany 1951). Algae that grow on rocks and in soil provide organic matter and nitrogen to the creation of soil. The chemistry of the underlying substrate has less impact on the algal community on rock than accessible moisture, the nature of the surface, and the degree of isolation. In different regions of the world with similar environmental circumstances, parallel algal species colonize rock. Algal resilience to temperature fluctuations might explain the similarities of algal growth on rocks in tropical and temperate regions (Zeitnder 1953).

The term "soil algae" refers to a wide range of biologically varied organisms, including algae that grow over the whole surface of soil and create a crust, algae that live on moist soil surfaces, and algae that live within the soil (Johansen and Shubert 2001; Starks and Shubert 1982). Underground soil algae in dry environments are seldom seen as solitary organisms, preferring to create continuous biological soil crusts with fungi, lichens, club mosses, and bryophytes (Evans and Lange 2003; Johansen and Shubert 2001; Belnap et al. 2003). A soil-algal crust is described as surface algae growing on sand grains in this study. Such surfaces offer varying degrees of ground stability, including structural, physical, and wind and water erosion resistance. Algae play a crucial part in the early stages of pedogenesis.

5 The Soil Algae Systems

Members of the major divisions (Bold and Wynne 1978) make up the algal flora of the soil: Cyanochloronta (blue-green algae), Chlorophycophyta (green algae), Chrysophycophyta (diatoms and yellow-green algae), Euglenophycophyta (euglenoids), and Rhodophycophyta (red algae). The Cyanochloronta and Chlorophycophyta account for the majority of soil algae, both in terms of quantity and kind; many belong to the Chrysophycophyta, and a few to the other two divisions. Although it is generally recognized that some taxa (species, genera, and in some instances orders) are unique to edaphic conditions, the methods of research and nomenclatural issues connected with soil algae are similar with freshwater species (Bold 1970). Although it was argued in 1963 that sufficient data for taxonomic treatment could be obtained from herbarium specimens, most phycologists now believe that proper taxonomic treatment of soil and freshwater microalgae should be based on prolonged study of unialgal and pure cultures under standard laboratory conditions (Bold 1970; Hoek 1963). The availability of live cultures of soil algae allows for a thorough examination of the morphological and physiological characteristics that have led to the classification of species. A number of culture collections have been created across the globe to guarantee that a live record of the variety of algae is accessible (Hindak 1970; Carr and Whitton 1973; Ettl 1976).

6 Distribution and Growth Conditions

Temperature loses relevance as a factor in determining distribution due to soil algae's resilience to extremes of heat and cold. The number of species present and the luxuriance of the soil algal flora are both influenced by nitrogen and moisture availability (Zeitnder 1953).

However, the chemical character of the substrate has less of an impact on the species composition of soil algal communities than certain physical factors that control soil moisture levels. In semiarid Southwest, soil texture has a significant impact on the occurrence of diatoms and blue-green species. Certain species have a better chance of surviving burial in the higher soil strata than others. Substrate, pH, water demand, and depth in soil are the four factors that most clearly impact the formation of soil algae (Shxelds and Francis 1962; LuNo 1945).

7 Soil Substrate

Pleurococcus in wet and *Microcoleus vaginatus* in dry environments are examples of soil algae communities that reside between the soil particles directly under the surface and those that occupy the soil surface or surface items, frequently producing a visible growth apparent to the human eye (Shxelds and Francis 1962). The quantity of essential mineral salts and moisture available on both cultivated and uncultivated soils is directly connected to the presence of algae. A million individual algae can be found in a gram of well-irrigated soil. In certain cases, the algal flora of woods and grasslands is more numerous than that of arable soil. Blue-green algae are more common in cultivated soils than in uncultivated soils in temperate climates (Tiffany 1951). The number of algal species in the rhizospheres of wheat, cotton, and sorghum in non-rice soils of India is higher than in control soils, with the species changing with the crop plant rather than soil fertility (Gonzalves and Valavicl 1960). Perennial and winter plants lead to the formation of large amounts of soil algae on sod-podzolic soils. Legumes encourage the growth of green algae, while cereal crops favor the growth of blue-green algae and diatoms (Shtma 1957).

8 Soil pH Affects Algae

The Nearly ubiquitous alkalinity of dry soil and the acidity of constantly moist soils make it challenging to distinguish the effects of pH and moisture difficult. The prevalence of the genera *Microcoleus*, *Schizothri*, *Scytonema*, and *Porphyrosiphon* in semiarid southwestern soils appears to be due to a combination of these two edaphic variables, as opposed to other genera on acid substrates (Fletcher and Martin 1948; Shxelds and Francis 1962; Coyle 1935). Despite the fact that most algae have a high pH tolerance, populations thrive in mildly acidic or alkaline environments (Blisks 1951; Durrell 1959; Coyle 1935). On acid and alkaline soils, however, the dominant species are frequently quite different (Tiffany 1951). While the pH value may not be immediately important, the chemical reactions in the substrate that are affected by pH may be (Lowe and Moyse 1930). Algal development and nitrogen fixation are commonly stimulated by natural humic acid, most likely because to the availability of iron (Allxsos et al. 1937).

pH is less important than moisture in influencing the spread of soil algae, according to some researchers. However, these findings appear to be based on laboratory cultures of large surface samples including spores and cells of forms that do not proliferate vigorously in soil but are favored by the changed pH, increased calcium carbonate concentration, and fluid culture medium (LuNo 1947).

9 Requirement of Water

Moisture appears to be the element that most commonly limits the spread of species not resistant to desiccation, notably the Chlorophyceae, despite the fact that micronutrient needs for green and blue-green forms differ (Allen and Arnon 1955; Allen 1956). Drought susceptibility varies greatly depending on the protoplast's physiologic characteristics and the existence of an enveloping sheath. The ideal moisture range for soil algae is similar to that of bacteria and actinomycetes, around 40–60%

of the soil moisture-holding capacity. A surplus of surface moisture is less harmful to algal growth than a surplus of moisture in the subsurface layers. Soil algae vary by season, especially surface forms, which proliferate quickly in favorable circumstances but perish in adverse ones (Petersen 1935; Bristol-RoAc 1926). There is usually no seasonal succession of algal species, simply a shift in populations in response to meteorological conditions, which is primarily due to moisture impact (LuNo 1947). The diatom flora on woodland and other soils may change periodically as a result of shade (LuNo 1945). There was no seasonal succession in the algal species present in garden soil kept free of weeds during a 5-year period, although fluctuations in numbers may be associated with changes in weather (LuNo 1947).

In Oklahoma, the composition of the algal population in the plow layer of farmland and protected prairie soil did not alter from the end of March until the end of May. Enrichment of soil cultures is also important. Using a mineral or organic solution may speed up the growth process; however, it had no effect on the species makeup in any other way (Wxllson and Forest 1957). The total number of seasonal variations in algal cells at the surface of soils in England has been estimated 105,550 per gram in the summer to 91 per gram in November, according to dilution method counts (Bristol-RoAc 1927a, b).

10 Soil Depth

All algal development takes place in the top few inches of the soil, with subsurface algae being transported down by natural forces but not multiplying to any significant degree after burial. Algal cells may be transported down for a distance of two meters by soil cultivation followed by rainwater action, although their numbers drop fast with depth (Petersen 1935; Durrell 1959). The algal flora in a farmed Oklahoma soil was comparable throughout the plow layer, albeit it was generally more luxuriant toward the surface (Wxllson and Forest 1957). Algae may live in the soil for up to 80 years, which explains the existence of algal cells in lower soil layers, despite the fact that the quantity transported down annually is not large (Petersen 1935). Cholodny's approach (Colodny 1930) of letting slides to remain in the soil in a vertical posture for a length of time and then staining for adhering organisms has been used by a number of researchers to determine the composition and relative abundance of the soil algal flora.

There are no algae species found in deeper soil layers that are not on the same soil's surface (Petersen 1932). In high light intensity, possibly motile species are negatively photoactic and move below the surface (Moore and Karrer 1919; Moore and Carter 1926; Petersen 1935). Algae may be more abundant at a depth of two to six inches than at the surface at periods of the year when conditions are adverse for development (Bristol-Roac 1927a, b).

Danish soils, on the other hand, have just 20,000 algal cells per gram below five cm, compared to a maximum of 3,000,000 above the surface (Petersen 1935). In a homogenous soil, however, the horizontal distribution of algae is uniform (Bristol-Roac 1927a, b). The underground algal flora of particular Missouri soils was characterized by the presence of *Protoderma viride* to a depth of one meter and the lack of blue-green surface forms (Moore and Karrer 1919). Other data suggest that algae may proliferate in soil in conditions of darkness and moderate wetness, as seen by a 500-fold rise in algae seeded into soil and kept at room temperature in complete darkness (Skinner 1932).

11 Physiological Ecology of Soil Alage

When it was initially realized that some groups of species were limited to soil in the late nineteenth century, algae were accepted as true soil microorganisms. Because aquatic forms are more easily observed and gradations between planktonic, hydro-terrestrial, and terrestrial microenvironments are slight, algae are thought to be misplaced freshwater creatures. Because of the realization that algae, as a component of the soil microflora, act as a reserve for plant nutrients, as organisms influencing soil structure and the activities of other organisms, and as agents for the incorporation of organic carbon and nitrogen through photosynthesis and dinitrogen fixation, the interpretation of edaphic algae as soil microorganisms is more widespread today.

Each alga's genotype determines the pace at which it responds to climatic and microenvironmental variables inside or on the soil. Little is known about how resident populations of algae adjust their activities to compensate for changes in nutrient availability, solar radiation quality and intensity, moisture, temperature, pH, ionic strength, oxidation-reduction potential, soil atmosphere composition, pressure, or the presence of associated microorganisms and plant roots.

Almost all ecological studies of algal development in soil have been indirect, with single or rare samplings of algal populations as well as related physical, chemical, and biological variables. As a result, laboratory studies and empirical data are used to study the eco-physiological characteristics of soil algae.

Despite the fact that no one has accompanied you, the characteristics of the individual plant and plant mass are listed in Table 1. Additionally, descriptions of newly discovered species, along with photographs and documentation of an additional set of physiological characteristics, are included.

12 Nutrition of Soil Algae

Most soil algae are likely photoautotrophs, meaning they get their energy from the sun and make all of their organic molecules from inorganic precursors. Other nutritional groups of soil algae include photoauxotrophs, which use solar radiation as an energy source but are unable to synthesize all required organic molecules from inorganic precursors. The extent to which the algal flora is reliant on organic growth

Table 1 Taxonomic requirements of microalgae from soil at the species level

Individual organism morphology:
Cell division type
Vegetative state (unicell, coenobe, filament, etc.)
The quantity and shape of plastids
Pyrenoids' presence, quantity, and shape
Zoospore morphology and behavior
Nuclei's number and position
Clone physiological characteristics:
Immunology that compares and contrasts
Production of extracellular enzymes
Facultative heterotrophy is a kind of heterotrophy that occurs as a result (photoheterotrophy and chemoheterotrophy)
Low oxygen pressures promote growth
Growth in microbiological medium that is complicated
Antibiotics and stains do not inhibit growth
The need for growth factors
Plant mass morphology and color on agar:
The plant mass's size and form
When nitrogen is depleted, the color changes
A colonial mucilage's presence or absence

factors from the soil cannot be correctly determined due to a lack of data, which stems in part from the fact that studying microorganism growth factor requirements requires tedious and meticulous procedures. Peranema, a colorless chemoheterotrophic (=organotrophic) alga, has been found in soil, despite the fact that the frequency of reports suggests that these algae are unlikely to be essential in energy or nutrient transformations. Treboux was the first to discover that a number of soil algae are facultative photoheterotrophs or chemoheterotrophs (1905), while investigations by Bristol-Roach (1926), Skinner and Gardner (1930), Ellner and Steers (1955), Samejima and Meyers (1958), Parker (1961), Parker et al. (1961), and more recently by Khoja and Whitton (1971) and Hoare et al. (1972) show their extensive prevalence across higher taxa. Taxonomic importance is attributed to the capacity to use simple sugars and organic acids in the light or dark (Table 1).

13 Oxidation-Reduction Potential

Part of blue-green algae's rapid growth and dinitrogen fixation may be attributed to low oxidation-reduction potentials seen in wet soils. After a rainfall, eukaryotic algae's activities may be limited since potentials have been reduced. With regard to a perennial *Scytonema ocellatum* Lyngbye from Indian grasslands, Singh (1942) found that optimal development took place between 20% and 40% of the soil's water holding capacity and between −410 and −456 millivolts in terms of moisture equivalents (MECs). The distribution of anaerobic and microaerophilic growth capacities is best approximated through pure culture experiments across different species and taxa.

14 Texture of Soil

Little is known about how soil texture affects algal distribution, even though there have been studies that link the micro-distribution of xeric algae to the substrata of rock, gravel, and diaphanous material. There has been some evidence to suggest that the texture of mineral soils influences diatom distribution and blue-green algae, but Gonzalves and Gangla (1949b, c) found that texture had no effect on algae distribution in soils sown to dry and irrigated wheat.

15 Biotic Interactions of Algae

Biotic symbioses with algae have been described as amensal, competitive, commensal, and protocooperative as well as parasitic and neutral in nature. Studies were based on observations of planktonic communities or trials using soil algae in most cases. This symbiotic relationship between dinitrogen-fixing blue-green algae, other microbes, and rice is the most well-known one involving soil algae. Rice could grow in a nitrogen-free medium when associated with blue-green algae, as demonstrated by M. B. Allen (1956a), and Stewart (1967b), as well as Mayland and McIntosh (1966a), who showed that 15N assimilated as N2 by blue-green algae was transferred to higher plants, provided convincing proof that dinitrogen fixation by soil algae is a commensal process.

Pratt (1940) verified previous hypothesis that algae generated products that hindered the development of other algae and related bacteria (Akehurst 1931) by reporting the discovery of an antibiotic material derived from an alga. Chlorellin, a compound discovered in old *Chlorella* sp. cultures, was revealed to be an unsaturated, long-chain fatty acid that inhibited both the alga and a variety of bacteria. *Chlorella* old cultures accumulate autoinhibitory organic peroxides, as well. There was evidence that the algae *Nitzchia palea* (Kutz) W. Sm., *Chlorella pyrenoidsa* Chick, *Stichococcus bacillaris*, *Protosiphon botryoides* (Kiitz) Kelbs, and *Scenedesmus quadricauda* (Turp) Breb also secrete inhibitory chemicals.

Another 122 two-membered associations involving soil algae were investigated by Parker (1960) using the medium derived from the Texas soil, where the microorganisms had been isolated. Among the combinations, 53.1% stimulated the algal component, 14.7% inhibited it, and 32.2% had no effect on it, according to his findings. It was shown that changes in irradiance, photoperiod, and temperature had no effect on 15 different relationships and that most instances of protocooperation or commensalism occurred when an autotroph and a heterotroph were working together. A higher nitrogen supply from the bacterium's degradation of complex chemical compounds resulted in *Bracteacoccus* sp. growing faster in combination with an unidentified gram-negative than in axenic culture *Streptomyces* sp. Waksman & Henrici (1943), increased *Chlamydomonas* growth and motility. The reduced nitrogen supply was thought to be responsible for the alga's enhanced mobility. The actinomycete and alga both produced more akinetes and conidia when nitrogen was depleted. When an unknown fungus was present, growth of the bacterium *Phormidium* was suppressed by an unknown agent.

Any two soil algae can have a community interaction value (CI), which Parker and Turner (1961) determined. The CI was calculated as the percentage of all comparable microbial interactions in which different associative effects were observed as the similarity of the predicted habitat requirements of two algae. In general, the closer two algae are to having CI values of zero, the more similar their niche requirements will be one or more strains of Azotobacter Beijernick have been found to interact with soil algae and other free-living bacteria. From the extracellular mucilage of *Nostoc* species, *Gloeocapsa* species, and *Rivularia* species (Roth), *Azotobacter* species (together with *Bacillus* Cohn and *Clostridium* Prazmousski) have been isolated. Tapwater cultures of *Chlorella* and *Azotobacter* retrieved almost two times as much total nitrogen as cultures of the bacteria alone. *Azotobacter* was said to be promoted by green algae, according to Nakano (1917); however, Stokes (1940a, b) found a commensal effect. In spite of being able to use nitrogen fixed by *Azotobacter* sp., the bacteria were unable to use carbon from *Microcystis* Lemm sp.

Azotobacter populations may be suffering from a lack of nutrition due to the growth of blue-green algae in rice fields. Artificial symbiosis has been used to isolate and reproduce *Rhizobium* Frank species using green and blue-green algae.

Using a soil dialyzate as the medium's base, an unknown heterotrophic bacterium in combination with *Chlamydomonas* sp. increased the alga's growth sevenfold. *Streptomyces* sp. growth was promoted by photosynthetic oxygen generation in *Chlorella* sp., while the alga was hindered by alcohol synthesis in the actinomycete. In the same series of investigations, a protocooperative relationship was discovered between *Chlorella* sp. and *Corynebacterium* Lehmann & Neumann. *Scenedesmus dimorphus* Kutz, which in pure culture produces coenobes, is unicellular when cultivated with a heterotrophic soil bacteria, as shown by Trainor (1963). Protozoa, nematodes, snails, microarthropods, and earthworms all utilize soil algae as a source of energy, as do nematodes and snails, and microarthropods. Bacteria and fungus quickly break down dead algae. They found that breakdown rates differed between *Achromobacter* (=*Alcaligenes* Castellani & Calmers), *Bacillus*, *Flavobacterium* Bergey, and *Pseudomonas* Migula according to Watanabe and Kiyohara (1960). Saprophytic bacteria in the soil used algal cell walls, cytoplasm, and extracellular polysaccharides as energy sources after the death of algae.

After 22 weeks of incubation in soil in the laboratory, species of *Anacystis*, *Microcoleus*, *Plectonema*, and *Stichococcus* from a sandy loam, as well as *Anabaena flos*-aquae (Lyngb.) *Breb*, *Nostoc muscorum*, and *Chlorella pyrenoidosa* from different habitats, easily degraded; 60–80% of the radioactive carbon that was added to cells and extracellular components during pretreatments has been converted into the radioisotope 14CO2. The degradation of *A. flos*-aquae and *N. muscorum* wall and cytoplasmic preparations was decreased by complexing them with model humic acid-type phenolic polymers.

Systematic relationships among bacteria found on soil algal cell surfaces and extracellular sheaths are poorly understood. Shtina (1968) found that nonspore-forming gram-negative bacteria predominated among the 170 blue-green algae strains she examined. With time, she discovered that the bacterial population became less diverse, with just three species remaining after three months: *Pseudomonas radiobacter* (=*Agrobacterior* sp. Beijernick & van Deldea), *Flavobacterium* sp., and *Caulobacter* sp. Henrici & Johnson The difficulty in separating blue-green algae from bacteria shows how closely they are linked. Most mucilagenous sheaths include bacterial species of satellites, such as actinomycetes, *Caulobacter*, and Chromosadum Bergonzini, but the interaction of these bacteria is not well known. If you consider lichens to be parasitic, commensal, or even protocooperative, they represent one of the most well-studied biotic relationships between terrestrial algae and their environment. The systematics and ecology of lichens are beyond the scope of this review. To summarize, lichens and algae seem to have little ecological differences in many situations.

16 Soil Ecology of the Algal Community

As soil is formed and matured, the actions of algae play an important role. Initial synusium includes algae on newly exposed substrates such as volcanic deposits and rocks. Random dispersion determines whether types of algae will be the main colonists. Treub (1888) stated that blue-green algae of the species *Anabaena*, *Lyngbya*, *Symploca*, and *Tolypothrix* were the first colonists on recent Krakatoa deposits in Indonesia. Schwabe (1972) found over 100 species of algae following the volcanic creation of Surtsey off the coast of Iceland, including about 10% of S blue-greens, while others found moss protonema, lichens, and coccoid green algae, a phenomenon repeated on Katmai (off the coast of Alaska) (Alaska). Green algae, lichens, and mosses dominated on Iceland and in the Aleutian Islands, while blue-green algae colonized Krakatoa and were the most numerous landforms closest to the new island. Given that air currents are a reliable supply of algal propagules, the rapid colonization of these volcanic islands is not unexpected. Roof discoloration is evidence of algae's ability to travel through the air. For example, algae aid in the retention of water, which is the primary weathering agent, and create slightly acidic conditions via photosynthetically induced production of carbonic acid in primary substrates, which promote the weathering of silicate minerals. Scientists have found evidence that algae extract orthophosphate from insoluble compounds, including insoluble compounds such as apatite, using extracellular polysaccharides of algae. These findings were made by Aristovskaya and colleagues (1969), who found that extracellular polysaccharide actions were among the most important microbiological processes that promoted the release of nutrients from insoluble compounds into the environment. During the breakdown of kaolin, the growth of diatoms has been implicated. Dinitrogen fixation and chemical element release from rocks are attributed to the cliff-dwelling algae's activities as the reason for the fertile mountain valley soils. Nitrate deposits in the Western Pamir, according to Odintsova (1941), are the consequence of long-term nitrogen fixation by chasmolithic blue-green algae.

Rock-associated algae's floral composition seems to be influenced by insolation and substrate wetness. Species of *Scytonema*, which may grow up to 8 millimeters through limestone in wet temperate areas, invade the rock. Blue-green algae of the genus *Aphanocapsa*, *Gloeocapsa*, *Gloeothece*, *Nostoc* (Fritsch 1907), and *Scytonema* may colonize other rock surfaces, particularly smooth ones. For filamentous green and blue-green algae, as well as moss protonema, wet temperate zones with more acid rock surfaces promote growth. Shaded siliceous and calcareous rock frequently supports *Tolypothrix byssoidea* Kirchn., *Stigonema minutum* (Ag.) Hassal, and *Gloeocapsa* species, which may weaken the rock to the point that it can be readily peeled off by hand to a depth of 1–2 millimeters, depending on the kind of rock.

Extreme wetness, insolation, and temperature in dry areas may prevent succession from progressing beyond initial cryptogam colonization. The climax community is made up of a variety of microenvironments dominated by different algae species. Rain crusts, algal crusts, and lichen crusts are all types of epedaphic communities that may be found. In contrast to the rain crust, which is dynamic and microephemeral, an algal crust is a clearly visible and adhering plant mass. A wide variety of algal crusts can be found throughout the world's latitudes, from the steppes and deserts to the polar regions (such as the Arctic and Antarctic), with Anacystis montana (Lightfoot) being the most frequent component of algal crusts worldwide, according to Cameron and Blank (1966a), who follow the conservative categorization of Drouet. *Scytonema hofmanii*, *Aphanothece stagnina* (Spreng.), *Coccochloris stagnina* (= Drouet & Daily, A. montana (Hansg.), Nostoc muscorum Fuscus Porphyrosiphon, *Microcoleus vaginatus* (Vaucher) Mert., Gomont, M. paludosus Zanard, M. chthonoplastes (Kützing) Gomont; The protuberans palm tree (Sm. & Sow.) According to Bolyshev (1952), who classified blue-green and green algae according to the traditional classification system, the following genera (with species numbers in parentheses) were the most important in takyr crusts: *Protococcus grevillei* (Agardh), Crouan, *Protosiphon botryoides*, *Chlorella vulgaris*, and *Trochisia hirta* (Reinsch). As an example, *Phormidium* has ten species, *Lyngbya* has nine, and the remaining species including *Nostoc* has four. Other species include *Aphanothece* and *Chlamydomonas* and the other species are *Calothrix*, *Nodularia*, and *Palmella* as well as *Protococcus* and *Scenedesmus* and *Schizothrix*. Forest (1965) found *Symploca atlantica* Gomont, *Schizothrix calcicola*, and *Microcoleus cthonoplastes* in two samples of takyr crust. On Sri Lanka's sandy beaches, *Nostoc* spp. also produce algal crusts (Fritsch 1907). There are algal metabolites in and around algae and rain crusts that chelate iron and manganese that may produce desert varnish, an irregularly distributed dark brown or black surface covering up to 1 mm thick in dry areas.

It has been suggested that in desert microenvironments, lichen crusts are ecologically similar to algal crusts, and that they have taken their place as the climatic climax after the invasion of fungi. Killian (1953), on the other hand, found no evidence of lichenization on Saharan algae. As a result of invasions and development along large fractures in the algal crust and full conversion to a lichen, takyr crusts are often transformed from an algal crust to a lichen. When it comes to chlamydia communities, *Gloeocapsa* sp. rules in Middle Eastern deserts, while in dolomitic and calcareous ones, sarcinoid green and filaments of blue-green algae rule.

There are microgreenhouses created by diaphanous materials buried on the surface of desert soils because they reduce insolation, retain moisture, and mitigate temperature extremes. Because of these unique circumstances, the microflora in the area is more varied than it would be in other desert microenvironments. This "fensterflanzen" flora was discovered by Friedmann and Galun (1974) to be associated with diaphanous substrates in the Negev (Israel). Vogel (1955) used the term "fensterflanzen" to describe highly evolved angiosperm (*Aizoaceae* and *Liliaceae*) species with leaf anatomy similar to that of diaphanous material. *Scytonema ocellatum*, a parasite from Egypt, was also found there. He also found *Tolypothrix distorta* Kützing and *Nostoc muscorum* as well as *Microcoleus vaginatus* and *Anacystis montana* under diaphanous materials in Death Valley, California, according to Durrell (1962). While *Phormidium tenue* (Meneg.) Gomont thrived under salt crusts in Durrell's time, Shields et al. (1957) discovered *Palmogloea protuberans*, *Plectonema nostocorum* Bornet, and the uncommon *Microcoleus vaginatus*, *M. paludosus*, *Schizothrix hofmanii*, and *Nostrocea* species in the same place.

A major source of organic carbon and nitrogen during secondary community succession is algal bloom. Moisture retention and seedbed establishment are both aided by humus accumulation (Fletcher and Martin 1948; Aleksandrova 1951). Compared to bare soil, mature algal crusts reduce the rate of germination of broadcast-sown grass seed, but once established, seedlings develop faster when exposed to algae. As part of the secondary succession, algae help to stabilize soil structure by aggregating colloids using extracellular polysaccharides, including filamentous sheaths, individual mucilages, and colonial mucilages. Water infiltration and percolation are improved with the assistance of soil particle aggregation, which in turn serves to reduce water erosion. During algal development, soil samples from degraded farmland in Oklahoma had an average of 8.9% water, while soil samples that were devoid of algae contained just 1.3% water. Samples from India's top inch of soil contained 10–15% less water while supporting algal development than bare soil.

Experiments show that no algal species act particularly as pioneer algae in secondary succession, which supports the theory. However, agricultural disturbances and other human activities may cause seasonal changes within algal populations. Annual successions in flooded paddy soils and successions linked with the restoration of sodic-saline ("Usar") soils in India are the most well-known instances, which will be addressed later. Massive areas in Brazil are covered by *Porphyrosa notarisii* Kützing, whereas in India, the same species is frequently found as a co-dominant in grasslands and on otherwise barren soils, giving off a crimson or brick-red color. In India, seasonal dominants include yellow-brown, jelly-like masses of *Nostoc* commune, dark blue-green patches of *Cylindrospermum* sp., shiny dark-brown patches of *Scytonema ocellatum*, and a variety of colored growths of *N. muscorum*, *N. humifusum* Carm. ex Bornet & Flahault, and *Aulosira fertilissima* Rao on fallow fields.

During the first year or two after a severe drought, *Klebsormidium flaccidum* (A. Brown) Silva, Mattox & Blackwell becomes dominant in dry grasslands in England, killing most of the vascular flora. Green algae, mosses, and forbs were the first plants to emerge on burnt heath. According to Feher (1936, 1948), in central European conifer woods, green algae were succeeded by blue-green algae in the summer, while in dicot woods, the opposite was true. Feher's findings were the sole evidence of seasonal succession in algal communities of undisturbed soils. According to Feher, in northern Europe, the reverse was true in terms of the predominance of algal groups in relation to season and macrovegetation.

17 Agriculture and Algae Soil

Aside from the carbon and nitrogen input, algal development in soil has beneficial impacts on soil structure and erosion management. The manipulation of algal populations in agricultural soils has so far served three distinct but connected objectives. For example, using dinitrogen-fixing species to inoculate paddy soils with combined nitrogen, using aquatic and edaphic algal development as green manure, and inoculating unproductive soils with green or blue-green algae to enhance the soil's structure are all examples of green manure techniques. While algae's impacts on crops are mostly unclear, it is possible that their production of hormones and vitamin-like substances stimulates root development or benefits other microbes.

India turned out to be helpful. Oat yields were improved by 34% in the Soviet Union when soil was injected simultaneously with *Nostoc* and *Azotobacter*, a figure greater than the total of separate inoculation experiments. The development of tobacco in greenhouses has been shown to be inhibited by algae, while the growth of existing citrus seedlings is promoted. Inoculating cranberry bogs with algae, according to Wakabayashi (1925), may boost harvests by raising oxygen levels in the root zone. In greenhouse experiments, vaccination of tomato plants with

Tolypothrix tenuis resulted in higher vitamin C content in the fruits, on par with other *Azotobacter* strains (Aiyer et al. 1964). *Calothrix*-fertilized lettuce and chilies outperformed non-algal controls in terms of yield.

18 Soil Conditioning with Algae

Algae were employed as green manure on cotton fields, for example, in Tashkent, India, and Mexico. They were also utilized to enhance structure in Mexico and in central Washington state (Metting, unpublished).

For the reclamation of "Usar" lands, according to Singh (1950, 1961), artificial impoundments are created by building half-meter earthen walls around an area no larger than an acre. Following the first seasonal rains in June, the first algae to develop on loose soils include *Microcoleus* (*M. cthonoplastes*, *M. vaginatus*) and *Scytonema* (*S. ocellatum*, *S. hofmanii*, *S. javanicum*). On hard, compacted soils, *Porphyrosiphon notarisii* assumes its rightful position as the dominant species. By August, *Scytonema* dominates puddled soils, and the resting spores of any of the following algae germinate at this time as well, in this order: *Nostoc commune*, *N. muscorum*, *N. linckia*, *Camptylonema lahorense*, *Cylindrospermum licheniforme*, and *Cylindrospermum trilobum*. During the rainy season, this synusium of dinitrogen-fixing bacteria takes over the moist soil (August–November). When soils in impoundments become waterlogged for an extended period of time, *N. commune* is obviously the dominating alga. It is possible that *Aulosira fertilissima* or *Anabaena ambigua*, *Cylindrospermum gorakhpuernse* and *Wollea bhardwajii* will take over. Water-holding capacity increases by up to 40%, pH drops to 7.6, and exchangeable calcium rises by 20–30% after a year or two.

As a green manure, sugarcane, maize, and grassland soils benefit from natural algal succession. Because of the country's well-drained soils, sugarcane is cultivated there on well-drained silt loam or sandy loam. *Scytonema ocellatum* and *Microcoleus cthonoplastes* emerge after the first monsoon rains. First came *Aphanothece pallida*, then *Nostoc commune*, then *N. muscoruin* and *N. linckia*, and ultimately *Cylindrospermum licheniforme* and *C. maius* on drier soils, or *Aulosira fertilissima* on more wet soils, until it was eventually superseded by all of them. *C. licheniforme* thrives in maize-growing soils, whereas *P. notarisii*, *S. ocellatum*, *S. hofmanii*, *C. lahorense*, *C. licheniforme*, *C. maius*, or *C. muscicola* dominate grasslands. Raju has a comprehensive list of India's blue-green algae species (1972). *Chlamydomonas mexicana* is used in high amounts (1–5 kg/ha).

This method, called Lewin, has been suggested for use on newly tilled soils in dry areas of Mexico and the American Southwest in order to increase fertility while also improving structure. It seems from preliminary research that up to 90% of the algal photosynthate is converted to extracellular biopolymer under optimum circumstances, and cotton and potato yields increase by 5–15% depending on prior climatic and edaphic conditions. Initially, the inoculum increased to 50–200 kg/ha after 3–4 weeks, and water savings were up to 35–45%. Erosion was decreased as well as the salt table, and root rot incidence was reduced.

19 Influence of the Rhizosphere

Algae are influenced by crops as well. As previously said, agricultural methods have the ability to enhance or reduce algal variety and populations. The effects of the rhizosphere on soil algae differ depending on the kind of vascular plant (Aleksakhina 1972). When diatoms are combined with *Artemesia* Linn. and *Lasiagrostis* Link., as well as when diatoms and blue-green algae are grown in the rhizosphere with the roots of malt, cotton, or wheat, the results were positive. Gonzalves and Yalavigi (1960) found that the effects were strongest during anthesis, when the variety of green and blue-green algae was comparable to that seen in soil supplemented with farmyard manure or (NH4)2SO4. Both Hadfield (1960) and Rauatt et al. (1960) found a detrimental rhizosphere impact on algae in tea bushes and wheat.

20 Pesticides and Soil Algae Management

Several research projects have looked at the effect of insecticides (Table 2) and herbicides on soil algae development. In situ and culture experiment results have been mixed, with growth inhibition and stimulation described, as well as no impact on growth. However, there are a few things to keep in mind when interpreting these findings. The impact of pesticides on non-target microorganisms may be influenced by factors such as soil variability, farming methods, plant cover, and formulation. As a consequence, findings gained in culture cannot be extrapolated to field settings without caution with the exception of fungicides, for which no in situ research has yet been done. At least, a few studies have been performed in soil and laboratory culture on the three main pesticide families widely used in agriculture (Table 3).

Herbicides generally have a detrimental impact on soil algae growth and reproduction, although the effects are typically more apparent in the lab. When propanil was applied to soil rather than cultured algae, it had a lower inhibitory effect on the algae. Additionally, herbicides may be less or more harmful to algae when given in

Table 2 Insecticides' impact on soil algae

Insecticides	References	
Lindane (diatoms in rice paddies)	Ishizawa and Matsuguchi (1966)	Negative effects
Chlorpyrifos (blue-green algae in rice)	Sivasithamparam (1970)	Positive effects
Diazinon (blue-green algae)	Sethunathan and MacRae (1969)	
Carbaryl (blue-green algae in rice)	Ishizawa and Matsuguchi (1966)	Neutral effects
Parathion	Naumann (1970b)	

Table 3 Effects of fungicides on cultured algal growth

Fungicide	Reference	Effects
Captan (*Euglena gracilis*)	Lazaroff (1967)	Negative effects
Captan (*Stichococcus bacillaris*)	Marton (1973)	
Ceresan (*Anabaena* sp.)	Venkataraman and Rajyalkshimi (1972)	
Sichlone and most other quinones	Sikka et al. (1971), Zweig et al. (1972)	
Various algae	Schwelitz et al. (1974)	
Dichlorophen (*Chlorella pyrenoidosa*)	Gupta and Saxena (1974)	
Difolatan (*Westiellopsis prolifica, Tolypothrix tenuis*)	Gangawane and Saler (1979)	
Dithane (*S. bacillaris*)	Marton (1973)	
Hexacap (*Nostoc* sp., *W. prolifica*)	Gangawane and Saler (1979)	
MBC (*W. prolifica*)	Gangawane and Saler (1979)	
MBC (*C. pyrenoidosa*)	Carton (1976)	
Nabam (*Euglena gracilis*)	Lazaroff (1967), Moore (1970)	
Nabam (*Anabaena* sp., *Chlorella* sp.)	Audus (1970)	
Sodium benzoate (*S. bacillaris*)	Marton (1973)	
Vapam (*E. gracilis*)	Moore (1970)	
Zinels (*Anabaena* sp., *Nostoc* sp.)	Venkataraman and Rajyalakshimi (1972)	
Difolatan (*Aulosira fertilissima*)	Gangawane and Saler (1979)	Positive effects
Ceresan (*Anabaena* spp., *A. fertilissima, Tolypothrix tenuis*)	Venkataraman and Rajyalakshimi (1972)	Neutral effects
Ethirimol (*C. pyrenoidosa*)	Teal (1974)	
Hexacap (*A. fertilissima, Calothrix* sp., *T. tenuis*)	Gangawane and Saler (1979)	
MBC (*A. fertilissima, Calothrix* sp., *Nostoc* sp., *T. tenuis*)	Gangawane and Saler (1979)	
Zineb (*A. fertilissima, Tolypothrix tenuis*)	Venkataraman and Rajyalkshimi (1972)	

combination rather than individually. The purity of the algal culture may also be used in the lab to gauge a chemical's impact on algal growth. Herbicides, fungicides, insecticides, soil fumigants, and pollutants all have a negative, positive, or neutral effect on soil algae, as shown in the studies listed below. More may be stated about pesticides' potential impact on closely related edaphic forms by adding findings of culture work with species of *Anabaena, Anacystis, Chlamydomonas, Chlorella*, and *Euglena* from aquatic environments (Table 4).

Many algae absorb and breakdown pesticides in addition to being inhibited or stimulated. Temperature and pH have a significant impact on uptake. There is no evidence to suggest that absorbing a pesticide in situ by algae alters its toxicity or speeds up its breakdown. In addition to the direct estimate of biomass, several techniques have been used to quantify culture studies. Agar plate tests and the study of

Table 4 Fumigants and pollution impact on soil algae (all effects are negative)

Fumigant	References
Allyl alcohol (indigenous populations)	Nauman (1970a, b, 1971, 1972)
Dazomet (indigenous populations)	Nauman (1970a, b, 1971, 1972)
Formalin (indigenous populations)	Nauman (1970a, b, 1971, 1972)
Methamsodium (indigenous populations)	Nauman (1970a, b, 1971, 1972)
Pollutant Fe, S wastes	Kildema and Roos (1977)
PCBs	Wright (1978)
Tyrene (phenylethylene)	Munjko and Brbic (1977)

bioluminescence by *Photobacterium* Beijernick sp., the intensity of which is related to algae's photosynthetic oxygen production, are examples of paper disc assays on soil or agar media. Herbicides' effects on algae have also been used as bioassays to test whether or not they might be utilized to control weeds.

21 Soil Algae Culture

Preliminary information on the ecology of an algal form may be obtained with the assistance of enrichment cultures. If you are doing taxonomic study before moving on to other areas of algal biology, such as physiology and genetics, you will need pure cultures of the algae. Distinct groupings are often formed depending on the cultural circumstances present. Extracts from soil or peat may help the culture medium perform better. Soil algae may be cultured on sterile quartz sand in nutrient solution in Erlenmeyer flasks, slanting the sand with flasks tilted. Plant-associated microbiome network models provide new ways to improve illness management. An ecological gradient from an aquatic to a barely wet environment may be seen along one side of the solution, with part of the solution submerged and portion above it. Algae cells and spores that do not actively proliferate in soil are encouraged by a pH change, a greater calcium carbonate content, or the fluid culture medium in laboratory cultures of large surface samples. After a few days, the dominant forms tend to fade away and be replaced by more common and less delicate ones. Bold and Pringsheim have done a thorough study and addition of algal cultivation methods.

22 Conclusion and Future Perspective

Crop diseases and agricultural materials are now at great risk due to climate change, polluted surroundings, and other external pressures. The health and function of the host may be adversely affected by plant-associated microorganisms. Attempts to use beneficial microorganisms in the field, on the other hand, have been ineffective. Plant-associated microbe interactions, ecological implications, and plant-microbial metabolic dynamics have only a limited amount of current understanding. Genomics has greatly aided plant-bacterial interaction research, but it still falls short of explaining how plants interact with diseases (Bender 2005). Plant-bacterial interactions are better understood because to methods like metagenomic sequencing and amplicon sequencing (Quirino et al. 2010) as well as databases (PHI-based) and metabolomics. Genetic variation in plants and the environment influence plant-microbe ecological ecosystems (Winnenburg et al. 2008). Various genotypes have varying physiological and immunological responses, resulting in unique microbial communities on different hosts. Exudates from plant roots (such as sugars, amino acids, organic acids, nucleotides, flavonoids, antimicrobial chemicals, and enzymes) help to form local communities and attract disease infections.

We discovered that particular species in a large microbial community had a substantial impact on the community's overall structure. We could identify the "hub species" and "milestone species," which are potential microbial assemblages for disease management, using a systems framework for microbial network analysis. Plant-associated microbiome network models provide new ways to improve illness management.

Since humans stopped being nomads and started farming, plants have been an essential component of our nutrition. Since then, feeding the world's ever-increasing population has been a constant problem. Eutrophication is being caused by overuse of fertilizer (Smith et al. 1999), and genetically engineering plants is a time- and money-consuming procedure. Plant-microbe interactions have seen extensive research into the function of microorganisms in the last decade (Vorholt 2012; Bulgarelli et al. 2013). The outcomes of comprehensive research indicate that using beneficial microorganisms is a superior long-term approach. A significant role in disease transmission and control will be played by increasing crop production (Reid 2011). Symbiosis between microbes and plants is one of three kinds of plant-microbe interaction that has been researched. Pathogenesis, plant-to-rhizobacteria interactions, and plant-to-rhizobacteria interactions are all examples of this.

Studies on plant abiotic stress have made significant use of omics technologies including transcriptomics, proteomics, and metabolomics, as well as bioinformatics (Soda et al. 2015). Proteomics is a branch of molecular biology. The majority of the research on plant protein patterns in response to abiotic stressors may spur the development of novel approaches to stress management (Gupta et al. 2013). Another method used to investigate the collection is microbial metabolomics by-products produced by microorganisms, such as metabolites. A metabolomics-based rhizosphere approach by Lee et al. utilized a method to investigate plant-microbe interactions to remove polychlorinated biphenyls. Food security, agricultural practices, and food species conservation have all been put at risk by pathogens and their development. Therefore, understanding them has become a top priority, particularly in relation to emerging novel pathogens and their significance. Up until now, plant pathosystems have been considered one gene or protein at a time for effective analysis; however, the genomic era has ushered in a comprehensive investigation into plant-pathogen interactions. Generally accepted pathogen genomes have been sequenced and analyzed, revealing surprising features of pathogen development influenced by their biochemistry. Reconstruction at the genomic scale is based on the reconstruction of metabolic pathways on a genomic scale for metabolite analysis and interpretation concentrations in certain situations. A successful version of GSRM has been created for more than one hundred species of creatures (including but not limited to).

23 Future Perspective

Knowledge of plant-microbiome and microbe-microbiome interactions will be useful in the future for controlling the microbiome to reduce disease incidence and improve gross plant production. The beneficial microbes associated with plants may also serve as counterparts to diseases in the microbial ecosystem, thus stabilizing the environment, increasing biodiversity, and reducing pathogen outbreaks while also increasing plant production. Another "Green Revolution" may be on the way if a successful plant-microbe cooperation can be developed in the future.

References

Adachi M, Hamda K (1971) Chlorella assay of the activity of herbicides. Weed Res 1970:54–58

Addison DA, Bardsley CE (1968) Chlorella vulgaris assay of the activity of soil herbicides. Weed Sci 16:427–429

Adesalu TA, Nwankwo DI (2005) Studies studied on the Phytoplankton of Olero creek and parts of Benin River, Nigeria. Ekologia 3(2):21–30

Adesalu TA, Olugbemi OM (2015) Soil algae: a case study of two vegetable farmlands in Lagos and Ogun states, Southwest Nigeria. IFE J Sci 17(3):765

Adesalu TA, Abiola TO, Bofia TO (2008) Studies on the epiphytic algae associated with two floating aquatic Macrophytes in a sluggish non-tidal Polluted Creek in Lagos, Nigeria. Asian J Sci Res 1(4):363–373

Agardh CA (1812) Disposito Algarum Suecial. Lundae

Agardh CA (1817) Synopsis Algarum Scandinaviae. Lundae

Aiyer RS (1965) Comparative algological studies in rice fields in Kerala state. Agric Res J Kerala 3:1–4

Aiyer RS, Venkataraiman GS, Sundara Rao WVB (1964) Effect of nitrogen-fixing blue-green algae and Azotobacter chroococcum on vitamin C content of tomato fruits. Sci Cult 30:557

Akehurst SC (1931) Observations on pond life, with special reference to the possible causation of swarming of phytoplankton. J Roy Microscop Soc London 51:237–265

Akiyama M (1961) Aerial and terrestrial algae in San-in region, Honshu, Japan. Bull. Shimane Univ Nat Sci 10:75–89

Akiyama M (1965) Some soil algae from Japan. Bull Shimane Univ Nat Sci 15:96–117

Akiyama M (1966) Some algal vegetation of y-irradiated field and natural strongly radioactive district in Japan. Bull Shimane Univ Nat Sci 16:20–31
Aleksakhina TI (1971) Characteristics of soil algae in different forest types. Bot Zurn (Moscow & Leningrad) 56:1658–1669
Aleksakhina TI (1972) Soil algae in the rhizosphere of predominant plants in forest biogeocenoses. Ekologiya 3:45–52
Aleksandrova IV (1951) The process of humus formation in several primitive mountain soils. Pocvovedenie 1951:604–616
Aleksiyeva G (1960) Types of composition of soil algae in some typical soils to the east of Plovdiv (Bulgaria). Proc Sci Higher Inst Agric Vasil Kolavov 8:10–15
Alexander VA (1975) Nitrogen fixation by blue-green algae in polar and subpolar regions. In: Stewart WDP (ed) Nitrogen fixation by free-living microorganisms. IBP Programme 6. Cambridge Univ. Press, Cambridge
Alexander VA, Schell DM (1973) Seasonal and spatial variation of nitrogen fixation in the Banow Alaska tundra. Arctic Alp Res 5:77–88
Ali S, Sandhu GR (1972) Blue-green algae of the saline soils of the Punjab. Oikos 23:228–272
Ali S, Rajoka MI, Sandhu GR (1978) Blue-green algae of different rice growing soil series of the Punjab, India. Pak J Bot 10:197–208
Alimagno BV, Yoshida T (1977) In situ acetylene-ethylene assays of biological nitrogen fixation in lowland rice soils. P1 Soil 47:239–244
Allen MB (1952) The cultivation of Myxophyceae. Arch Mikrobiol 17:34–53
Allen MB (1956) Photosynthetic nitrogen fixation of blue-green algae. Sci Mon 88:100–106
Allen MB (1956a) Photosynthetic nitrogen fixation by blue-green algae. Sci Month 83:100–106
Allen MB (1956b) Excretion of organic compounds by Chlamydomonas. Arch Mikrobiol 24:163–168
Allen TFH (1971) Multivariate approaches to the ecology of algae on terrestrial rock surfaces in North Wales. J Ecol 59:803–826
Allen MB, Arnon DI (1955) Studies on nitrogen-fixing blue-green algae. II. The sodium requireemnt of dnabaena cylindrica. Physiol Plant 8:653–660
Allen MB, Arnon DI (1955a) Studies on nitrogen fixing blue-green algae. II. The Na requirements of Anabaena cylindrica Lemm. Physiology P1. (Copenhagen) 8:653–660
Allen MB, Arnon DI (1955b) Studies on nitrogen-fixing blue-green algae. I. Growth and nitrogen fixation by Anabaena cylindrica Lemm. P1. Physiology (Lancaster) 30:366–372
Allen MM, Stanier RY (1968) Selective isolation of blue-green algae from water and soil. J Gen Microbiol 51:203
Allxsos FE, Morris HJ (1930) Nitrogen fixation by blue-green algae. Science 71:221–223
Allxsos FE, Hoover SR, Morris HJ (1937) Physiological studies with the nitrogen-fixing alga, Nostoc muscorum. Bot Gaz 98:433–463
Belnap J, Büdel B, Lange OL (2003) Biological soil crust: characteristics and distribution. Ecol Stud 150:3–30. https://doi.org/10.1007/978-3-642-56475-8_1
Bender CL (2005) The post-genomic era: new approaches for studying bacterial diseases of plants. Australas Plant Pathol 34(4):471–474
Bharti VS, Dotaniya ML (2017) Managing soil fertility through microbes: prospects, challenges and future strategies
Blisks LR (1951) Physiology and biochemistry of algae. In: Smith GM (ed) Manual of phycology. Chronica Botanica Co., Waltham, pp 263–291
Bold HC (1970) Some aspects of the taxonomy of soil algae. In: Fredrick JF, Klien RM (eds) Phylogeny and morphogenesis in the algae, vol 175. Ann. New York Acad. Sci, p 601416
Bold HC, Wynne MJ (1978) Introduction to the algae. Structure and reproduction. Prentice Hall, Englewood Cliffs, New Jersey
Booth WE (1941) Algae as pioneers in plant succession and their importance in erosion control. Ecology 22:38–46

Bristol-RoAc BM (1926) On the relation of certain soil algae to some soluble carbon compounds. Ann Bot 40:149–201
Bristol-RoAc BM (1927a) On the carbon nutrition of some algae isolated from soil. Ann Bot 41:509–517
Bristol-Roac BM (1927b) On the algae of some normal English soils. J Agric Sci 17:563–588
Bulgarelli D, Schlaeppi K, Spaepen S, van Themaat EVL, Schulze-Lefert P (2013) Structure and functions of the bacterial microbiota of plants. Annu Rev Plant Biol 64:807–838
Carr NG, Whitton BA (eds) (1973) The biology of blue-green algae. Botan. Mono. No. 9. Blackwell Scientific Publ, Oxford
Chodat F (1928) Sur la specifitive des Stichococcus du sol du Parc National. Comp Rend Soc Phys Hist Nat G~n6ve 45:26–27
Colodny N (1930) Ober eine neue Methode zur Untersuchung der Bodenmikroflora. Arch Mikrobiol 1:620–652
Coyle EE (1935) Algae of some Ohio soils. Abs. Doe. Diss. Ohio State Univ 17:217–227
Dotaniya ML, Kushwah SK (2013) Nutrient's uptake ability of various rainy season crops grown in a vertisol of Central India. Afr J Agric Res 8(44):5592–5598
Droue (1943) Myxophyceae of eastern California and western Nevada. Field Mus Nat Hist Bot Ser 20:145–176
Durot M, Bourguignon P-Y, Schachter V (2009) Genome-scale models of bacterial metabolism: reconstruction and applications. FEMS Microbiol Rev 33(1):164–190
Durrell LW (1959) Algae in Colorado soils. Am Mid Nat 61:322–328
Durrell LW, Shields LM (1961) Characteristics of soil algae relating to crust formation. Trans Am Micros Soc 80:73–79
Elwell HM, Slosser JW, DANmL HA (1939) Revegetative and gully-control experiments in the Red Plains Region. Soil Conserv 5:17–20
Eneje R, Donatus EA (2009) Algal compost effect on soil nutrient standard aggregate and stability. Nigeria J Soil Sci 19(2):19
EttI H (1976) Die Gattung Chlamydomonas Ehrenberg. Beihefte Nova Hedwigia, Heft, p 49
Evans RD, Lange OL (2003) Biological soil crusts and ecosystem nitrogen and carbon dynamics. Ecol Stud 150:263–279. https://doi.org/10.1007/978-3-642-56475-8_20
Fletcher JOELE, Martin WP (1948) Some effects of algae and molds in the raln-crust of desert soils. Ecology 29:95–100
Foot GE (1947) Nitrogen fixation by blue-green algae. Endeavor 6:172–175
Fritsch FE (1907) The role of algal growth in the colonization of new ground and in the determination of scenery. Geogr J 30:531–548
Fuller WH, Rogers RN (1952) Utilization of the phosphorus of the algal cells as measured by the Neubauer technique. Soil Sci 74:417–430
Gartner G (1976) List of the culture collection of algae at the Institute for Botanische Systematik und Geobotanik der Universitat Innsbruck. Ber Naturwiss Med Vereins Innsbruck 63:67–89
Gonzalves ELLAA, Valavicl VS (1960) Algae in the rhizospheres of some crop plants. In: Kachroo P, Led I (eds) Proc. Symp. Algol., Indian Council of Agr. Res, New Delhi, pp 335–342
Gupta B, Saha J, Sengupta A, Gupta K (2013) Plant abiotic stress: 'Omics' approach. J Plant Biochem Biotechnol 1:e108
Haraison WH, Aiyer PAS (1913) The gases of swamp rice soils: their composition and their relation to the crop. Mem Dept Agr India (Chem Ser) 3:65–106
Hindak F (1970) Culture collection of algae at laboratory of algology in Trebon. Arch. Hydrobiol. Supp. 39. Algol Stud 3:86–126
Hoek CV (1963) Nomenclatural typification of some unicellular and colonial algae. Nova Hedwigia 6:277–296
Hoffman L (1998) Algae of terrestrial habitats. Botan Rev 55:77–105
Johansen JR (1993) Cryptogamic crusts of semiarid and arid lands of North American. J Phycol 29:140–147
Johansen JR, Shubert LE (2001) Algae in soil. Nova Hedwigia, Beih 123:297–306

Lowe CW, Moyse AV (1934) An investigation of some Manitoba soils for the presence of soil algae. Trans Roy Soc Canada 28:119–152
Lukešová A (1993) Productivity of algocenoses in zonal arable soils of steppe and foreststeppe. Eurasian Soil Sci 31:406–410
Lukešová A (2001) Soil algae in brown coal and lignite post mining areas in Central Europe (Czech, Republic and Germany) restore. Ecology 9:341–350
Lukešová A, Hoffmann L (1996) Soil algae from acid rain impacted forest areas of the Krušné hory Mts. 1. Algal communities. Vegetation 125:123–136
LuNo JWG (1945) Observations on soil algae. I. The ecology, size and taxonomy of British soil diatoms. New Phytol 44:196–219
LuNo JWG (1947) Observations on soil algae. II. Notes on groups other than diatoms. New Phytol 46:35–60
Magnúsdóttir S, Heinken A, Kutt L, Ravcheev DA, Bauer E, Noronha A (2017) Generation of genome-scale metabolic reconstructions for 773 members of the human gut microbiota. Nat Biotechnol 35:81–89
Martin TL (1949) The occurrence of algae in some virgin Utah soils. Soil Sci Soc Am Proc 4:249–250
Meena VD, Dotaniya ML, Rajendiran S, Coumar MV, Kundu S, Rao AS (2013) A case for silicon fertilization to improve crop yields in tropical soils. Proc Natl Acad Sci India Sect B Biol Sci 84(3):505–518
Metting B (1981) The systematics and ecology of soil algae. Botan Rev 47:195–312
Moore GT, Carter N (1926) Further studies on the subterranean algal flora of the Missouri Botanical Garden. Ann Mo Bot Gard 13(101–117):123
Moore GT, Karrer JL (1919) A subterranean algal flora. Ann Mo Bot Gard 6:281–307
Narasimhan K, Basheer C, Bajic VB, Swarup S (2003) Enhancement of plant-microbe interactions using a rhizosphere metabolomics-driven approach and its application in the removal of polychlorinated biphenyls. Plant Physiol 132(1):146–153
Nogales J, Palsson BØ, Thiele I (2008) A genome-scale metabolic reconstruction of Pseudomonas putida KT2440: iJN746 as a cell factory. BMC Syst Biol 2(1):79
Osborne BO (1950) Range cover tames the raindrop. U. S. Soil Conserv. Spry., Oper. & Res. Bur
Petersen JB (1932) The algal vegetation of Hammer Bakker. Bot Tidsskr 42:1–48
Petersen JB (1935) Studies on the biology and taxonomy of soil algae. Dansk Bot Arkiv 8:1–180
Prescott GW (1956) A guide to the literature on ecology and life histories of the algae. Bot Rev 22:167–240
Quirino B, Candido E, Campos P, Franco O, Krüger R (2010) Proteomic approaches to study plant–pathogen interactions. Phytochemistry 71(4):351–362
Reid A (2011) Microbes helping to improve crop productivity. Microbe 6(10):435
Russell EJ (1952) Soil conditions and plant growth, 8th edn. Longmans, Green and Co., London, p 518
Salanoubat M, Genin S, Artiguenave F, Gouzy J, Mangenot S, Arlat M (2012) Genome sequence of the plant pathogen Ralstonia solanacearum. Nature 415(6871):497–502
Schneider DJ, Collmer A (2010) Studying plant-pathogen interactions in the genomics era: beyond molecular Koch's postulates to systems biology. Annu Rev Phytopathol 48:457–479
Shtma EA (1957) The development of soil algae in sod-podzolic soils. Pochvovedenie 1957:12–18
Shukla M, Patel RH, Verma R, Deewan P, Dotaniya ML (2013) Effect of bio-organics and chemical fertilizers on growth and yield of chickpea (Cicer arietinum L.) under middle Gujarat conditions. Vegetos 26(1):183–187
Shxelds LM (1957) Algal and lichen floras in relation to nitrogen content of certain volcanic and arid range soils. Ecology ~8:661–663
Shxelds LM, Drouet F (1962) Distribution of terrestrial algae within the Nevada Test Site. Am J Bot 49:547–554
Shxelds LM, Mitchell C, Drouet F (1957) Alga-and lichen-stabilized surface crust as a soil nitrogen source. Am J Bot 44:489–498

Simpson AJG, Reinach F, Arruda P, Abreu F, Acencio M, Alvarenga R (2000) The genome sequence of the plant pathogen Xylella fastidiosa. Nature 406(6792):151–157
Singh JS (2013) Plant growth promoting rhizobacteria: potential microbes for sustainable agriculture. Resonance 18(3):275–281
Singh JS (2014) Cyanobacteria: a vital bio-agent in eco-restoration of degraded lands and sustainable agriculture. Clim Chang Environ Sustain 2:133–137
Singh JS (2015) Microbes: the chief ecological engineers in reinstating equilibrium in degraded ecosystems. Agric Ecosyst Environ 203:80–82
Singh JS, Pandey VC (2013) Fly ash application in nutrient poor agriculture soils: impact on methanotrophs population dynamics and paddy yields. Ecotoxicol Environ Saf 89:43–51
Singh JS, Singh DP (2012) Reforestation: a potential approach to mitigate the excess CH4 buildup. Ecol Manag Restor 13(3):245–248
Singh JS, Singh DP (2013) Plant Growth Promoting Rhizobacteria (PGPR): microbes in sustainable agriculture. In: Malik A, Grohmann E, Alves M (eds) Management of microbial resources in the environment. Springer, Dordrecht, pp 307–319
Singh JS, Strong PJ (2016) Biologically derived fertilizer: a multifaceted bio-tool in methane mitigation. Ecotoxicol Environ Saf 124:267–276
Singh JS, Pandey VC, Singh DP, Singh RP (2010) Influence of pyrite and farmyard manure on population dynamics of soil methanotroph and rice yield in saline rain-fed paddy field. Agric Ecosyst Environ 139:74–79
Singh JS, Abhilash PC, Singh HB, Singh RP, Singh DP (2011a) Genetically engineered bacteria: an emerging tool for environmental remediation and future research perspectives. Gene 480:1–9
Singh JS, Pandey VC, Singh DP (2011b) Efficient soil microorganisms: a new dimension for sustainable agriculture and environmental development. Agric Ecosyst Environ 140:339–353
Singh JS, Singh DP, Dixit S (2011c) Cyanobacteria: an agent of heavy metal removal. In: Maheshwari DK, Dubey RC (eds) Bioremediation of pollutants. IK International Publisher, New Delhi, pp 223–243
Singh JS, Kumar A, Rai AN, Singh DP (2016) Cyanobacteria: a precious bio-resource in agriculture, ecosystem, and environmental sustainability. Front Microbiol 7:529
Skinner CE (1932) The soil as a habitat for growth of green algae. Soil Sci 84:25–28
Smith VH, Tilman GD, Nekola JC (1999) Eutrophication: impacts of excess nutrient inputs on freshwater, marine, and terrestrial ecosystems. Environ Pollut 100(1):179–196
Soda N, Wallace S, Karan R (2015) Omics study for abiotic stress responses in plants. Adv Plants Agric Res 2(1):00037
Starks TL, Shubert LE (1982) Colonization and succession of algae and soil algae interactions associated with disturbed areas. J Phycol 18:99–107. https://doi.org/10.1111/j.1529-8817.1982.tb03162.x
Starks TL, Elliot Shubert L, Trainor FR (1981) Ecology of soil algae: a review. Phycologia 20(1):65–80
Stokes JL (1940a) The influence of environmental factors upon the development of algae and other microorganisms in the soil. Soil Sci 49:171–184
Stokes JL (1940b) The relation of algae to the nitrogen economy of the soil. Chron Bot 6:202–203
Thynne E, McDonald MC, Solomon PS (2015) Phytopathogen emergence in the genomics era. Trends Plant Sci 20(4):246–255
Tiffany LH (1951) Ecology of freshwater algae. In: Smith GM, Led L (eds) Manual of phycology. Chroniea Botanica Co, pp 293–311
Timm CM, Campbell AG, Utturkar SM, Jun S-R, Parales RE, Tan WA (2015) Metabolic functions of Pseudomonas fluorescens strains from Populus deltoides depend on rhizosphere or endosphere isolation compartment. Front Microbiol 6:1118
Trueb M (1888) Notice sur la nouvelle flora de Krakatau. Anti Jard Bot Buitenzorg 7:221–223
Vorholt JA (2012) Microbial life in the phyllosphere. Nat Rev Microbiol 10(12):828–840
Waksman SA (1952) Soil Microbiology. Wiley, New York, p 356

Winnenburg R, Urban M, Beacham A, Baldwin TK, Holland S, Lindeberg M et al (2008) PHI-base update: additions to the pathogen–host interaction database. Nucleic Acids Res 36(Suppl 1):D572–D576

Wxllson D, Forest HS (1957) An exploratory study on soil algae. Ecology 38:309–313

Zeitnder ALFONS (1953) Beitrag zur Kenntnis von Mikroklima und Algenvegatation des nackten Gesteins in den Tropen. Ber Schweiz Bot Ges 63:5–26

Commercial Applications of Soil Algae

V. Sivamurugan, P. Priya, N. Shanthi, S. Murugesan, and T. V. Poonguzhali

Abstract Algae are rich sources of polysaccharides such as alginates, carrageenan, and agar. Commercial applications of algal biomass are primarily in energy production, new functional foods, and medicines. Soil algae help to fix atmospheric CO_2 in soil and are primarily present on the surface of the soil to prevent soil erosion. Seaweed contains essential nutrients and bioactive compounds that have pending or potential applications in animal and human health and wellness. Various seaweed and seaweed extracts have been proven to have antibacterial, anti-diabetic, antiviral, antifungal, and antibacterial capabilities. Using algae for the extraction or bioconversion of pollutants such as nutrients and xenobiotics in polluted water and CO_2 in polluted air currents is known as phycoremediation. Seaweed, which is commonly used in wastewater treatment, also helps to lower or eliminate hazardous heavy metal levels. The chapter delineates the contribution of algae toward sustainable development and commercial aspects of algae in fuel, medicine, and food applications.

Keywords Algal metabolites · Energy · Bioremediation · Theragnostic compounds · Carbon emission · Algae biomass

V. Sivamurugan
PG and Research Department of Chemistry, Pachaiyappa's College, Chennai, India

P. Priya
PG and Research Department of Zoology, Pachaiyappa's College, Chennai, India

N. Shanthi · S. Murugesan (✉)
PG and Research Department of Botany, Pachaiyappa's College, Chennai, India

T. V. Poonguzhali
PG and Research Department of Botany, Queen Mary's College, Chennai, India

A. Mishra, A. Varma (eds.), *Soil Algae*, https://doi.org/10.1007/978-981-97-9873-5_7

1 Introduction

Algae are the primary source of energy and many other fascinating secondary metabolites. There are interesting prospects for biomass energy production, novel functional foods, and medicines due to its nutraceutical value. It has been characterized by its availability, color, and metamorphosis. Algae are rich sources of polysaccharides such as alginates, carrageenan, and agar. The biological community of microorganisms that lives in association with soil particles up to a certain depth on the soil crust forms the cryptobiotic soil crust (Jane and Duan 2015). Soil algae help to fix atmospheric CO_2 in soil and are primarily present on the surface of the soil to prevent soil erosion. Cyanobacteria and Chlorophyceae algae are the first to house soil surface. The four major divisions of algae present on the soil are Cyanophyta, Chlorophyta, Xanthophyta, and Bacillariophyta (Khan and Rao 2019). Algae are a rich collection of photosynthetic organisms that provide major organizational roles and functional environmental services that can play a significant role in the pillars of sustainable development.

Algae-produced biomass is a renewable energy source and alternative fuel. Algae have a long history of commercial use in biodiesel, food, and cosmetics industries. Algae, particularly for biodiesel production, have emerged as one of the most promising alternatives (Avagyan 2008). Algal oil yields are orders of magnitude higher than ordinary oil seed yields (Sheehan et al. 2006; Riesing 2007). Algae are also advantageous since they provide secondary metabolites for the production of novel theragnostic compounds (Ślusarczyk et al. 2021). These "algae" are said to include all macro- and micronutrients, including around 65% of proteins, phycocyanins, vitamin B complex, vitamin E, chlorophyll, carotenoids, and superoxide dismutase, as well as several minerals, making it one of the most nutritious diets on the planet (Wang et al. 2008). The theragnostic potential of soil algae refers to the potential theragnostic mechanism of a single soil-derived algal medicine that performs dual activities, such as locating the disease and providing remedial treatment, specifically in the identified affected spot. Seaweeds have the potential to improve socioeconomic conditions. Seaweed contains essential nutrients and bioactive compounds that have pending or potential applications in animal and human health and wellness. Various seaweed and seaweed extracts have been proven to possess antibacterial, anti-diabetic, antiviral, antifungal, and antibacterial capabilities, as well as abscess prevention properties and HIV in studies (Fedorov et al. 2013). They have been demonstrated to reduce blood cholesterol and blood pressure, and they are all high in antioxidants, making them ideal candidates for active food consumption (Holdt and Kraan 2011). Phycoremediation utilizes algae to remove or transform pollutants such as excess nutrients and xenobiotics from contaminated water as well as polluted air streams (e.g., exhaust from gas and cement factories) into harmless gaseous components (Olguin 2003).

Seaweed, which is commonly utilized for the treatment of wastewater, also helps to lower or eliminate hazardous heavy metal levels (He and Chen 2014). Heavy metal elimination from effluent is a serious threat to the environment today and is gaining major attention due to the obvious destructive nature of heavy metals to the environment and living creatures, particularly when they exceed regulatory standards (Adamu et al. 2015). This chapter focuses on the commercial value of soil algae and its use as a greener alternative to fossil fuels, as well as the therapeutic and environmental fields, which are briefly explored in different parts.

2 Role of Algae in Sustainable Development

Phycology has the potential to play an important role in the new field of sustainability science. Although "plants" frequently mentioned in public debates about climate change, biofuel production, and green technology, only experts in the field are familiar with the role and potential sustainability of "algae" (Reisser 2010). The major problem globally is the rising cost of food due to the inability to produce food suitable for the growing population and other related foods and food crops that can be used in biofuel production. Traditional knowledge and delicious algal foods can aid the development of Western algae-based methods for human consumption in various parts of the world (Mouritseno 2013). Traditional uses include the use of cochayuyo (*Durvillaea antarctica*) in Chile (Cardenas Alvarez and Munoz Arias 2008) and the production of flour from "llaska" (*Cladophora crispata*) by local people on the Peruvian-Chile border (Cardenas) (Alvarez and Munoz Arias 2008; Canales et al. 2008). Seaweed contains polysaccharides, minerals, essential amino acids, and various vitamins. They also contain bioactive compounds such as sulphated polysaccharides, proteins, lipids, and polyphenols, which have pending or potential applications in animal and human health and wellness.

The use of algae for capturing carbon emissions, extracting bioactive chemicals, and producing agricultural raw materials, along with its potential for energy production, has become a popular field of interest, for example, the Algal Carbon Conversion (ACC). The Flagship of the National Research Council (NRC) aims to provide the Canadian industry with a cost-effective solution that produces quantitatively to divert CO_2 emissions from algal biomass, which can be processed into biofuels and other products. In addition, the potential for biofuels based on grain to convert fossil fuels, particularly transport, is considered relatively limited (Global Fuels Magazine 2007). To switch to biofuels in a controlled manner, as the government's aim stated, a strong productive and environmentally friendly agricultural sector must be developed. Consumption of grain in the world will also increase as the population grows. The global population is projected to increase from 6.5 billion to 9 billion in 2050, and the demand for grain in developing countries is expected to double, with an annual average of 1.1 billion tons by 2050. Global interest in using biofuels as fuels has increased due to possible global climatic changes facing regulatory systems and energy conservation.

Oil is undoubtedly the lifeblood of today's world, and the hydrocarbon economy has intensified, causing many conflicts in oil-producing areas. Oil and its extraction are required to provide "cheap" fuel for internal combustion engines, which are essential for transportation. In addition, a few oil marks have developed for a variety of uses, from candle wax to plastics to essential drugs (such as aspirin, psychiatric drugs, antihistaminics, or antimicrobials): a test of global limitations and global instability.

The limits for the ecological sustainability of algal biofuels have not yet reached those of other food stocks in terms of testing and selection. Many institutions and researchers have proposed indicators of bioenergy sustainability (Roundtable on Sustainable Biomaterials 2010; Global Bioenergy Partnership 2011). McBride et al. (2011) and others, such as the International Organization for Standardization, work on them (ISO 2010). Most bioenergy indicators and standards are derived from the nutrients found in the vegetation, including crops like corn, switch grass, and forest products (CSBP 2012). The National Research Council of the United States (NRC 2012) published a report on the potential environmental impact and metrics for algal biofuels development, as well as a list of critical issues that may be sustainable, but it did not identify significant potential or a set of measurable environmental indicators. As a result, technological advancements in algae biofuels are accelerating despite the lack of specific tools for identifying and evaluating their long-term efficacy.

Given the significant role of algae in our ecosystem, it's important to highlight that they are often not actively included in global discussions on biodiversity (Norton et al. 1996; Connan 2004), and there was little or no effort on their conservation worldwide (Brodie 2009; Mollo and Thomas-Bourgneuf 2009). However, since the beginning of this century, literature on algae and environmental concerns has proliferated, especially in algae-based bioenergy. Annual scientific analysis reveals that the number of papers published with search terms "algae" and "bioenergy" has increased significantly over the past three decades. Although there is no competition for agricultural land between food and algae biofuels, there is competition for fertilizer use, which is needed for both successful agricultural (terrestrial) and algae (water) production. The commercialization of algal biofuel production as a sustainable and renewable resource will require significant use of phosphorus (P), now found in the non-renewable mineral phosphate rock. This is disputed over its long-term effectiveness (Rhodes 2010). The re-digestion of phosphorus in manure and other plant species may provide a solution to the supply of P as fertilizer.

Many countries around the world make many efforts for the sustainable development of various sections of society through past performance. Their agriculture is deteriorating as a result of global climate change and inadequate rainfall. Alternatively, our biotechnology scientist can highlight algae and its function as a key solution to many of the emerging challenges. Algae play an important role in solving many problems in emerging sciences. Microalgae and marine plants receive very little attention from humans compared to plants and animals. Recently, numerous countries have been advocating for research on algae biofuels to foster community growth and promote sustainable development within society.

3 Contribution of Biotechnology in Algal Metabolites Production

By 2050, the world's population will reach ten million. Agriculture uses a lot of arable land, so they face a big problem, especially global climate change, and urbanization is an important challenge for future agriculture (Foley et al. 2011). Growth in agriculture, fishing, and the production of fossil fuels have not attained sustainable solutions for human development. There are novel solutions and additional resources required to address these complex issues. Because of advances in agricultural technologies, genetics, and genomics, agriculture has experienced significant advances in production, efficiency, and segregation of products in recent years. Apart from the fact that applications for algae have been around since the beginning of time, the attempt to exploit these organisms as industrial resources is a new goal.

Algae-based processes are a relatively new field of application compared to conventional agricultural crops, and all existing algal species are unethical. Years of basic research in algal biochemistry and physiology can accelerate the use of algae in biotechnology. The emergence of genetic sequencing projects and functional genetic tools has recently helped efforts to advance understanding of the symptoms of various diseases, revealing additional aspects of algal metabolism associated with industrial applications (Kirst et al. 2012; Radakovits et al. 2012; Fabris et al. 2014).

Algae are now used in a few industrial applications. Recent research has elaborated on the shift in emphasis from algae-based bioenergy to high-value bioproducts, as well as the algae-based biorefineries model (Laurens et al. 2017). Various bio-compounds that can be used as active substances, such as carotenoids, phycocyanins, polyphenols, fatty acids, and polyunsaturated compounds, are developed and produced with great interest (Plaza et al. 2008). Microalgae are also gaining momentum as the next-generation bio-factory equivalent to long-term production cells for a variety of products (Rasala and Mayfield 2015; Vavitsas et al. 2018), following in the footsteps of smaller platforms such as yeasts and bacteria. Microalgal bio-factories have the potential to become more cost-effective and sustainable platforms to produce specialized plant-based goods (Vavitsas et al. 2018). Advances in technologies such as automation, synthetic biology, and phenomics may be used to exploit the already vulnerable microalgae in the future.

Many different species of algae produce carbohydrates, lipids, proteins, pigments, and a range of second-line metabolites that are currently commercialized in traditional agriculture (Koyande et al. 2019). Bioactive chemicals are active molecules in the human body that have functional properties. A growing number of scientific studies showing the beneficial effects of these chemicals on health have aroused interest in the formation of bioactive compounds from natural sources (Herrero 2013). In researching new molecules that work in pharmacies, natural products are important. For the benefit of the people in the future, today's globalization of the need for nutrition should focus on research into the testing of solid second metabolites from microbial sources with high nutritional value (Garcial et al.

2017). *Chlorella*, *Haematococcus*, *Dunaliella*, and *Spirulina* are just a few species of algae that provide a wide variety of secondary metabolites to promote optimal nutrition (Gouveia et al. 2008). Alkaloids, flavonoids, glycosides, terpenoids, and phenazines are secondary metabolites generated by algae (Alassali et al. 2016). Proteins, carbohydrates, lipids, polysaccharides, polyols, and phycobiliproteins are all found in algae (Borowitzka and Vonshak 2017). In the healthcare industry, these second metabolites can be used in a variety of ways (Wells et al. 2014). Secondary metabolites in algae have been developed under a variety of stress circumstances (Ranga Rao et al. 2016). Secondary metabolites in microalgae have been studied pharmacologically (De Morais et al. 2015). Secondary metabolites are being studied against bacteria, and cytotoxicity against microalgae is being tested.

Macroalgae is well known as a source of secondary metabolites that hold great promise for therapeutic benefit. In vitro research is merely the initial step in a long process; nevertheless, in vivo research and clinical trials are the most informative phases of the metabolite's true strengths and limits when used as a novel medication. Minerals (7–36%), lipids (1–5%), polysaccharides (15–76%), and proteins (5–47%) are all prevalent in macroalgae (Anyanwu et al. 2018). Iodine, calcium, iron, copper, and magnesium are all found in high concentrations in macroalgae. Polyunsaturated fatty acids (PUFAs), including omega-3 and omega-6 acids, help to prevent cardiovascular disease, osteoarthritis, and diabetes, and they constitute the lipid composition of macroalgae. Other fatty acids found in green macroalgae include alpha-linolenic acid (C18:3), eicosapentaenoic acid (EPA, C20:5), and arachidonic acid (C20:4). In contrast, red and brown macroalgae contain similar fatty acids. Additionally, macroalgae are rich in micronutrients, such as vitamins (particularly B3, B9, B12, C, and E) and antioxidants like polyphenols, the most important of which are phlorotannins. Other notable components include carotenoids and chlorophyll. Furthermore, the *Dictionary of Natural Products* (DNP) website lists over 4000 metabolites classified as originating from red, brown, and green macroalgae. Macroalgae are high in macromolecules and micronutrients, both of which have biological and metabolic functions.

Mycosporine-like amino acids (MAAs) are intracellular molecules that protect aquatic organisms from UV radiation. Structurally, they consist of amino acid substitutions, as well as a chromophore composed of either a cyclohexenimine or cyclohexenone with absorption wavelengths ranging from 310 to 362 nm (Bhatia et al. 2011). Scytonemin is a photoprotective dimer made up of indolic and phenolic subunits that weigh 544 Da. It was discovered as a soluble yellow pigment in the exo-polysaccharide sac of certain earth cyanobacterial species (Wada et al. 2013). Because of the presence of the new chlorosulfolipid, an alga named *Poterioochromonas malhamensis* has been demonstrated to block the action of the enzyme protein tyrosine kinase. Apoptosis impacts cell signaling by activating enzymes that express the protein kinase C family 82, and many cyanobacteria are promising producers of beneficial substances and can destroy cancer cells through the mechanism of apoptosis. These compounds are particularly effective against lymphocytic leukemia and lung cancer. Cryptophycin, an anti-cancer metabolite obtained from *Nostoc* ATCC 53789 84, and scytonemin, a protein serine inhibitor identified from *Stigonema*, may produce a great anti-inflammatory and anti-inflammatory medication. All organisms possess 40–50 distinct oligosaccharides. These oligosaccharides exhibit biochemical features characterized by glycosidic linkages and additional modifications such as acyl groups, amino acids, or sulfates. Polysaccharides serve as sources of carbon and energy, released during both normal and stressful physical activities. They are thickeners and gelling agents, respectively (Kraan 2012). Antibacterial, anticoagulant, antimutagenic, radioprotective, anti-oxidative, antiulcer, anticancer, and anti-inflammatory properties are also found in them (Kraan 2012; Misurcova et al. 2015). For example, *Dunaliella salina* and *Botryococcus braunii* may store up to 60% of their final lipids as triglycerides (TAGs), which could be a helpful fat source for biodiesel production (Scott et al. 2010). Algae are more prolific than plants, with the potential to produce 10 to 100 times more oil per hectare, though this has yet to be commercialized (Greenwell et al. 2010).

Emerging technology accelerates the manufacturing process of these hooks by offering information and guidance on their proper use, delivering effective solutions tailored to address the most pressing issues in the community. It covers many aspects of biology, genetics, and the metabolism of all living organisms and the environment. All of this is a natural or genetically engineered approach that helps to improve both biodiversity and ecosystems with relatively short-term productivity in the future. Internal performance in almost all industries, including those needs of human society such as food, medicine, and mass production, all have a small impact on the environment and low production costs. (Fabris et al. 2020). The development of highly efficient aluminum purifiers (Lam et al. 2018) will allow for the cooperation of a wide range of products while minimizing waste and increasing productivity, which is why we are developing economically efficient systems now. The scientist says that algae biotechnology will make a huge difference in the current industry. The result is a sustainable balance and efficiency of the algae economy.

4 Algae as a Source of Alternative Energy

Algae are slowly becoming recognized as one of the most promising long-term sources of biomass and oils for fuel, feed, food, and other coproducts (Hyka et al. 2013), as well as an alternative fuel feedstock that could alleviate the fuel vs. food (Hyka et al. 2013; Wanchat et al. 2014). Microalgae oil productivity is higher than that of traditional crops (Rudras and Susan 2010). Microalgae has a higher oil content than other rivals, ranging from 20 to 50 percent; on an area basis, microalgae produce 15–300 times more oil for biodiesel generation than traditional crops. When compared to other renewable energy sources such as solar, geothermal, wind, tidal energy, and so on, algae growth can produce more biofuel with less water and fertile land (Ullah et al. 2014). Microalgae cultivation requires less water than conventional energy oil crops. The algal biofuel technology includes the selection of specific species for production and extraction of valuable coproducts (Ono and

Cuello 2006). The algae are bio-engineered for achieving advanced photosynthetic efficiencies through continued development of production systems (Pulz and Scheinbenbogan 1998).

4.1 Types of Energy

Energy is the power we use for transportation, for heat and light in our homes, and for manufacturing all kinds of products. There are two sources of energy: renewable and nonrenewable energy (Energy Information Administration 2007).

4.2 Nonrenewable Sources of Energy

The majority of the energy we use comes from fossil fuels, such as natural gas, coal, and petroleum. Although uranium is a nonrenewable resource, it is not a fossil fuel. It is converted into a fuel and then passed through nuclear power facilities. These common resources are used up as soon as they are depleted. The extraction of these fuels can be hazardous to the biomes from which they come. To generate energy, fossil fuels go through a process known as burning. Sulfur dioxide and carbon monoxide are released during combustion, which can produce acid rain and global warming.

4.3 Renewable Sources of Energy

Renewable energy sources can be used over and over again. Wind, geothermal energy, solar energy, hydropower, and biomass are examples of renewable resources. Nonrenewable sources produce far more pollution throughout the harvesting and manufacturing of this resource.

4.4 Algal Biomass

Algal biomass is a renewable resource with the potential to meet a small percentage of the world's energy requirements (Elbehri et al. 2013). Research and development aimed at improving biomass production has focused on enhancing cultivation, harvesting, and biomass conversion technologies. There are goals to achieve the lowest carbon emissions, recycle inorganic nutrients, minimize energy inputs at each stage of production, and integrate algal biomass production systems into the existing energy infrastructure as seamlessly as possible.

4.5 *Algae Fuel*

Algal industry techniques are integrated processes for CO_2 capture that contribute to solving the global warming problem while producing valuable byproducts such as lipids (oils), carbohydrates, proteins, and various feedstocks that can be converted into biofuels and other useful materials. Microalgae are probable to be an essential raw material for amino acids, vitamins, and yields of valuable by-products. The production of microalgae is the more gainful business in the biotechnological process. It is wasteless and environmentally safe. Oil productivity of microalgae is greater compared to conventional crops. The oil content of microalgae is between 20% and 50%, which is greater than other competitors. Microalgae have a higher photon conversion efficiency than terrestrial plants, ranging from 3–8 to 0.5%. Microalgae is an alternate fuel feedstock that avoids the fuel versus food dilemma, and it may be grown as a nutrient source in wastewater. The wastewater treatment function covers the costs of algae growing and harvesting for biofuel production.

Algae fuel, or algal biofuel, is another form of fossil fuel that uses microalgae as its source of natural deposits (Scott et al. 2010). Some of the unique characteristics of algal fuels are: they can be grown with negligible impact on freshwater resources (Yang et al. 2011), they can be synthesized using ocean and wastewater, and they are biodegradable and relatively harmless to the environment if spilled (Green Dream 2003). Algal fuel is highly favorable and workable related to other biofuels, as they do not have to produce structural compounds and they can convert higher fractions of biomass to oil compared to other cultivated crops (Low Cost Algae Production System Introduced 2007). We can produce several fuel types from algae, such as bioethanol, biodiesel, methane, aircraft fuel (kerosene), bio-butanol, biogas, and green diesel.

4.6 *Bioethanol Production from Microalgae*

Because they have the same chemical and physical features, we consider bioethanol to be a substitute for conventional petroleum (Naik et al. 2010). In order to reduce our reliance on fossil fuels, bioethanol has been recognized as the most appropriate liquid transportation fuel (Bardhan et al. 2019a). To reduce pollutant emissions, bioethanol can be blended with existing gasoline or utilized as pure ethanol. However, because of food security and dwindling arable land area concerns, the main use of terrestrial crops such as maize and sugarcane for making bioethanol has sparked public outrage (Ge et al. 2011). Non-food feedstock, particularly microalgal biomass, is one of the most promising renewable feedstocks for third-generation biofuel production in this context (Vassilev and Vassileva 2016). Compared to other first- and second-generation bioethanol feedstocks, microalgae was proven to be a superior feedstock for producing bioethanol (Chisti 2007, 2008). When dried microalgae biomass with pre-extracted lipid was compared to dried microalgae biomass without lipid extraction, the ethanol concentration was 60 percent greater (Choi et al. 2010). Similarly, some microalgae species may store huge amounts of carbohydrates in their cells, such as triacylglycerol esters and starch. These carbohydrates can be employed as a carbon source or a substrate for bioethanol production during fermentation (Harun et al. 2010). Proteins, as well as carbohydrates and fats, can be stored within cells under nitrogen limitation or starvation (González-Fernández and Ballesteros 2012). The complicated nitrogen compounds are broken down into proteins by microalgae. Carbohydrates can be accumulated as a result of changes in salinity, light intensity, and temperature. Microalgae also have a low hemicellulose content and lack lignin, making hydrolysis and fermentation yields more efficient (Ueno et al. 1998). Microalgae biomass is the most sought-after biomass for alternative energy sources due to its high carbohydrate content, rapid growth rate, lack of rivalry with regular plant crops, and simpler processing stages compared to lignocellulosic biomass (Martín-Juárez et al. 2017; Phwan et al. 2018; Tan et al. 2019). The bioethanol yields got are comparable to the yields from sugary and lignocellulosic feedstock, showing that microalgae biomass is a feasible alternative substrate for commercial-scale bioethanol production (Choi et al. 2010).

4.7 Biodiesel

Microalgae oil contains more free fatty acids than oil from terrestrial plants. These free fatty acids can be esterified using acid catalysts prior to transesterification during the biodiesel production process from microalgae oil. Biodiesel made from algal lipid is non-toxic and biodegradable to a large extent. Different ways for converting biomass to biofuel exist, and they are essentially classed as thermal, chemical, and biological approaches (Volk et al. 2000). The third-generation biofuel, according to many academics, is biodiesel made from microalgae. To overcome the constraints of first- and second-generation biofuels, microalgae can be a sustainable renewable energy source for biodiesel. Although microalgal biodiesel is becoming more cost-effective, true economic parity with petroleum diesel has yet to be achieved. With diesel presently costing $3.08 per gallon on average around the world (Diesel Prices 2021), a corresponding market price for algal diesel seems unavoidable. To achieve this, algal biomass yields must be steadily increased from 12 to over 30 g dw/m^2/day (assuming that all integrated systems based on algal biomass processing are subject to significant cultivation variability). Ideally, the energy return on investment (EROI) for harvesting algae from ponds should exceed 20, meaning that only 5% of the energy content of the algae should be expended during harvesting. Additionally, the lipid extraction and conversion efficiency to biodiesel must be improved to minimize expenses in the defined process (Barry et al. 2015; Barsanti and Gualtieri 2018). Microalgal biodiesel prices can be further reduced by focusing on optimizing lipid content in high biomass-yielding microalgal strains and valorization of the algal biomass, as this leads to a more significant cost reduction (Bellou et al. 2014; Zhu 2015).

4.8 *Bio-methane*

Biogas is one of the most promising biofuels, with the potential to alleviate some of the growing concerns about fossil fuels, such as the energy crisis and weather change (Bhale et al. 2009; Zabed et al. 2017). In the production of biogas, microalgae has proven to be efficient, practical, and cost-effective (Ward et al. 2014; Saratale et al. 2018). Because of their ability to assimilate large amounts of carbon dioxide and the possibility of blending microalgae cultivation with flue gas emissions or biogas upgrading, which involves removing carbon dioxide (as biogas) to increase methane percentage, microalgae are particularly well suited for combined nutrient removal from wastewater treatment and carbon dioxide sequestration (Xu et al. 2015). Anaerobic digestion of leftover biomass can also yield biogas (Gonzalez et al. 2018). Biomethane can be produced through biochemical conversion of biomass, followed by gas upgrading, or through thermochemical conversion of solid biomass through gasification, followed by gas cleaning, methanation as a synthesis process, and biogas upgrading.

4.9 *Biogas*

Microalgae biomass develops a suitable alternate origin of biogas. However, the use of algae for biogas production is associated with many difficulties (Giordano and Wang 2018). The efficiency of biogas production has improved when algae are combined with conventional feedstocks (Miao et al. 2014). Glucose and glycine derived from algae have been exposed to thermochemical treatment under neutral, acidic, and alkaline conditions to establish their decomposition mode and their future usage in the concurrent fermentation of bio-hydrogen and bio-methane (Chen et al. 2020). In combination with environmental bioremediation, water-blooming cyanobacteria can be utilized for biogas processing as well. During methanogenesis, the concentration of microcystin decreases significantly, indicating that anaerobic digestion of cyanobacteria could develop biogas from renewable energy while biodegrading microcystin (Chen et al. 2020).

4.10 *Biohydrogen*

Hydrogen is a sustainable energy source produced by photosynthetic organisms, with a higher energy content of roughly 122 kJ/g, nearly 2.75 times that of hydrocarbon fuels (Argun et al. 2008). As a result, it has been regarded as a feasible alternative to fossil fuels and as a carrier of energy. Biological processes, either directly or indirectly dependent on solar light, or fermentation processes and other thermochemical procedures employed to convert the biomass, enable microalgal hydrogen production. Microalgae produce molecular hydrogen in the presence of the enzyme hydrogenase, which exclusively accepts water as an electron donor (Vijayaraghavan et al. 2009). *Chlorella vulgaris* is the most widely utilized untreated substrate for hydrogen production among the many species. The maximum hydrogen yield was obtained from *Chlorella vulgaris* (Wieczorek et al. 2014) and *Scenedesmus* sp. (Yang et al. 2010). According to studies, using immobilized cells for hydrogen production is more appealing than using free cells (Rashid et al. 2013).

4.11 Bio-butanol

The potential for biodiesel and bioethanol production from microalgae has been well documented and investigated in the literature (Lage and Gentili 2018), but microalgal bio-butanol has received less attention (Wang et al. 2016). Bio-butanol is superior to bioethanol since it can be blended with fuels at much higher ratios without requiring modifications to engines (Bankar et al. 2013). However, bio-butanol production is severely restricted due to low product yields from Acetone-Butanol-Ethanol (ABE) fermentation, which is the biochemical route used to produce this biofuel (García et al. 2011). Thus, before bio-butanol production (whether from microalgae or any other feedstock) becomes an economically viable alternative to the most dominant bioethanol, the low ABE fermentation yields should be improved.

4.12 Aircraft Fuel (Kerosene)

In terms of jet fuel mixing options, ASTM currently only certifies a 50/50 blend (Bwapwa et al. 2017). Jet fuels generated from microalgae have already been tested in commercial and military aircraft with great success. ASTM has certified the use of biomass-based jet fuel for commercial flights since July 2011. The ASTM certification is only valid for blending with conventional jet fuel at a 50/50 volume ratio (B50). The blended gasoline is said to be ultra-pure. It must not crystallize at low temperatures and must produce more energy than is required for aircraft propulsion.

The main issue with algae-based jet fuel is finding species that can produce a large amount of bio-oil made up of hydrocarbons that are molecularly extremely similar to petroleum oil. This will allow jet fuel to be produced using the same downstream procedures as conventional jet fuel. Because of its composition and culture possibilities in the marine environment, *Nannochloropsis* sp. is one type of species that has the potential to create algae jet fuel. The inclusion of algae-based jet fuel in Jet A1 has the potential to make the 50/50 blended jet fuel more ecologically friendly, but it is still not economically viable due to the high cost of producing algae-based jet fuel (Bwapwa et al. 2019). The continued development of biotechnological and bioengineering methods to enhance the production of microalgae and extraction of oil from algae is one of the best solutions for the aviation industry to grow towards a sustainable and green future.

4.13 Green Diesel

Green diesel is an alternative fuel created from the hydrogenation of oils from renewable feedstocks like algae. Contrary to biodiesel, green diesel is produced by hydrogenation of fats and oils, and its chemical composition is very similar to that of petroleum-based diesel. Many life cycle assessment-based viability studies have already been done for the production of green diesel from microalgae. The production of green diesel via open raceway pond, both in dry and wet routes, has less negative net energy balance and comparatively higher net energy balance than the photo bioreactors. Comparison says that open raceway pond dry route has slightly higher value for net energy balance than the wet route. Even with the best feasible route (open raceway pond dry route), the total energy use is almost 5 times more than the energy produced, with a negative Net Energy Balance of 4.07 MJ & very low net energy balance value of 0.20. Study concludes that R & D in green diesel production from microalgae has yet to go a long way and has a huge scope to further lower its input energy demand for biofuel production (Namita and Krishnan 2016).

4.14 Bio-oil

Bio-oil production from microalgae, which has been known as a type of algae, has done by using the pyrolysis technique (Jena and Das 2011). Microalgal bio-oil is important for obtaining clean fuels and sustainable chemicals. It has been possible that, after upgrading, microalgal bio-oil has been used as biofuel (Li et al. 2019). Upgrading has been necessary because the bio-oil has possessed some undesired properties such as high oxygen content, low high heating value, and high viscosity (Saber et al. 2016). For upgrading, it had been used hydrogenation under high pressures after pyrolysis (Feng et al. 2017). Nowadays, it has been tried two routes to upgrade microalgal bio-oil quality directly during pyrolysis. Using that route, it has been obtained high-quality biofuel in one step. The first one has been known as thermal cracking with zeolites (Zainan et al. 2018) while the second one has been known as co-pyrolysis with several materials (Peng et al. 2015).

4.15 Solar Energy

Microalgae, as a very efficient converter of solar power, has been used in mass culture for both biomass and high-value product production. Solar radiation is often too high relative to the photosynthetic capacity of microalgae, thus leading to photo saturation and photo inhibition and also to over-warming, eventually significantly increasing mortality (Parlevliet and Moheimani 2014; Collet et al. 2014). In recent years, besides synthetic photovoltaic devices, bio-photovoltaics (BPVs, also known as biological solar cells) have emerged as an environmentally friendly and low-cost approach to harvesting solar energy and converting it into electrical current. These solar cells utilize the photosynthetic properties of microorganisms, such as algae, to convert light into electric current that can provide electricity. Thus, shading the microalgae with solar panels appears to be a feasible strategy for both enhancing productivity and supplying local electricity for the process during hotter periods (Parlevliet and Moheimani 2014). When photovoltaic (PV) panels were employed as a source of electricity in the plant, Jez et al. (2017) exhibited an increase in economic competitiveness for microalgae biofuels. It is also a viable method for manufacturing algal biofuels in isolated places (usually deserts) when the electric infrastructure is unavailable or difficult to reach. The excellence of microbial fuel cell by utilizing microalgae cultivation is because it is efficient and eco-friendly. It naturally changes chemical energy into electricity. It also plays an important role in overcoming air pollution (Ping-Zhong et al. 2016). Scientists showed a method to encase algae protein in liquid droplets, leading to an extraordinary improvement in the light-harvesting and energy conversion characteristics of algae by nearly three times. Phycobiliproteins can harvest light energy from the entire spectral range of light wavelengths, including those that chlorophylls cannot absorb well, and turn it into electricity.

5 Theragnostic Potential of Soil Algae

Algal cells contain protein bodies, which contain proteins that can bound carbohydrates are called lectins (Skjånes et al. 2013). The only microorganism that has produced potential lectin is algae, hence termed algal lectin (Singh and Thakur 2015). It plays a hopeful role in various applications of pharmacognosy such as identification of any changes in glycan synthesis related to disease, glycan aberrations, detection of antigen in blood and bodily fluid, biological markers for detection of disease-causing agents such as parasites, bacteria, viruses, and fungi. Based on the functions of algal cells, the production of algal lectins varies. The soil derived cyanobacteria that produce lectins are *Nostoc* sp., *Microcystis* sp., *Oscillatoria* sp., *Scytonema* sp., and *Lyngbya* sp. by Chlorophycean members such as *Spirogyra* sp., *Chlorella* sp., *Desmococcus* sp., *Scenedesmus* sp., and *Chlamydomonas* sp.

5.1 *Therapeutic Activities of Microalgae*

Soil microalgae are the simple photosynthetic, autotrophic microorganisms, which form the primary producers. They produce primary and secondary bioactive compounds and other molecules of high values, such as antioxidants, vitamins, carotenes, polysaccharides, etc., which have interesting aspects of industrial and medical importance. In addition, eco-friendly nanoparticles produced using living and dead microalgae have medicinal values. Algal-derived nanoparticles (NPs) such as gold (Au), silver (Ag), zinc oxide (ZnO), copper oxide (CuO), iron oxide (Fe_3O_4 or γ–Fe_2O_3) are widely used in applications of biosensing and therapy.

5.2 *Primary Metabolites*

5.2.1 Polysaccharides

Algal cell walls contain huge amount of polysaccharides (Alba and Kontogiorgos 2018), which are polymers of carbohydrates called polycarbohydrates. It has a lot of medicinal properties like antifungal, antibacterial, antiviral, antioxidant, anti-inflammatory, anticancer, immunostimulatory, antidiabetic, antiproliferative, anti-coagulant, anti-aging, and anti-glycatochlorella ellipsoidea interact with RAW264.7 cells of the macrophage murine, which triggers nuclear factor-kappa B (NF-B) and mitogen-activated protein kinase pathways and produce a large amount of diverse nitric oxide and cytokines. Green algae produces polysaccharide glucuronorhamnan, which absorbs the virus particles and downrigger protein kinase B signaling of the host there by suppressing the multiplication and infection of enterovirus 71 (Nigam et al. 2021).

5.2.2 Protein-Based Therapeutics

Algal proteins such as glycoprotein extracted from *Chlorella vulgaris* strain CK22 have antimetastatic immunopotentiation and are efficacious in curing tumors and cancer. Toll-like receptor of the glycoprotein recognizes bacteria. ARS-2 water-soluble protein moiety of *C. vulgaris* is anti-tumorous glycoprotein (Hasegawa et al. 2002). Green algal strains of very colder regions produce antifreeze proteins, which prevent protein and cell damage and recrystallization by binding with ice crystals. Algal antifreeze proteins have biomedical applications (Skjånes et al. 2013). Algal proteins produce antibodies in the chloroplast, which are capable of producing 3D structures because of their protein folding ability. Algal human antibodies were produced for treating human pulmonary emphysema in San Diego (Aditya et al. 2016). Cyanovirin present in *Nostoc* is a useful therapeutic protein compound for treating viruses like HIV and H1N1 (De Morais et al. 2015).

5.2.3 Lipids

The cell membrane is made of fatty acids like glycolipids and phospholipids or stored as TAG. Oxidative stress is cured by using TAG. Essential fatty acids and polyunsaturated fatty acids are produced in large quantities by some algae, which is the essential nutrient of food. Deficiency of these nutrients leads to severe illness to the organism. Essential fatty acids such as linoleic acid, α-linolenic acid, γ-linolenic acid, omega-3 fatty acid, and acetylenic lipids are present in the soil. Microalgae has medicinal values such as antibiotics, depression, phototoxic, immune-suppressive agent, ADHD, hypertension, anti-cancer, antitumor, anti-inflammatory, anti-thrombic, anti-arrhythmic, hypolipidemic, and vasodilatory activity, renal problems, anti-diabetic, anti-arthritic, antibacterial, antifungal, antiviral against HIV, antimicrobial, ulcer, vasodilatory, rheumatoid, Crohn's, and chronic pulmonary heart diseases, and mental illness. *Parietochloris incisa* produces arachidonic acid-rich TAG globules that, when integrated with the cell membrane, are effective against cold stress. Linolenic acid of *Dunaliella* sp. and *Chlorococcum* sp. produces antibodies against microorganisms. Large amount of sterols produced by *Pyramimonas* sp. and *Tetraselmissuecica* are necessary components of cell membrane and signal transduction. 3 PUFA of *Pavlova* sp. and *Thalassiosira* is used in food instead of fish oil (De Morais et al. 2015). Lipids of *N.commune* inhibit cholesterol and fatty acid synthesis metabolism by lowering the regulatory activity of sterol regulatory element binding protein (Rasmussen et al. 2008).

5.2.4 Photosynthetic Pigments

Microalgal chlorophyll improves blood circulation, pressure, digestion, immunity, and anti-cancerous. Chlorophyll a and b are present in large amounts in Chlorophycean members of soil algae; because of their photosensitization activity, used cancer photodynamic therapy. It is also injected into the cells for regeneration of damaged tissues. Reactive oxygen species release enough energy and are thus used as therapeutics for malignant and carcinogenic effect diseases (Riswi et al. 2015).

5.2.5 Secondary Metabolites

Soil microalgae are the natural source of pigments for pharmaceutical products that cure neurodegenerative diseases, tumors, and neuronal disorders. Fucoxanthin carotenoid present in diatoms is anti-inflammatory and causes apoptosis in the cancer cells of human beings. Histamine produced by the phycocyanin of cyanobacteria is anti-inflammatory. Carotenoids and Lutein therapy are carried out to treat cataracts, macular degeneration, reactive oxygen species treatment, and atherosclerosis. β-Carotenoids of *Dunaliella saina* have anticancer activity and rehabilitate degenerative effects in treatment of angina pectoris (Michele et al. 2015).

High-value carotenoid astaxanthin produced by *Haematococcus pluvialis* has anti-diabetic mechanism, antioxidant, anti-inflmmatory, and anticancer efficacy (Mularczyk et al. 2020). β-Carotenoids and lutein extracts of *Chlorella ellipsoidea* and *C.vulgaris* inhibit colon cancer.

5.2.6 Phytosterols

Esters of algal phytosterol are used to prepare the drug statins, which are anti-hypercholesterolemia and for coronary heart disease caused by low-density lipoprotein cholesterol. Microalgae-derived phytosterols such as stigmasterol of *Schizochytrium* interact with ACAT2 intestinal gene and down-regulate the process of intestinal cholesterol absorption. Phytosterol absorbs cholesterols from gastrointestinal tracts and performs cholesterol-lowering activity (Luo et al. 2015).

5.2.7 Phenolic Compounds

Soil microalgae are rich in polyphenols, which consist of various types of phenols and flavonoids and have antioxidant properties. Gallic acid and protocatechuic acid have effective antioxidant efficacy. Phenols of *Spirulina maxima, Chlorella* sp., *Navicula* sp. *Nannochloropsis* sp. protect the liver from damage through radical scavenging activity. Phenols of microalgae exhibit anticancer activity, anti-proliferative activity, and anti-metastatic activity. Ellagic acid, in combination with cisplatin, cures murine colon cancer. Likewise, ellagic acid and cisplatin, in combination with doxorubicin, heal hepatocellular carcinomas (Ferdous and Yusof 2021). Soil microalgae derived lectin has a miraculous property of dual functional mechanism:

5.2.8 Lectins

Microalgal lectins have a normal function of greater affinity and specificity to bind with sugar, i.e., sugar binding specificity and activity. Lectin has specificity in binding with glycoconjugate receptors of foreign agents. The glycoconjugate receptor may be glycans of polysaccharides, proteins, lipids, glycosylated serine-hydroxyproline (resemblance of cell wall protein), polynucleotide: adenosine glycosidase, glycosyltransferase (glycogen of serum protein), glycopeptides, aberrant glycans, and so on. Foreign glycans are well recognized by the lectin binding site. This recognition by the algal lectins is because of the abundant availability of non-plant-specific lectins, i.e., glycans of animals and human beings. It conjugates with a glycan that is not present in the algae and plays the role of a specific defense lectin (Willy and Els 1998).

5.2.9 First Mechanism

Algal lectin efficiently detects the microorganisms or infectious agents and binds with their glycans or glycoconjugate receptors to form bioadhesives or glycoconjugates, or lectin-binders without modifying the glycan to which it binds. It performs its natural function of exogenous lectin receptor because of its affinity with foreign glycans. Algal lectins undergo evolutionary adaptation and detect the host glycans and that of infectious agents from animal or man glycans. Chitin-binding lectins bind with the cell wall of fungi. Microalgal lectin identifies microbial glycans from recombinant glycans of man and animals, glycan synthesis because of disease, and forms specific and accurate glycosylation. These binding activities form the prerequisite for further therapeutic activity (Willy and Els 1998).

5.2.10 Second Mechanism

The second function of the dual mechanism of micro-algal lectin is therapeutic potential. Lectins do efficient restoration activity through lectin-direct targeting of drug delivery systems (Fohona and Bi-Botti 2016).

5.2.11 Lectins in Pathogens

Lectins are reported on the cell surface of many Gram-positive and Gram-negative bacteria and also in viruses called viral lectins. These lectins have the property of saccharide specificity. These lectins are bound with particular carbohydrate moieties (Hassan et al. 2020). Efficient strategy is required for the effective binding of microalgal-derived lectins with the glycan moiety of the infectious agents. To tackle this problem, advanced biotechnological tools are to be implied to achieve targeted recovery.

5.2.12 Nanotechnology

Algal lectin-conjugated nanoparticles perform the dual functions of disease deduction, targeted drug delivery, and monitoring of recovery responses with higher efficacy (Fohona and Bi-Botti 2016).

5.2.13 Biosensors and Microarray

Specific biosensors are capable of bio-recognition of diseases. Glycan integrated biosensors are employed to find microbial lectins, which is an important prerequisite. Targeting can be achieved by using a parallel array of lectins integrated with sensors to prepare glycan biosensor chips. Electrochemical and optical transducers

Lectin-selective biosensor chips provide real time monitoring of interaction affinity. Lectin electrodes have greater affinity for binding to glycans of bacteria (Hassan et al. 2020). Microalgal lectin as a dual mechanism. First, it naturally analyzes and targets diseases; second, it delivers drugs onto the infected surface for therapeutic activity. Thus, algal lectinology is a powerful theragnostic potential of soil algae. Soil algae such as *Nostoc ellipsosporum, Microcystis aeruginosa*, and *Griffithsia* sp. produce anti-HIV and anti-HCV lectins, cyanovirin-N, and microvirin, which are highly effective antiviral lectins (Hassan et al. 2020).

6 Environmental Remediation

As garbage from cities, towns, and factories is poured into aquatic and terrestrial ecosystems without treatment, rapid urbanization and industrialization have prompted scientific communities to address a number of environmental challenges. Aquatic ecosystems are especially vulnerable to these environmental threats since practically all contaminants end up in natural water bodies. Although numerous physical, chemical, and biological approaches for the bioremediation of aquatic contaminants have been proposed, each method has its own set of benefits and drawbacks. The traditional chemical approaches for decontaminating wastewaters are either ineffective or prohibitively expensive. Microalgae are now being used to safely remediate wastewater. Because of their high surface-to-volume ratio, high removal capability toward toxic pollutants, higher bio-sorption capacity, adaptability to different environments, and cost-effectiveness, the exploited use of microalgae for wastewater remediation is advantageous over other wastewater decontamination techniques. Algal species are gaining popularity as a means of removing organic (hydrocarbons) and inorganic (heavy metals) contaminants from soils (Gomez & Sartaj, 2013). A number of algae species have been identified as having the ability to clean up polluted soils and wastewaters.

Bioremediation is the most common technique used to treat wastewater because of its low cost and environmentally friendly nature. Almost every wastewater-treating industry opted bioremediation (Gomez et al. 2008). The chemical and physical removal of contaminants is expensive and less effective than bioremediation. Phycoremediation is the use of macro or microalgae to remove or biotransform contaminants from wastewater, such as nutrients and xenobiotics (Ahmad et al. 2013). The microalgae are versatile in converting the contaminants into non-hazardous resources, enabling the treated water to be recycled, reused, or safely discharged (Rao et al. 2011). Not like the conventional wastewater treatment process, where the slime is used to debase natural carbonaceous issues to their more straightforward form, algae can absorb natural toxins into cell constituents. Microalgae-based treatment framework is additionally one of the suitable answers for taking care of the ecological issues; for example, global warming, the expansion of the ozone gap, and the atmosphere changed because of its capacity to devour a high amount of carbon dioxide in photosynthesis procedure to deliver oxygen and glucose (Ahmad et al. 2019). This technology is low in cost, and it is an effective approach to remove excess nutrients, and contaminants in wastewater and generating possibly useful biomass (Sengar et al. 2011).

Algae consume waste as a nutrient and digest contaminants enzymatically. Phycoremediation is utilized as a secondary or tertiary treatment method for industrial or mine wastewater treatment due to the high concentration of heavy metals and their toxicity (Sharma and Pandey 2014). The xenobiotics and heavy metals are known to be detoxified, transformed, or volatilized by algal metabolism (Sangar et al. 2011). They can take up various kinds of nutrients like nitrogen and phosphorus (Olguin 2003). They can utilize various organic compounds containing nitrogen and phosphorus from their carbon sources.

Heavy metals accumulated by microalgal cells will be involved in carbon metabolism in the cells. The produced carbon compounds can be utilized in the production of biofuels (Piccini et al. 2019). Some other researchers, such as Sivasubramanian and Muthukumaran (2012), Kotteswari and Murugesan (2012), Nordin et al. (2014), etc. have also documented some studies on phycoremediation. Recent research is focusing on wastewater treatment with microalgae and bacteria to mitigate CO_2 and to produce low-cost biomass for bioenergy and other value additions of biomass (Viswanaathan et al. 2022; Rugnini et al. 2019).

The main mechanism used by algae for phycoremediation of contaminated soils is very much related to its surface properties and its cell-wall constituents, which contain polymers, namely pectin, hemicelluloses, cellulose, lignin, and arabinogalactan proteins (Domozych et al. 2012). The phycoremediation of algae to accumulate and absorb heavy metals into its biomass has been investigated (Jeyakumar and Chandrasekaran 2014). Mechanisms such as chemisorption, ion exchange, surface precipitation, and covalent bonding are observed in phycoremediation (Munoz and Guieyssea 2006). The harvesting of CO_2 from the atmosphere into algal biomass and subsequent deposition of this carbon from algae to agricultural soil results in considerable positive changes in both physical and microbiological properties, therefore enhancing soil fertility status.

Microalgae-mediated CO_2 bio-mitigation can be more economic, cost-effective, and eco-friendly when it is incorporated into a wastewater treatment infrastructure (Collotta et al. 2018). The choice of microalgae to be used in wastewater treatment is determined by their robustness against wastewater and by their efficiency to grow in it and take up nutrients from wastewater (Olguin 2003).

Contaminants often result from domestic, agricultural, and industrial effluents, and their physical and chemical compositions differ according to their origins (Felix Wollmann et al. 2019). The most frequently found are metals, pesticides (Emparan et al. 2019), and nutrients, such as nitrate (Menció et al. 2016) and phosphate (Schaum 2018). The occurrence of emerging organic and inorganic pollutants, such as microplastics (Koelmans et al. 2019), pharmaceuticals (Muñoz et al. 2009), flame retardants (Donna Sutherland et al. 2019), personal care products (Norvill et al. 2016), and hazardous and noxious substances, has been increasing since conventional wastewater treatment plants (WWTP) not yet equipped and suitable to remove these new contaminants. The contaminants can be divided into different classes, such as pesticides, personal care products, drugs, and flame retardants (Norvill et al. 2016). The main technologies for performing the removal or degradation of emerging contaminants include membrane separation processes (e.g., ultrafiltration, nano filtration, and reverse osmosis), advanced oxidation processes (e.g., photocatalysis, UV/H_2O_2, electro-Fenton, and ozonation), adsorption (e.g., activated carbon, carbon nanotubes, and graphene), and bioremediation (e.g., wetlands and microalgae) (Teodosiu et al. 2018; Fischer et al. 2019; Rigueto et al. 2020).

Organic pollutants (pesticides, phenolics, petroleum hydrocarbons) can also be converted/degraded by microalgae into less/no hazardous chemicals (Egberomoh and Fagade 2016). Algae can degrade azo dyes through an induced form of azoreductase.They promote thereductive cleavage of the azo bonds and release the aromatic amines that will be degraded in CO_2 and H_2O (Lellis et al. 2019; Jamee and Siddique 2019). There are reports of the treatment of wastewater using *Chlorella* spp. and *Scenedesmus* spp. that degrade or remove paracetamol, acetylsalicylic acid, caffeine, and diclofenac (Escapa et al. 2016), levofloxacin (Xiong et al 2017) among others, with efficiencies above 90% (Escapa et al. 2017a, b, c; Xiong et al. 2018; Tripathi et al. 2019). Microalgae for emerging pollutants bioremediation stands out due to considerable improvements in treatment efficiency and the possible use of biomass for the creation of high-value products (Escapa et al. 2017a, b, c; Sutherland and Ralph 2020).

The proper disposal of phycoremediation in aquatic plants is a new topic of study that has to be explored (Newete and Byrne 2016). The employment of hyper accumulating aquatic plants like algae for arsenic treatment technology has both environmental and economic benefits; however, the accumulation and removal of the metalloid is insufficient for the successful application of this technology if it is not properly controlled and disposed of (Rahman and Hasegawa 2011).

Harvesting techniques have large energy demands and operating expenses; hence, solutions to simplify this process have been researched (Pal et al. 2019). As a result, novel microalgae culture methods have emerged, such as the immobilization method on alginate beads and the development of bioreactors that utilize microalgae biofilm, and have been linked to wastewater treatment (Gross & Wen, 2014; Li et al. 2024).

Some heavy metals are toxic and carcinogenic, even at minute concentrations. Bio-sorption using seaweed is obviously a promising method with naturally existing raw material and a higher efficiency with a low-cost investment in the treatment.

Algal/cyanobacterial-based biofilms are getting increasingly popular as an alternative to suspension-based culture systems because of their low-cost algal harvesting (Blanken et al. 2014), simple, energy-efficient absorption of the essential nutrients, nitrogen and phosphorus, followed by simple and reliable separation of the algae biomass from the wastewater bulk (Shi et al. 2014; Li et al. 2015). When compared to the same biomass grown in suspension cultures, high algal concentrations can be reached with a significantly reduced amount of media in algal biofilm systems (Berner et al. 2015). Despite the obvious benefits of biofilm-based technology for low-cost nutrient removal and low-energy biomass production, widespread adoption of algalbiofilm-based treatment of municipal, industrial, and agricultural waste streams has been limited thus far (Kesaano and Sims 2014). Reverse osmosis (RO) technology is being used worldwide for full-scale municipal wastewater reclamation to produce high-quality recycled water to meet the increasing water supply demand (Shannon et al. 2008). The need to investigate cost-effective treatment techniques for the ROC in order to reduce the environmental and health dangers associated with its disposal or reuse is developing.

Nanotechnology provides an opportunity for upgrading waste water treatment. Moreover, it has many advantages, like cost effectiveness, reusability, and being highly efficient in removing and recovering pollutants due to its high surface area. The main types of nanomaterials that can be synthesized from algae are either metal nanoparticles or metal oxide nanoparticles (Asmathunisha and Kathiresan 2013). Traditionally, algal systems have been used as a tertiary wastewater treatment technique. They have also been mentioned as a possible subsequent therapy option. Algae can be utilized in treatment processes for a variety of objectives, including the elimination of coliform bacteria, lowering chemical and biochemical oxygen demand, removing N and/or P, and removing heavy metals, to name a few (Rao et al. 2012).

After the pollutants are removed, the remaining biomass can be utilized to make bioproducts, which adds another benefit to the process (Xiong et al. 2018; Gojkovic et al. 2019). So for effective phycoremediation treatment, there should be regular harvest and safe disposal of the biomass (Rai 2008) to avoid the release of the absorbed or adsorbed metals back into their source when the plant dies and decomposes. After treatment, biomass can be used to produce biofuels on account of high levels of proteins and carbohydrates, depending on the species (Dasan et al. 2019), considering full use and completion of a cycle closure, thereby realizing the concept of biorefinery (Ubando et al. 2020).

The important challenges of the present-day phycoremediation technologies include the space requirement for the cultivation of the algae and the remediation rate of the system. Still, technology needs to develop to produce algae in higher concentrations in the outdoor environment to tackle the challenge. Many bioactive compounds can be produced from algae biomass, which has the larger application in pharmaceutical sector. However, due to the lower production of biomass, the earlier is not accomplished, affecting the economic feasibility. Algal species are gaining popularity as a means of removing organic (hydrocarbons) and inorganic (heavy metals) contaminants from soils.

7 Personal Care Applications of Soil Algae

The term "PERSONAL CARE" comes from the word cosmetics used by a person to enhance his personality, beauty, or appearance, and to encourage attraction in his daily life. Different types of personal care products used toothpaste, ear buds, wet or dry wipes, tissue paper, cotton swabs, sanitary pads, colognes, insect repellent, diapers, and much more. Apart from this, there are toothpaste, delicacies, biting toys, trash cans, demetting, eye and ear cleaners, shiny footwear, hair washer, shampoos, lotions, soap, towels, perfumes, urine stool, bags, nappies, deodorants, insect repellents are available as personal care products for domestic cattle, dogs, cats, and birds in a modern sophisticated lifestyle. In addition, a variety of these products are produced by various factories and companies and are available in many of today's offline and online markets.

As a result of various environmental changes such as pollution, stress, and other environmental changes occurring due to climate change and heavy rainfall, such as extreme heat waves and heavy rainfall and temperatures, depletion of the ozone layer, and ultra violet radiation.

7.1 Biomass Production

Algae grown at different pressures cause the production of metabolites. Based on demand, the culture can be subjected to appropriate stress conditions in order to produce the bioactives you want (Skjånes et al. 2013). Growing microalgae under severe stress conditions creates the accumulation of secondary rich metabolites (De Luca et al. 2021).

7.2 Algal Oil

Lipid-rich soil alloys are subject to disruption by traditional printing presses for oil extraction. This method is expensive, and the oil available is clean. The supercritical fluid extraction method is also used, which is very expensive due to the high technology (Barasa et al. 2021). Oil is extracted from liquids such as *Scenedesmus, Spirulina, Chlorella, Arthrospira,* and *Desmodesmus* (De Luca et al. 2021 & US Patent # 20120149075). Microalgae-based additives are added to obtain the required algal oil (Table 1).

7.3 Algal Soap Bar

A soil algae biomass rich in oil content, such as *Chlorella* and *Spirulina*, will be further processed, and all algal cells will undergo saponification to produce soap. In other words, algal fatty acid esters and glycerolipids require less alkaline hydrolysis to obtain fatty acid salts. Then a metal halide such as sodium chloride, and potassium chloride, which are alkaline and dissolved in water, is added to the fatty acid salts for their precipitation. Alkaline hydrolysis is performed using sodium hydroxide and aminomethylpropanol, which are used to produce bar soap. These soaps can

Table 1 Microalgae derived additives are added to get desired algal oil

Microalgae additives	Cosmetic products	Cosmetic property	Cosmeceutical property	References
Fatty acid salts	Algal soap bar Algal liquid soap Skin deodorant Fragrance products Oral care products Lens disinfectants	Detergents Coloration Film former Non-sensitizing Non-irritating	Normal skin barrier Skin protection Cosmetic biocide Antibacterial effect to oral bacteria Anti *Acanthamoeba*—disinfecting contact lens.	De Luca et al. (2021), Tanaka et al. (2017), www.personalcarecouncil.org
Lipids	Gel	Emollient Colour and fragrance carrier Lipid Bio-factories Nanoparticles nanospheres, nanosomes	Skin softness and smoothness Moisturizing agent	De Luca et al. (2021)
Poly unsaturated fatty acids	Different cosmetic formulations	Creams or emulsions	Anti-oxidant Anti-inflammatory Anti-cancer	De Luca et al. (2021)
Wax	Lipstick Eye makeup sticks Eye shadow Deodorant sticks	Stick texture, rigidity, hardness,	Personal care applications	De Luca et al. (2021), Blasio and Balzano (2021)
Glycerophospholipids	Skin care products—skin hydration	Emulsifying agent Stabilizing oil-water emulsion. Surfactant to adjust viscosity and crystallization	Skin care applications	De Luca et al. (2021)

(continued)

Table 1 (continued)

Microalgae additives	Cosmetic products	Cosmetic property	Cosmeceutical property	References
Lecithins	Oilments	Emulsifying agent	Water based emulsifying agent in cosmetic ingredients	De Luca et al. (2021)
Glycerolipids	Bath products Cleaning products Eye makeup Fragrance Foot powder Facial makeup Personal care products Skin suntan	Emollient	Antioxidant Antimicrobial	De Luca et al. (2021)
Sphingolipids	Skin care product	Moisturizer	Epidermal barrier repair	De Luca et al. (2021)
Carotenoids β-carotene, zeaxanthin, canthaxanthin, Astaxanthin	Skin and hair care products Sun tan lotion Natural dye	Moisturizer	Antioxidant UV-rays damage protection Colour enhancer Recover skin tone, elasticity and wrinkles	De Luca et al. (2021), Gupta et al. (2019)
Chlorella whole algal cells	Skin cream Face gel Shampoo	Moisturizer Keratin	Protection of collagen and elastin fiber Anti-wrinkling agent Restores skin firmness Anti –aging Skin moisturizer Anti –photo–ging Strong, shiny, long hair	Gupta et al. (2019), Pooja et al. (2020) Vaibhav and Sahasrabuddhe (2018)
Alkenones (Sustainable wax)	Lipsticks	Ease bending More firmness	High color intensity for skin Low friction	Blasio and Balzano (2021)

(continued)

Table 1 (continued)

Microalgae additives	Cosmetic products	Cosmetic property	Cosmeceutical property	References
Algal oil	Make up Bath care Cream Sun care products	Moisturizer Best emollient	Anti-microbial activity Freshen skin barrier Skin moisturizer Anti-ageing	U.N. FAO (2010), Birjandi et al. (2020)
Oleic acid	Lotions soaps	Long-lasting moisturizer	Skin nourishment	Basily et al. (2018)
Linoleic acid	Lotion, oil, soap and sunscreen	Emollient and thickening agent	Skin-barrier repair Anti-oxidant Anti-inflammatory Lowers pain Comedos and eczemas formation Freshen Sebaceous gland	Basily et al. (2018)
Mycosporine-like amino acid Mycosporine-serinol and porphyra-334	Cream	Sunscreen	Anti-oxidant Anti-skin aging UVA and UVB damage protection Photoprotection Procollagen C proteinase and elastin enhancer	de la Coba et al. (2019)
Nostoc commune whole algal cells	Cream	Moisturizer	Anti-inflammatory Skin nourishment, Softener and whitening	Vaibhav and Sahasrabuddhe (2018)
Nostoc flagelliformenm whole algal cells	Cream	Sunscreen	UVA and UVB damage protection	Vaibhav and Sahasrabuddhe (2018)
Chlorella vulgaris And *Spirulina* whole algal cells	Facial cleanser	Cleansing agent	Skin cleanser	Vaibhav and Sahasrabuddhe (2018)

(continued)

Table 1 (continued)

Microalgae additives	Cosmetic products	Cosmetic property	Cosmeceutical property	References
Spirulina whole algal cells	Cream	Moisturizer Pigmentation	Anti-aging, remove dark circle, replenishment of skin Skin coloration Skin cleanser Anti-acene	Vaibhav and Sahasrabuddhe (2018), Ragusa et al. (2021)
Chlorophyll	Cream Lotion	Natural dye	Antioxidant Anti-inflammatory Anti-wrinkle Skin damage repair Collagen repair	Agustina et al. (2020)
Vitamin E	Skin cream	Anti-oxidant	Photoprotection	Skjånes et al. (2013)

be made in different ways. Microalgae-based additives are added to obtain the desired algal oil (US Patent # 20120149075) (Table 1).

7.4 Algal Liquid Soap

Algal biomass is subjected to water hydrolysis to extract algal-oil-in-cell emulsion concentrate. The remaining amount of the preferred KOH base is added. After alkaline hydrolysis fatty acid salts were obtained. In addition, glycerin, ethylenediamine, and cocoamidopropyl betaine are added to improve clarity, moisture, solubility, purity, and aroma is added to algal liquid soaps. Microalgae *Chlorella, Spirulina*-based additives are added to obtain the algal oil you wish (US Patent # 20120149075) (Table 1).

7.5 Algal Lotion

Algal lotion is made using a water and oil phase, where glycerin and distilled water are applied to the water phase. Algal metabolites such as lipids, oleic acid, linoleic acid, phospholipids, and saccharolipids are used in the oil phase. The water phase is added to the oil phase with constant stirring until the emulsion state. Other bioactive components of microalgae are added and mixed in the same way to produce algal lotions (Basily et al. 2018; Agustina et al. 2020). Microalgae-based additives are added to obtain the required algal oil (Table 1).

7.6 Algal Cream

Similarly, algal lotions are made using a category of water and oil. Oil found in microalgae has a negative effect on unsaturation, thus reducing oxidizing efficiency and increasing shelf life. Microalgae under fermentation convert glycans into lipid molecules. Microalgal oil has excellent emollient properties (Birjandi et al. 2020). The oil phase is added to the water phase and is stirred into a homogenizer to obtain the emission status. Bioactive components of microalgae are added to it and mixed evenly to form algal cream. (Kalasariya et al. 2021). Microalgae-based additives are added to obtain the required algal oil (Table 1).

7.7 Algal Gel, Body Wash as Well

Lipids and polysaccharides found in soil microalgae such as *Spirullina* provide the same gel and moisturizers (Hernandez et al. 2020). This gel is used as a cleansing gel, bath gel, sunscreen, and body wash (Ruma et al. 2021). Microalgae-based additives are added to obtain the required algal oil (Table 1).

7.8 Algal Foam

Phycobilliprotein and microalgae proteins such as *Arthrospira maxima*, *A.plantensis*, *Nannochloropsis gaditana* and *Tetraselmis impellucida* increase the size of the bubble, the solid emulsion of the optimal air-oil connector, and the formation of 500 m *amafuthaamafutha* oil and amakhulu Pascal 1200 mM emulsion. Microalgae-based additives are added to obtain the required algal oil (Table 1).

7.9 Algal Shampoo

Polysaccharides, lipids, peptides, vitamins, phycobilliproteins, and amino acids such as mycosporine in microalgae such as *Chlamydomonas*, *Spirulina*, and *Chlorella* are used as foam and gels in the production of algal shampoo (Ruma et al. 2021; Barasa et al. 2021). Microalgae-based additives are added to obtain the required algal oil (Table 1).

Table 2 Microalgae in pet care

Microalgae additives	Cosmetic product	Cosmetic property	Cosmeceutical property	References
Spirullina Polysaccharides Vitamins B-carotene	Oral care Skin care	Lustrous skin	Dental cure for dog Healthy and balanced skin Boost immunity to fight infection	https://www.greatpetcare.com/wellness/spirulina-for-dogs/ https://theinnocentpet.co.uk/health-benefits-of-spirulina-for-dogs https://holistichound.com/spirulina-for-cats-and-dogs/

7.10 Algal Eyeliner Sticks, Lips, and Lip Balms

Euglena gracilis and *Neochloris oleoabundans* certain microalgae exist in wet soil compaction of large amounts of wax by wax ester fermentation of polysaccharide paramylon (Blasio and Balzano 2021). This wax has good texture, hardness, and durability. Therefore, this wax is still distributed as a wax obtained by microalgae on eyebrows and lip balms (De Luca et al. 2021). Microalgae-based additives are added to obtain the required algal oil (Table 1). These microalgal supplements can still be distributed against certain targeted effects on various personal care products. Some of the freshwater algae recorded in this chapter are found in wet soil. In modern life, *Spirulina* has been extensively tested for personal animal care products (Table 2).

8 Opportunities and Challenges

A number of algae species have been identified as having the ability to clean up polluted soils and wastewaters. Despite advancements in alga-based removal methods, significant limits remain, necessitating further research in order to effectively use holistic phycoremediation removal strategies on a broad scale. Commercial microalgal wastewater treatment, on the other hand, would necessitate the integration of numerous separate unit processes, such as microalgal cultivation, biomass separation, and biomass valorization. Traditional knowledge and delicious algal foods can aid the development of western algae-based methods for human consumption. Algal biomass can be processed into biofuels and other products. As the world's population expands, so will the need for grain. The limits for the ecological sustainability of algal biofuels have not yet reached those of other food stocks in terms of testing and selection. Many institutions and researchers have proposed indicators of bioenergy sustainability. But they have not identified significant potential or a set of measurable environmental indicators. To create the target's bioactivities, cultures can be produced in a number of methods. The build-up of secondary rich metabolites occurs when microalgae are grown under stressful conditions. This is crucial in the manufacturing of biofuels. Traditional oil extraction presses damage lipid-rich soil alloys. The supercritical fluid extraction process is also employed, which is quite costly due to the high level of technology needed. The space needed for algae growth, as well as the system's remediation rate, are two important challenges for today's phycoremediation systems. To overcome the issue, technologies that produce algae in higher quantities in an outside environment must evolve. Algal biomass can produce a wide spectrum of bioactive compounds, which have numerous applications in the pharmaceutical business. However, due to declining biomass yield, the first phase is not finished, jeopardizing economic feasibility. Algal species are becoming increasingly popular as a method of eliminating organic (hydrocarbons) and inorganic (heavy metals) pollutants from soils.

9 Conclusion and Future Perspectives

In developing sciences, algae plays a critical role in resolving a variety of issues. Humans pay extremely less attention to little algae and sea plants compared to other plants and animals. Commercialization of algae biofuel production will necessitate a large amount of algae production. Soil algae can thrive under a number of stress circumstances, including changes in light intensity, temperature, salt, pH, and nutritional content. Other cutting-edge technologies, such as reactive permeable barrier systems, have indeed been recognized as having the chance to facilitate heavy metal removal in the future. Microalgae-based wastewater or effluent bioremediation research has progressed to the point that it now focuses on combining bioremediation with the development of economically viable break through bioproducts, thereby benefiting resource efficiency and serving as a multipurpose solution. Recent funding opportunities have focused on the bioeconomy and the development of alternative bioproducts, which will help this field of research process. Not only is further research needed to identify the chemicals involved, but it is also necessary to better understand how microalgae make nanoparticles. To make algal-based biofuels more promising, further research is needed to improve the technology used to convert biomass to biofuels and to develop better harvesting methods. When microalgae such as blue and green cosmetics goods become available in the future, personal care products will play a significant role in fostering human appreciation of attractiveness or acclaim.

References

Adamu CI, Nganje TN, Edet A (2015) Heavy metal contamination and health risk assessment associated with abandoned barite mines in Cross River State, southeastern Nigeria. Environ Nanotechnol Monit Manag 3:10–21

Aditya T, Bitu G, Mercy Eleanor G (2016) The Role of Algae in Pharmaceutical Development. Research & Reviews: J Pharm Nanotech 4:82–89

Agustina S, Aidha NN, Oktarina E (2020) The 6th international symposium on applied chemistry (ISAC). In: IOP conference series: materials science and engineering, vol 1011 (2021). IOP Publishing, p 012057

Ahmad F, Khan AU, Yasar A (2013) Comparative phycoremediation of sewage water by various species of algae. Proc Pak Acad Sci 50:131–139

Ahmed S, Griffin TS, Kraner D, Schaffner MK, Sharma D, Hazel M (2019) Environmental factors variably impact tea secondary metabolites in the context of climate change. Front Plant Sci 10:939

Alba K, Kontogiorgos V (2018) Seaweed polysaccharides (agar, alginate carrageenan). In: Varelis P (eds) Encyclopedia of food chemistry. Elsevier, Amsterdam, pp 240–250.

Alassali A, Cybulska I, Alkhori A, Przemyslaw Brudecki G, Thomsen N (2016) Methods for upstream extraction and chemical characterization of secondary metabolites from algae biomass-a mini review. Adv Tech Biol Med 4(1):1–16

Alvarez V, Muñoz Arias J (2008) Ethnobotanical Practices with Durvillaea antarctica and Cladophora crispata in Chile and Peru. Ethnobiol Conserv 18:78–89

Anyanwu RC, Rodriguez C, Durrant A, Olabi AG (2018) Micro-Macroalgae Properties and Applications. In Reference Module in Materials Science and Materials Engineering; Elsevier: Amsterdam, The Netherlands. pp. 1–28. ISBN 9780128035818.

Argun H, Kargi F, Kapdan IK, Oztekin R (2008) Biohydrogen production by dark fermentation of wheat powder solution: Effects of C/N and C/P ratio on hydrogen yield and formation rate. Int J Hydrogen Energy 33:1813–1819

Asmathunisha N, Kathiresan K (2013) A review on biosynthesis of nanoparticles by marine organisms. Colloids Surf B: Biointerfaces 103:283–287

Avagyan A (2008) Global prospects for microalgae production for biofuels and for the preservation of nature. Global Fuel Magazine, February, 22–27

Bankar SB, Survase SA, Ojamo H, Granström T (2013) Biobutanol: the outlook of an academic and industrialist. RSC Adv 3:24734

Barasa M, Kaustubha M (2021) The Budding Potential of Algae in Cosmetics. Springer. Pg. 181–200.

Bardhan P, Gupta K, Mandal M (2019a) Microbes as bio-resource for sustainable production of biofuels and other bioenergy products. In: New and Future Developments in Microbial Biotechnology and Bioengineering. Elsevier, pp. 205–222

Barry AN, Starkenburg SR, Sayre RT (2015) Strategies for optimizing algal biology for enhanced biomass production. Front Energy Res. 3:1

Barsanti L, Gualtieri P (2018) Is exploitation of microalgae economically and energetically sustainable? Algal Res 31:107–115

Basily S, Haidy NM, Mamdouh D, Guzine I, Abo El-Enin, S (2018) Extraction of algal lipid as a natural cosmetic component. Egyptian. Pharm J 17:13. 10.4103

Bellou S, Baeshen MN, Elazzazy AM, Aggeli D (2014) Microalgal lipids biochemistry and biotechnological perspectives. Biotechnol Adv 32:1476–1493

Berner F, Heimann K, Sheehan M (2015) Microalgal bioflms for biomass production. J Appl Phycol 27(5):1793–1804

Bhale PV, Deshpande N, Thombre SB (2009) Improving the low temperature properties of biodiesel fuel. Renew Energy 34:794–800

Bhatia S, Garg A, Sharma K, Kumar S, Sharma A, Purohit AP (2011) Mycosporine and mycosporine-like aminoacids: A paramount tool against ultra violet irradiation. Pharmacogn Rev 5(10):138–146

BirjandiNejad H, Blasco H, Moran B, Cebrian J, Woodger J, Gonzalez E, Pritts C, Milligan (2020) Bio-based algae oil: an oxidation and structural analysis. Int J Cosmet Sci 42(3):237–247

Blanken W, Janssen M, Cuaresma M, Libor Z, Bhaiji T, Wijffels RH (2014) Biofilm growth of Chlorella sorokiniana in a rotating biological contactor based photobioreactor. Biotechnol Bioeng 111:2436–2445

Blasio M, Balzano S (2021) Fatty acids derivatives from Eukaryotic microalgae, pathways and potential applications. Front Microbiol 12:718933
Borowitzka MA, Vonshak A (2017) Scaling up microalgal cultures to commercial scale. Eur J Phycol 52:407–418
Brodie J (2009) Algal Conservation and Its Global Importance. J Marine Biol 56(6):587–597
Bwapwa JK, Anandraj A, Trois C (2017) Possibilities for conversion of microalgae oil into aviation fuel: a review. Renew Sust Energy Rev 80:1345–54
Bwapwa JK, Anandraj A, Trois C (2019) Jet fuel blend from Algal Jet Fuel and Jet A1 in 50/50 volume ratio. Inter J Low-Carbon Technol 14:234–240
Canales M, Aguilar J, Gutiérrez M, & Rodríguez E (2008) Traditional Knowledge and Use of Llaska (Cladophora crispata) by Indigenous Communities on the Peruvian-Chile Border. J Ethnopharmacol 116(2):295–301
Cardenas Alvarez R, Munoz Arias V (2008) Ethnobotanical Uses of Cochayuyo (Durvillaea antarctica) in Chile. J Ethnobiol 23(4):189–197
Chen H, Wang X, Wang Q (2020) Micro algal biofuels in China: The past, progress and prospects. GCB Bioenergy 12:1044–1065
Chisti Y (2007) Biodiesel from microalgae. Biotechnol Advan 25(3):294–306
Choi SP, Nguyen MT, Sim SJ (2010) Enzymatic pretreatment of Chlamydomonas reinhardtii biomass for ethanol production, Bioresource Technol 101(10):330–5336
Collotta MP, Champagne WM, Tomasoni G (2018) Wastewater and waste CO_2 for sustainable biofuels from microalgae. Algal Res 29:12–21
Connan S (2004) Algae and Biodiversity: A Global Perspective. Global Biodiversity Conservation 12(3):145–156
Collet P, Lardon L, Hélias A, Bricout S, Lombaert-Valot I, Perrier B (2014) Biodiesel from microalgae—life cycle assessment and recommendations for potential improvements. Renew Energy 71:525–33
CSBP (Council on Sustainable Biomass Production) (2012) Sustainable Biomass Production Standards and Indicators. CSBP, Washington, DC
Dasan YK, Kee M, Lam YS, Lim J-W, Lee KT (2019) Life cycle evaluation of microalgae biofuels production: effect of cultivation system onenergy, carbon emission and cost balance analysis. Sci Total Environ 688:112
de la Coba F, Aguilera J, Korbee N, de Gálvez MV, Herrera-Ceballos E, Álvarez-Gómez F, Figueroa FL (2019) UVA and UVB Photoprotective capabilities of topical formulations containing Mycosporine-like amino acids (MAAs) through different biological effective protection factors (BEPFs). Mar Drugs 17(1):55
De Luca M, Pappalardo I, Limongi IR, Viviano E, Radice RP, Todisco S, Martelli G, Infantino V, Vassallo A (2021) Lipids from microalgae for cosmetic applications. Cosmetics 8(2):52
De Morais MG, Vaz BDS, De Morais EG, Costa JA (2015) Biologically active metabolites synthesized by microalgae. BioMed Res Int 835761
Domozych DS, Ciancia M, Fangel JU, Mikkelsen MD, Ulvskov P, Willats WGT (2012) The cell walls of green algae: a journey through evolution and *diversity*. Front Plant Sci 3:82
Diesel Prices (2021) US Gallon. Available online at: https://www.globalpetrolprices.com/USA/diesel_prices/ (accessed May 6, 2021)
Egberomoh G, Fagade O (2016) Microalgal-bacterial consortium in polyaromatic hydrocarbon degradation of petroleum-based effluent. J Bioremed Biodegr 7(4):359
Elbehri A, Segerstedt A, Liu P (2013) Biofuels and the Sustainability Challenge: A Global Assessment of Sustainability Issues, Trends and Policies for Biofuels and Related Feedstocks. Rome: Trade and Markets Division, Food and Agriculture Organization of the United Nations; ISBN 978-92-5-107414-5
Emparan Q, Harun R, Danquah MK (2019) Role of phycoremediation for nutrient removal from wastewaters: a review. Appl Ecol Environ Res 17:889–915
Energy Information Administration. Annual Energy Review [Internet]. [Accessed: 2007]

Escapa C, Coimbra R, Paniagua S, García A, Otero M (2016) Comparative assessment of diclofenac removal from water by different microalgae strains. Algal Res 18:127–134
Escapa C, Coimbra RN, Paniagua S, García AI, Otero M (2017a) Comparison of the culture and harvesting of Chlorella vulgaris and Tetradesmus obliquus for the removal of pharmaceuticals from water. J Appl Phycol 29(3):1179–1193
Escapa C, Coimbra RN, Paniagua S, García AI, Otero M (2017b) Paracetamol and salicylic acid removal from contaminated water by microalgae. J Environ Manag 203:799–806
Escapa C, Coimbra RN, Nuevo C, Vega S, Paniagua S, García AI, Calvo LF, Otero M (2017c) Valorization of microalgae biomass by its use for the removal of paracetamol from contaminated water. Water 9(5):312
FAO Aquatic Biofuels Working Group, U. N (2010) Cosmetics in algae-based biofuels: applications and co-products
Fabris M, George J, Kuzhiumparambil U, Lawson CA, Jaramillo Madrid AC, Abbriano RM (2020) Extrachromosomal genetic engineering of the marine diatom Phaeodactylum tricornutum enables the heterologous production of monoterpenoids. ACS Synth Biol 9(3):598–612
Fabris M, Matthijs M, Carbonelle S, Moses T, Pollier J, Dasseville R (2014) Tracking the sterol biosynthesis pathway of the diatom Phaeodactylum tricornutum. New Phytol 204:521–535
Fedorov S, Ivanova L, Petrov A, Sokolov M (2013) Bioactive Properties of Seaweed Extracts and Their Applications in Medicine and Health Sciences. Marine Biology Res 49(3):112–120
Feng J, Yang Z, Hse CY, Su Q, Wang K, Jiang J, Xu J (2017) In situ catalytic hydrogenation of model compounds and biomass-derived phenolic compounds for bio-oil upgrading. Renew Ener 105:140–148
Ferdous UT, Yusof ZNB (2021) Medicinal Prospects of Antioxidants From Algal Sources in Cancer Therapy. Front Pharmacology 12:157
Fischer A, Van Wezel AP, Hollender J, Cornelissen E, Hofman R, Van der Hoek JP (2019) Development and application of relevance and reliability criteria for water treatment removal efficiencies of chemicals of emerging concern. Water Res 161:274–287
Fohona CS, Bi-Botti CY (2016) Review Current status of lectin-based cancer diagnosis and therapy AIMS Molecular Science 4(1):1–27
Foley JA, Ramankutty N, Brauman KA, Cassidy ES, Gerber JS, Johnston M, Zaks DPM (2011) Solutions for a cultivated planet. Nature 478(7369):337–342
García V, Päkkilä J, Ojamo H, Muurinen E, Keiski RL (2011) Challenges in biobutanol production: how to improve the efficiency? Renew Sustain Energy Rev 15(2):964–980
Ge L, Wang P, Mou H (2011) Study on saccharification techniques of seaweed wastes for the transformation of ethanol. Renew Ener 36(1):84–89
Gojkovic Z, Lindberg RH, Tysklind M, Funk C (2019) Northern green algae have the capacity to remove active pharmaceutical ingredients. Ecotox Environ Safe 170:644–656
Gomez E, Johnson M, Ramirez S (2008) Bioremediation: A sustainable solution for wastewater treatment. J Environ Manag 86(2):334–342
Gomez F, Sartaj M (2013) Field scale ex-situ bioremediation of petroleum contaminated soilunder cold climate conditions. Int Biodeterior Biodegrad 85:375–382
González-Fernández C, Ballesteros M (2012) Linking microalgae and cyanobacteria culture conditions and key-enzymes for carbohydrate accumulation. Biotechnol Adv 30:1655–1661
Gonzalez LM, Correa DF, Ryan S, Jensen PD, Pratt S, Schenk PM (2018) Integrated biodiesel and biogas production from microalgae: Towards a sustainable closed loop through nutrient recycling. Renew Sustain Energy Rev 82:1137–1148
Giordano M, Wang Q (2018) Microalgae for Industrial Purposes, In: Vaz Jr. S. (eds) Biomass and Green Chemistry, Springer, Cham 133–167
Global Fuels Magazine (2007) The Limitations of Grain-Based Biofuels for Fossil Fuel Replacement in Transportation. Global Fuels 12(3):45–52
Global Bioenergy Partnership (2011) Sustainability Indicators for Bioenergy. Food and Agriculture Organization of the United Nations, Rome, Italy

Greenwell HC, Laurens LML, Shields RJ, Lovitt RW, Flynn KJ (2010) Placing Microalgae on the Biofuels Map: An Overview of the Literature. Biofuels 1(6):763–783

Green Dream (2003) Backed by MPs, Eastern Daily Press.

Gross M, Wen Z (2014) Yearlong evaluation of performance and durability of a pilot-scale Revolving Algal Biofilm (RAB) cultivation system. Bioresour Technol 171:50–58

Gouveia L, Batista P, Raymundo A, Bandarra N (2008) Spirulina maxima and Diacronemavlkianum microalgae in vegetable gelled desserts. Nutr Food Sci 38:492–501

Gupta PL, Rajput M, Oza T (2019) Eminence of microbial products in cosmetic industry. Nat Prod Bioprospect 9:267–278

Harun R, Singh, M, Forde, GM, Danquah MK (2010) Bioprocess engineering of microalgae to produce a variety of consumer products. Renew Sustain Energy Rev 14:1037–1047

Hasegawa T, Matsuguchi T, Noda K, Tanaka K, Kumamoto S, Shoyama Y, Yoshikai Y (2002) Toll-like receptor 2 is at least partly involved in the antitumor activity of glycoprotein from Chlorella vulgaris. Int Immunopharmacol 2:579–589

Hassan SU, Donia A, Sial A, Zhang X and Bokhari H (2020) Glycoprotein- and Lectin-Based Approaches for Detection of Pathogens. Pathogens (Basel, Switzerland). 9(9):694

He J, Chen JP (2014) A comprehensive review on biosorption of heavy metals by algal biomass: materials, performances, chemistry, and modeling simulation tools. Bioresour Technol 160:67–78

Hernandez EM (2020) Pharmaceutical and Cosmetic Use of Lipids. In Bailey's Industrial Oil and Fat Products; Wiley: Hoboken, NJ, USA. pp. 1–28.

Herrero M (2013) The roles of livestock in developing countries. Animal 7(Suppl 1):3–18

Holdt SL, Kraan S (2011) Bioactive compounds in seaweed: Functional food applications and health benefits. Phycologia 50(6): 577–592

Hyka P, Lickova S, Pribyl P, Melzoch K, Kovar K (2013) Flow cytometry for the development of biotechnological processes with microalgae. Biotechnology Advan 31:2–16

International Organization for Standardization (2010) Sustainability Criteria for Bioenergy (ISO 13065). ISO, Geneva, Switzerland

Jamee R, Siddique R (2019) Biodegradation of synthetic dyes of textile effluent by microorganisms: an environmentally and economically sustainable approach. Eur J Microbiol Immunol 9(4):114–118

Jeyakumar RS, Chandrasekaran V (2014) Adsorption of lead (II) ions by activated carbons prepared from marine green algae: equilibrium andkinetics studies. Int J Ind Chem 5:1–9

Jane S, Duan L (2015) Cryptobiotic Soil Crusts and Their Role in Soil Ecosystems. J Soil Bio 34(2): 123–132

Jena U, Das KC (2011) Comparative evaluation of thermochemical liquefaction and pyrolysis for bio-oil production from microalgae. Energy and Fuels 25(11):5472–5482

Jez S, Fierro A, Dibenedetto A, Aresta M, Busi E, Basosi R (2017) Comparative life cycle assessment study on environmental impact of oil production from micro-algae and terrestrial oilseed crops. Bioresour Technol 239:266–75

Kalasariya HS, Yadav VK, Yadav KK, Tirth V, Algahtani A, Islam S, Gupta N, Jeon BH (2021) Seaweed-Based Molecules and Their Potential Biological Activities: An Eco-Sustainable Cosmetics. Molecules 26:5313

Kesaano M, Sims RC (2014) Algal bioflm based technology for wastewater treatment. Algal Res 5:231–240

Khan M, Rao S (2019) Diversity and Ecological Roles of Soil Algae. Inter J Phycol 27(3):45–56

Kirst H, Garcia-Cerdan JG, Zurbriggen A, Ruehle T, Melis A (2012) Truncated Photosystem Chlorophyll Antenna Size in the Green Microalga Chlamydomonas reinhardtii upon Deletion of the TLA3-CpSRP43 Gene. Plant Physiol.160:2251–2260

Koelmans AA, Nor NHM, Hermsen E, MerelKooi SM, De Mintenig J, France (2019) Microplastics in freshwaters and drinking water: critical review and assessment of data quality. Water Res. https://doi.org/10.1016/j.watres.02.054

Kotteswari M, Murugesan S (2012) Potential use of cyanobacterium in bioremediation of dairy effluent. Glob J Appl Environ Sci 2(3):157–163

Koyande AK, Chew KW, Rambabu K, Tao Y, Chu DT, Showa PL (2019) Microalgae: a potential alternative to health supplementation for humans. Food Sci Hum Wellness 8:16–24

Kraan S (2012) Algal polysaccharides, novel applications and outlook, in Carbohydrates – Comprehensive Studies on Glycobiology and Glycotechnology, ed C.-F. Chang (Croatia: InTech) 489–532

Lam T, Vermuë GP, Eppink MH, Wijffels MHM RH, Van Den Berg C (2018) Multi-product microalgae biorefineries: from concept towards reality. Trends Biotechnol 36:216–227

Lage LP, Gentili FG (2018) The potential of microalgae as a source for biodiesel and bioethanol production: A review. Renew Sustain Energy Rev 81:767-778

Laurens LML, Chen-Glasser M, Mcmillan JD (2017) A perspective on renewable bioenergy from photosynthetic algae as feedstock for biofuels and bioproducts. Algal Res 24:261–264

Lellis B, Fávaro-Polonio CZ, Pamphile JA, Polonio JC (2019) Effects of textile dyes on health and the environment and bioremediation potential of living organisms. Biotechnol Res Innov 3(2):275–290

Li F, Srivatsa SC, Bhattacharya S (2019) A review on catalytic pyrolysis of microalgae to high -quality bio-oil with low oxygeneous and nitrogenous compounds. Renew Sust Ener Rev 108:481–497

Li T, Lin GY, Podola B, Melkonian M (2015) Continuous removal of zinc from wastewater and mine dump leachate by a microalgal bioflm PSBR. J Hazard Mater 297:112–118

Li J, Li T, Sun D, Guan Y, Zhang Z (2024) Chapter Two - Treatment of agricultural wastewater using microalgae: A review. In: Gadd GM, Sariaslani SBT-A in AM (eds) Adv Appl Microbiol Academic Press, pp 41–82

Low Cost Algae Production System Introduced, Energy-Arizona, August 2007.

Luo X, Su P, Zhang W (2015) Advances in Microalgae-Derived Phytosterols for Functional Food and Pharmaceutical Applications. Marine drugs 13(7):4231–4254

Martín-Juárez J, Markou G, Muylaert K, Lorenzo-Hernando A, Bolado S (2017) Breakthroughs in bioalcohol production from microalgae: solving the hurdles. In: Microalgae-Based Biofuels and Bioproducts. Woodhead Publishing. pp.183–207

McBride A, Dale VH, Baskaran L, Downing M, Eaton L, Efroymson RA, Garten CT, Kline KL, Jager HI, Mulholland PJ, Parish ES (2011) Indicators to support environmental sustainability of bioenergy systems. Ecological Indicators 11(5):1277–1289

Menció A, Mas-Pla J, Otero N, Regàs O, Boy-Roura M, Puig R, Bach J, Domènech C, Zamorano M, Brusi D (2016) Nitrate pollution of groundwater; all right, but nothing else? Sci Total Environ 539:241–251

Michele Greque de Morais, Bruna da Silva Vaz, Etiele Greque de Morais, Jorge Alberto Vieira Costa (2015) Biologically Active Metabolites Synthesized by Microalgae. BioMed Research International. 2015(1):835761

Mollo E, Thomas-Bourgneuf L (2009) Global Efforts in Algal Conservation and Its Significance to Biodiversity. International J Phycol 18(4):88–94

Mouritseno G, Dawczynski C, Duelund L, Jahreis G, Vetter W, Schroder M (2013) On the human consumption of the red seaweed dulse (Palmaria palmata (L.) Weber & Mohr). J Applied Phycol 25: 1777–1791

Misurcova L, Orsavova J and Ambrozova JV (2015) Algal polysaccharides and health in Polysaccharides Bioactivity and Biotechnology, eds K. G. Ramawat and J. M. Merillon (Cham: Springer International Publishing Switzerland). 110–144

Miao H, Wang S, Zhao M, Huang Z, Ren H, Yan Q, Ruan W (2014) Codigestion of Taihu blue algae with swine manure for biogas production. Energy Convers Manag 77:643–649

Mularczyk M, Michalak I, Marycz K (2020) Astaxanthin and other Nutrients from Haematococcus pluvialis-Multifunctional Applications. Mar Drugs 18(9):459

Munoz R, Guieyssea B (2006) Algal-bacterial processes for the treatment of hazardous contaminants: a review. Water Res 40:2799–2815

Muñoz I, RamosA MJG, Agüera AR, Fernández-Alba JF, García-Reyes AM-D (2009) Chemical evaluation of contaminants in wastewater effluents and the environmental risk of reusing effluents in agriculture. TrAC Trends Anal Chem 28:676–694

Naik S, Goud VV, Rout, PK, Dalai AK (2010) Production of first and second generation biofuels: A comprehensive review. Renew Sustain Energy Rev 14:578–597

Namita P, Krishan KP (2016) Life cycle assessment of green diesel production from microalgae. Renew Ener 86:623-632

National Research Council (NRC) (2012) Sustainable Development of Algal Biofuels in the United States. The National Academies Press, Washington, DC

Newete S, Byrne M (2016) The capacity of aquatic macrophytes for phytoremediation and their disposal with specific reference to water hyacinth. Environ Sci Pollut Res Int 23:10630–10643

Norton TA, Hixon MA, Dobson ADW (1996) The Role of Algae in Marine Ecosystem Diversity and Conservation. Marine Ecology Progress Series 145(2):209–222

Nordin N, Yusof N, Samsudin S (2014) Microalgae biomass production and nitrateremoval from landfill leachate. Proceeding of international conference on research, implementation and education of mathematics and sciences 2014, Yogyakarta StateUniversity, 18–20 May 2014

Norvill Z, Shilton A, Guieysse B (2016) Emerging contaminant degradation and removal in algal wastewater treatment ponds: identifying the research gaps. J Hazard Mater 313:291–309

Olguin EJ (2003) Phycoremediation: key issues for cost effective nutrient removal processes. Biotechnol Adv 22:81–90

Ono E, Cuello JL (2006) Feasibility assessment of microalgal carbon dioxide sequestration technology with photobioreactor and solar collector. Biosystems Eng 95(4):597–606

Pal P, Singh N, Kumar A (2019) Innovative solutions for simplifying the harvesting process in microalgae-based bioenergy production: Addressing energy demand and operational costs. Renew Sust Ener Rev 101:116–124

Parlevliet D, Moheimani NR (2014) Efficient conversion of solar energy to biomass and electricity. Aquat Biosyst 10(1):4

Peng X, Ma X, Lin Y, Guo Z, Hu S, Ning X, Zhang Y (2015) Co-pyrolysis between microalgae and textile dyeing sludge by TG–FTIR: kinetics and products. Energy Convers Manag 100:391–402

Phwan THNN, Sulaiman MM, Ali A, Annuar S (2018) Microalgae as a renewable feedstock for biofuels: Challenges and opportunities. Renew Sustain Ener Rev 94:312–327

Piccini M, Raikova S, Allen MJ, Chuck CJ (2019) A synergistic use of microalgae and macroalgae for heavy metal bioremediation and bioenergy production through hydrothermal liquefaction. Sustainable Energy & Fuels 3(1):292–301

Ping-Zhong F, Lian-Dong Z, Xiao-Xi Q, Zhao-Hua L (2016) Water Footprint of Biodiesel Production from Microalgae Cultivated in Photobioreactors. J Environ Eng 142:4016067

Pooja RA, Sachin PB, Vishal AC, Namrata DB, Ritesh DB (2020) Formulation and evaluation of photo-aging facegel by using algae powder extract (Chlorella vulgaris). IJCIRAS 2(12):2581–5334

Plaza M, Cifuentes A and Ibáñez E (2008) In the search of new functional food ingredients from algae. Trends Food Sci Technol 19:31–39

Pulz O, Scheinbenbogan K (1998) Photobioreactors: design and performance with respect to light energy input. Adv Biochem Eng Biotechnol 59:123–152

Radakovits R, Jinkerson RE, Fuerstenberg SI, Tae H, Settlage RE, Boore JL (2012) Draft genome sequence and genetic transformation of the oleaginous alga Nannochloropis gaditana. Nat Commun 3:686

Ragusa I, Nardone GN, Zanatta S, Bertin W, Amadio E (2021) Spirulina for skin care: a bright blue future. Cosmetics 8(1):7

Rahman M, Hasegawa H (2011) Aquatic arsenic: phytoremediation using floating macrophytes. Chemosphere 83:633–646

Rai PK (2008) Heavy metal pollution in aquatic ecosystems and its phytoremediation usingwetland plants: an eco-sustainable approach. Int J Phytoremediation 10(2):131–158

Ranga Rao A, Sarada R, Ravishankar GA, Phang SM (2016) Industrial production of micro algal cell-mass and bioactive constituents from green micro alga-Botryococcus braunii. In: Liu J, Sun Z, Henri G (Eds.). Recent Advances in Micro algal Biotechnology. OMICS Group Incorporation, 731 Gull Ave, Foster City, CA 94404, USA. 103–126

Rao PH, Kumar RR, Raghavan BG, Subramanian VV, Sivasubramanian V (2011) Application of phycoremediation technology in the treatment of wastewater from a leather processing chemical manufacturing facility. Water SA 37:7–14

Rao DG, Senthilkumar R, Byrne A, Feroz S (2012) Wastewater treatment: advanced processes and technologies. In: Thanikal V (ed) Anaerobic fxed bed reactor for treatment of industrial wastewater Joseph. CRC Press, Boca Raton, pp 335–354

Rasala BA, Mayfield SP (2015) Photosynthetic biomanufacturing in green algae; production of recombinant proteins for industrial, nutritional, and medical uses. Photosynth Res 123:227–239

Rashid N, Lee K, Han J, Gross M (2013) Hydrogen production by immobilized Chlorella vulgaris: optimized pH, carbon source and light. Bioprocess Biosyst Eng 36:867–872

Rasmussen HE, Blobaum KR, Park YK, Ehlers SJ, Lu F, Lee JY (2008) Lipid extract of Nostoc commune var. sphaeroides Kutzing, a blue-green alga, inhibits the activation of sterol regulatory element binding proteins in HepG2 cells. J Nutr Mar 138(3):476–81

Rhodes CJ (2010) Biofuel from algae: salvation from peak oil? In Seaweeds and their Role in Globally Changing Environments (pp. 229–248). Springer Netherlands.

Rigueto CVT, Nazari MT, De Souza CF, Cadore JS, Brião VB, Piccin JS (2020) Alternative techniques for caffeine removal from wastewater: an overview of opportunitiesand challenges. J Water Proc Eng 35:101231

Riesing S (2007) Comparative Study on Algal and Oil Seed Yields for Biofuel Production. Bioenergy Research J 14(1):88–96

Reisser W (2010) The future is green: On the biotechnological potential of green algae. In Seaweeds and their Role in Globally Changing Environments (pp. 373–383). Springer Netherlands.

Roundtable on Sustainable Biomaterials (2010) Indicators of Bioenergy Sustainability. Roundtable on Sustainable Biomaterials, Geneva, Switzerland

Rudras M, Susan M (2010) Comparative Study of Microalgae Oil Productivity vs. Traditional Crops for Biofuel Production. Renewable Energy Rev 15(2):654–660

Riswi H, Shinta Candra D, Afra M, Selvi R, Ihda Zuyina RS (2015) Preliminary Study the Potency of Macroalgae in Yogyakarta: Extraction and Analysis of Algal Pigments from Common Gunungkidul Seaweeds. Procedia Chemistry 14:373–380

Rugnini L, Tredici MR, Panzavolta S (2019) Microalgae and bacterial consortia for wastewater treatment: Potential for CO2 sequestration and biomass production. Algal Research 43:101658

Ruma ASK Sudhakar RSR, Baredar P (2021) The Budding Potential of Algae in Cosmetics. Springer. Pg. 201–226

Saber M, Nakhshiniev B, Yoshikawa K (2016) A review of production and upgrading of algal bio-oil. Renew Sust Energy Rev 58:918–930

Sangar RMS, Singh KK, Singh S (2011) Showed the application of phycoremediation technology in the treatment of sewage water to reduce pollution load. Indian J Sci 2:33–39

Saratale RG, Kumar G, Banu R, Xia A, Periyasamy S, Saratale GD (2018) A critical review on anaerobic digestion of microalgae and macroalgae and co-digestion of biomass for enhanced methane generation. Bioresour Technol 262:319–332

Schaum C (2018) Phosphorus: polluter and resource of the future-removal and recovery from wastewater. Water Intell Online 17:450–457

Scott A, Davey MP, Dennis JS (2010) Biodiesel from algae: challenges and prospects. Curr Opin Biotechnol 21(3):277–286

Sengar RMS, Singh KK, Singh S (2011) Application of phycoremediation technology in the treatment of sewage water to reduce pollution load. Indian J Sci Res 2:33–39

Shannon MA, Bohn PW, Elimelech M, Georgiadis JG, Marinas BJ, Mayes AM (2008) Science and technology for water purifcation in the coming decades. Nature 452(7185):301–310

Sharma P, Pandey S (2014) Status of phytoremediation in world scenario. Int J Environ Bioremed Biodegrad 2(4):178–191

Sheehan J, Dunahay T, Benemann J, Roessler P (2006) A Look Back at the U.S. Department of Energy's Aquatic Species Program: Biodiesel from Algae. National Renewable Energy Laboratory, Golden, CO.

Shi J, Podola B, Melkonian M (2014) Application of a prototype-scale twin-layer photobioreactor for efective N and P removal from diferent process stages of municipal wastewater by immobilized microalgae. Bioresour Technol 154:260–6

Singh RS, Thakur, SRP (2015) Bansal Algal lectins as promising biomolecules for biomedical research. Crit Rev Microbiol 41(1):77–88

Sivasubramanian V, Muthukumaran M (2012) Large scale phycoremediation of oil drilling effluent. J Algal Biomass Util 3:5–1

Skjånes K, Céline R, Peter L (2013) Potential for green microalgae to produce hydrogen, pharmaceuticals and other high value products in a combined process. Crit Rev Biotechnol 33(2):172–215

Ślusarczyk J, Adamska E, Czerwik-Marcinkowska J (2021) Fungi and algae as sources of medicinal and other biologically active compounds: a review. Nutrients 13:3178

Sutherland D, Ralph P (2019) Microalgal bioremediation of emerging contaminants - opportunities and challenges. Water Res 164:114921

Sutherland DL, Ralph PJ (2020) 15 years of research on wastewater treatment high ratealgal ponds in New Zealand: discoveries and future directions. N Z J Bot 4:1–24

Tan ZY, Yang X, Li S, Zhang F, Yu (2019) Advances in microalgae-based biofuels: Challenges and future directions. Biomass Bioener 127:1–18

Tanaka A, Era M, Obata Y (2017) Role of fatty acid salts as anti Acanthamoeba agents fordisinfecting contact lens. Biocontrol Sci 22(3):153–161. www.personalcarecouncil.org

Teodosiu C, Gilca AF, Barjoveanu G, Fiore S (2018) Emerging pollutants removal through advanced drinkingwater treatment: a reviewon processes and environmental performances assessment. J Clean Prod 197:1210–1221

Tripathi R, Gupta A, Thakur IS (2019) An integrated approach for phycoremediationofwastewater and sustainable biodiesel production by green microalgae, Scenedesmussp. ISTGA1. Renew Energy 135:617–625

Ubando AT, Felix CB, Chen WH (2020) Biorefneries in circular bioeconomy: a comprehensive review. Bioresour Technol 299:122585

Ueno Y, Kurano N, Miyachi S (1998) Ethanol production by dark fermentation in the marine green alga, Chlorococcum littorale. J Ferment Bioeng 86:38–43

Ullah K, Ahmad M, Sofia VKS, Lu P, Harvey A, Zafar M, Sultana S, Anyanwu CN (2014) Algal biomass as a global source of transport fuels: Overview and development perspectives. Prog Nat Sci: Mater Intern 24(4):329–339

Vaibhav V, Sahasrabuddhe S (2018) 'BLUE' is the new 'GREEN' for cosmetic industry. IJRTI 3(10):134–144

Vavitsas K, Fabris M, Vickers CE (2018) Terpenoid metabolic engineering in photosynthetic microorganisms. Genes. 9(11):e520

Vassilev SV, Vassileva CG (2016) Non-food feedstocks for third-generation biofuels: Microalgal biomass as a renewable source. Renew Sustain Ener Rev 57:277–288

Vijayaraghavan K, Karthik R, Nal SK (2009) Hydrogen generation from algae: A review. J Plant Sci 5:1–19

Viswanaathan S, Perumal PK, Sundaram S (2022) Integrated Approach for Carbon Sequestration and Wastewater Treatment Using Algal–Bacterial Consortia: Opportunities and Challenges. Sustainability 14(3):1075

Volk L, Abrahamson E, White E, Neuhauser E, Gray C, Demeter C, Lindsey J, Jarnefeld D, Aneshansley R, Pellerin S, Edick S (2000) Developing a willow biomass crop Enterprise for Bioenergy and Bioproducts in the United States. Proceed Bioener 12–16

Wada N, Sakamoto T, Matsugo S (2013) Multiple roles of photosynthetic and sunscreen pigments in cyanobacteria focusing on the oxidative stress. Metabolites. 3:463–483

Wanchat S, Thapat S, Athikom B, Gheewala SH (2014) Life cycle cost of biodiesel production from microalgae in Thailand. Energy for Sustainable Develop 18:67–74

Wang Y, Li X, Cheng Z, Huang J (2008) Nutritional Composition and Health Benefits of Algae. J Nutritional Sci 26(4):223–234.

Wang Y, Ho SH, Cheng CL, Guo WQ, Nagarajan D, Ren NQ, Lee DJ, Chang JS (2016) Perspectives on the feasibility of using microalgae for industrial wastewater treatment. Bioresour Technol 222:485–497

Ward A, Lewis D, Green F (2014) Anaerobic digestion of algae biomass: A review. Algal Res 5:204–214

Wells TE, Johnson SL, Thompson MA (2014) Applications of Secondary Metabolites in Healthcare: An Overview of Their Potential Uses and Benefits. J Pharma Sci 103(7):2112–2121

Wieczorek N, Kucuker MA, Kuchta K (2014) Fermentative hydrogen and methane production from microalgal biomass (Chlorella vulgaris) in a two-stage combined process. Appl Energy 132:108–117

Willy PJ, Els VDJM (1998) Plant Lectins: Versatile Proteins with Important Perspectives in Biotechnology. Biotechnol Genet Eng Rev 15(1):199–228

Wollmann F, Dietze S, Ackermann J-U, Bley T, Walther T, Steingroewer J, Krujatz F (2019) Microalgae wastewater treatment: biological and technological approaches. Eng Life Sci 19:860–871

Xiong JQ, Kurade MB, Jeon BH (2017) Biodegradation of levofloxacin by an acclimated freshwater microalga, Chlorella vulgaris. Chem Eng J 313:1251–1257

Xiong JQ, Kurade MB, Jeon BH (2018) Can microalgae remove pharmaceutical contaminants from water? Trends Biotechnol 36(1):30–44

Xu J, Zhao Y, Zhao G, Zhang H (2015) Nutrient removal and biogas upgrading by integrating freshwater algae cultivation with piggery anaerobic digestate liquid treatment. Appl Microbiol Biotechnol 99:6493–6501

Yang Z, Guo R, Xu X, Fan X, Li X (2010) Enhanced hydrogen production from lipid-extracted microalgal biomass residues through pretreatment. Int J Hydrogen Energy 35:9618–9623

Yang J, Xu M, Zhang X, Hu Q, Sommerfeld M, Chen Y (2011) Life-cycle analysis on biodiesel production from microalgae: water footprint and nutrients balance. Bioresource Technol 102(1):159–165

Zhu L (2015) Biorefinery as a promising approach to promote microalgae industry: an innovative framework. Renew Sust Energ Revi 41:1376–1384

Zabed HM, Boyce AN, Sahu JN, Faruq G (2017) Evaluation of the quality of dried distiller's grains with solubles for normal and high sugary corn genotypes during dry–grind ethanol production. J Clean Prod 142:4282–4293

Zainan NH, Srivatsa SC, Li F, Bhattacharya S (2018) Quality of bio-oil from catalytic pyrolysis of microalgae Chlorella vulgaris. Fuel 223:12–19

Potential Use of Extracellular Substances from Soil Algae

S. Murugesan, V. Sivamurugan, and D. Kathiresan

Abstract The numerous forms of extracellular polymeric substances (EPS) produced by soil algae, as well as their applications, are covered in different sections in this chapter. Algal biomass is mostly used to isolate EPS such as ulvans, carrageenan, alginates, proteins, and lipids. The commercial highlights of EPS are examined in detail in a separate section. In the last section of the chapter, frequent applications of EPS in bioremediation for wastewater treatment, i.e., removal of emerging toxins, at the laboratory scale, pilot scale, and industrial level are highlighted.

Keywords Alginates · Bioremediation · Carrageenans · Extracellular polymeric substance · Lipids · Soil algae

1 Introduction

Extracellular polymeric substances (EPS) are a kind of polymeric carbohydrates formed by various microorganisms through biofilm formation, which provides protection and structural stability to the microorganisms. Algae can produce more than 70% of carbon-based EPS with respect to its total weight. EPS are macromolecules especially comprised of polysaccharides or proteins. These polysaccharides are best known as exopolysaccharides; they are multifunctional in nature and have various applications in pharmaceutical and food industries (Wingender et al. 2012). Natural products derived from algae possess commercial importance (Borowitzka 2013; Michalak et al. 2015). Many types of algae have the potential to generate active

S. Murugesan
PG and Research Department of Botany, Pachaiyappa's College, Chennai, India

V. Sivamurugan (✉)
PG and Research Department of Chemistry, Pachaiyappa's College, Chennai, India

D. Kathiresan (✉)
PG and Research Department of Botany, Saraswathi Narayanan College, Madurai, India

A. Mishra, A. Varma (eds.), *Soil Algae*, https://doi.org/10.1007/978-981-97-9873-5_8

biological substances such as proteins, polysaccharides, polyphenols, pigments, and fatty acids. EPS help microorganisms stick to plant roots or soil particles and entrap nutrients and moisture. Multiple functions of EPS are summarized in Fig. 1 (Costa et al. 2018).

EPS mainly function as a defense system that protects microorganisms from various physical and community stress conditions such as pH, temperature, light, seasonal changes, and drought. They also help microorganisms to have symbiotic relationship with plants or any other living organisms to entrap nutrients from hosts.

For instance, cyanobacteria are rich in carbohydrates, carotenoids, and phycobiliproteins, which are important ingredients in many commercial products. EPS extracted from *Pseudomonas fluorescens* PGM37 were found to have more moisture retention capacity and suitable for cosmetics and pharmaceutical products. Sacran is a sulfated polysaccharide isolated from *Aphanothece sacrum* and can be used as a moisturizing agent. Microalgae biomass rich in starch, carbohydrate, and lipids are utilized as raw materials for making biodegradable plastics either in pure form or it is blended with traditional petrochemical-based raw material or biopolymers like corn starch.

Dixoniella grisea, a red microalga, can produce EPS mainly composed of polysaccharides and proteins and is explored as a lubricant for tribological applications (Gavalás-Olea et al. 2021). EPS generated from algae have a potential to bioaccumulate heavy metals as a complex to reduce pollution in sewage or during wastewater treatment. Due to the anionic nature of functional groups, (a) EPS act as

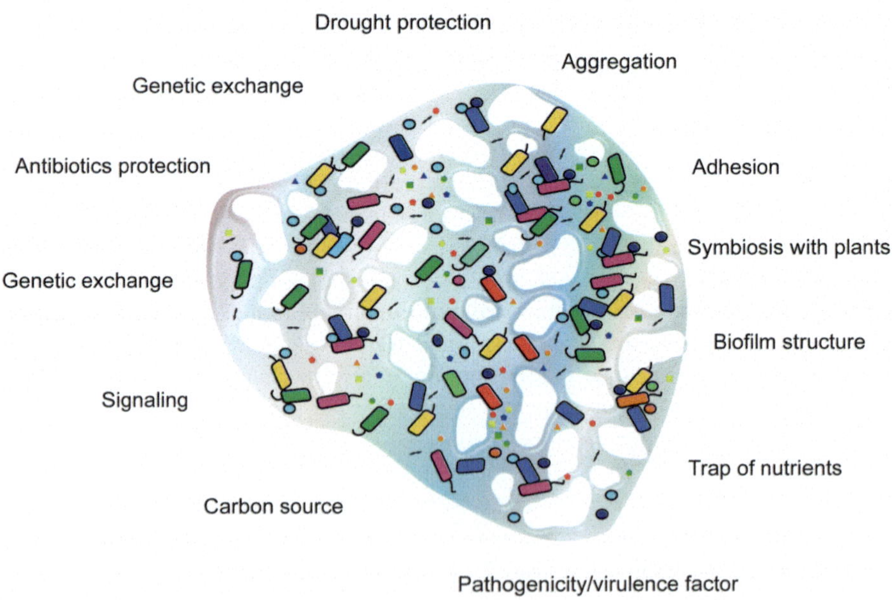

Fig. 1 Multifunctional nature of extracellular polymer substances (EPS). (Credit: Costa et al. (2018). This is an open-access article distributed under the terms of the Creative Commons Attribution License—CC BY)

complexing agent for heavy metals (Micheletti et al. 2008), (b) exhibit tolerance toward salts, surfactants, and alpha-hydroxy acids, and (c) act as bioemulsifier for hydrocarbons (Banerjee et al. 2021).

1.1 Generation of EPS

Quantitative and qualitative production of EPS is improved by optimizing stress conditions (Maršálek and Rojíčkovâ 1996). The production of EPS depends on the environment of the algae, and biosynthesis is triggered by external parameters such as acidity, alkalinity, heat, light, nutrients, seasons, and other environmental factors. The quantitative composition of EPS varies with respect to algal species, life cycle, physiological stage, and nutritional availability (Bafanaa 2013). Through the optimization of physiological conditions, EPS such as organic acids, amino acids, peptides, proteins, sugars, oligo, and polysaccharides can be synthesised in commercial algal cultivation (Angelis et al. 2012).

Maršálek and Rojíčkovâ (1996) evaluated the EPS composition produced by *Scenedesmus quadricauda, Chlorella kessleri,* and *Raphidocelis subcapitata* under 2,5-dichlorophenol- and under glyphosate-induced toxic organic compound stress, cadmium chloride-induced heavy metal stress, and NaCl-induced osmotic stress. The study showed that the production of proteins and carbohydrates increased tremendously. Further, selective production of carbohydrate was observed under osmotic stress, whereas in contrast, under heavy metal stress, the production of proteins increased. Similarly, protein production increased under 2,5-dichlorophenol- and glyphosate-induced toxic organic compounds stress.

More recently, dos Reis et al. (2022) studied the cadmium- and cobalt-induced growth on microalgae *Raphidocelis subcapitata* (Chlorophyceae). The study indicated that heavy metal contamination results in pigment and oxygen-evolving complex. Increase in fatty acid composition have been observed for biodiesel application using microalgae *Scendesmus quadricauda* among 23 microalgae species tested from Qaron Lake, El Fayoum, Egypt. The study showed the production of increased fatty acid under wastewater stress (Ahmed 2021). Sylwestrzak et al. (2021) reported on the effect of glyphosate concentration on marine microalgal species and the results showed an increase in cyanobacteria production with an increase in the concentration of glyphosate.

Mishra and Jha (2009) explored NaCl-induced osmotic stress on the EPS production of *Dunaliella salina* strain. The results showed that at a concentration of 5 M NaCl, *D. salina* synthesized four monosaccharides (glucose, galactose, fructose, and xylose) as observed by HPLC analysis and also amino acids, aliphatic acids, and aromatic compounds as determined by FT-IR analysis. The exposure of nanoparticle on microorganism showed enhanced EPS production and qualitative enhancement of EPS observed with respect to stressor. For instance, the EPS production under Cd and iron oxide nanoparticles exposure studied on *Halomonassp* as a model algae for bioremediation for heavy metal removal (Cao et al. 2022). In vivo studies revealed enhanced production of EPS proteins under Cd/iron oxide NPs exposure, whereas Cd NPs alone induced more production of EPS polysaccharides. Enhanced production of exopolysaccharides (EPS) was observed when bacterial strain was grown under heavy metal nanoparticles. In a study, *Bacillus cereus* KMS3-1 strain was grown in 100 ppm of heavy metal such as Pb (II), Cu (II), and Cd (II) nanoparticles (Mathivanan et al. 2021). EPS production was increased by Pb (II), and bioaccumulation of heavy metals was observed due to interaction of metals with EPS functional groups.

Qian et al. (2021) explored the effect of surface-functionalized polystyrene (PS) nanoparticles on the total EPS and protein production in activated sludge. As per the study, PS nanoparticles showed negative impact on EPS production and substantially reduced EPS protein production. The presence of polar functional group in PS-NH_2 resulted in cell wall damage and reduced the EPS protein content. Qian et al. (2022) and Otero and Vincenzini (2003) explored the effect of nitrogen on the generation of EPS. Any microorganism can interact with other organism through adhesion, cohesions, and genetic material transfer (Costa et al. 2018). Soil algae can symbiotically interact with plants and other microorganisms. Liu et al. (2016) reviewed various types of extracellular metabolites such as polysaccharides, proteins, organic acids, lipids, phytohormones, and allelopathic chemicals as well as their applications. The generation of various types of EPS, their commercial aspects, and their role in heavy metal accumulation have been discussed in the following sections.

2 Extracellular Substances

2.1 *Ulvans*

Ulvan is a polysaccharide found in the cell walls of green macroalgae. It is a member of the family Ulvaceae and belongs to the genus *Ulva*. It is predominantly composed of a sulfonic acid group that includes rhamnose, uronic acids (glucuronic acid and iduronic acid), and xylose. Among the members of the genus *Ulva*, ulvan contributes to between 9% and 36% of the dry weight of the biomass (Lahaye and Robic 2007). Cellulose, xyloglucan, and glucuronan are examples of other cell wall polysaccharides found in *Ulva* species. According to Lahaye et al. (1997), these polysaccharides account for up to 45% of the dry biomass in certain species. According to Lahaye et al. (1997), xyloglucan and glucuronan are soluble polysaccharides that constitute a portion of the cell wall polysaccharides. Ulvan is another polysaccharide that falls into this category. Out of the four cell walls that may be discovered within the organism, the *Ulva* cell wall is the only one that includes both rhamnose and iduronic acid (Quemener et al. 1997). The influence that rhamnose has on biosynthetic pathways in the dermis (Adrien et al. 2017) and on plant immunity (Trouvelot et al. 2014) is something that has piqued the curiosity of researchers.

Uronic acids, which include glucuronic and iduronic acids, as well as their sulfate esters, are essential components found in glycosaminoglycans (GAGs) of mammalian organisms (Fig. 2). Ulvan, like GAGs, possesses a repeating disaccharide structure that is mostly composed of a uronic acid connected to a sulfated neutral sugar. This structure gives Ulvan the potential to modulate the processes and functions of mammalian polysaccharides (Lahaye et al. 1999). According to Paradossi et al. (1999), the Ulvan backbone is primarily made up of α- and β-(1,4)-linked monosaccharides, which include rhamnose, xylose, glucuronic acid, and iduronic acid. These monosaccharides are characterized by their distinctive repeating disaccharide units.

Both in vitro and in vivo tests conducted on animal and plant systems have demonstrated that Ulvan possesses a significant amount of biological activity. In mammals, sulfated polysaccharides, which include free glycosaminoglycans (GAGs) and proteoglycans (protein linked GAGs), are involved in a wide range of biological processes and share structural similarities with ulvan (Scharnweber et al. 2015). These polysaccharides play a significant role in several biological processes. Ulvan is possibly applied in biomaterial science sector (wound dressings, tissue engineering, biofilm prevention, and excipients); utilized as nutraceuticals (anticoagulant), excipients (Adrien et al. 2017), antioxidants (Le et al. 2019); and possesses antihyperlipidemic (Rizk et al. 2016), antimicrobial (Junter et al. 2016), antiviral (Aguilar-Briseño et al. 2015; Bagasra and Lischner 1998; Cheng et al. 2013), anticancer (Ahmed et al. 2017; Ahmed and Ahmed 2014), and immunomodulatory properties

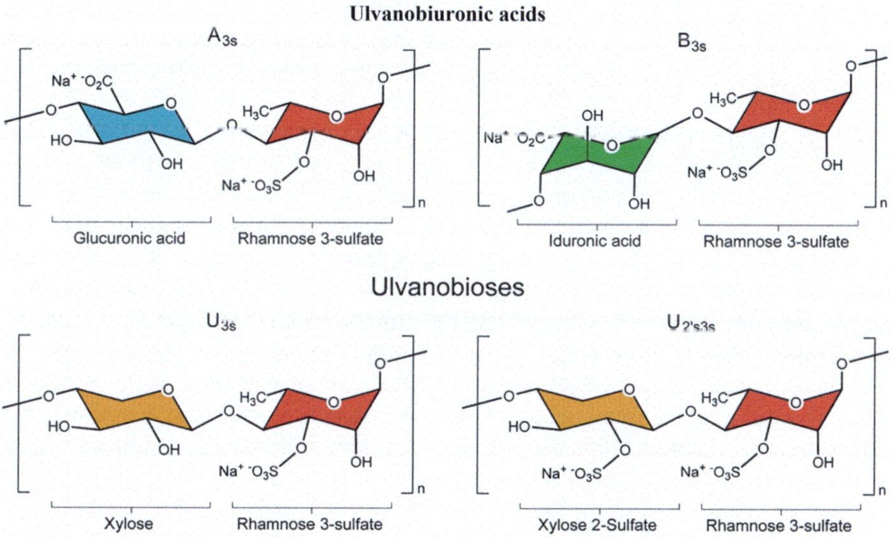

Fig. 2 The primary disaccharide polymeric units that make up ulvan have a distinct nomenclature and structure. Ulvanobiuronic acid A3s have glucuronic acid (blue) connected to rhamnose 3 sulfate (red), whereas B3s have rhamnose 3-sulfate but iduronic acid (green) instead of glucuronic acid. Rhamnose 3 sulfate is linked to xylose in ulvanobioses (orange). As observed in U2′s, 3 s, xylose can include a sulfate group. (Credit: Kidgell et al. (2019). This is an open access article distributed under the CC BY-NC-ND license. Published with permission from Elsevier)

(Bussy et al. 2019). The pharmacokinetics and bioavailability of such a complicated macromolecule, on the other hand, need to be thoroughly researched before it can be employed in a therapeutic capacity. In addition to this, Venkatesan et al. (2015) presented evidence that ulvan has the potential to be utilized in the fields of agricultural and functional foods. According to Cunha and Grenha (2016), they also contribute to the development of technologies in the fields of horticulture, agriculture, and plant physiology at the same time.

The development of material applications containing ulvan has come up along with the advancement of this material from nanofibers (Kikionis et al. 2015), membranes and films (Guidara et al. 2019), matrix tablets (Vlachou et al. 2018), emulsions (Shao et al. 2017), nanoparticles (González-Ballesteros et al. 2019), nanocomposites, and hydrogels (Morelli et al. 2016).

2.2 Carrageenan

Carrageenans are natural compounds with high molecular weight linear sulfated polysaccharide extracted from various genera and species of red algae (Rhodophyceae) (Bui et al. 2019) (Fig. 3). *Chondrus crispus, Gigartina stellata*, and *Eucheuma* are a few of them (Robertson 2016).

Zablackis and Santos (1986) divided carrageenans into kappa, iota, alpha, beta, lambda, and theta families. These families are distinguished from one another by the amount of sulfate groups present in the repeating units and the position of those groups. *j*-, *i*-, and *k*-carrageenans are found in abundance in nature and are frequently utilized as viscosity intensifiers in the cosmetics and food sectors (Tye 1989). A significant portion of carrageenans is composed of 1→3-linked b-d-galactose and 1→4-linked a-d-galactose, with galactose residues partially sulfated at positions 2 and/or 6 and/or 3,6-anhydrided (Clark and Ross-Murphy 1984) (Fig. 3). According to Kulkarni et al. (2016), the units are connected by 1,3- and 1,4-alternating glycosidic connections. According to Sedayu et al. (2019), the molecular weight of carrageenan ranges from 100 to 1000 kDa. Specifically, carrageenan contains negatively charged sulfate groups, which are capable of forming strong interactions with the amine groups found in proteins, particularly casein.

According to Necas and Bartosikova (2013), carrageenan finds use in a diverse range of industries, including the field of drug research, the cosmetics industry, the pharmaceutical industry, the food industry, and the textile industry. Gelling, thickening, stabilizing, and protein-binding agents are some of the many applications that they have seen in the food sector (Weiner 2014) ; (Venugopal 2011). Kappa (κ), iota (ι), and lambda (λ) are the forms of carrageenans that are utilized in the food sector with the highest frequency, as stated by Sedayu et al. (2019). Pharmaceutical applications include drug delivery devices, matrices, and excipients (Akbar and Shakeel 2019). Carrageenan are also used in many controlled-release tablets either alone or in combination with polysaccharides (Picker 1999). Carrageenan also possesses antiviral and anti-tumoral activities (Mishima et al. 1998; Rinaudo 2008;

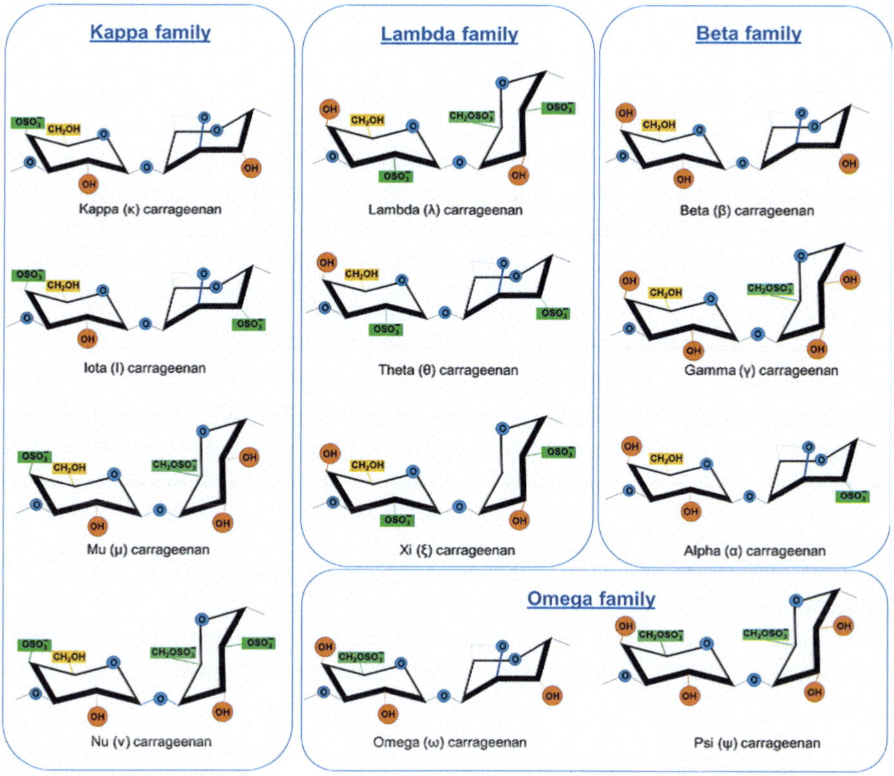

Fig. 3 Basic carrageenan forms: iota (*I*), kappa (*κ*), lambda (*λ*), mu (*μ*), nu (*Υ*), theta (*θ*), xi (*έ*), alpha (*α*), beta (*β*), gamma (*γ*), omega (*ω*), and psi (*ψ*). (Credit: Pradhan and Ki (2023). Published with permission from Elsevier)

Raposo et al. 2014). Both original polysaccharides and their derivatives are widely used in many areas of pharmaceutical formulations, and in some cases, they play a fundamental role as that in delivery systems. Various formulation of carrageenan like nanoparticles (Rodrigues et al. 2012), pellets (Ghanam and Kleinebudde 2011), hydrogel beads (Selvakumaran et al. 2016), films (Bajpai et al. 2015), matrices (Panyoyai et al. 2015), and other devices (Pawar et al. 2014) have been widely investigated as drug delivery carriers. Thus, carrageenan has many applications in various important fields, as well as novel investigations steadily increasing attraction of researchers to discover new applications.

2.3 Alginates

Alginate is a naturally occurring hydrocolloid extracted from the cell wall of brown seaweeds (Kloareg and Quatrano 1988). Due to its rheological properties, such as gelling, viscosifying, and stabilizing dispersions, natural alginate is widely employed in various industries, including textiles, agri-food, paper, cosmetics, biomedical, and pharmaceuticals (Draget et al. 2006). Alginates are primarily isolated from brown seaweed species such as *Ascophyllum nodosum*, *Laminaria* spp., *Lessonia nigrescens*, *Ecklonia maxima*, *Macrocystis pyrifera*, and *Durvillea antarctica*, where they account for up to 40% of the dry weight (Rinaudo 2007).

Alginic acid is a complex organic molecule composed of homogeneous (poly-M and poly-G) or heterogeneous (MG) block configurations of monomers of 1,4-β-d-mannuronic acid (M) and 1,4-α-l-guluronic acid (G). The percentage of the three types of blocks—GG, MM, and MG—determines the physical properties of alginates (Fig. 4). Alginates with a high proportion of M blocks have higher viscosity, while those with a high proportion of G blocks have better gelling ability (Pereira and Cotas 2020). Alginates are primarily found as gels comprising magnesium, calcium, sodium, and barium ions, with a dry matter content ranging from 20% to 60% (Arunkumar 2017).

Alginate has a number of benefits, including ease of production, biocompatibility, biodegradability, and nontoxicity. In the pharmaceutical industry, alginate has a

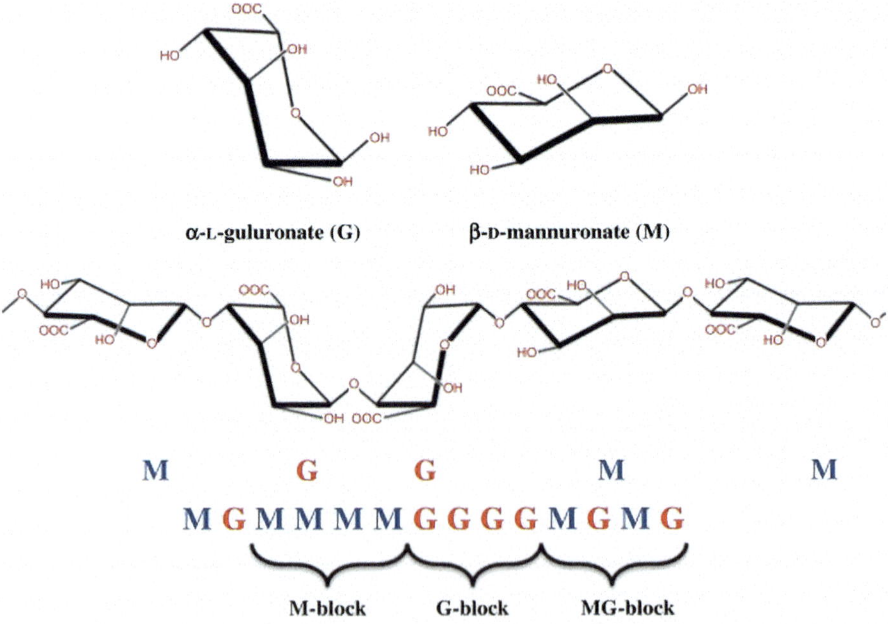

Fig. 4 Chemical structure of alginates. (Credit: Paredes Juárez et al. (2014). This is an open-access article distributed under the Creative Commons Attribution License—CC BY)

variety of uses, including gel formation, stabilizing agents, wound healing, tissue engineering, stem cells, cell encapsulation, localized drug delivery, and protein delivery (as a matrix for three-dimensional tissue cultures, adjuvants for antibiotics and antiviral agents, and cell transplantation for the treatment of diabetes and neurodegenerative diseases (Urtuvia 2017)). Additionally, alginate is utilized in various industries, including textiles, food, paper, pharmaceuticals, and cosmetics (Szekalska et al. 2016). In the paper industry, alginate serves as an adhesive and preservative. It is also used as a glazing agent in fish freezing to prevent oxidation reactions.

Alginates have traditionally been used as excipients in pills and for the treatment of stomach ulcers, gastric reflux, and heartburn due to their absorptive, swelling, and hemostatic properties (Draget and Taylor 2011). These properties also make alginates beneficial for wound healing (Lee and Mooney 2012; Bhatnagar et al. 2014; Cassidy et al. 2018). Calcium alginates, in particular, have the ability to absorb bile acids and lipids, which can help lower cholesterol and lipid levels (Georg Jensen et al. 2013; Kasahara et al. 2019). Although many alginate-based hydrogels have not yet been successfully tested in clinical trials, several new and promising alginate gels are currently being developed, suggesting potential new treatment options in the near future.

Currently, about 50% of imported alginate is used in the textile industry, 30% in food production, 6% in paper production, 5% in welding rod production, and 5% in pharmaceuticals. Given its extensive use across various industries, alginate is one of the most widely exploited seaweed polysaccharides. As demand for alginate, especially sodium alginate, continues to grow worldwide, it is essential to identify and study new raw sources for alginate extraction to meet this increasing demand (Global Alginate Market Size Industry Report 2021–2028).

2.4 Proteins

Protein is thought to be abundant in seaweed and microalgae. Protein levels in some seaweed and microalgae species are comparable to those found in traditional protein sources such as meat, eggs, soy beans, and milk (Gouveia et al. 2008). Until date, macroalgae proteins, peptides, and amino acids have been extracted and separated mostly on a laboratory scale.

Algae-based protein production has a few benefits over typical high-protein plants in terms of production and nutritional content. However, the most significant disadvantage of employing marine vegetation as a protein source is that the protein content of raw materials can range from 3% to 47% (Harnedy and FitzGerald 2013), depending on animal species, geographical region, and growing season (Fleurence 1999).

In general, the protein content associated with winter is high due to decreased photosynthetic activity, which results in limited carbohydrate generation and storage. Brown seaweed also has a lower protein level than green or red seaweed. When compared to terrestrial crops like soya (0.6–1.2 tonnes/ha/year), pulse legumes (1–2 tonnes/ha/year), and wheat (1.1 tonnes/ha/year), seaweed and microalgae provide more protein per region (2.5–7.5 tonnes/Ha/year and 4–15 tonnes/Ha/year, respectively) (Van Krimpen et al. 2013).

Single-cell protein (SCP) refers to protein derived from unicellular and multicellular algae, offering both production and nutritional benefits. Dried SCP cells, typically produced from bacteria, fungi, or microalgae, are used as protein sources in food and animal feed. In addition, SCP from microalgae is a significant source of bioactive compounds, making it a promising option for future human diets (Sathasivam et al. 2019).

Algae does not needed clean water or arable land for flourishing. Thus, freeing up resources could be used to produce more food or other income crops (Van Krimpen et al. 2013). Algae, on the other hand, are prone to severe oxidative and free-radical stress as a result of their harsh and phototropic environment (Sampath-Wiley et al. 2008). This has resulted in the formation of natural immune systems, such as the creation of pigments (carotene, chlorophyll, and phycobiliprotein), and polyphenols (catechins, flavonols, and phlotannins), which can provide consumers with health advantages (Wang et al. 2009).

Several factors, including harvest access and rights, season and algae-growing area, and the availability of growing production methods for the breakdown of proteins in algae, limit the widespread use of marine vegetation and microalgae, which is further limited by the fact that current processes for algal protein separation are time-consuming and ineffective (Wijffels and Barbosa 2010).

Microalgae and seaweed are rich in protein and contain all essential amino acids, making them valuable sources of nutrition (Fleurence et al. 2012). Algae can be particularly beneficial for athletes who require high protein intake, especially vegans who may not find sufficient protein in eggs and dairy whey protein (Du 2015). For thousands of years, indigenous peoples have utilized marine plants and microalgae as nutritious food sources (Chakdar et al. 2012).

Two notable bioactive algal protein families, lectins and phycobiliproteins, have various industrial applications. Phycobiliproteins, for example, are widely used in fluorescence labeling, flow cytometry, fluorescent microscopy, and fluorescent immunohistochemistry (Aneiros and Garateix 2004). Due to their excellent nutritional profile, it is estimated that over 30% of the world's algal production is used in animal feed, and about 50% of *Spirulina* biomass is primarily utilized as a dietary supplement (Becker 2004). Incorporating microalgae into chicken feed can improve health, productivity, and nutritional value. Various species of algae, such as *Chlorella* sp., *Arthrospira* sp., *Porphyridium* sp., and *Haematococcus* sp., are used for these purposes (Holman and Malau-Aduli 2013).

In pig diets, protein derived from *Arthrospira maxima, A. platensis*, and *C. vulgaris* has been shown to replace up to 33% of soy protein with no adverse effects (Gatrell et al. 2014). Herders, out of all the animals tested for algae, have the best chance of digesting high fiber content and releasing algal proteins efficiently (Holman and Malau-Aduli 2013). Microalgae, in addition to being a food source for zooplanktons, play a vital role in aquatic animal rearing by maintaining pH, limiting bacterial development, and enhancing the quality of the breeding habitat (Lio-Po et al. 2005). Many dietary sources, including meat, egg, fish, blood, and vegetable sources, such as rice, soybeans, wheat, peas, broccoli, garlic, and algae, have been found to contain bioactive peptides (Fan et al. 2014).

Algal proteins can produce bioactive peptides (BAPs) and other biologically valuable molecules with various health benefits, including antioxidant, anti-proliferative, anti-inflammatory, antihypertensive, antidiabetic, anti-atherosclerotic, anticoagulant, and antibacterial properties (Pimentel et al. 2019). By 2020, the market for algae protein exceeded USD 690 million and is projected to grow at a compound annual growth rate (CAGR) of over 6% from 2021 to 2027.

Growing concerns about health awareness among consumers is an important factor that will encourage the sharing of the maritime protein industry. The seaweed protein supplements promote the levels of minerals and vitamins in human body and regulate estradiol and estrogen levels. Seaweed is an excellent source of omega-3 fats and vitamin B12, which improves heart health and nerve health; therefore, it is widely used in diet and nutritional supplements, which has a positive impact on all market predictions. It is interesting to note that there is increased demand for processed foods because of rapid urban growth and due to which workforce will increase that relies on demand for food and beverages, eventually having a positive impact on algal protein business. This relates to consumers' concerns about eating a healthy diet that includes low-calorie, low-sugar, and protein-rich foods, necessitating customer selection of these proteins. The food and beverage sector will be driven by the growing demand for organic food products and the strong demand for high-protein beverages, reviving demand for algal protein production for a limited time.

Despite the fact that algae have been recognized as a significant source of nutrients, there are conflicting findings about algae digestion and protein bioavailability. A comprehensive evaluation of various algae sources, cultivation methods, harvesting systems, production techniques, growing demand, components used, and the safety of the final products is needed to better understand the actual energy usage of algae-based proteins and products. Additionally, more research is necessary to investigate the bioavailability of algal proteins in vivo.

2.5 Lipids

Algae are photosynthetic creatures that are widely spread and contain a diverse range of lipids as a result of their wide range of environments. Phospholipids, glycolipids (glycosylglycerides), and non-polar glycerolipids (neutral lipids) such as plant extracts and betaine, as well as other aberrant lipids, are all found in algal lipids. Microalgae can produce a variety of lipids, including triacylglycerols, phospholipids, glycolipids, and phytosterols, which are essential for energy storage and various metabolic processes such as signal transmission, translation, intercellular communication, extraction, and vesicle transport. While typical lipid molecules found in microalgae are also present in higher plants, certain unique lipids are specific to microalgae and not found in other species (Stonik and Stonik 2018). Under optimal conditions, microalgae can accumulate lipids that make up 70–80% of their biomass weight, making them a more prolific oil producer compared to the most commercially viable plants (Dolganyuk et al. 2020).

Microalgae also produce compounds such as lipids, sterols, hydrocarbons, and waxes. Among the lipid classes, glycerolipids are abundant and well expressed in microalgae, consisting of a glycerol backbone with one, two, or three attached fatty acids (FAs) (Klok et al. 2014). Fatty acids are a significant component of microalgae biomass, comprising between 5% and 60% of the dry cell weight (Ben-Amotz et al. 1985). Glycerolipids are categorized into storage lipids, which primarily include triacylglycerols (TAGs) and structural lipids, which are important for cell and organelle membranes. Membrane lipids usually consist of two fatty acid groups and a polar head group attached to the glycerol backbone, which are vital for maintaining cell and organelle function.

The primary storage lipids, TAGs, are formed with three fatty acid chains linked to a glycerol molecule (Ordog et al. 2016). Microalgae-based cosmetics are a revolutionary technique to broaden the search for innovative natural components in environmentally friendly biomass. Algae are a long-term source of bioactive and nutraceutical compounds that can be used for nutritional purposes. Low fat content, high PUFAs (both n-3 and n-6), and anti-hypercholesterolemic, antioxidant, anticancer, antidiabetic, antihypertensive, anti-inflammatory, antifungal, and antibacterial characteristics of C28 and C29 sterols make macroalgae low-calorie meals. Glycolipids and phospholipids (Breuer et al. 2013) are the most common polar lipids, whereas the main class of lipids retains triacylglycerols (TAGs), in which the three fatty groups are formed using the glycerol structure (Ordog et al. 2016). Microalgae-derived cosmetics are an intriguing technique to broaden the search for innovative natural components in environmentally friendly biomass.

Seaweed generally has a low lipid content (Pati et al. 2016). However, compared to other plant and animal-derived foods, it is an excellent source of polyunsaturated fatty acids (PUFAs), which have been shown to benefit blood pressure, brain function, and the nervous system (Notowidjojo 2021). While large quantities of PUFAs have traditionally been found in cold-water regions where *Chlorophyta* species grow, recent findings indicate that warm-water *Chlorophyta* species also have high levels of fatty acids, including oleic acid (Notowidjojo 2021). Among the different types of seaweed, *Enteromorpha clathrata* has the highest lipid content at 4.6%, followed by *Codium tomentosum* at 2.53% (Pati et al. 2016).

Seaweed lipids make up a small percentage of phytochemicals that, although their importance, are underappreciated, owing to their low content, considerable structural variety and complexity, and little understood biological function. They are well known as omega-3 polyunsaturated fatty acid (PUFA) reservoirs, which have been linked to a variety of health advantages (van Ginneken et al. 2011). Seaweed, on the other hand, contains complex lipids like phospholipids (PLs) and glycolipids (GLs) that have unique features like omega-3 fatty acids (FA) like eicosapentaenoic acid (EPA) and docosahexaenoic acid (DHA) that aren't present in earthy plants (Kagan et al. 2015). Bioavailability, oxidation resistance, and omega-3 PUFA content are all higher in marine PLs than in lipids from other sources. They are more effective than global PLS at supplying omega-3 PUFA meals, as evidenced by a few comparison trials (Drouin et al. 2021). GLs from seaweed that include long-chain PUFA (20 or more carbon atoms) offer potential biotechnological applications, unlike their terrestrial equivalents (Lopes et al. 2021).

Complex lipids have recently been identified as promising phytochemicals with intrinsic bioactive qualities such as antioxidant, anticancer, anti-inflammatory, and antibacterial activities (Lopes et al. 2019), suggesting potential applications in medicine, nutrition, and cosmeceuticals. Based on the above, it is reasonable to anticipate that advancements in high-production technologies and increased consumption of algal lipids, fatty acids, and sterols will enhance their incorporation into efficient diets.

3 Commercial Aspects of Extracellular Polymeric Substances

Exopolysaccharides (EPS) are high-molecular-weight biopolymers released into the environment by microorganisms like microalgae during their development and proliferation (Delbarre-Ladrat et al. 2014; Zheng and Xiao 2016). They can either loosely adhere to the cell wall or be discharged into the environment (Trabelsi et al. 2016). Based on the binding force with the cell, EPS are generally classified as loosely bound (LB-EPS) or firmly bound (TB-EPS) (Wu 2013).

Exopolysaccharides (EPS) have garnered significant interest, leading to the development of promising pharmaceutical candidates with a range of activities, including antibacterial, antiviral, free radical scavenging, antioxidant, anti-tumor, anticancer, wound healing, immunomodulatory, and anticoagulant properties. EPS are also useful in coating capsules and tablets (Sun et al. 2012; Nguyen and Nguyen 2014; Parwani et al. 2020). Various *Porphyridium* species, including *P. cruentum*, produce EPS with hypocholesterolemic effects and the ability to alter intestinal morphology in rats (Dvir et al. 2009). Additionally, sulfated EPS from *A. platensis* exhibit antithrombogenic, antiatherogenic, and anticoagulant properties. Exopolymers from three *Anabaena* species and *T. tenuis* have been shown to reduce activated partial thromboplastin time and prothrombin time by 16–41% and 12–65%, respectively. Polysaccharides from *Aphanothece sacrum* demonstrate anti-inflammatory effects (Talyshinsky et al. 2002). Research by Gudmundsdottir et al. (2015) found that exopolysaccharides from certain cyanobacteria modulate immune system responses in vitro, while UV protection studies have shown EPS pigments derived from *Nostoc* to be effective (Ehling-Schulz et al. 1997; Ehling-Schulz and Scherer 1999).

Microalgae produce large amounts of EPS for cell defense in challenging environments (Angelis et al. 2012). EPS play a role in cell-to-cell interactions, adhesion, and biofilm formation (Dertli et al. 2015) and help protect cells from environmental stress (Prajapat and Patel 2013). They can absorb and retain water, creating a gelatinous coating around cells and regulating water and heavy metal absorption (Adhikary and Sahu 1998; Pereira et al. 2011, 2013). Polysaccharides may also influence allelochemical activity (El-Sheekh et al. 2012). The formation of both intracellular and extracellular polysaccharides can alter the adaptive response of microalgae such as *Anabaena* PCC 7120 and *Chlorella vulgaris* to microcystin toxins from *Microcystis aeruginosa* extracts (Yang et al. 2014). At low concentrations, EPS from *P. cruentum* and *Rhodella maculata* exhibit excellent drag-reducing properties, suggesting their potential use as additives to enhance ship velocity and reduce fuel consumption (Gasljevic et al. 2007; Chen et al. 2010).

Cyanobacterial EPS have diverse applications, and they can be used as bioflocculants, bioadhesives, soil conditioners, biopolymers, biofilm developers, food supplements, and medicinal bioactive substances and in bioremediation (Chen et al. 2003; Rossi and De Philippis 2015). Biofilms, primarily formed by cyanobacteria, release negatively charged EPS into water, which bind positively charged heavy metals and cause them to settle (Bender and Phillips 2004). *Anabaena* species produce extracellular flocculants with rapid flocculation abilities and strong temperature resistance (Choi et al. 1998; Freire-Nordi et al. 2005). Coelho et al. (2015) investigated the formation of polyhydroxybutyrate (PHB) by *Spirulina* sp. LEB 18 under various nutritional conditions, finding that higher carbon source concentrations enhanced biopolymer accumulation.

Cyanobacteria produce a substantial amount of exopolysaccharides (EPS), which play a crucial role in soil structure and fertility, contributing significantly to nitrogen fixation (60%). These hydrophilic polymers enhance the soil's ability to absorb and retain water, increase soil carbon content (18%), and serve as a reservoir for nutrients. The hydrophilic nature of EPS improves soil water uptake and retention, which collectively enhances soil characteristics and supports seed germination and plant growth, thereby benefiting biocrusts and plant communities (Lan et al. 2014; Rossi et al. 2017).

EPS can bind various positively charged metal ions, which are essential for physiological processes such as nitrogen fixation and growth, or can act as toxins against predators due to the presence of functional groups like carbonyl, carboxyl, hydroxyl, and sulfate (Potts 1999). The biosorption capacity of *Nostoc linckia* EPS for Co(II) and Cr(VI) ions depends on contact time, pH, and initial metal ion concentration (Mona and Kaushik 2015). EPS from *Chlorella stigmatophora* can complex with Zn(II) and Cd(II) ions, potentially due to negative surface charges from uronic acids or sulfate groups (Kaplan et al. 1987). The capsular polysaccharides (CPS) of *Chroococcus paris* are capable of extracting Cu, Cd, and Zn from water (Li et al. 2001). Additionally, *Aphanothece* sp. and *Rivularia* sp. exhibit species-specific Ni adsorption capabilities, while sulfate groups in the EPS of the green alga *Chlorella stigmatophora* can bind and remove zinc and cadmium (Asthana et al. 1995).

Cyanobacteria's EPS can cling to a solid surface and therefore can act as bioadhesives. Biofilm development is aided by the sticky properties of the polymeric compounds generated by cyanobacteria (Tsuneda et al. 2003). As a result, biofilm

EPS help living communities survive in a wide range of environments (Neu and Lawrence 2016). The role of *Microcystis aeruginosa* EPS in heteroaggregation between the organism's cells and CeO nanoparticles was explored by Yang et al. (2018). Biofouling by cyanobacterial EPS results in antifouling paint detoxication as well as increased corrosion, conditioning, and colonization of metal surfaces by a variety of species (Scott et al. 1996). To avoid the formation of slime layers, antifouling technology can be utilized.

Exopolysaccharides (EPS) from cyanobacteria play a significant role in bioremediation, constituting about 25% of their total biomass. These EPS help remove and degrade environmental toxicants and contaminants through various mechanisms (Pisciotta et al. 2010). Abed and Köster (2005) explored the use of cyanobacteria and chemotrophic bacteria consortia for bioremediating wastewater and oil-contaminated sediments. Their research found that the *Oscillatoria-Gammaproteobacteria* consortium effectively degrades hydrocarbons such as phenanthrene, dibenzothiophene, pristane, and n-octadecane. Additionally, cyanobacterial EPS serve as biosorbents for removing organic contaminants like dyes and pesticides (Bhatnagar et al. 2012). Furthermore, the cyanobacterial sheath has recently been identified as a protective barrier against harmful biomineralization processes (Benning et al. 2004).

Anabaena sp. EPS has a beneficial flow feature known as pseudoplasticity, which makes it valuable in commercial applications when mixed with other materials (Moreno et al. 2000). *Anabaena* EPS dispersions have a viscosity that is comparable to commercially available food-grade polysaccharides and is very near to that of xanthan gum. It is used as a stabilizer in the food industry. Due to the solubilization of molecular weight EPS in the culture, it was recently claimed that the high viscosity of *C. capsulata* cultures impeded the free diffusion of copper ions into the culture (De Philippis and Micheletti 2009).

In the food sector, it is used as a stabilizer.

In the food industry, EPS can be used as a thickener and gel to improve food quality and texture (Cheng et al. 2011). EPS is largely used in industry as a gelling and thickening agent to suspend or stabilize the aqueous phase. EPS is produced by a variety of species and has a wide range of industrial applications (Sandford et al. 1984). As a result of the recent growth in demand for natural polymers in many commercial applications, new algae have been exploited to synthesize high-value products like pigments, osmoprotectants, metabolites, fatty acids, and proteins. In EPS, novel algae could be employed as biosurfactants and/or bioemulsifiers (Zheng and Xiao 2016; Shen et al. 2018).

Algal EPS can also be used in agriculture to minimize soil erosion. Cyanobacteria and algae produce EPS, which capture and bind soil particles, causing soil aggregates to grow in size. Higher soil aggregates become heavier, have a larger surface area, and are more difficult to blow away or wash away by water (Belnap and Gillette 1998). Biological soil crusts, also known as microbial mats, are formed when cyanobacteria, algae, and soil come together on the earth's surface. Cyanobacterial crusts improve water retention time by reducing bulk density and soil porosity, as well as organic matter and cation exchange capacity (CEC) (Mager 2009, 2010).

Microalgae can synergistically influence quorum sensing in cyanobacteria and other prokaryotes, leading them to colonize or develop on substrates and enhance soil stability (Morales et al. 2010). Cyanobacterial exopolysaccharides (EPS) absorb water, which helps regulate water movement in the soil, creating a microenvironment that improves water retention (Potts 2001). Wetting expands the polysaccharides in the mucilage, increasing cyanobacterial colony size and reducing water drainage, thereby enhancing soil water-holding capacity and reducing evaporation losses (Malam Issa et al. 2009).

EPS, including sheaths, capsules, and slime, also protect cyanobacterial cells from UV radiation. UV exposure stimulates the synthesis of the extracellular polysaccharide matrix in *Nostoc commune* (Wright et al. 2005), and UV-screening pigments concentrate in the sheath and extracellular matrix, forming a protective barrier against UV damage (Fleming and Castenholz 2007, 2008). Additionally, mycosporine-like amino acids (MMAs) found in the sheaths of several cyanobacterial strains suggest a role in storing UV-absorbing compounds to shield cells from harmful UV radiation (Adhikary 1998; Adhikary and Sahu 1998).

EPS can also affect the motility of gliding cyanobacteria, with slime secretion providing the necessary propulsion for movement (Li et al. 2001). While experimental and ecological observations support several ecological functions of cyanobacterial EPS, their role in protecting cells from desiccation is well documented (Potts 2004; Wright et al. 2005; Tamaru et al. 2005). Cyanobacteria from arid environments, such as desert soils and dry surfaces, often produce significant amounts of extracellular polysaccharides, underscoring their importance in surviving harsh conditions (Crispim and Gaylarde 2005; Zhang et al. 2006).

Polysaccharides can also function as prebiotics, promoting the growth of beneficial gut bacteria (Vidanarachchi et al. 2009). Rich polysaccharide extracts from *Padina gymnospora* and *Ulva lactuca* have been shown to accelerate tomato development under low-cost inorganic farming conditions (Hernández-Herrera et al. 2016). Application of these crude polysaccharide extracts to plants has been found to enhance root growth, yields, seed germination, and disease resistance in various crops (Laporte 2007; Chandía and Matsuhiro 2008).

4 Removal of Emerging Contaminants (ECS) at Pilot and Industrial Scale

Emerging contaminants (ECs) are synthetic organic chemicals that have gained attention due to their potential risks to human health and aquatic ecosystems, or because they have recently been detected in aquatic environments or drinking water due to advancements in analytical techniques. The long-term effects of many new pollutants are not well understood, and regulatory requirements are often lacking.

Common emerging pollutants include pharmaceuticals, personal care products, agri-chemicals, pesticides, and flame retardants, with over 200 distinct substances documented so far (Ahmed et al. 2017; Villar-Navarro et al. 2018).

These pollutants typically enter rivers through human activities, such as the direct discharge of raw or treated municipal wastewater, landfill leachate, urban and rural runoff, and industrial waste discharge (Tran and Gin 2017; Ali et al. 2018; Tran et al. 2018). Reports indicate that the concentrations of emerging pollutants in wastewater, surface water, groundwater, and drinking water are generally in the nanogram to low microgram per liter range (Kelly and Brooks 2018; Nakayama et al. 2019).

Traditional wastewater treatment facilities (WWTPs) are not designed to effectively remove emerging pollutants from influent. Various technologies have been explored for treating and removing these pollutants, including activated sludge, constructed wetlands, chemical precipitation, solvent extraction, electrocoagulation, and anaerobic bed reactors (Sutherland and Ralph 2019). Additionally, the practice of disposing of emerging pollutant-laden sludge on land at many WWTPs can lead to leaching of contaminants into groundwater or surface runoff, further contaminating ecosystems. In contrast, microalgal-based wastewater treatment systems have demonstrated lower capital and operating costs, natural disinfection properties, and enhanced efficiency in nutrient pollution remediation (Craggs et al. 2012). Microalgae can bioremediate emerging pollutants.

4.1 Microalgal Bioadsorption of Emerging Contaminants

Emerging pollutants are adsorbed onto cell wall components or organic chemicals released by microalgal cells, a process known as bioadsorption (Saavedra et al. 2018). Research on heavy metal removal has primarily focused on microalgal bioadsorption systems, with adsorption rates varying from 0% to 100% (Guo et al. 2016; Ali et al. 2018).

Microalgal Biouptake of Emerging Contaminants

Biouptake involves the active transport of contaminants into the cell, where they bind to intracellular proteins and other substances. Microalgal cells can take up emerging contaminants through three mechanisms: (1) passive diffusion, (2) passive-facilitated diffusion, or (3) energy-dependent/active uptake. The latter requires energy from the cell to transport the contaminants across the cell membrane. Biouptake is less commonly studied, with only 10% of research indicating its involvement in the treatment of emerging contaminants. For instance, biouptake was responsible for about 23% of 17-ethinyl estradiol removal by the green alga *Desmodesmus subspicatus* (Maes et al. 2014) and roughly 42% of triclosan removal by the green alga *Nannochloris* sp. (Bai and Acharya 2016). Biouptake is particularly useful for treating dilute concentrations of emerging pollutants where other treatment methods may be cost-prohibitive.

4.2 *Microalgal Biodegradation*

Microalgal biodegradation involves the metabolic breakdown of complex chemicals into simpler components. This can occur via metabolic degradation, where the contaminant serves as a carbon source or electron donor/acceptor, or co-metabolism, where the contaminant is degraded by enzymes that also process other substrates (Tiwari et al. 2017). Biodegradation can occur intracellularly, extracellularly, or through a combination of both, with extracellular degradation often followed by intracellular processing of the breakdown products (Tiwari et al. 2017). The process involves Phase I and Phase II enzymes that enhance the hydrophilicity of pollutants through hydrolysis, oxidation, or reduction (Xiong et al. 2018). Biodegradation can transform contaminants into less hazardous compounds, making it a promising approach for remediation.

Photodegradation and Volatilization

When microalgae cannot bioremediate emerging pollutants through bioadsorption, biouptake, or biodegradation, photodegradation and volatilization become alternative methods. Photodegradation involves two main processes: photolysis and photooxidative destruction (Abo et al. 2016). Photolysis occurs when a contaminant absorbs light, leading to chemical changes and degradation. Photooxidative degradation, on the other hand, involves the contaminant reacting with hydroxyl radicals or other oxidants produced from dissolved organic molecules or nitrate under light (Abo et al. 2016). Combining photodegradation with nutrient removal in wastewater treatment systems using microalgae has proven successful (Hom-Diaz et al. 2017a, b; Villar-Navarro et al. 2018).

Volatilization refers to the release of volatile organic compounds from the liquid phase into the atmosphere. In microalgal treatment systems, high aeration rates from mixing devices (e.g., paddle wheels, bubble lift columns, stirrers), along with increased sunlight and temperatures, can enhance the removal of volatile pollutants. For instance, volatilization of hydrophobic compounds like musk fragrances was observed to be more effective during summer operations when sunlight and temperatures were higher (Matamoros et al. 2015).

4.3 *Membrane Filtration*

Membrane filtration is an attractive method for removing emerging pollutants due to its high removal efficiency, low energy requirements, ease of scaling up, rapid kinetics, and minimal carbon footprint. Membrane-based technologies have been used for years to filter microorganisms, organic materials, pollutants, and other particles from water and wastewater (Gómez-Espinosa and Arizmendi-Cotero 2019). Various pressure-driven membrane technologies, including nanofiltration (NF), ultrafiltration (UF), reverse osmosis (RO), and forward osmosis (FO), have been explored for emerging pollutants removal. Hybrid systems that combine membrane filtration with adsorption offer additional benefits by integrating the strengths of both methods (Dhangar and Kumar 2020).

4.4 Full-Scale Systems

Matamoros and Salvadó (2013) assessed the removal of various emerging contaminants using a pilot-scale high-rate algal pond (HRAP) with an organic loading rate of 7–29 g COD/m^3/d. The study categorized contaminants into four classes based on removal efficiency: (i) class A (>90% removal, e.g., caffeine, ibuprofen); (ii) class B (60–90% removal, e.g., naproxen, bisphenol A, tributyl phosphate); (iii) class C (40–60% removal, e.g., diclofenac, benzotriazole); and (iv) class D (<30% removal, e.g., methyl paraben, 2,4-D and tetracycline) (de Godos et al. 2012).

While much research on microalgal bioremediation has been conducted in laboratory settings under batch culture conditions with long hydraulic retention times (HRT), scaling up to full-scale systems presents challenges. Batch treatment may be suitable for low-volume point sources of emerging contaminants, such as industrial discharges, but municipal wastewater treatment plants require continuous processing due to the large volume and daily flow of wastewater. Semi-continuous or continuous algal-based treatment systems are necessary for managing emerging pollutants in municipal wastewater streams.

Two studies, one conducted outdoors and one indoors, have demonstrated the effectiveness of pilot-scale HRAPs and built treatment wetlands containing both plants and algae in removing emerging pollutants, including pharmaceuticals and personal care products (PCPs). These systems employed a semi-continuous culture method, changing a portion of the culture medium daily (e.g., 12.5% exchange every 8 days). Despite the longer HRT in laboratory cultures, removal rates for diclofenac and salicylic acid were comparable between outdoor HRAPs and laboratory batch cultures. Further research is needed to validate the performance of microalgal-based treatment systems for emerging contaminants across seasonal variations in outdoor HRAPs with mixed microalgal species.

4.5 Laboratory Scale

Bioreactors—Open Pond

Meneses-Jácome et al. (2016) describe open ponds as shallow artificial pools, with depths of approximately 0.03–0.07 meters, designed for cultivating microalgae without stirring. These are among the simplest bioreactors. Factors such as cell suspension mixing, light diffusion, CO_2 supply, pH, temperature, and nutrient availability all influence algal growth (Sawant et al. 2018). Open ponds are widely used for treating industrial and agricultural wastewaters due to their low investment cost, ease of operation, and minimal operational expenses (Norvill et al. 2016, 2017; Wang et al. 2016; Sawant et al. 2018).

Raceways enhance the open-pond system by incorporating mixing equipment, typically a paddle wheel, to ensure better microalgae suspension and system homogenization, leading to improved production and stability (Maity 2015). Photobioreactors (PBRs), which come in various shapes and sizes, and raceway pond systems are among the most common methods for cultivating photoautotrophic algae (Gouveia et al. 2016). High-rate algal ponds (HRAPs) are a specific type of raceway pond.

Laboratory-scale studies using open ponds and bubble column photobioreactors have shown high removal efficiencies for ciprofloxacin, metoprolol, triclosan, and salicylic acid (over 90%); moderate removal for tramadol and carbamazepine (50–90%); and very low removal for trimethoprim and ciprofloxacin (10%) (Tolboom et al. 2019).

Bubble Column Photobioreactors

Tubular photobioreactors (PBRs) are closed systems composed of transparent tubes that allow natural or artificial light to penetrate (Singh and Sharma 2012). These PBRs can be configured in vertical, horizontal, or helical arrangements (Christenson and Sims 2011). They create a controlled environment designed to isolate specific microalgal strains, thereby enhancing productivity, biomass quality, and enabling the exploration of a broader range of strains. Despite the high investment and operational costs, and the challenges associated with constructing and managing these closed systems, the benefits include improved O_2 release, more efficient residual gas mixture handling, and enhanced photosynthetic efficiency through increased gas flow rates (Zhu et al. 2014; Gouveia et al. 2016).

4.6 High-Rate Algal Ponds (HRAPs): A Pilot-Scale Study

High-rate algal ponds (HRAPs) are shallow raceway reactors where bacteria and microalgae coexist. In these systems, heterotrophic bacteria digest organic matter using oxygen produced by microalgal photosynthesis, eliminating the need for additional aeration (Passos et al. 2014). While microalgae-based wastewater treatment systems have been explored since the 1950s for nutrient removal, heavy metal extraction, and pathogen reduction, research on their capacity to remove organic pollutants is limited (Abdel-Raouf et al. 2012). These systems utilize both abiotic processes (such as sorption, volatilization, and photodegradation) and biotic processes (including biodegradation, microalgae absorption, and metabolization) to eliminate microcontaminants (Abargues et al. 2013). Consequently, HRAPs are considered a viable option for treating wastewater containing emerging organic contaminants, offering the additional advantages of producing microalgal biomass, requiring no aeration, and using less land compared to other engineered natural treatment technologies, such as constructed wetlands (CWs) and waste stabilization ponds (WSPs).

5 Conclusion and Future Perspectives

There has been limited fundamental research on the structural and functional diversity of cyanobacterial exopolysaccharides (EPS). The structure, physicochemical properties, and biosynthesis of these EPS remain largely unexplored, creating substantial knowledge gaps. Consequently, further investigation is essential to both advance fundamental understanding and explore the potential applications of these unique substances across various biotechnology fields. Based on current knowledge, cyanobacterial EPS show promise for use as bioflocculants, as soil conditioners, in bioremediation of heavy metals and other toxins, and within the food and pharmaceutical industries.

Despite their demonstrated structural stability and favorable rheological properties, additional research is needed to fully evaluate their practical utility in these sectors.

The lack of specific information on the genes involved for EPS production, biosynthetic routes, and regulatory variables, on the other hand, limits their prospective uses in a range of emerging biotech domains. As a result, further research is needed at the molecular level to understand their biosynthesis routes. Similarly, the link between cyanobacterial biomass and EPS generation is poorly understood. More research in this area could aid in the identification of industrial-scale biomass production methods that are economically viable. After EPS production, the residual biomass can be utilized as a valuable resource for creating various value-added products and biofuels.

Additionally, developing integrated methods for synthesizing and extracting EPS from cyanobacteria could enhance the commercial viability of this bioprocess by improving material productivity and efficiency. While the structural complexity of these polysaccharides offers potential advantages, further research is required to understand their structure, rheological properties, and molecular stability under different environmental conditions before they can be established as a new source of natural biopolymers.

Cyanobacterial EPS are important in soil neogenesis. Innovative methods for combining cyanobacterial EPS with soil-fixing chemicals to form durable biocrusts are urgently needed to combat desertification. Finally, adopting a multi-product approach in researching cyanobacterial EPS could uncover their full potential and hidden benefits.

To optimize the bioremediation capabilities of microalgal species for emerging contaminants, further research is needed to enhance bioadsorption, develop more effective biodegrading enzymes, and improve growing conditions. Combining microalgal treatment of emerging pollutants with nutrient removal offers a cost-effective and practical solution for reducing contamination in water bodies.

References

Abargues MR, Ferrer J, Bouzas A, Seco A (2013) Removal and fate of endocrine disruptors chemicals under lab-scale post-treatment stage. Removal assessment using light, oxygen and microalgae. Bioresour Technol 149:142–148

Abdel-Raouf N, Al-Homaidan AA, Ibraheem IBM (2012) Microalgae and wastewater treatment, Saudi. J Biol Sci 19:257–275

Abed R, Köster MMJ (2005) The direct role of aerobic heterotrophic bacteria associated with cyanobacteria in the degradation of oil compounds. Int Biodeter Biodegr 55:29–27

Abo R, Kummer NA, Merkel BJ (2016) Optimized photodegradation of Bisphenol A in water using ZnO, TiO2 and SnO2 photocatalysts under UV radiation as a decontamination procedure. Drink Water Eng Sci 9(2):27–35

Adhikary SP (1998) Polysaccharides from mucilegenous envelope layers of cyanobacteria and their ecological significance. J Sci Ind Res 57:454–466

Adhikary SP, Sahu JK (1998) UV-protecting pigment of the terrestrial cyanobacterium Tolypothrixbyssoidea. J Plant Physiol 153:770–773

Adrien A, Bonnet A, Dufour D, Baudouin S, Maugard T, Bridiau N (2017) Pilot production of ulvans from Ulva sp. and their effects on hyaluronan and collagen production in cultured dermal fibroblasts. Carbohydr Polym 157:1306–1314

Aguilar-Briseño JA, Cruz-Suarez LE, Sassi JF, Ricque-Marie D, Zapata-Benavides P, Mendoza Gamboa E, Rodríguez-Padilla C, Trejo-Avila LM (2015) Sulphated polysaccharides from Ulva clathrata and Cladosiphonokamuranus seaweeds both inhibit viral attachment/entry and cell-cell fusion, in NDV infection. Mar Drugs 13(2):697–712

Ahmed EA (2021) Studies on some phytoplankton and phytobenthos from Lake Qaroun, El-Fayoum Egypt. Egypt J Phycol 22:49–72. https://doi.org/10.21608/egyjs.2021.89705.1004

Ahmed OM, Ahmed RR (2014) Anti-proliferative and apoptotic efficacies of ulvan polysaccharides against different types of carcinoma cells in vitro and in vivo. J Cancer Sci Ther 6:202–208

Ahmed MB, Zhou JL, Ngo HH, Guo W, Thomaidis NS, Xu J (2017) Progress in the biological and chemical treatment technologies for emerging contaminantremoval from wastewater: a critical review. J Hazard Mater 323:274–298

Akbar A, Shakeel A(2019) Marine polysaccharides: advances and multifaceted applications Shakeel Ahmed and Aisverya Soundararajan. Pan Stanford Publishing Pte. Ltd.29–52

Ali ME, El-Aty AMA, Badawy MI, Ali RK (2018) Removal of pharmaceutical pollutants from synthetic wastewater using chemically modified biomass of green alga Scenedesmus obliquus. Ecotoxicol Environ Saf 151:144–152

Aneiros A, Garateix A (2004) Bioactive peptides from marine sources: pharmacological properties and isolation procedures. J Chromatogr B 803:41–53

Angelis S, Novak AC, Sydney EB, Soccol VT, Carvalho JC, Pandey A, Noseda MD, Tholozan JL, Lorquin J, Soccol CR (2012) Co-culture of microalgae, cyanobacteria, and macromycetes for exopolysaccharides production: process preliminary optimization and partial characterization. Appl Biochem Biotechnol 167:1092–1106

Arunkumar K (2017) Extraction, isolation, and characterization of alginate. In: Industrial applications of marine biopolymers. CRC press, Boca Raton, pp 19–35

Asthana RK, Chatterjee S, Singh P (1995) Investigations on nickel biosorption and its remobilization. Process Biochem 30:729–724

Bafanaa A (2013) Characterization and optimization of production of exopolysaccharide from Chlamydomonas reinhardtii. Carbohydr Polym 95:746–752

Bagasra O, Lischner HW (1998) Activity of dextran sulfate and other polyanionic polysaccharides against human immunodeficiency virus. J Infect Dis 158:1084–1087

Bai X, Acharya K (2016) Removal of trimethoprim, sulfamethoxazole, and triclosan by the green alga Nannochloris sp. J Hazard Mater 315:70–75

Bajpai SK, Dehariya P, Singh Saggu SP (2015) Investigation of moisture sorption, permeability, cytotoxicity and drug release behavior of carrageenan/poly vinyl alcohol films. J Macromol Sci Part A 52:243–251

Banerjee A, Sarkar S, Govil T, González-Faune P, Cabrera-Barjas G, Bandopadhyay R, Salem DR, Sani RK (2021) Extremophilic exopolysaccharides: biotechnologies and wastewater remediation. Front Microbiol 12:721365. https://doi.org/10.3389/fmicb.2021.721365

Becker W (2004) Microalgae in human and animal nutrition. In: Richmond A (ed) Handbook of microalgal culture: biotechnology and applied phycology. Wiley-Blackwell, Hoboken, p 312

Belnap J, Gillette DA (1998) Vulnerability of desert soil surfaces to wind erosion: impact of soil texture and disturbance. J Arid Environ 39:133–142

Ben-Amotz A, Tornabene T, Thomas W (1985) Chemical profile of selected species of microalgae with emphasis on lipids. J Phycol 21(1):72–81

Bender J, Phillips P (2004) Microbial mats for multiple applications in aquaculture and bioremediation. Bioresour Technol 94:229–238

Benning LG, Phoenix VR, Yee N, Tobin MJ (2004) Molecular characterization of cyanobacterial silicification using synchrotron infrared micro-spectroscopy. Geochim Cosmochim Ac 68:729–741

Bhatnagar MS, Ganguly PJ, Bhatnagar A (2012) Rheology and composition of a multi utility exopolymer from a desert borne cyanobacterium Anabaena variabilis. J Appl Phycol 24:1387–1384

Bhatnagar M, Parwani L, Sharma V, Ganguly J, Bhatnagar A (2014) Exopolymers from *Tolypothrix tenuis* and three *Anabaena* sp. (Cyanobacteriaceae) as novel blood clotting agents for wound management. Carbohydr Polym 99:692–699

Borowitzka MA (2013) High-value products from microalgae—their development and commercialisation. J Appl Phycol 25:743–756

Breuer G, Lamers P, Martens D, Draaisma R, Wijffels R (2013) Effect of light intensity, pH, and temperature on triacylglycerol (TAG) accumulation induced by nitrogen starvation in Scenedesmus obliquus. Bioresour Technol 143:1–9

Bui VTNT, Nguyen BT, Nicolai T, Renou F (2019) Mobility of carrageenan chains in iota- and kappa carrageenan gels. Colloids Surf A Physicochem Eng Asp 562:113–118

Bussy F, Matthieu LG, Salmon H, Delaval J, Berri M, Pi NC (2019) Immunomodulating effect of a seaweed extract from *Ulva armoricana* in pig: specific IgG and total IgA in colostrum, milk, and blood. Vet Anim Sci 7:1–5

Cao X, Xu L, Chen YP, Decho AW, Cui Z, Lead JR (2022) Contribution, composition, and structure of EPS by in vivo exposure to elucidate the mechanisms of nanoparticle-enhanced bioremediation to metals. Environ Sci Technol 56:896–906. https://doi.org/10.1021/acs.est.1c05326

Cassidy YM, McSorley EM, Allsopp PJ (2018) Effect of soluble dietary fibre on postprandial blood glucose response and its potential as a functional food ingredient. J Funct Foods 46:423–439

Chakdar H, Jadhav SD, Dhar DW, Pabbi S (2012) Potential applications of blue green algae. J Sci Ind Res 71:13–20

Chandía N, Matsuhiro B (2008) Characterization of a fucoidan from Lessonia vadose (Phaeophyta) and its anticoagulant and elicitor properties. Int J Biol Macromol 42:235–240

Chen L, Li D, Liu Y (2003) Salt tolerance of MicrocoelusvaginatusGom., a cyanobacterium isolated from desert algal crust, was enhanced by exogenous carbohydrates. J Arid Environ 55:645–646

Chen B, You W, Huang J, Yu Y, Chen W (2010) Isolation and antioxidant property of the extracellular polysaccharide from Rhodella reticulata. World J Microbiol Biotechnol 26:833–840

Cheng YS, Zheng Y, Labavitch JM, VanderGheynst JS (2011) The impact of cell wall carbohydrate composition on the chitosan flocculation of chlorella. Process Biochem 46(10):1927–1933

Cheng YS, Zheng Y, Labavitch JM, VanderGheynst JS (2013) Virus infection of Chlorella variabilis and enzymatic saccharification of algal biomass for bioethanol production. Bioresour Technol 137:326–331

Choi CW, Yoo S, Oh I, Park SH (1998) Characterization of an extracellular flocculating substance produced by a planktonic cyanobacterium, Anabaena sp. Biotechnol Lett 20:643–646
Christenson L, Sims R (2011) Production and harvesting of microalgae for wastewater treatment, biofuels, and bioproducts. Biotechnol Adv 29:686–702
Clark AH, Ross-Murphy SB (1984) Structural and mechanical properties of biopolymer gels. Adv Polym Sci 83:57
Coelho VC, De silva CK, Terra AL, JAV C, De Morais MG (2015) Polyhydroxybutyrate production by Spirullina sp. LEB 18 grown under different nutrient concentrations. Afr J Microbiol Res 24:1586–1584
Costa OYA, Raaijmakers JM, Kuramae EE (2018) Microbial extracellular polymeric substances: ecological function and impact on soil aggregation. Front Microbiol 9:1636. https://doi.org/10.3389/fmicb.2018.01636
Craggs R, Sutherland D, Campbell H (2012) Hectare-scale demonstration of high rate algal ponds for enhanced wastewater treatment and biofuel production. J Appl Phycol 24(3):329 337
Crispim CA, Gaylarde CC (2005) Cyanobacteria and biodeterioration of cultural heritage: a review. Microb Ecol 49:1–9
Cunha L, Grenha A (2016) Sulfated seaweed polysaccharides as multifunctional materials in drug delivery applications. Mar Drugs 14:42
de Godos I, Muñoz R, Guieysse B (2012) Tetracycline removal during wastewater treatment in high-rate algal ponds. J Hazard Mater 229:446–449
dos Reis LL, Alho LD, de Abreu CB, Gebara RC, da Silva Mansano A, Melão MD (2022) Effects of cadmium and cobalt mixtures on growth and photosynthesis of Raphidocelissubcapitata (Chlorophyceae). Aquat Toxicol 244:106077. https://doi.org/10.1016/j.aquatox.2022.106077
De Philippis R, Micheletti E (2009) Heavy metal removal with exopolysaccharide producing cyanobacteria. In: Wang LK, Chen JP, Hung YT, Shammas NK (eds) Heavy metals in the environment. CRC Press, Boca Raton, pp 89–122
Delbarre-Ladrat C, Sinquin C, Lebellenger L, Zykwinska A, Colliec-Jouault S (2014) Exopolysaccharides produced by marine bacteria and their applications as glycosaminoglycan like molecules. Front Chem 2:85
Dertli E, Mayer MJ, Narbad A (2015) Impact of the exopolysaccharide layer on biofilms, adhesion and resistance to stress in Lactobacillus johnsonii FI9785. BMC Microbiol 15:8
Dhangar K, Kumar M (2020) Tricks and tracks in removal of emerging contaminants from the wastewater through hybrid treatment systems: a review. Sci Total Environ 738(336):140320
Dolganyuk V, Belova D, Babich O, Prosekov A, Ivanova S, Katserov D, Patyukov N, Sukhikh S (2020) Microalgae: a promising source of valuable bioproducts. Biomol Ther 10:8
Draget KI, Taylor C (2011) Chemical, physical and biological properties of alginates and their biomedical implications. Food Hydrocoll 25:251–256
Draget KI, Moe ST, Skjak-Bræk G, Smidsrød O (2006) Alginates. In: Stephen AM, Phillips GO, Wiliams PA (eds) Food polysaccharides and their applications, vol 14. CRC Press, Boca Raton, pp 159–1178
Du J (2015) Research on functionality sports nutrition and health food security issues based on circulation. Open Cybern Syst J 9:1945–1949
Dvir I, Stark AH, Chayoth R, Madar Z, Arad SM (2009) Hypocholesterolemic effects of nutraceuticals produced from the red microalga Porphyridium sp. in rats. Nutrients 1:156–167
Drouin G, Johnson R, Smith P, Lee M (2021) Comparison of Omega-3 Fatty Acids from Marine and Plant Sources in Clinical Trials. J Nutrit Sci 13(2):45-60
Ehling-Schulz M, Scherer S (1999) UV protection in cyanobacteria. Eur J Phycol 34:329–338
Ehling-Schulz M, Bilger W, Scherer S (1997) UV-B-induced synthesis of photoprotective pigments and extracellular polysaccharides in the terrestrial cyanobacterium Nostoc commune. J Bacteriol 179:1940–1945
El-Sheekh MM, Khairy HM, El-Shenody R (2012) Algal production of extra and intra-cellular polysaccharides as an adaptive response to the toxin crude extract of Microcystis aeruginosa. Iran J Environ Health Sci Eng 9:10

Fan X, Bai L, Zhu L, Yang L, Zhang X (2014) Marine algae-derived bioactive peptides for human nutrition and health. J Agric Food Chem 62:9211–9222

Fleming ED, Castenholz RW (2007) Effects of periodic desiccation on the synthesis of the UV-screening compound, scytonemin, in cyanobacteria. Environ Microbiol 9:1448–1455

Fleming ED, Castenholz RW (2008) Effects of nitrogen source on the synthesis of the UV screening compound, scytonemin, in the cyanobacterium Nostoc punctiforme PCC73102. FEMS Microbiol Ecol 63:301–308

Fleurence J (1999) Seaweed proteins: biochemical, nutritional aspects and potential uses. Trends Food Sci Technol 10(1):25–28

Fleurence J, Morançais M, Dumay J, Decottignies P, Turpin V, Munier M, Garcia-Bueno N, Jaouen P (2012) What are the prospects for using seaweed in human nutrition and for marine animals raised through aquaculture? Trends Food Sci Technol 27:57–61

Freire-Nordi CS, Vieira AAH, Nascimento OR (2005) The metal binding capacity of Anabaena spiroides extracellular polysaccharide: an EPR study. Process Biochem 40:2215–2224

Gasljevic K, Hall K, Chapman D, Matthys EF (2007) Drag-reducing polysaccharides from marine microalgae: species productivity and drag reduction effectiveness. J Appl Phycol 20:299–310

Gatrell S, Lum K, Kim J, Lei X (2014) Nonruminant nutrition symposium: potential of defatted microalgae from the biofuel industry as an ingredient to replace corn and soybean meal in swine and poultry diets. J Anim Sci 92:1306–1314

Gavalás-Olea A, Siol A, Sakka Y, Köser J, Nentwig N, Hauser T, Filser J, Thöming J, Lang I (2021) Potential of the red alga Dixoniella grisea for the production of additives for lubricants. Plan Theory 10:1836

Georg Jensen M, Pedersen C, Kristensen M, Frost G, Astrup A (2013) efficacy of alginate supplementation in relation to appetite regulation and metabolic risk factors: evidence from animal and human studies. Obes Rev 14:129–144

Ghanam D, Kleinebudde P (2011) Suitability of κ-carrageenan pellets for the formulation of multiparticulate tablets with modified release. Int J Pharm 409:9–18

Global Alginate Market SizelIndustry Report, 2021–2028. Available online: https://www.grandviewresearch.com/industryanalysis/alginate-market. Accessed 31 Mar 2021

Gómez-Espinosa RM, Arizmendi-Cotero D (2019) Role of membrane on emerging contaminant removal. In: Handbook of environmental chemistry, vol 66. Springer Verlag, pp 157–174

González-Ballesteros N, Rodríguez-Argüelles MC, Prado-López S, Lastra M, Grimaldi M, Cavazza A, Nasi L, Salviati G, Bigi F (2019) Macroalgae to nanoparticles. study of *Ulva lactuca* L. role in biosynthesis of gold and silver nanoparticles and of their cytotoxicity on colon cancer cell line. Mater Sci Eng C 97:498–509

Gouveia L, Batista AP, Sousa I, Raymundo A, Bandarra N (2008) Microalgae in novel food products. In: Konstantinos N, Papadopoulos PP (eds) Food chemistry research development. Nova Science Publishers, New York, pp 75–112

Gouveia L, Graça S, Sousa C, Ambrosano L, Ribeiro B, Botrel EP, Neto PC, Ferreira AF, Silva CM (2016) Microalgae biomass production using wastewater: treatment and costs: scale up considerations. Algal Res 16:167–176

Gudmundsdottir AB, Omarsdottir S, Brynjolfsdottir A, Paulsene BS, Olafsdottir ES, Freysdottir J (2015) Exopolysaccharides from *cyanobacterium aponinum*from the Blue Lagoon in Iceland increase IL-10 secretion by human dendritic cells and their ability to reduce the IL 17+ RORt+/ IL-10+ FoxP3+ ratio in CD4+ T cells. Immunol Lett 163(2):157–152

Guidara M, Yaich H, Richel A, Blecker C, Boufi S, Attia H, Garna H (2019) Effects of extraction procedures and plasticizer concentration on the optical, thermal, structural and antioxidant properties of novel ulvanfilms. Int J Biol Macromol 135:647–658

Guo WQ, Zheng HS, Li S, Du JS, Feng XC, Yin RL, Wu QL, Ren NQ, Chang JS (2016) Removal of cephalosporin antibiotics 7-ACA from wastewater during the cultivation of lipid accumulating microalgae. Bioresour Technol 221:284–290

Harnedy PA, FitzGerald RJ (2013) Extraction of protein from the macroalga Palmariapalmata. LWT Food Sci Technol 51(1):375–382

Hernández-Herrera RM, Santacruz-Ruvalcaba F, Zañudo-Hernández J, Hernández Carmona G (2016) Activity of seaweed extracts and polysaccharide–enriched extracts from Ulva lactucaand Padina gymnosporaas growth promoters of tomato and mung bean plants. J Appl Phycol 28:2549–2560

Holman B, Malau-Aduli A (2013) Spirulina as a livestock supplement and animal feed. J Anim Physiol Anim Nutr 97:615–623

Hom-Diaz A, Norvill ZN, Blanquez P, Vicent T, Guieysse B (2017a) Ciprofloxacin removal during secondary domestic wastewater treatment in high rate algal ponds. Chemosphere 180:33–41

Hom-Diaz A, Jaen-Gil A, Bello-Laserna I, Rodríguez-Mozaz S, Vicent T, Barcelo D, Blanquez P (2017b) Performance of a microalgal photobioreactor treating toilet wastewater: pharmaceutically active compound removal and biomass harvesting. Sci Total Environ 592:1–11

Issa OM, Défarge C, Trichet J (2009) Microbiotic soil crusts in the Sahel of western Niger and their influence on soil porosity and water dynamics. Catena 77:48–55

Junter GA, Thébault P, Lebrun L, Biomater A (2016) Polysaccharide-based antibiofilm surfaces. Acta Biomater 30:13–25

Kaplan D, Christiaen D, Arad SM (1987) Chelating properties of extracellular polysaccharides from Chlorella spp. Appl Environ Microbiol 53:2953–2956

Kasahara F, Kato T, Idota Y, Takahashi H, Kakinuma C, Yano K, Arakawa H, Hara K, Miyajima C, Ogihara T (2019) Reduction effect of calcium alginate on blood triglyceride levels causing the inhibition of hepatic and total body accumulation of fat in rats. Biol Pharm Bull 42:365–372

Kelly KR, Brooks BW (2018) Global aquatic hazard assessment of ciprofloxacin: exceedances of antibiotic resistance development and ecotoxicological thresholds. Prog Mol Biol Transl Sci 159:59–77

Kidgell JT, Magnusson M, de Nys R, Glasson CRK (2019) Ulvan: a systematic review of extraction, composition and function. Algal Res 39:101422. https://doi.org/10.1016/j.algal.2019.101422

Kikionis S, Ioannou E, Toskas G, Roussis V (2015) Electrospunbiocomposite nanofibers of ulvan/PCL and ulvan/PEO. J Appl Polym Sci 132:1–5

Kloareg B, Quatrano RS (1988) Structure of the cell walls of marine algae and eco physiological functions of the matrix polysaccharides. Oceanogr Mar Biol Annu Rev 26:259–315

Klok A, Lamers P, Martens D, Draaisma R, Wijffels R (2014) Edible oils from microalgae: insights in TAG accumulation. Trends Biotechnol 32(10):521–528

Kulkarni VS, Shaw C, Kulkarni VS, Shaw C (2016) Use of polymers and thickeners in semisolid and liquid formulations. In: Essential chemistry for formulators of semisolid and liquid dosages. Academic Press, pp 43–69

Kagan ML, Levy A, Leikin-Frenkel A (2015) Comparative study of tissue deposition of omega-3 fatty acids from polar-lipid rich oil of the microalgae Nannochloropsis oculata with krill oil in rats. Food & Function 6(1):185–191

Lahaye M, Kaeffer B (1997) Seaweed dietary fibres: structure, physico-chemical and biological properties relevant to intestinal physiology. Sci Aliment 17:563–584

Lahaye M, Robic A (2007) Structure and functional properties of ulvan, a polysaccharide from green seaweeds. Biomacromolecules 8:1765–1774

Lahaye M, Brunel M, Bonnin E (1997) Fine chemical structure analysis of oligosaccharides produced by an ulvan-lyase degradation of the water-soluble cell-wall polysaccharides from Ulva sp. (Ulvales, Chlorophyta). Carbohydr Res 304:325–333

Lahaye M, Cimadevilla EAC, Kuhlenkamp R, Quemener B, Lognoné V, Dion P (1999) Chemical composition and ^{13}C NMR spectroscopic characterisation of ulvans from Ulva (Ulvales, Chlorophyta). J Appl Phycol 11:1–7

Lan S, Zhang Q, Wu L, Liu Y, Zhang D, Hu C (2014) Artificially accelerating the reversal of desertification: cyanobacterial inoculation facilitates the succession of vegetation communities. Environ Sci Technol 48:307–305

Laporte D (2007) Structurally unrelated oligosaccharides obtained from marine macroalgae differentially stimulate growth and defense against TMV in tobacco plants. J Appl Phycol 19:79–88

Le B, Golokhvast KS, Yang SH, Sun S (2019) Optimization of microwave-assisted extraction of polysaccharides from *Ulva pertusa* and evaluation of their antioxidant activity. Antioxidants 8:129–133

Lee KY, Mooney DJ (2012) Alginate: properties and biomedical applications. Prog Polym Sci 37:106–126

Li P, Harding SE, Liu Z (2001) Cyanobacterial exopolysaccharides: their nature and potential biotechnological applications. Biotechnol Genet Eng 18:375–374

Lio-Po GD, Leaño EM, Peñaranda MMD, Villa-Franco AU, Sombito CD, Guanzon NG (2005) Anti-luminous vibrio factors associated with the 'green water'grow-out culture of the tiger shrimp Penaeus monodon. Aquaculture 250:1–7

Liu L, Pohnert G, Wei D (2016) Extracellular metabolites from industrial microalgae and their biotechnological potential. Mar Drugs 14:1–19

Lopes D, Melo T, Meneses J, Abreu MH, Pereira R, Domingues P, Lillebø AI, Calado R, Rosário Domingues M (2019) A new look for the red macroalga *Palmaria palmata*: a seafood with polar lipids rich in EPA and with antioxidant properties. Mar Drugs 17:533

Lopes D, Rey F, Leal MC, Lillebø AI, Calado R, Domingues MR (2021) Bioactivities of Lipid Extracts and Complex Lipids from Seaweeds: Current Knowledge and Future Prospects. Marine Drugs. 19(12):686.

Maes HM, Maletz SX, Ratte HT, Hollender J, Schaeffer A (2014) Uptake, elimination, and biotransformation of 17a-ethinylestradiol by the freshwater alga Desmodesmussubspicatus. Environ Sci Technol 48:12354–12361

Mager DM (2009) Extracellular polysaccharides from cyanobacterial soil crusts and their role in dryland surface processes. [Ph.D. Thesis]. Manchester Metropolitan University, Manchester, p 273

Mager DM (2010) Cyanobacterial soil crusts: analysing resilience in Kalahari sand soils. In: Degenovine KM (ed) Semi-arid environments: agriculture, water supply and vegetation. NOVA Scientific Publishers, Inc, Hauppauge

Maity SK (2015) Opportunities, recent trends and challenges of integrated biorefinery: part II. Renew Sust Energ Rev 43:1446–1466

Maršálek B, Rojíčkovâ R (1996) Stress factors enhancing production of algal exudates: a potential self-protective mechanism? Z Naturforsch C 51:646–650. https://doi.org/10.1515/znc-1996-9-1008

Matamoros V, Gutierrez R, Ferrer I, García J, Bayona JM (2015) Capability of microalgae based wastewater treatment systems to remove emerging organic contaminants: a pilot-scale study. J Hazard Mater 288:34–42

Mathivanan K, Chandirika JU, Mathimani T, Rajaram R, Annadurai G, Yin H (2021) Production and functionality of exopolysaccharides in bacteria exposed to a toxic metal environment. Ecotoxicol Environ 208:111567. https://doi.org/10.1016/j.ecoenv.2020.111567

Meneses-Jácome A, Diaz-Chavez R, Velásquez-Arredondo HI, Cárdenas-Chávez DL, Parra R, Ruiz-Colorado AA (2016) Sustainable energy from agro industrial wastewaters in Latin America. Renew Sust Energ Rev 56:1249–1262

Michalak I, Dmytryk A, Wieczorek PP, Rój E, Łęska B, Górka B, Messyasz B, Lipok J, Mikulewicz M, Wilk R, Schroeder G, Chojnacka K (2015) Supercritical algal extracts: a source of biologically active compounds from nature. J Chem 2015:597140. https://doi.org/10.1155/2015/597140

Micheletti E, Colica G, Viti C, Tamagnini P, De Philippis R (2008) Selectivity in the heavy metal removal by exopolysaccharide-producing cyanobacteria. J Appl Microbiol 105:88–94

Mishima T, Murata J, Toyoshima M, Fujii H, Nakaiima M, Hayashi T (1998) Inhibition of tumor invasion and metastasis by calcium spirulan (Ca-SP), a novel sulfated polysaccharide derived from a blue-green alga, *Spirulina platensis*. Clin Exp Metastasis 16(6):541

Mishra A, Jha B (2009) Isolation and characterization of extracellular polymeric substances from micro-algae Dunaliella salina under salt stress. Bioresour Technol 100:3382–3386. https://doi.org/10.1016/j.biortech.2009.02.006

Mona S, Kaushik A (2015) Chromium and cobalt sequestration using exopolysaccharides produced by freshwater cyanobacterium Nostoc linckia. Ecol Eng 82:121–125

Morales VL, Parlange JY, Steenhuis TS (2010) Are preferential flow paths perpetuated by microbial activity in the soil matrix? A review. J Hydrol 393:29–36

Morelli A, Betti M, Puppi D, Bartoli C, Gazzarri M, Chiellini F (2016) Enzymatically crosslinked ulvan hydrogels as injectable systems for cell delivery. Macromol Chem Phys 217:581–590

Moreno J, Vargas MA, Madiedo JM, Munoz J, Rivas J, Guerrero MG (2000) Chemical and rheological properties of extracellular polysaccharide produced by the cyanobacterium Anabaena sp. ATCC 33047. Biotechnol Bioeng 67:283–290

Matamoros V, Salvadó V (2013) Evaluation of a coagulation/flocculation-lamellar clarifier and filtration-UV-chlorination reactor for removing emerging contaminants at full-scale wastewater treatment plants in Spain. J Environ Management 117:96–102

Nakayama SF, Yoshikane M, Onoda Y, Nishihama Y, Iwai-Shimada M, Takagi M, Kobayashi Y, Isobe T (2019) Worldwide trends in tracing poly-and perfluoroalkyl substances (PFAS) in the environment, vol 121. TrAC Trends Anal Chem, Chem, p 115410

Necas J, Bartosikova L (2013) Carrageenan: a review. Vet Med 58(4):187–205

Neu TR, Lawrence JR (2016) The extracellular matrix – an intractable part of biofilm systems. In: Flemming HC, Neu TR, Wingender J (eds) The perfect slime – microbial extracellular polymeric substances (EPS). IWA Publishers, London, pp 25–60

Nguyen DTA, Nguyen THK (2014) Detection on antioxidant and cytotoxicity activities of exopolysaccharides isolated in plant-originated Lactococcus lactis. Biomed Pharmacol J 7:33–38

Norvill ZN, Shilton A, Guieysse B (2016) Emerging contaminant degradation and removal in algal wastewater treatment ponds: identifying the research gaps. J Hazard Mater 5(313):291–309

Norvill ZN, Toledo-Cervantes A, Blanco S, Shilton A, Guieysse B, Muñoz R (2017) Photodegradation and sorption govern tetracycline removal during wastewater treatment in algal ponds. Bioresour Technol 232:35–43

Notowidjojo L (2021) Seaweed as novel food for prevention and therapy for life style related disease. World Nutr J 6:1–5. https://doi.org/10.25220/WNJ.V05.i1.0001

Ordog V, Stirk W, Alint PB, Aremu A, Okem A, Lovasz C, Molnar Z, van Staden J (2016) Effect of temperature and nitrogen concentration on lipid productivity and fatty acid composition in three chlorella strains. Algal Res 16:141–149

Otero A, Vincenzini M (2003) Extracellular polysaccharide synthesis by Nostoc strains as affected by N source and light intensity. J Biotechnol 102:143–152

Panyoyai N, Bannikova A, Small DM, Kasapis S (2015) Controlled release of thiamin in a glassy κ-carrageenan/glucose syrup matrix. Carbohydr Polym 115:723–731

Paradossi G, Cavalieri F, Pizzoferrato L, Liquori AM (1999) A physico-chemical study on the polysaccharide ulvan from hot water extraction of the macroalga Ulva. Int J Biol Macromol 25:309–315

Paredes Juárez GA, Spasojevic M, Faas MM, de Vos P (2014) Immunological and technical considerations in application of alginate-based microencapsulation systems. Front Bioeng Biotechnol 2(AUG). https://doi.org/10.3389/fbioe.2014.00026

Parwani L, Srivastava M, Singh J (2020) Uncovering potential of cyanobacteria in wound healing. Res J Biotech 15(6):170–179

Passos F, Mariona HM, García J, Ferrer I (2014) Long-term anaerobic digestion of microalgae grown in HRAP for wastewater treatment. Effect of microwave pretreatment. Water Res 49:351–359

Pati MP, Sharma S, Das NL, Panda CR (2016) Uses of seaweed and its application to human welfare: a review. Int J Pharm 8(10):12–20

Pawar HV, Boateng JS, Ayensu I, Tetteh J (2014) Multifunctional medicated lyophilised wafer dressing for effective chronic wound healing. J Pharm Sci 103:1720–1733

Pereira L, Cotas J (2020) Introductory chapter: alginates-A general overview. In: Alginates recent uses of this natural polymer. IntechOpen, London

Pereira S, Micheletti E, Zille A, Santos A, Moradas-Ferreira P, Tamagnini P, Philippis RD (2011) Using extracellular polymeric substances (EPS)-producing cyanobacteria for the bioremediation of heavy metals: do cations compete for the EPS functional groups and also accumulate inside the cell? Microbiology 157:451–458

Pereira SB, Mota R, Santos CL, De Philippis R, Tamagnini P (2013) Assembly and export of extracellular polymeric substances (EPS) in cyanobacteria. Adv Bot Res 65:235–279

Picker KM (1999) Matrix tablets of carrageenans. I. A compaction study. Drug Dev Ind Pharm 25:329–337

Pimentel FB, Alves RC, Harnedy PA, FitzGerald RJ, Oliveira MBPP (2019) Macroalgal derived protein hydrolysates and bioactive peptides: enzymatic release and potential health enhancing properties. Trends Food Sci Technol 93:106–124

Pisciotta JM, Zou Y, Baskakov IV (2010) Light-dependent electrogenic activity of cyanobacteria. PLoS One 5(5):e10821

Potts M (1999) Mechanisms of desiccation tolerance in cyanobacteria. Eur J Phycol 34:319–328

Potts M (2001) Desiccation tolerance: a simple process? Trends Microbiol 9:553–559

Potts M (2004) Nudist colonies: a revealing glimpse of cyanobacterial extracellular polysaccharide. J Phycol 40:1–3

Pradhan B, Ki JS (2023) Biological activity of algal derived carrageenan: a comprehensive review in light of human health and disease. Int J Biol Macromol 238:124085. https://doi.org/10.1016/j.ijbiomac.2023.124085

Prajapat J, Patel A (2013) Food and health applications of exopolysaccharides produced by lactic acid bacteria. Adv Dairy Res 1:107

Qian J, He X, Wang P, Xu B, Li K, Lu B, Jin W, Tang S (2021) Effects of polystyrene nanoplastics on extracellular polymeric substance composition of activated sludge: the role of surface functional groups. Environ Pollut 279:116904. https://doi.org/10.1016/j.envpol.2021.116904

Qian L, Ye X, Xiao J, Lin S, Wang H, Liu Z, Ma Y, Yang L, Zhang Z, Wu L (2022) Nitrogen concentration acting as an environmental signal regulates cyanobacterial EPS excretion. Chemosphere 291:132878. https://doi.org/10.1016/j.chemosphere.2021.132878

Quemener B, Lahaye M, Bobin-Dubigeon C (1997) Sugar determination in ulvans by a chemical-enzymatic method coupled to high performance anion exchange chromatography. J Appl Phycol 9:179–188

Rafika C, Lamia T, Rym BD, Omeya EA, Ali Y, Khamissa G, Jihen BA, Hela O, Ouada HB (2011) Evaluation of cytotoxicity and biological activities in extracellular polysaccharides released by cyanobacterium Arthrospira platensis. Braz Arch Biol Technol 54:831–838

Raposo MF, de Morais AM, de Morais RM (2014) Influence of sulphate on the composition and antibacterial and antiviral properties of the exopolysaccharide from Porphyridiumcruentum. Life Sci 101:56–63

Rinaudo M (2007) Seaweed polysaccharides. In: Kalmerling JP (ed) Comprehensive glycoscience from chemistry to systems biology, vol 2. Elsevier, London, pp 691–735

Rinaudo M (2008) Main properties and current applications of some polysaccharides as biomaterials. Polym Int 57:397–430

Rizk MZ, Aly HF, Matloub AA, Fouad GI (2016) The anti-hypercholesterolemic effect of ulvan polysaccharide extracted from the green alga Ulva fasciata on aged hypercholesterolemic rats. Asian J Pharm Clin Res 9:165–176

Robertson GL (2016) Food packaging: principles and practice. CRC press

Rodrigues S, da Costa AMR, Grenha A (2012) Chitosan/carrageenan nanoparticles: effect of cross-linking with tripolyphosphate and charge ratios. Carbohydr Polym 89:282–289

Rossi F, De Philippis R (2015) Role of cyanobacterial exopolysaccharides in phototrophic biofilms and in complex microbial mats. Life 5(2):1218–1238

Rossi F, Li H, Liu Y, De Philippis R (2017) Cyanobacterial inoculation (cyanobacterisation): perspectives for the development of a standardized multifunctional technology for soil fertilization and desertification reversal. Earth Sci Rev 171:28–43

Saavedra R, Muñoz R, Taboada ME, Vega M, Bolado S (2018) Comparative uptake study of arsenic, boron, copper, manganese and zinc from water by different green microalgae. Bioresour Technol 263:49–57

Sampath-Wiley P, Neefus CD, Jahnke LS (2008) Seasonal effects of sun exposure and emersion on intertidal seaweed physiology: fluctuations in antioxidant contents, photosynthetic pigments and photosynthetic efficiency in the red alga Porphyraumbilicaliskützing (Rhodophyta, Bangiales). J Exp Mar Biol Ecol 361:83–91

Sandford PA, Cottrell IW, Pettit DJ (1984) Microbial polysaccharides: new products and their commercial applications. Pure Appl Chem 56:879–892

Sathasivam R, Radhakrishnan R, Hashem A, Abd_Allah EF (2019) Microalgae metabolites: a rich source for food and medicine. Saudi J Biol Sci 26:709–722

Sawant SS, Khadamkar HP, Mathpati CS, Pandit R, Lali AM (2018) Computational and experimental studies of high depth algal raceway pond photo-bioreactor. Renew Energy 118:152–159

Scharnweber D, Hubner L, Rother S, Hempel U, Anderegg U, Samsonov SA, Pisabarro MT, Hofbauer L, Schnabelrauch M, Franz S, Simon J, Hintze V (2015) Glycosaminoglycan derivatives: promising candidates for the design of functional biomaterials. J Mater Sci Mater Med 26:232

Scott C, Fletcher RL, Bremer GB (1996) Observations on the mechanisms of attachment of some marine fouling blue-green algae. Biofouling 10:161–173

Sedayu BB, Cran MJ, Bigger SW (2019) A review of property enhancement techniques for carrageenan-based films and coatings. Carbohydr Polym 216:287–302

Selvakumaran S, Muhamad II, Razak SIA (2016) Evaluation of kappa carrageenan as potential carrier for floating drug delivery system: effect of pore forming agents. Carbohydr Polym 135:207–214

Shao P, Ma H, Zhu J, Qiu Q (2017) Impact of ionic strength on physicochemical stability of o/w emulsions stabilized by *Ulva fasciata* polysaccharide. Food Hydrocoll 69:202–209

Shen L, Li ZF, Wang JJ, Liu A, Li ZH, Yu RL, Wu XL, Liu YD, Li JK, Zeng WM (2018) Characterization of extracellular polysaccharide/protein contents during the adsorption of Cd(II) bySynechocystis sp. PCC6803. Environ Sci Pollut Res 25(21):20713–20722

Singh RN, Sharma S (2012) Development of suitable photobioreactor for algae production–a review. Renew Sust Energ Rev 16(4):2347–2353

Stonik VA, Stonik IV (2018) Sterol and Sphingoid Glycoconjugates from microalgae. Mar Drugs 16:514

Sun L, Wang L, Zhou Y (2012) Immunomodulation and antitumor activities of different molecular weight polysaccharides from Porphyridiumcruentum. Carbohydr Polym 87(1206):1210

Sutherland DL, Ralph PJ (2019) Microalgal bioremediation of emerging contaminants-opportunities and challenges. Water Res 164:114921

Sylwestrzak Z, Zgrundo A, Pniewski F (2021) Ecotoxicological studies on the effect of Roundup® (glyphosate formulation) on marine benthic microalgae. Int J Environ Res Public Health 18:3

Szekalska M, Pucibowska A, Szymanska E, Ciosek P, Winnicka K (2016) Alginate: current use and future erspectivein pharmaceutical and biomedical applications. Int J Polym Sci 16:1–17

Talyshinsky MM, Souprun YY, Huleihel MM (2002) Anti-viral activity of red microalgal polysaccharides against retroviruses. Cancer Cell Int 2:1–7

Tamaru Y, Takani Y, Yoshida T, Sakamoto T (2005) Crucial role of extracellular polysaccharides in desiccation and freezing tolerance in the terrestrial cyanobacterium Nostoc commune. Appl Environ Microbiol 71:7327–7333

Tiwari B, Sellamuthu B, Ouarda Y, Drogui P, Tyagi RD, Buelna G (2017) Review on fate and mechanism of removal of pharmaceutical pollutants from wastewater using biological approach. Bioresour Technol 224:1–12

Tolboom SN, Carrillo-Nieves D, de Jesús R-AM, de la Cruz Quiroz R, Barceló D, Iqbal HM, Parra-Saldivar R (2019) Algal-based removal strategies for hazardous contaminants from the environment—a review. Sci Total Environ 665:358–366

Trabelsi L, Chaieb O, Mnari A, Abid-Essafi S, Aleya L (2016) Partial characterization and antioxidant and antiproliferative activities of the aqueous extracellular polysaccharides from the thermophilic microalgae Graesiella sp. BMC Complement Altern Med 16:210
Tran NH, Gin KYH (2017) Occurrence and removal of pharmaceuticals, hormones, personal care products, and endocrine disrupters in a full-scale water reclamation plant. Sci Total Environ 599:1503–1516
Tran NH, Reinhard M, Gin KYH (2018) Occurrence and fate of emerging contaminants in municipal wastewater treatment plants from different geographical regions-a review. Water Res 133:182–207
Trouvelot S, Héloir MC, Poinssot B, Gauthier A, Paris F, Guillier C, Combier M, TrdáL DX, Adrian M (2014) Carbohydrates in plant immunity and plant protection: roles and potential application as foliar sprays. Front Plant Sci 5:592
Tsuneda S, Aikawa H, Hayashi H, Yuasa A, Hirata A (2003) Extracellular polymeric substances responsible for bacterial adhesion onto solid surface. FEMS Microbiol Lett 223:287–292
Tye RJ (1989) Industrial and non-food uses for carrageenan. Carbohydr Polym 10(4):259–280
Urtuvia V (2017) Bacterial alginate production: an overview of its biosynthesis and potential industrial production. World J Microbiol Biotechnol 33(11):198
van Ginneken VJT, Helsper JPFG, de Visser W, van Keulen H, Brandenburg WA (2011). Polyunsaturated Fatty Acids in Various Macroalgal Species from North Atlantic and Tropical Seas. Lipids in Health and Disease 10:104–111
Van Krimpen M, Bikker P, Van der Meer I, Van der Peet-Schwering C, Vereijken J (2013) Cultivation, processing and nutritional aspects for pigs and poultry of European protein sources as alternatives for imported soybean products. Wageningen UR Livestock Research, Lelystad, p 48
Venkatesan J, Lowe B, Anil S, Manivasagan P, Kheraif AAA, Kang KH, Kim SK (2015) Seaweed polysaccharides and their potential biomedical applications. Starch-Stärke 67:381–390
Venugopal V (2011) Marine polysaccharides—food applications. CRC Press, Boca Raton
Vidanarachchi J, Iji P, Mikkelsen L, Sims I, Choct M (2009) Isolation and characterization of water-soluble prebiotic compounds from Australian and New Zealand plants. Carbohydr Polym 77:670–676
Villar-Navarro E, Baena-Nogueras RM, Paniw M, Perales JA, Lara-Martín PA (2018) Removal of pharmaceuticals in urban wastewater: high rate algae pond (HRAP) based technologies as an alternative to activated sludge based processes. Water Res 139:19–29
Vlachou M, Tragou K, Siamidi A, Kikionis S, Chatzianagnostou AL, Mitsopoulos A, Ioannou E, Roussis V, Tsotinis A (2018) Modified in vitro release of the chronobiotic hormone melatonin from matrix tablets based on the marine sulfated polysaccharide ulvan. J Drug Deliv Sci Technol 44:41–48
Wang T, Jonsdottir R, Ólafsdóttir G (2009) Total phenolic compounds, radical scavenging and metal chelation of extracts from icelandic seaweeds. Food Chem 116:240–248
Wang Y, Ho SH, Cheng CL, Guo WQ, Nagarajan D, Ren NQ, Lee DJ, Chang JS (2016) Perspectives on the feasibility of using microalgae for industrial wastewater treatment. Bioresour Technol 222:485–497
Weiner ML (2014) Food additive carrageenan: part II: a critical review of carrageenan in vivo safety studies. Crit Rev Toxicol 44:244–269
Wijffels RH, Barbosa MJ (2010) An outlook on microalgal biofuels. Science 329:796–799
Wingender J, Neu TR, Flemming HC (2012) Microbial extracellular polymeric substances: characterization, structure and function. Springer Science & Business Media
Wright DJ, Smith SC, Joardar V, Scherer S, Jervis J, Warren A, Helm RF, Potts M (2005) UV irradiation and desiccation modulate the three-dimensional extracellular matrix of Nostoc commune (Cyanobacteria). J Biol Chem 280:40271–40281
Wu Z (2013) Effect of ultrasonic power density on extracting loosely bound and tightly bound extracellular polymeric substances. Desalination 329(15):35–40

Xiong JQ, Kurade MB, Jeon BH (2018) Can microalgae remove pharmaceutical contaminants from water? Trends Biotechnol 36(1):30–44

Yang J, Deng X, Xian Q, Qian X, Li A (2014) Allelopathic effect of Microcystis aeruginosa on Microcystis wesenbergii: microcystin-LR as a potential allelochemical. Hydrobiologia 727:65–73

Yang Y, Hou J, Wang P, Wang C, Wang X, You G (2018) Influence of extracellular polymeric substances on cell-NPs hetero aggregation process and toxicity of cerium dioxide NPs to *Microcystis aeruginosa*. Environ Pollut 242:1206–1206

Zablackis E, Santos GA (1986) The carrageenan of Catenella nipae Zanard., a marine red alga. Bot Mar 29:319–322

Zhang YM, Wang HL, Wang XQ, Yang WK, Zhang DY (2006) The microstructure of microbiotic crust and its influence on wind erosion for a sandy soil surface in the Gurbantunggut Desert of Northwestern China. Geoderma 132:441–449

Zheng Y, Xiao R (2016) Overview of microalgal extracellular polymeric substances (EPS) and their applications. Biotechnol Adv 34(7):1225–1244

Zhu LD, Hiltunen E, Antila E, Zhong JJ, Yuan ZH, Wang ZM (2014) Microalgal biofuels: flexible bioenergies for sustainable development. Renew Sust Energ Rev 30:1035–1046

Role of Soil Algae in the Maintenance of Soil Fertility and Soil Health

Satish Kumar Sen, Praveen Jain, Rishi Kumar Singh, and Akhilesh Kumar Singh

Abstract The undiscriminating exploitation of different chemical fertilizers/pesticides for boosting agricultural productivity has given rise to severe issues such as accumulation of poisonous pollutants. Such issues cause ecological imbalance among soil, plant, as well as microbiota, thereby affecting the fertility and health of the soil. However, sustainable agriculture is straightforwardly dependent on soil health. The soil health involves the capability of the soil in establishing ecological quality, sustaining bio-productivity, as well as maintaining health of all living things. Soil health is closely linked to sustainable agriculture as soil microbe variety and activity have been the key constituents of soil health. In this context, the algae (microalgae and cyanobacteria) of soil perform a key functions toward maintaining the nutrient balance (in case of nitrogen, phosphorus, potassium, and carbon) as well as soil health. These maintain the fertility and health of soil in both living and dead conditions. This chapter is an attempt to highlight the major role of soil algae toward the sustenance/maintenance of soil health, fertility, and agricultural productivity.

Keywords Algae · Agriculture microalgae · Cyanobacteria · Fertilizer · Soil · Health · Microbe · Sustainable agriculture

S. K. Sen
Department of Botany, Government VYT PG College, Durg, Chhattisgarh, India

P. Jain
Department of Botany, Government Chandulal Chandrakar Arts and Science PG College Patan, Durg, Chhattisgarh, India

R. K. Singh
School of Biosciences, IMS Institute University Courses Campus, Ghaziabad, Uttar Pradesh, India

A. K. Singh (✉)
Department of Biotechnology, School of Life Sciences, Mahatma Gandhi Central University, Motihari, Bihar, India
e-mail: akhileshsingh@mgcub.ac.in

A. Mishra, A. Varma (eds.), *Soil Algae*, https://doi.org/10.1007/978-981-97-9873-5_9

1 Introduction

The soil involves three crucial components: (i) soil health, (ii) soil quality, and (iii) soil fertility (Fig. 1). The healthy soil serves as a dynamic living platform for providing numerous ecological amenities, for instance, maintaining water characteristic as well as plant productivity, regulating soil nutrient recycling disintegration, and so on. Soil health is an inherent feature of a soil, which alters as a result of preferred exploitation of that soil by human beings (Tahat et al. 2020). Further, a large number of physical, chemical, as well as biological factors have been accountable toward the soil fertility (Cheik and Jouquet 2020; Jiang et al. 2020). There is need to maintain these soil parameters in order to keep it healthy, which is possible by adopting various agricultural strategies. The biological characteristics of soil have been found to be extremely sensitive toward the minor-scale alterations taking place in a soil over physical and chemical characteristics of soil. The deviations of physical, chemical, as well as biological characteristics from the natural status have been found to be regulated through buffering capability as well as resilience of soil. When such alterations are not overcome through soil resilience, soil loses its original state, thereby causing soil degradation. In other words, any substantial alteration in physical, chemical, and biological characteristics away from the range of crop tolerance limits causes soil unhealthy for crop cultivation. Therefore, soil with all characteristics within the acceptable range is considered healthy. Soil health is closely linked to sustainable agriculture as soil microbe diversity and activity have been the key constituents of soil health. The status of soil organisms together with crops requires to be taken into account to ensure the long-standing sustainability of cropping

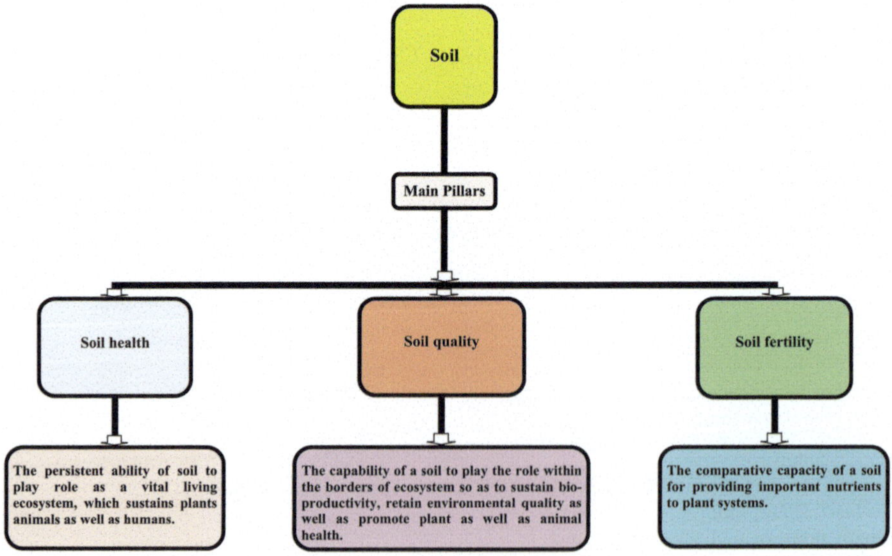

Fig. 1 Core constituents of soil

systems (Pelosi and Römbke 2018; De Vries and Wallenstein 2017). Agricultural sustainability involves the capability of a crop production system to unceasingly yield food without ecological degradation (Tahat et al. 2020). Development and use of sustainable farming with viewpoint of environmental and nutritional balance for feeding human population have been the issues for scientists in the twenty-first century (Prasanna et al. 2013). This is supported by the fact that the human population has been forecasted to reach nine billion by 2050, which can cause a 70% enhancement toward food production demand (FAO 2009). Such higher food demands with the lack of new agronomic land expansion in the future would need doubling crop productions by sustainable approaches (Tahat et al. 2020). Various agronomic soils depict shortage of key nutrients like nitrogen, phosphorus, potassium, and iron. This causes the exploitation of about 200 million metric tons of chemical fertilizers (FAO 2009; Rashid et al. 2016). Further, fulfilling the agricultural demand through rigorous exploitation of chemical fertilizers and pesticides results in land degradation with ecological pollution in numerous agroecosystems. This has imposed a detrimental impact on humans, animals, and aquatic ecosystems (Devarinti 2016). The foundation of sustainable farming has been critically reliant on soil. The soil is the main constituent of crop yield with a complex life reservoir, which requires nutrients toward its long stability as well as productivity (Pankhurst and Lynch 1995). In addition, the soil acts as the biggest sink for organic carbon and aids in the control of greenhouse gas fluxes. Therefore, it is essential to comprehend the soil ecology for ecological sustainability along with reestablishment of habitations (Chen et al. 2007). The function of soil microbiota is to enhance organic substances like carbon, nitrogen, phosphates, and potassium, which in turn increase soil health and fertility (Singh et al. 2011).

Earth acts as a reservoir of greater than one trillion microbial species. The microbes like the bacterial, fungal, actinomycetes, and algal species have been found closely associated as well as grow well in the topmost soil part called biological soil crust (BSC). Such microbes perform a chief function toward enhancing soil fertility together with crop productivity (Manjunath et al. 2016). Among soil microbes, soil algae are photosynthetic microbes that serve as a soil biofertilizer in sustainable organic agriculture. There are four diverse kinds of algae present in soil, that is, green (chlorophyta), blue-green (cyanobacteria), yellow-green (xanthophyta), as well as diatoms (bacillariophyta) (Paul and Clark 1989). Soil algae increase the soil characteristics like texture, organic content, and airing (Ibraheem 2007). They fix nitrogen, improve soil fertility, and play a significant role in plant succession (Hameed 2007). The algae like microalgae have been found to serve as the primary producers in the food chain. The soil microalgal species liberate nutritive and organic substances including active constituents, which can act as food for other organisms such as bacteria and fungi (Abinandan et al. 2019). Further, decomposition of microalgae leads to addition of enhanced organic content along with liberation of macronutrients as well as micronutrients to the soil.

2 Importance of Soil Algae: Microalgae/Cyanobacteria

Microalgae are composed of eukaryotic (green algae and so on) as well as prokaryotic (blue-green algae/cyanobacteria) microscopic algae. They are receiving increasing attention owing to their simple unicellular structure, higher photosynthetic efficiency, capability for heterotrophic mode of nutrition, adaptability toward domestic and industrial wastewater, amenability for metabolic engineering, and prospect to generate valuable co-products. Cyanobacteria are gram-negative prokaryotic photoautotrophic bacteria that depict oxygenic photosynthesis where they produce oxygen in the presence of light energy (Singh et al. 2017; Singh and Mallick 2017). These are present in plenty in various soils. They perform significant functions in building up as well as maintaining soil fertility (Prasanna et al. 2009; Sahu et al. 2012). These depict a substantial positive impact on soil health and plant tolerance against abiotic/biotic stresses (Zulpa et al. 2003; Prasanna et al. 2009). Cyanobacterial species have great adaptability for different types of extreme ecological conditions like temperature (0–70 °C), salinity (~20 dS m^{-1}), high and low pHs, drought, and heavy metals like Cd, Pb, and so on, as detailed in Tahat et al. (2020). The enhanced tolerance of plants toward desiccation and drought stress can be due to the occurrence of modified vegetative cells like spores and akinetes in cyanobacterial cells (Perez et al. 2016). The cyanobacterial species are spread globally and found to increase the growth and development of the crops by the following means (Ibraheem 2007): (i) enhancement of soil fertility, (ii) generation of different bioactive products, and (iii) depicting greater effectiveness in biosorption of heavy metals (bioremediation). Overall, the microalgae play an important role in the management of soil fertility and health. The microalgae-mediated soil management involves the economic benefits (reduced input cost), nutrient cycling, nitrogen (N_2) fixation, phosphorus bioavailability, water storage/movement, environmental protection, prevention of pollution/land degradation, particularly by decreasing the utilization of agro-chemicals, recycling of nutrients, and re-establishment of soil fertility via reclamation.

3 Role of Microalgae/Cyanobacteria in Soil Fertility

Soil fertility involves three interconnected as well as mutually dependent constituents: (i) physical fertility, (ii) chemical fertility, and (iii) biological fertility (Fig. 2). Agricultural lands have been vulnerable to the loss of soil fertility. This is because of the rigorous farming approaches that can cause 30% of the cultivable land to undergo soil degradation. Among microbiota, microalgae alone contribute 27% of the overall biomass in agricultural land (Megharaj 2001). Further, cyanobacteria have been recognized as the most appropriate class of microbes toward soil bioindicator of land utilization (Zancan et al. 2006). Table 1 depicts the microalgal/cyanobacterial potential for enhancing/boosting soil fertility.

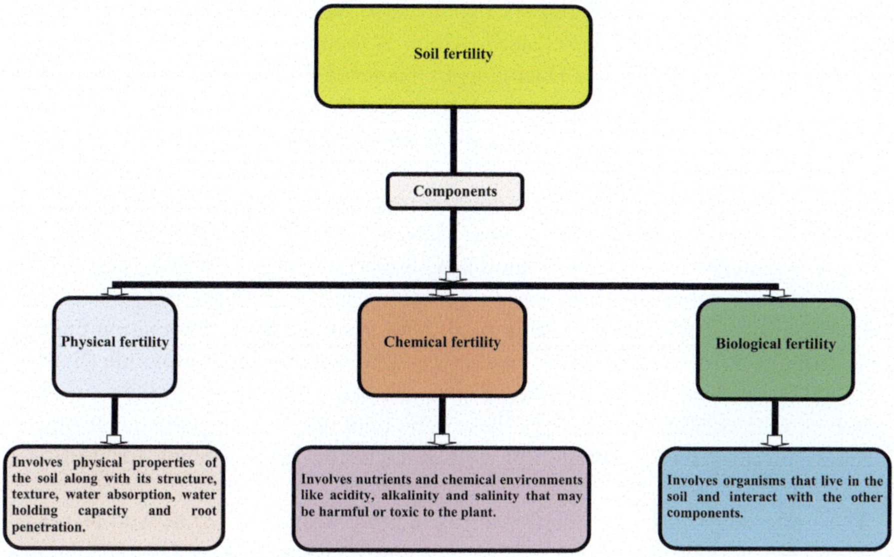

Fig. 2 Main components of soil fertility

Table 1 Potential of microalgae/cyanobacteria for enhancing fertility of soil

S. no.	Innoculants of cyanobacterial/microalgae	Improvement pertaining to soil fertility	References
1.	*Oscillatoria* sp. and so on	Formation of microbial crusts with high carbon/nitrogen content	Acea et al. (2003)
2.	*Nostoc* sp.	Enhancement in the stability of aggregate	Malam Issa et al. (2007)
3.	*Anabaena doliolum* HH209 and so on	Enhancement in carbon/nitrogen Mineralization by promoting Microbial activities	Nisha et al. (2007)
4.	*Scytonema javanicum* and so on	Greater than fivefold enhancement of organic carbon/nitrogen by more than fivefolds	Wang et al. (2009)
5.	*Tribonema minus*, *Choricystis minor*, and so on	Enhancement in organic matter, water drop infiltration, and so on	Lichner et al. (2013)
6.	*Anabaena* sp. and *Providencia* sp.	Iron concentration was enhanced by more than twofold to threefold	Manjunath et al. (2016)
7.	*Chroococcus* sp. and so on	Improvement in microbial biomass carbon	Renuka et al. (2016)
8.	*Chlorella vulgaris*	Increased nitrogen and enzyme activities	Renuka et al. (2016)

The cyanobacteria are the utmost essential N_2-fixing organism in agricultural soils (Rodrigo and Eberto 2007). The majority of cyanobacterial species, which fix N_2, are perhaps heterocystous (Granhall and Henriksson 1969). Nevertheless, non-heterocystous cyanobacterial species are also capable of fixing N_2. The N_2 so fixed via the cyanobacteria is released, which is then re-assimilated by the higher plant systems (Stewart 1970). During cyanobacterial N_2 fixation, the atmospheric N_2 is reduced to ammonia. However, O_2 liberated through the photosynthesis process in the same cell is found to be harmful for N_2 fixation. Therefore, approaches for avoiding O_2 involve temporal separation of N_2 fixation and O_2 evolution (as in various non-heterocyst's species) to spatial separation as well as cellular differentiation into N_2-fixing heterocysts species. Heterocysts are terminally differentiated cells having anaerobic condition internally. Such anaerobic condition permits the process of N_2 fixation to carry on. For instance, the cyanobacterial species like *Nostoc* sp. are capable of colonizing plant root systems. Such association enables the transfer of nutrients as well as metabolites with the help of hormogonia (Svircev et al. 1997). Further, the cyanobacterial species need solar energy for the fixation of carbon as well as N_2. Consequently, these depict not only higher potential as biofertilizers, but also their exploitation will reduce fuel requirement for fertilizer production.

The important soil nutrients after nitrogen are phosphorus (P) as well as potassium (K) for agricultural plants cultivation/crop production. Usually, the topmost layer of soil comprises organic P ranging from 20% to 80% of overall P (Steffens et al. 2010). It has been found that soil microalgal species gather inorganic P and produce polyphosphates. The polyphosphates are the easily accessible type of P for plant systems (Mukherjee et al. 2015). Furthermore, cyanobacterial species can also secrete enzymes, which decompose the soil inorganic P for enhancing its bioavailability to plant systems (Whitton et al. 1991). Under P-deficient environments, cyanobacterial species have been able to dissolve the mineral rock having P by phthalic acid formation (Yandigeri et al. 2011). For instance, the exploitation of cyanobacterial biofilms in non-flooded environments has been found to enhance N_2 fixation and P solubilization. This in turn enhances the growth and development of agricultural plants (Prasanna et al. 2015). Figure 3 summarizes the potential roles of microalgae together with cyanobacteria in boosting fertility apart from improving soil health.

4 Role of Microalgae Including Cyanobacteria in Soil Health

The soil health involves the balance between organisms inside the soil, among soil organisms, as well as their environment (Brady and Weil 1999). Thus, soil health is of utmost importance for the survival of agricultural plants. Apart from role in soil fertility, the algae are also found to play an important role in maintaining soil health and thereby act as an important bio-agent in sustainable agriculture. Owing to the improved impact upon the structure, stability, nutrient content, as well as moisture availability of soil, the algae increase the microhabitat of the soil heterotrophic community. This also assists in the colonization of other organisms like lichens/mosses and establishment as well as survival of annual/vascular plants under dry environments. Also, the enhanced organic carbon content supplied by cyanobacteria results

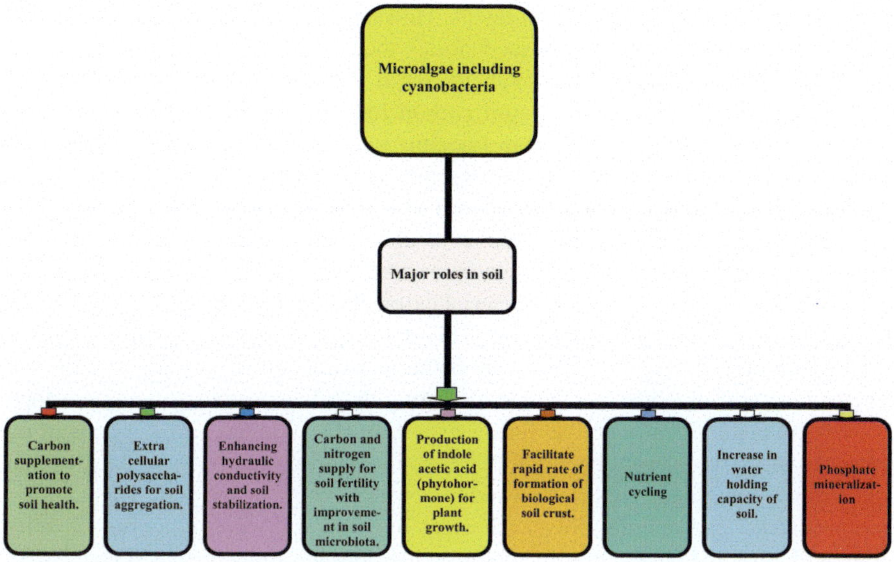

Fig. 3 Various potential roles of microalgae including cyanobacteria in enhancing soil fertility and health

in the supportive growth for huge as well as diverse population of soil microfauna. Hence, well-developed BSC, which is dominated by cyanobacterial species, is utilized as an indicator of soil health including biodiversity. For instance, the collective use of cyanobacterial formulations along with natural substance additives will assist in the quicker rate of formation of BSCs over the natural processes and increase soil stability as well as water-retaining capacity in degraded soils (Park et al. 2017). In other words, microalgal crusts transform soil by enhancing its nutrient content and water availability, inhibiting soil erosion, and stimulating the growth of other plants (Stal 2007). Further, algae have been the source of soil organic content. The death and decomposition of microalgal cells not only yield organic substances but also add macro- and micronutrients to the soil particles. Such organic substances/nutrients are utilized by the plants/microbial communities, which is closely connected with the cation exchange capacity (CEC) of the soil. The CEC is directly linked to the whole amount of cations that a specific soil can hold. This in turn is directly connected with the soil capability to retain plant nutrients. Since organic content has contribution toward cation exchange, the greater the organic content of the soil, the higher the CEC. Thus, organic substance improves the soil fertility and health via supporting the growth of microbes. Apart from this, cyanobacteria are observed to enhance the soil humus content through decomposing and generation of exopolysaccharides.

Cyanobacteria helpful in providing resistance to soil against wind/water erosion through the mechanism of soil particle binding either mechanically or chemically. The mechanically mediated soil particle binding involves intertwined filaments that develop a network in the upper soil layers. However, chemically mediated soil particle binding involves sticky polysaccharides that tie soil particles. Owing to the improved soil stabilization, the cyanobacterial species are accountable for the decrease in water eutrophication, soil carbon losses, CO_2, and other greenhouse gases into the atmosphere, therefore boosting soil carbon contents. On the other hand, cyanobacteria improve water availability via enhancing infiltration as well as retaining surface moisture longer over bare soils (Chamizo et al. 2012). Because of the capability of polysaccharide sheaths to absorb huge water, cyanobacteria are capable of transforming soil characteristics like hydrophobicity, water retention, and so on. For instance, the cyanobacterial extracellular polysaccharides hold water movement in soil through absorption. This develops a soil microenvironment for water regulation (Potts 2001). The cyanobacterial extracellular polysaccharides also act as a reservoir of water storage for managing the adverse environments. Further, this can be metabolically active during rewetting (De Winder 1990). The wetting enhances the size of cyanobacterial colonies by swelling the polysaccharides that occur in the mucilage. This in turn reduces water runoff in the soil. This enables cyanobacterial extracellular polysaccharides to boost soil water-retaining capacity along with diminishing the transevaporation loss into the atmosphere (Mager 2010).

Microalgal and cyanobacterial enzymes are exploited to increase soil fertility. The phosphatases catalyze the breakdown of organophosphates in soil, thereby enabling P available to plant systems. For instance, the formation of cyanobacterial alkaline phosphatase in P-deficient environments was found to be greater. This enzyme hydrolyzes the phosphate residues of polyphosphate (Grzesik and Romanowska-Duda 2015). This subsequently facilitates in boosting the nitrogenase activity enzyme (Reed et al. 2011). Likewise, cyanobacterial hydrolytic enzymes depict fungicidal activities like protease, carboxymethyl cellulase, cellobiase, and filter paperase. These enzymes have been accountable for the speedy nutrient recycling of carbon, nitrogen, and phosphate that are essential for the growth and development of plant system (Burns et al. 2013). A substantial enhancement was observed pertaining to the activities of dehydrogenase, urease, as well as phosphatase with cyanobacterial inoculants. Such enhanced enzymatic activities caused enhanced mineralization of organic substances (Rao and Burns 1990). On the other hand, phytohormones are essential biomolecules usually present in plants and microbes like microalgae and so on. These phytohormones have been found to regulate the cellular processes. They aid in overcoming the abiotic stress in plants and retain growth. The cyanobacteria/microalgae that form a symbiotic association with plants produce hormones like auxins, cytokinin, abscisic acid, gibberellins, as well as ethylene (Hartung and Gimmler 1994; Sergeeva et al. 2002; Tarakhovskaya et al. 2007). However, the synthesis of microalgal phytohormones has been found to be reliant on the plant amino acids as precursor molecules (Zhao 2012). For instance, formation of cyanobacterial auxins was merely detected with the tryptophan amino acid accumulation excreted by the plants (Sergeeva et al. 2002). Usually, the phytohormones synthesized by cyanobacteria are transported into the plants by their roots along with other nutrients (Mazhar et al. 2013). The phytohormones also have the capability to detoxify/modify the lethal substances, which prevent plants/algal growth. For instance, phytohormones aid in modifying the phytotoxicity of heavy metal and stop algal growth (Piotrowska-Niczyporuk et al. 2012). Figure 3 throws light on the potential roles of microalgae/cyanobacteria in improving the health of the soil apart from maintaining soil fertility.

5 Conclusion

The enhancement in global population leads to an enhanced demand for food production. This may be merely achieved by boosting crop yields. One potential approach involves the development as well as exploitation of sustainable agriculture that is based on biological systems having well-known activity in nutrition, fortification, as well as growth promotion. Among these biological systems, microalgae/cyanobacteria have been attaining specific interest. These are the source of important nutrients/products having diverse bioactivities with ability to considerably enhance agricultural productivities. They also play a key function in improving the soil health/fertility needed for sustained agriculture. Microalgal/cyanobacterial bioactive substances can control many plant response mechanisms like nutrition, protection (toward biotic and abiotic stresses), and growth promotion of plants. Hence, the exploitation of microalgae/cyanobacteria can be a sustainable as well as feasible substitute to chemical fertilizers, pesticides, and growth promoters. In spite of these advantages, the exploitation of microalgae/cyanobacteria in agriculture is still restricted as their biomass production and processing prices are still reasonably high. Also, a deep understanding of the interactions among these microalgae/cyanobacteria with higher plants is required. However, one promising way is to promote the biorefinery model. This model will permit the cost-effective exploitation of microalgae/cyanobacteria in the agricultural field.

References

Abinandan S, Subashchandrabose SR, Venkateswarlu K, Megharaj M (2019) Soil microalgae and cyanobacteria: the biotechnological potential in the maintenance of soil fertility and health. Crit Rev Biotechnol 39(8):981–998

Acea MJ, Prieto-Fernández A, Diz-Cid N (2003) Cyanobacterial inoculation of heated soils: effect on microorganisms of C and N cycles and on chemical composition in soil surface. Soil Biol Biochem 35:513–524

Brady NC, Weil RR (1999) The nature and properties of soils, 12th edn. Prentice Hall Inc, Upper Saddle River

Burns RG, DeForest JL, Marxsen J, Sinsabaugh RL, Stromberger ME, Wallenstein MD, Weintraub MN, Zoppini A (2013) Soil enzymes in a changing environment: current knowledge and future directions. Soil Biol Biochem 58:216–234

Chamizo S, Cantón Y, Lázaro R, Solé-Benet A, Domingo F (2012) Crust composition and disturbance drive infiltration through biological soil crusts in semiarid ecosystems. Ecosystems 15:148–161

Cheik S, Jouquet P (2020) Integrating local knowledge into soil science to improve soil fertility. Soil Use Manag 36:561–564

Chen MM, Zhu YG, Su YH, Chen BD, Fu BJ, Marschner P (2007) Effects of soil moisture and plant interactions on the soil microbial community structure. Eur J Soil Biol 43:31–38

De Vries FT, Wallenstein MD (2017) Below- ground connections underlying above- ground food production: a framework for optimising ecological connections in the rhizosphere. J Ecol 105:913–920

De Winder GBM (1990) Ecophysiological strategies of drought-tolerant phototrophic micro-organisms in dune soils. PhD thesis, University of Amsterdam, Amsterdam, p 94

Devarinti SR (2016) Natural farming: eco-friendly and sustainable? Agrotechnology 5:147

FAO (2009) Food and Agriculture Organization of the United Nations. Rome: How to feed the World in 2050. Available from www.fao.org/fileadmin/templates/wsfs/docs/expert_paper/How_to_Feed_the_World_in_2050.pdf

Granhall U, Henriksson E (1969) Nitrogen-fixing blue-green algae in Swedish soils. Oikos 20:175–178

Grzesik M, Romanowska-Duda Z (2015) Ability of cyanobacteria and green algae to improve metabolic activity and development of willow plants. Pol J Environ Stud 24:1003–1012

Hameed SMM (2007) Studies on nitrogen fixing cyanobacteria. MSc Thesis, Botany Department, Faculty of Science, Beni-Suef University, Beni-Suef

Hartung W, Gimmler H (1994) A stress physiological role for abscisic acid in lower plants. Prog Bot 55:157–173

Ibraheem IBM (2007) Cyanobacteria as alternative biological conditioners for bioremediation of barren soil. Egypt J Phycol 8:99–116

Jiang N, Tang C, Hata T, Courcelles B, Dawoud O, Singh DN (2020) Bio- mediated soil improvement: the way forward. Soil Use Manag 36:185–188

Lichner L, Hallett PD, Drongova Z, Czachor H, Kovacik L, Mataix-Solera J, Homolak M (2013) Algae influence the hydrophysical parameters of a sandy soil. Catena 108:58–68

Mager DM (2010) Cyanobacterial soil crusts: analysing resilience in Kalahari sand soils. In: Degenovine KM (ed) Semi-arid environments: agriculture, water supply and vegetation. NOVA Scientific Publishers, Inc, Hauppauge

Malam Issa O, Défarge C, Le Bissonnais Y, Marin B, Duval O, Bruand A, D'Acqui LP, Nordenberg S, Annerman M (2007) Effects of the inoculation of cyanobacteria on the microstructure and the structural stability of a tropical soil. Plant Soil 290:209–219

Manjunath M, Kanchan A, Ranjan K, Venkatachalam S, Prasanna R, Ramakrishnan B, Hossain F, Nain L, Shivay YS, Rai AB, Singh B (2016) Beneficial cyanobacteria and eubacteria synergistically enhance bioavailability of soil nutrients and yield of okra. Heliyon 2:e00066

Mazhar S, Cohen JD, Hasnain S (2013) Auxin producing non-heterocystous cyanobacteria and their impact on the growth and endogenous auxin homeostasis of wheat. J Basic Microbiol 53:996–1003

Megharaj M (2001) Healthy levels of soil algae lift plant growth. In: Farming ahead, December edn. CSIRO, Canberra

Mukherjee C, Chowdhury R, Ray K (2015) Phosphorus recycling from an unexplored source by polyphosphate accumulating microalgae and cyanobacteria -a step to phosphorus security in agriculture. Front Microbiol 6:1421

Nisha R, Kaushik A, Kaushik CP (2007) Effect of indigenous cyanobacterial application on structural stability and productivity of an organically poor semi-arid soil. Geoderma 138:49–56

Pankhurst CE, Lynch JM (1995) The role of soil microbiology in sustainable intensive agriculture. Adv Plant Pathol 11:229–247

Park CH, Li XR, Zhao Y, Jia RL, Hur JS (2017) Rapid development of cyanobacterial crust in the field for combating desertification. PLoS One 12:e0179903

Paul EA, Clark FE (1989) Soil microbiology and biochemistry. Academic Press, Inc, San Diego

Pelosi C, Römbke J (2018) Enchytraeids as bioindicators of land use and management. App Soil Ecol 123:775–779

Perez R, Forchhammer K, Salerno G, Maldener I (2016) Clear differences in metabolic and morphological adaptations of akinetes of two *Nostocales* living in different habitats. Microbiology 162:214–223

Piotrowska-Niczyporuk A, Bajguz A, Zambrzycka E, Godlewska-Żyłkiewicz B (2012) Phytohormones as regulators of heavy metal biosorption and toxicity in green alga *Chlorella vulgaris* (Chlorophyceae). Plant Physiol Biochem 52:52–65

Potts M (2001) Desiccation tolerance: a simple process? Trends Microbiol 9:553–559

Prasanna R, Jaiswal P, Nayak S, Sood A, Kaushik BD (2009) Cyanobacterial diversity in the rhizosphere of rice and its ecological significance. Indian J Microbiol 49:89–97

Prasanna R, Sharma E, Sharma P, Kumar A, Gupta V, Pal RK, Shivay YS, Nain L (2013) Soil fertility and establishment potential of inoculated cyanobacteria in rice crop grown under non-flooded conditions. Paddy Water Environ 11:175–183

Prasanna R, Babu S, Bidyarani S, Kumar A, Triveni S, Monga D, Mukherjee AK, Kranthi S, Gokte-Narkhedkar N, Adak A, Yadav K, Nain L, Saxena AK (2015) Prospecting cyanobacteria-fortified composts as plant growth promoting and biocontrol agents in cotton. Exp Agric 51:42–65

Rao DLN, Burns RG (1990) The effect of surface growth of blue-green algae and bryophytes on some microbiological, biochemical, and physical soil properties. Biol Fertil Soils 9:239–244

Rashid MI, Mujawar LH, Shahzad T, Almeelbi T, Ismail IMI, Oves M (2016) Bacteria and fungi can contribute to nutrients bioavailability and aggregate formation in degraded soils. Microbiol Res 183:26–41

Reed SC, Cleveland CC, Townsend AR (2011) Functional ecology of free-living nitrogen fixation: a contemporary perspective. Annu Rev Ecol Evol Syst 42:489–512

Renuka N, Prasanna R, Sood A, Ahluwalia AS, Bansal R, Babu S, Singh R, Shivay YS, Nain L (2016) Exploring the efficacy of wastewater-grown microalgal biomass as a biofertilizer for wheat. Environ Sci Pollut Res 23:6608–6620

Rodrigo V, Eberto N (2007) Seasonal changes in periphyton nitrogen fixation in a protected tropical wetland. Biol Fertil Soils 43:367–372

Sahu D, Priyadarshani I, Rath B (2012) Cyanobacteria-as potential biofertilizer. CIBTech J Microbiol 1:20–26

Sergeeva E, Liaimer A, Bergman B (2002) Evidence for production of the phytohormone indole-3-acetic acid by cyanobacteria. Planta 215:229–238

Singh JS, Pandey VC, Singh DP (2011) Efficient soil microorganisms: a new dimension for sustainable agriculture and environmental development. Agric Ecosyst Environ 140:339–353

Singh AK, Sharma L, Mallick N, Mala J (2017) Progress and challenges in producing polyhydroxyalkanoate biopolymers from cyanobacteria. J Appl Phycol 29:1213–1232.

Singh AK, Mallick N (2017) Advances in cyanobacterial polyhydroxyalkanoates production, FEMS Microbiol Lett 364:fnx189.

Stal LJ (2007) Cyanobacteria. In: Seckbach J (ed) Algae and cyanobacteria in extreme environments, Cellular origin, life in extreme habitats and astrobiology, vol 11. Springer press, pp 659–680

Steffens D, Leppin T, Luschin-Ebengreuth N, Min Yang Z, Schubert S (2010) Organic soil phosphorus considerably contributes to plant nutrition but is neglected by routine soil-testing methods. J Plant Nutr Soil Sci 173:765–771

Stewart WDP (1970) Algal fixation of atmospheric nitrogen. Plant Soil 32:555–588

Svircev Z, Tamas I, Nenin P, Drobac A (1997) Co-cultivation of N_2-fixing cyanobacteria and some agriculturally important plants in liquid and sand cultures. Appl Soil Ecol 6:301–308

Tahat MM, Alananbeh KM, Othman YA, Lescovar DI (2020) Soil health and sustainable agriculture. Sustainability 12:4859

Tarakhovskaya ER, Maslov YI, Shishova MF (2007) Phytohormones in algae. Russ J Plant Physiol 54:163–170

Wang W, Liu Y, Li D, Hu C, Rao B (2009) Feasibility of cyanobacterial inoculation for biological soil crusts formation in desert area. Soil Biol Biochem 41:926–929

Whitton BA, Grainger SL, Hawley GR, Simon JW (1991) Cell-bound and extracellular phosphatase activities of cyanobacterial isolates. Microb Ecol 21:85–98
Yandigeri MS, Meena KK, Srinivasan R, Pabbi S (2011) Effect of mineral phosphate solubilization on biological nitrogen fixation by diazotrophic cyanobacteria. Indian J Microbiol 51:48–53
Zancan S, Trevisan R, Paoletti MG (2006) Soil algae composition under different agro-ecosystems in North Eastern Italy. Agric Ecosyst Environ 112:1–12
Zhao Y (2012) Auxin biosynthesis: a simple two-step pathway converts tryptophan to indole-3-acetic acid in plants. Mol Plant 5:334–338
Zulpa G, Zaccaro MC, Boccazzi F, Parada JL, Storni M (2003) Bioactivity of intra and extracellular substances from cyanobacteria and lactic acid bacteria on "wood blue stain" fungi. Biol Control 27:345–348

Toxic Compounds of Free-Living and Symbiotic Soil Algae

Maya P. Stoyneva-Gärtner, Georg Gärtner, and Blagoy Uzunov

Abstract The chapter provides a comprehensive review of the versatile toxic compounds produced by soil algae. Toxin production was proved chemically or genetically for at least 67 species and 3 varieties from 41 genera of soil algae (34 cyanoprokaryotes and 7 diatoms), from which more than 216 strains have been investigated. Four more genera are supposed to be toxigenic according to their phylogenetic relations or due to the existence of toxic aquatic counterparts. Soil algae have yielded a plethora of toxic compounds with remarkable structural diversity: altogether 430 toxins (including variants and the odoriferous nontoxic geosmin) from 11 main classes and 77 subclasses were isolated with antimicrobial or cytotoxic activity, 320 of which from 9 classes and 56 subclasses were novel. They were discovered in 18 cyanoprokaryote genera, and the most prolific producers were heterocytous *Nostoc*, *Fischerella*, *Scytonema*, and *Hapalosiphon*, in particular. Along with ecotoxicological concerns and potentially hazardous effects on human and ecosystem health, some of these toxins are of potentially significant importance for application in pharmacology, medicine, cosmetics, and biotechnology.

Keywords Algal toxins · Cyanobacteria · Cyanotoxins · Edaphic algae · Geosmin · Terrestrial algae

M. P. Stoyneva-Gärtner · B. Uzunov (✉)
Department of Botany, Faculty of Biology, Sofia University "St. Kliment Ohridski", Sofia, Bulgaria
e-mail: mstoyneva@uni-sofia.bg; buzunov@uni-sofia.bg

G. Gärtner
Institut für Botanik der Universität Innsbruck, Innsbruck, Austria
e-mail: georg.gaertner@uibc.ac.at

A. Mishra, A. Varma (eds.), *Soil Algae*, https://doi.org/10.1007/978-981-97-9873-5_10

1 Introduction

The term "soil algae" ("terrestrial algae," or also "edaphic algae") embraces diverse algae, which inhabit the terrestrial environments, occurring commonly on soil surface or beneath it, penetrating its deeper layers (Metting 1981; Ettl and Gärtner 2014; Gärtner et al. 2021). Since the first description of soil *Nostoc commune* by Dillenius (1741) and floristic studies of Gräbner (1895), much work has been done to elucidate the biodiversity of the peculiar edaphic environment with increasing number of genera and species discovered (Ettl and Gärtner 2014). By number of recorded species, two main groups can be outlined: (1) blue-green algae (known also as cyanobacteria due to their prokaryotic character), united in the division Cyanoprokaryota, and (2) green algae from eukaryotic divisions Chlorophyta and Streptophyta, among which species of Chlorophyta prevailed (Metting 1981; Mühlsteinová and Hauer 2013; Ettl and Gärtner 2014; Whitford and Duval 2019; Hofbauer and Gärtner 2021; Gärtner et al. 2021). Apart from being primary producers, terrestrial algae can release exopolysaccharides and often develop in crusts, thus contributing to enhancement of soil stability and water retention, at the same time providing microhabitats for other organisms (Metting 1981; Mazor et al. 1996; Yeager et al. 2004; Warren et al. 2019). In addition, Cyanoprokaryota are unique among photosynthetic organisms in their ability to fix atmospheric nitrogen (Whitford and Duval 2019). Therefore, the enrichment by blue-green algae of arable lands was considered beneficial, and different approaches for their introduction in soils have been developed (e.g., Venkataraman 1961, 1972; Joshi et al. 2017). However, over the years, lots of evidence have accumulated that algae can produce toxic compounds (phycotoxins) which are chemically diverse and have different targets of action, with the possibility to affect dangerously human and ecosystem health (Evangelista et al. 2007; Meriluoto et al. 2017). For a long time, algal toxicity has been associated with aquatic habitats with growing knowledge of the significant impact of global warming and anthropogenic pressure on their rising global spread in freshwaters and coastal zones (Meriluoto et al. 2017; Janssen 2019). But, over time, a lot of evidence has accumulated on the presence of toxigenic free-living and symbiotic algae and different toxins in other extreme and aeroterrestrial habitats, including soils (Gärtner et al. 2021; Hofbauer 2021; Ivanov et al. 2021).

Nowadays, the diversity of toxic compounds and their producers in soils is of primary importance in view of the current population growth, the scarcity of arable land and the worldwide problem of degraded soils, whose global estimates vary between 1 and 6 billion hectares, the global estimates for which varied between 1 and 6 billion ha (Campbell 2011; Gibbs and Salmon 2015). Moreover, the concernment about soil contamination is growing since many studies reported the transport of phycotoxins in crops, with different pathways of penetration through roots or leaves (Bouaïcha and Corbel 2016; Gärtner et al. 2021). The ubiquity of soil algae makes them important for wide human exposure, and the understanding of their effects on humans and other organisms is gaining stronger attention (Cox et al. 2005; Gärtner et al. 2021; Hofbauer 2021). Although a few studies were conducted, they demonstrated the adverse effects of soil phycotoxins and associated health risks after inhalation of dust, which contained toxic algae from soil algal crusts or after drinking water, overlying such crusts (e.g., Cox et al. 2009; Metcalf et al. 2012; Chatziefthimiou et 2014; Facciponte et al. 2018; Nowruzi et al. 2012, 2018). Some soil algae, temporarily present in the air (airborne algae), such as Chlorophyta, may be allergenic, and some of them, such as Cyanoprokaryota, may not only be toxic but also allergenic (Hofbauer 2021). In terrestrial ecosystems, some lichens containing toxic algae can be a source of hepatotoxins for grazing animals, and in this way the toxins can enter the food chain (Kaasalainen et al. 2012; Ivanov et al. 2021). On the other side, the interest in terrestrial algae and their toxicity grows along with the rising safe requirements and application of soil algae as conditioners or fertilizers in new eco-friendly processes, or due to their direct utilization or implementation in different novel products used in humans' life (Berry 2008; Ersmark et al. 2008; Prakash and Nikhil 2014; Ditmann et al. 2015; Mitchell et al. 2017; Singh et al. 2017; Demay et al. 2019; Śliwińska-Wilczewska 2019; Stoyneva-Gärtner et al. 2020; Gärtner et al. 2021; Mazur-Marzec et al. 2021; Pagarete et al. 2021, among the many others). Therefore, this chapter aims to provide a summary of the knowledge on free-living and symbiotic soil algae, capable of toxin production and of their versatile toxic compounds.

With the lack of a single approach to the classification of phycotoxins, in this review, only chemically defined compounds are included, covering not only the well-known but also some unusual and less-studied toxins. In order to keep the text within reasonable borders, and considering the enormous amount of information given in the chapter references, we do not provide all details on chemical characteristics of the toxic compounds, methods of their extraction, exact dozes, values, and modes of their action, as well as data on their usage in human life. It is necessary to stress also that one and the same algal species can have both nontoxic and toxic strains and that strains of different toxicity may coexist in the same environment (Meriluoto et al. 2017). In addition, it became known that some species do not produce a single class of compounds, and some of their strains can produce simultaneously different groups of secondary metabolites and their congeners (Welker and Döhren 2006; Janssen 2019; Monteiro et al. 2021). Moreover, one and the same strain can produce different variants of the same compound or different classes of chemicals. To exemplify, we provide data on two heterocytous filamentous cyanoprokaryotes: (1) in culture, the single toxigenic *Nostoc* strain UK18, isolated from 29 different Nostoc 16S rRNA gene haplotypes from a lichen *Peltigera leucophlebia*, generated 19 microcystin (MC) variants (Kaasalainen et al. 2009); (2) a single strain of *Scytonema wolleanum* (as *S. mirabile* BY-8-1), isolated from the roof of abandoned house at Mt Tantalus, Oahu, Hawaii, and then mass cultured in laboratories, generated 38 phycotoxins (cyanotoxins) from 4 distinct classes of compounds (Patterson et al. 1994). However, in this review, for clearer presentation, each toxin and the relevant producing genera, species or strains are enlisted separately. Algal nomenclature follows AlgaeBase (Guiry and Guiry 2021) with additional checking in DiatomBase (Kociolek et al. 2022), with synonyms provided at the first appearance in the text.

2 Results

2.1 *Toxic Amino Acids*

The neurotoxic diamino acid, β-methylamino-L-alanine, commonly abbreviated as BMAA, was isolated for the first time from seeds of *Cycas circinalis* (Vega and Bell 1967), but afterward, its production by the *Cycas* root-symbiont heterocytous filamentous genus *Nostoc* (Cyanoprokaryota) was proved (Cox et al. 2003, 2005; Murch et al. 2004; Meriluoto et al. 2017). This led to examination and finding of BMAA in two more symbiotic heterocytous cyanoprokaryotes, namely *Trichormus azollae* from the leaves of the floating fern *Azolla filiculoides*, which can develop amphibially on wet soils (Draganov and Stoyneva 1992; Stoyneva 1995) and is commonly used in Southeast Asia as green manure of the rice fields, and *Nostoc* sp. from the petiolar tissues of the large-leafed angiosperm *Gunnera kauaiensis* (Cox et al. 2003). Subsequent analyses covered other 11 symbiotic and 4 free-living strains of *Nostoc*, together with 26 strains of other free-living cyanoprokaryotes from 20 genera of all main orders of Cyanoprokaryota, collected from diverse aquatic and terrestrial habitats (Cox et al. 2005). They revealed that 73% of symbiotic and 97% of the free-living strains produced BMAA, and moreover, BMAA was produced in all main cyanoprokaryote orders and in 95% (20/21) of the genera tested. Considering soils, besides *Nostoc*, BMAA-producing free-living strains were found in two heterocytous genera *Chlorogloeopsis* (*C. fritschii* PCC 6912, collected from an Indian soil) and *Calothrix* (strain PCC 7103) and in two non-heterocytous genera *Phormidium* (no strain number indicated) and *Plectonema* (strain PCC 73110) (Cox et al. 2005). The habitats for the last three strains were mentioned as "unknown," but since these genera have been reported from both soils and waters, they were enlisted among potential aeroterrestrial toxin producers (Gärtner et al. 2021). Among the studied 11 symbiotic strains, BMAA was discovered in the *Nostoc* strains from lichen (*Peltigera*), hornwort (*Anthoceros*), two other cycads (*Cycas*, *Encephalartos*), and in the flowering plants (*Gunnera*) but was not detected in *Nostoc* LBG1 and PCC 73102, isolated from the liverwort *Blasia* and from the cycad *Macrozamia*, respectively (Cox et al. 2005). Later, BMAA and its three isomers (commonly abbreviated as DAB, BAMA, and AEG) were found in cultures obtained from "salt" crusts, collected in dry, vegetation-free undrained desert basins of northwest Qatar, formed by diverse cyanoprokaryotes (predominantly non-heterocytous filamentous *Microcoleus* and *Phormidium*, but also coccal *Chroococcus* and *Pleurocapsa*), and a range of possible producers was speculated (Cox et al. 2009; Metcalf et al. 2012, 2015). It is noteworthy to mention that in part of these cultures in order to confine the BMAA production to cyanoprokaryotes, eukaryotic growth was inhibited through addition of nitrogen and cycloheximide to BG11 medium (Cox et al. 2009). Conducted tests proved the neurotoxic character of both isomers BMAA and DAB, and the weak toxicity of AEG, but the toxicity of BAMA awaits its investigation (Metcalf et al. 2015). Production of BMAA was proved for lab-cultured *Leptolyngbya* PCC 73110 and for field *Leptolyngbya* sp., forming terrestrial mats on the Swedish west coast (Jiang et al. 2013, 2014). For some years blue-green algae were thought to be the sole producers of BMAA and its isomers, but since 2014 their production by algae from two other taxonomic groups, such as dinoflagellates (Pyrrhophyta) and diatoms (Bacillariophyceae, Ochrophyta) from marine and freshwater habitats, has been reported (Jiang et al. 2014; Lage et al. 2014, 2016; Metcalf et al. 2021a; Violi et al. 2019). Among the diatoms, discovered as a novel source for BMAA, AEG, and DAB, are *Achnanthes*, *Aulacoseira* (Syn. *Melosira* p.p.), *Cyclotella* (except for AEG), *Fragilaria*, *Navicula*, and *Tabellaria*, which have been recorded with different frequency in soil samples (Ettl and Gärtner 2014). Therefore, in this review, we enlist them among potential toxigenic soil algae. However, it is worth mentioning that BMAA production by diatoms is notably less when compared with cyanoprokaryotes (Violi et al. 2019).

Domoic acid is a water-soluble cyclic amino acid, known as a kainic acid-type neurotoxin, affecting molluscan shellfish, birds, marine mammals, and humans, and responsible for so-called amnesic shellfish poisoning (Evangelista et al. 2007; Dhar et al. 2015). For a long-time, diatoms of *Pseudo-nitzschia* genus were most frequently associated with domoic acid production, but Bates (2000) suggested such ability in another diatom genus, *Amphora*. Further studies, including molecular detection, revealed that domoic acid can be produced by *Amphora coffeaeformis*, currently synonymized as *Halamphora coffeaeformis* (Kociolek et al. 2022). Since this species had often been misidentified, the identity of the toxigenic species was considered uncertain (Sala et al. 1998). *Amphora coffeaeformis* s.l. was recorded in hot springs, and species from *Amphora* were frequently found in Arctic soil samples, which led to the including of the genus in the list of toxigenic aeroterrestrial and extremophilic algal genera (Gärtner et al. 2021).

2.2 Toxic Isocyanides

Isocyanides (called also isonitriles or carbilamines) are organic compounds with a functional group, in which carbon is bounded to a single nitrogen atom, and are considered building blocks in large multimolecular assemblies (Scheuer 1992; Patil et al. 2020). Isocyanides have been known for more than 200 years, but only in 1957 the first report on the isolation of a naturally occurring isocyanide, xanthocillin, from the mold *Penicillium notatum* was published (Scheuer 1992). Subsequently they have been found in different organisms, and it has been thought that most of the marine isocyanides are terpenoid (characterized by unpleasant odor and some ichthyotoxic properties), but some of the terrestrial ones originate from α-amino acids (Scheuer 1992). Nowadays over 26,000 isocyanides are known, but yet their toxicity is less investigated (Patil et al. 2020). Six novel mildly cytotoxic and antimicrobial isocyanides, named mirabilene isonitriles A-F, have been isolated from *Scytonema wolleanum* (strain *S. mirabile* BY-8-1) (Carmeli et al. 1990b).

2.3 Toxic Cyanopeptides and Polyketides

Cyanopeptides comprise a large group of natural peptides, mostly non-ribosomal oligopeptides, produced by Cyanoprokaryota (Janssen 2019). Many of these metabolites exhibit hepatoxicity and neurotoxicity and also a broad spectrum of other biological activities of ecotoxicological concern (Radau 2005; Janssen 2019). Although the screening of 146 axenic strains from the Pasteur Culture Collection (PCC) demonstrated that the majority (75%) of them, and the filamentous ones specifically, contained non-ribosomal peptide synthetase genes (Christiansen et al. 2001), studies of cyanopeptides in aeroterrestrial and extremophilic algae are scarce (Gärtner et al. 2021).

Microcystins (MCs) are considered to be the most often recorded toxins, produced by Cyanoprokaryota, and are the most widely studied metabolites of these algae, especially in aquatic habitats (Janssen 2019; Svirčev et al. 2019; Uzunov et al. 2021b). Although these heptapeptides traditionally were classified as hepatotoxins, evidence of their potential to possess adverse neurological activities was collected (Meriluoto et al. 2017; Metcalf et al. 2021b). Regarding soil algae, most data demonstrated or presumed the microcystin (MC) production in heterocytous filamentous strains of genera *Fischerella*, *Hapalosiphon*, *Nostoc*, and *Westiellopsis*, as shown below. The first MC, presented by the type LA in particular, was found in a strain of *Hapalosiphon hibernicus* BZ-3-1 from Hawaiian soil (Prinsep et al. 1992b). MC production was recorded also in some strains of *Nostoc*, symbiotic in terrestrial lichens, liverworts, and flowering plants (Oksanen et al. 2004; Cox et al. 2005; Kaasalainen et al. 2009; Fidor et al. 2019). For example, from *Nostoc* sp. IO-102-I, a symbiont collected from the bipartite lichen *Protopannaria pezizoides*, at Sysmä, Finland, three variants of MC-LR (i.e., [MeAsp^3, DMAdda^5, Mdha^7], [Asp^3, ADMAdda^5, Mdha^7], and [MeAsp^3, ADMAdda^5, Mdha^7]) and two variants of MC-XR (i.e., [MeAsp^3, ADMAdda^5, Mdha^7] and [MeAsp^3, ADMAdda^5, Mdha^7]) were isolated (Oksanen et al. 2004), whereas, as it was mentioned above, 19 variants were discovered in its closely related strain UK18, collected at a location of 500 km apart, in the commune of Autti within the area of reindeer husbandry in northern Finland (Kaassalainen et al. 2009). Among them, three highly toxic MCs were detected, [ADMAdda^5]MC-LR being the major one, accounting for ca. 60% of the total intensity of the MC peaks, followed by [DMAdda^5]MC-LR and [DAsp^3, ADMAdda^5]MC-LR, which comprised ca. 20%. Despite the differences in the composition, the total MC concentration of both strains was almost similar (Oksanen et al. 2004; Kaasaalainen et al. 2009). Besides the fact that Kaasalainen et al. (2009) for the first time provided direct evidence of production in situ of MCs in lichens, they proved the presence of the MC synthetase gene *mcyE* in the studied strain. Subsequent screening of 803 lichen specimens from five continents revealed the presence of *mcyE* or the toxin itself in 12% of them (Kaasalainen et al. 2012). Moreover, they demonstrated very high diversity of the *mcyE* gene (30 diverse genotypes of 98 sequences), and 52 chemical variants of MCs were found, many of which being only rarely reported from free-living cyanoprokaryotes (Kaasalainen et al. 2012). In the same study, lichens containing toxins were found from all geographical regions, and 20 MC variants with intensity >10% were discovered at least in one sample. In lichens of the *Nephroma* lichen guild, the most common variant was [ADMAdda[5]] MC-RR, whereas in other lichens the most common main variant was [Leu[1]] MC-LR (Kaasalainen et al. 2012). By contrast, among free-living blue-green algae these two variants were infrequently found (Kaasalainen et al. 2012). Free-living *Nostoc* from hydroterrestrial Antarctic mats was supposed as a producer of MC-LR, MC-FR, and MC-RR and of eight novel MC congeners with specific modifications containing glycine (Wood et al. 2008; Puddick et al. 2015), and *Nostoc paludosum* from Arctic biocrusts was presumed as responsible for the production of MC-LR (Chrapusta et al. 2015). MCs in concentration of 1.5 ppb were detected in *Nostoc microscopicum* CCALA #124 by enzyme-linked immunosorbent assay (ELISA) (Batsalova et al. 2016). Biosynthetic gene clusters (BGC) coding MC production (*mcy*) were found in *Fischerella* sp. PCC 9339 (Shih et al. 2013; Micaleff et al. 2015). Among non-heterocytous cyanoprokaryotes, two genera, *Microcoleus* and *Phormidium*, were supposed as MC-producing agents in algal soil mats, collected from salt flats, dry river beds, and hot deserts of Qatar (Cox et al. 2009; Metcalf et al. 2012). The sole supposed MC producer from the coccal cyanoprokaryotes was *Chroococcus* sp. from the Arctic soil crusts (Chrapusta et al. 2015).

A recently published review on toxic algae from aeroterrestrial and extremophilic environments suggested that future studies will reveal cyanotoxins in soils from hypersaline environments (Gärtner et al. 2021). Almost simultaneously appeared the first record of a new toxin producer, the boring *Perforatum tunneli* gen. et sp. nov. from the wind-tidal sand flats of the worldwide largest laguna, Laguna Madre (USA), connected with Laguna Tamaulipas (Mexico) (Zimba et al. 2021). The strain of this non-heterocytous filamentous species produced different toxins, already known from marine and freshwater algae, one of which was the MC variant [Asp[3], Dha[7]] MYST-LY (Zimba et al. 2021).

Nodularins (NODs) are cyclic pentapeptides, which represent the second group of hepatotoxins, and are commonly released in brackish and freshwater environments by the heterocytous filamentous genus *Nodularia* (Meriluoto et al. 2017). Although in low amounts, nodularin (NOD) was found in free-living *Nostoc* sp. CENA543 from an extreme environment in Brazil (benthos of saline-alkaline lake in Pantanal wetland area) (Jokela et al. 2017). The synthesis of NOD and [L-Har2]-NOD was confirmed by genetic and chemical methods in two terrestrial symbiotic strains of *Nostoc*, such as *Macrozamia serpentina* 73.1 and *Macrozamia riedlei* 65.1 (Gehringer et al. 2012). During the abovementioned screening of 803 lichen specimens from five continents, three different NOD variants (with NOD-R being the main among them) were detected in an Argentinian and two Kenyan specimens (Kaasalainen et al. 2012). Search for the causative toxigenic agent showed that cyanoprokaryote 16SrRNA or *mcyE* gene sequences from lichens, which produce NOD, were spread only among lichen-symbiotic *Nostoc* sequences (Kaasalainen et al. 2012). All these data pointed *Nostoc* as a rare non-*Nodularia* potential NOD-producer (Fidor et al. 2019; Gärtner et al. 2021).

Interestingly, cyanoprokaryotes can produce not only hepatotoxins but also their antitoxins, nostocyclopeptides (NCPs) and nostopeptolides (NPs), which can block the transport of NODs and MCs to hepatocytes (Jokela et al. 2010; Fidor et al. 2019; Gärtner et al. 2021). NCPs are low-molecular non-ribosomal peptides, which possess a unique imino linkage in the macrocyclic ring and are produced only by *Nostoc* (Fidor et al. 2021). The first NCPs (found in two variants, A1 and A2) and their coding BGC were identified in the terrestrial symbiotic *Nostoc* sp. ATCC 53789, collected from a lichen at Arron Island, Scotland (Golakoti et al. 2001; Luesch et al. 2003; Becker et al. 2004). Later, a gene cluster, encoding nostocyclopeptide (NCP) A in the same strain was confirmed by Liu et al. (2014a), and NCPs were reported from *Nostoc* sp. KVJ2 symbiotic in *Blasia*, isolated from Norway (Liaimer et al. 2016), as well as in a free-living soil *Nostoc* sp. ASN_M, isolated from Iran (Nowruzi et al. 2012). More recently, the whole genome sequence of *Nostoc edaphicum* CCNP1411, collected from the region of Gulf of Gdansk (Baltic Sea), revealed the presence of four new NCP analogues (the linear E1-L, E2-L, and E4-L and the cyclic E3) besides the already known cyclic variants A1 and A2 (Fidor et al. 2020). Despite this strain being mentioned by authors as isolated from "brackish waters," *N. edaphicum* was discussed together with other toxigenic aeroterrestrial algae (Gärtner et al. 2021). Here, this opinion is shared, considering that the species has been originally described from salty soils of Crimea, Ukraine (Kondratyeva 1962; Komárek 2013) and that in the database of the National Center for Biotechnology Information (NCBI 2022) the strain CCNP1411 (NCBI:txid1472755) has been assigned to the "Terrabacteria group." NPs are cyclic hybrid non-ribosomal peptides/polyketides, branched acylated octapeptides with a heptapeptide lactone structure (Golakoti et al. 2000; Welker and Döhren 2006). Their coding BGC was first identified in *Nostoc* sp. GSV224, collected from Indian soil (Luesch et al. 2003). Further screening of 116 freshwater and soil strains from eight cyanoprokaryote genera revealed that the nostopeptolide (NP) encoding genes (the *4-mPro* biosynthetic genes in particular) were quite common for *Nostoc* (Liu et al. 2014a). In soil algae, the same study identified genes, encoding NP A in *Nostoc* sp. GSV224 and in *Nostoc punctiforme* PCC 73102. Chemically, the following variants of NPs have been identified in soil algae: (1) A1-A3 in the free-living soil *Nostoc* sp. GSV224 (Golakoti et al. 2000; Hofmann et al. 2003); (2) L1-L4 in the symbiotic *Nostoc* sp. UK2aImI, isolated from a lichen (Liu et al. 2014a); and (3) NP A and NP 1052 in *Nostoc punctiforme* PCC 73102, taken as a model because of its facultative symbiotic relations with diverse plants and fungi (Liaimer et al. 2015; Guljamov et al. 2017; Dehm et al. 2019). The genome mining in filamentous heterocytous *Brasilonema* sp. CT11, isolated from a black mat in a bamboo forest in Costa Rica, revealed a BGC with 100% similarity to NP A2 (Saha et al. 2020).

Nostocyclyne A is a novel cyclophane, extracted as a major component from *Nostoc* sp. TAU IL-220, which grew abundantly in the ground of a greenhouse in Neve Monosson, Israel (Ploutno and Carmeli 2000). This polyketide was active against *Staphylococcus aureus* and *Bacillus subtilis* and weakly (5% at 100 μg/mL) inhibited the photosynthesis in a green alga (Ploutno and Carmeli 2000).

The polyketide nosperin, a distinct member of the pederin family of toxic amides with two tetrahydropyran rings, was previously thought exclusive to animal–bacteria association as being made by bacteria that live in rove beetles of the genus *Paederus* (Kampa et al. 2013). Nosperin was discovered genetically and proved chemically in symbiotic *Nostoc* from the terrestrial lichen *Peltigera membranacea* (Kampa et al. 2013). This unexpected finding of the first trans-ATP KS BGC from Cyanoprokaryota not only expands the structural range and known distribution of this family of natural products but also suggests a role associated with symbiosis (Kampa et al. 2013).

Fischerellins (FRs), natural antimicrobial agents and growth inhibitors, were discovered in the benthic freshwater *Fischerella musicola* UTEX 1829 and also in terrestrial *Fischerella ambigua* UTEX 1903 and *Fischerella tisserantii* Kom. 1964/-17, suggesting that the synthesis of such allelochemicals might be characteristic of the genus (Gross et al. 1991). Fischerellin (FR) A is a natural heterocyclic product, unique and unrelated to all known cyanobacterial metabolites, which contains an enediyne moiety and two heterocyclic ring systems, that is a potent photosystem-II inhibitor acting against cyanoprokaryotes (e.g., *Synechococcus* PCC 6911) and other photoautotrophic organisms, and has interesting activity against fungi and plants (algae) (Hagmann and Juttner 1996; Srivastava et al. 1998, 1999; Demay et al. 2019; Al-Yosef and Amina 2021). FR B, a 2-pyrrolidinone with a polyunsaturated side chain, with similar algicidal activity, was subsequently isolated and characterized from *Fischerella muscicola* (Papke et al. 1997). Both FRs A and B were isolated also from *Fischerella muscicola* UTEX LB1829 and *Westiellopsis* sp. SAG 20.93 (Kim et al. 2012). Up to now, FRs A and B are the only fully characterized enediyne metabolites from Cyanoprokaryota (Hagmann and Juttner 1996; Papke et al. 1997). Members of enediyne family share a unique enediyne core, which has a polyketide origin proved by sequencing of relevant BGC, and is the cause of their potent cytotoxicity (Van Lanen and Shen 2008; Liang 2010; Shen et al. 2015).

Anabaenopeptins (ABPs) are a class of cyclic non-ribosomal hexapeptides with a conserved ureido bond and a D-Lys moiety (Harada et al. 1995; Rouhiainen et al. 2010; Monteiro et al. 2021). ABPs are protein (including protease) inhibitors, some of which are capable of inhibiting the carboxypeptidase activated thrombin activatable fibrinolysis inhibitor (TAFIa), important in the fibrinolysis process (i.e., breakdown of blood clots), and thus exerting antithrombotic effect (Chlipala et al. 2011; Schreuder et al. 2016). Some ABPs demonstrated potent allelochemical activity against cyanoprokaryotes, such as *Nostoc* and *Microcystis* (Sedmak et al. 2008; Liaimer et al. 2016; Guljamow et al. 2017). Firstly, they were isolated from the freshwater heterocytous filamentous genus *Anabaena* (Harada et al. 1995), and later their production by different, mainly filamentous, aquatic cyanoprokaryotes was demonstrated (Saha et al. 2020; Shishido et al. 2017; Monteiro et al. 2021). The first anabaenopeptin (ABP) BGC was identified also in freshwater *Anabaena* sp. 90 with two encoded alternative starter modules organized in separate bimodular proteins (with a high substrate specificity for L-Arg/L-Lys and L-Tyr, respectively) that allow the specific biosynthesis of different ABP variants (Rouhiainen et al. 1995,

2010). The two starter modules were found also in other 11 freshwater *Anabaena* strains, in which ABPs A1-F were found in different combinations (Rouhiainen et al. 2010). However, only a single module was detected in the ABP BGC of *Nostoc punctiforme* PCC 73102 and of the aquatic *Nodularia spumigena* CCY 9414, respectively (Rouhiainen et al. 2010). Further genome mining demonstrated the widespread of ABP BGC in Cyanoprokaryota (Shishido et al. 2017). According to Monteiro et al. (2021), 50 variants of ABPs were identified in marine cyanoprokaryotes, when compared with their 43 and 34 types in freshwater and soil strains, respectively. In soil algae, likewise in freshwaters, ABPs were found in strains of *Anabaena* s.l., currently assigned to genus *Dolichospermum*, namely *D. circinale* (strain *Anabaena circinalis* 90, variants A, B and C, Fujii et al. 1995; Monteiro et al. 2021) and *D. flos-aquae* (strain *A. flos-aquae* NRC 525-17, variants A and B, Harada et al. 1995; Monteiro et al. 2021). Altogether 14 new ABPs (803A, 803Б, 842, 849, 850, 856, 858, 863, 864, 872, 878A, 878B, 883, 892, 897, 906) were identified by Sanz et al. (2015) in different combinations in *Desmonostoc* sp. CENA 386 (849, 850, 863, 864, 878A, 878B, 892, 906), *Nostoc* sp. CENA 352 (842, 856, 858, and 872), and *Nostoc* sp. CENA 358 (883 and 897). They were found also in *Nostoc* ASN_M (808, 828, 842, 844, and 858, Nowruzi et al. 2012), *Nostoc* sp. ATCC 53789 (with novel congeners SA9, SA10, SA11, and SA12, Schreuder et al. 2016), and *Nostoc punctiforme* PCC 73102 (ABPs A2-F and NZ857, Guljamow et al. 2017; Dehm et al. 2019). ABPs (specifically their new variants KVJ827, KVJ841, and KVJ811) were found also in the strain *Nostoc* sp. KVJ2, symbiont in *Blasia*, isolated from Norway (Guljamow et al. 2017). Interestingly, ABPs were detected in its high-dense cultures but were virtually absent in the extracts from the conventional cultures (Rouhiainen et al. 2010). Therefore, Gulijamow et al. (2017) supposed that this explains why they were not detected earlier by Liaimer et al. (2016). In the genome of KVJ2 was identified BGC close to that of *N. punctiforme* PCC 73102 (Gulijamow et al. 2017). The structures of KVJ827, KVJ841, and KVJ811 were very similar to the ABP variants from *Anabaena* sp. TAU strain NZ-3-1 and *N. punctiforme* PCC 73102 (Grach-Pogrebinsky and Carmeli 2008). In the last strain the specific ABP nostamide A was also discovered (Gulijamov et al. 2017). Specific ABPs were found also in strains of non-heterocytous *Plectonema* sp. (plectonemid A, Müller et al. 2005) and *Schizothrix* sp. TAU IL-208-2-2, collected from a soil sample in Givat Brener nursery, Israel (schizopeptin 791, Reshef and Carmeli 2002). However, not long ago schizopeptin 791 was found also in *Nostoc* sp. CENA543 together with ABP 807 and novel nostamides B-E (Shishido et al. 2017).

The large family of ABPs includes also some tryptophan-containing variants, which long were attributed only to marine theonellid sponge from Mozambique (mozamide A, Schmidt et al. 1997). Variants of this type had antiproliferative activity against HeLa cell lines, which varied from weak to moderate (Saha et al. 2020). Currently, the first specific tryptophan-containing variants (803A, B) in algae were detected in *Brasilonema* sp. (strains CENA360 and CENA382) (Sanz et al. 2015). In another strain, *Brasilonema* sp. CT11, genome studies revealed a BGC with maximum similarity (100%) with ABP 908/915 (Saha et al. 2020). By chemical analysis in the same strain, three novel tryptophan-containing congeners (788, 802, and 816) were found together with two other homologues, i.e., ABPs 802A and 802B (Saha et al. 2020).

Simultaneous production of typical ABPs and their shorter structural homologues, the cyclic tetrapeptides namalides, was detected in *Nostoc* sp. CENA543 (Shishido et al. 2017). Namalides, firstly identified in marine sponges and then recorded in the freshwater heterocytous *Sphaerospermopsis torques-reginae*, are carboxypeptidase A inhibitors, and likewise the more common hexapeptides can exhibit protease inhibitory activity (Cheruki et al. 2012; Shishido et al. 2017).

ABPs and other cyanopeptides, such as the linear tetra- to heptapeptides microginins, as well as their structural analogues, nostoginins together with banyasides, are commonly known from aquatic environments (Ploutno and Carmelli 2002, 2005a, b; Welker and Döhren 2006). However, currently all of them were detected in 33% (45/133) *Nostoc* and *Nostoc*-like strains isolated from grasslands of the Great Hungarian Plain (Riba et al. 2020). Microginins inhibit angiotensin-converting enzyme (ACE), and currently their use is suggested in the treatment of COVID-19 patients, who have injured ACE2-expressing organs (Mazur-Marzec et al. 2021). Nostoginins and banyasids, firstly isolated from water blooms of *Nostoc* (Ploutno and Carmelli 2002), are close to the linear tetrapeptides aeruginosins (AERs) (Welker and Döhren 2006). They are also better investigated in strains from aquatic environments in comparison with soil inhabitants (Welker and Döhren 2006; Fidor et al. 2019; Gärtner et al. 2021). AERs exhibit potent inhibitory activity against trypsin and other trypsin-like serine proteases (e.g., thrombin) and also have shown toxicity against the freshwater crustacean *Thamnocephalus platyurus* (LC50 = 22.4 μM), which is higher than the toxicity reported for MC-LR (LC50 = 10.8 μM) (Murakami et al. 1995; Welker and Döhren 2006; Kohler et al. 2014). In soil algae, AERs were first detected in *Nostoc* KVJ2 symbiotic in *Blasia* (Liaimer et al. 2016). The peculiar aeruginosin (AER) 865 was discovered in soil *Nostoc* sp. Lukešová 30/93 from a forest of Krušné mountains in the Czech Republic (Kapuścik et al. 2013). According to these authors, this is the first peptide of AER-type, which contains a fatty acid along with a carbohydrate moiety, and is characterized by strong anti-inflammatory activity, but no cytotoxicity. The AER 865 along with ten additional newly identified congeners (variants 737, 753, 767, 823, 837, 849, 851, 879A, 879B, and 893) were produced in different combinations by strains of *Brasilonema* and *Nostoc*, isolated from leaf surfaces of a Brazilian forest (Sanz et al. 2015), as follows: *Brasilonema* sp. CENA360 (737, 753, 767, 823, 837, 849, 851, 865, 879A, 879B, 893), *Brasilonema* sp. CENA361 (851, 865), *Brasilonema* sp. CENA381 (737, 753, 767, 823, 851, 865, 879A, 879B, 893), *Brasilonema* sp. CENA382 (737, 753, 767, 851, 865, 893), *Nostoc* sp. CENA352 (737, 753, 767, 823, 837, 849, 851, 865, 879A, 893), *Nostoc* sp. CENA358 (737, 767, 823, 837, 849, 851, 865, 879B, 893), and *Nostoc* sp. CENA369 (851, 865). Currently, AER (microcin SF608) was identified among the toxins of the salt-sand boring *Perforatum tunneli* (Zimba et al. 2021).

Other cyanopeptides, nostosines, are low-molecular linear tripeptides with anti-trypsin activity, discovered in *Nostoc* sp. FSN, isolated from a paddy field in Iran (Liu et al. 2014b).

Cyanopeptolins (CPPs) represent a large class of hybrid non-ribosomal/polyketide synthesized cyclic depsipeptides (peptidolactones), composed of six amino-acid rings and a unique 3-amino-6-hydroxy-2-piperidone (Ahp) residue, frequently produced by different Cyanoprokaryota (Fidor et al. 2019; Mazur-Mazurec et al. 2021). Most CPPs, isolated from aquatic species, are inhibitors of serine proteases (such as trypsin, chymotrypsin, thrombin, and elastase), and some of their variants are acutely toxic to small aquatic crustaceans (Mazur-Marzec et al. 2018, 2021). Not long ago 13 CPPs (962, 969, 978, 985, 990, 992, 999, 1006, 1013, 1018, 1020, 1027, 1048) were isolated from *Nostoc edaphicum* CCNP1411, 12 of which were new variants, and also showed trypsin and chymotrypsin inhibitory activity (Mazur-Marzec et al. 2018). Altogether 11 new CPPs (984A, 984B, 998A, 998B, 998C, 1002, 1004, 1012, 1018, 1026, 1032) were identified by Sanz et al. (2015) in different combinations in *Desmonostoc* strains CENA 371 (i.e., 984B, 998A, 1002, 1004, 1012, 1018, 1026, 1032) and CENA 386 (984A-B, 998A-C, 1002, 1004, 1012, 1018, 1026, 1032). From the soil *Scytonema hofmannii* PCC7110, three novel CPPs, namely hofmannolin and scyptolins (A and B), were isolated, and for scyptolins, the porcine pancreatic elastase inhibition was shown (Matern et al. 2001, 2003a, b). Hapalosin is another cyclic depsipeptide able to reverse P-glycoprotein multiple drug resistance (MDR), isolated firstly from *Hapalosiphon welwitschii* UH IC-52-3, collected from an Australian soil sample and mass cultivated, and UTEX51830 (Stratmann et al. 1994a). The hapalosin BGC, comprising five genes (*hapA-E*), was identified within the genomes of soil *Fischerella* sp. PCC 9431 and *Westiella intricata* UH HT-29-1 (Micaleff et al. 2015). In *Brasilonema* CT11 the finding of an interesting hybrid BGC that showed just 60% similarity to the BGC of the hapalosin, allowed to suppose the discovery of a new cyclic peptide (Saha et al. 2020). Eight new CPPs, namely insulapeptolides A-H, identified in soil *Nostoc insulare* 54.79, and their variant D in particular, have shown selective inhibition of human leukocyte elastase when applied at nanomolar range (Mehner et al. 2008; Fidor et al. 2019). Elastase inhibitors with similar structure, nostopeptines A and B, were extracted from *Nostoc minutum* NIES-26, sampled at Ishigaki, Japan (Okino et al. 1997; Ahmad et al. 2020). Although Okino et al. (1997) mentioned this strain as a freshwater one, obtained from a culture collection, Fidor et al. (2019) assigned it to terrestrial algae, and in NCBI database this strain (NCBI:txid1844469) was included in the "Terrabacteria group" (NCBI 2022). Considering that according to Komárek (2013) *N. minutum* develops on wet soils, we share the opinion of Fidor et al. (2019) and include it among the toxigenic soil algae.

Apratoxins (APRs) are a family of natural products, mainly cyclodepsipeptides, characterized by a thiazoline unit and an extensive polyketide-derived moiety as part of the macrocyclic structure, that are formed by a polyketide synthase/non-ribosomal peptide synthase pathway and are known in seven variants (A-G) (Grindberg et al. 2011; Tan 2012). The first APRs and their BGC were isolated from marine non-heterocytous *Moorea producens* (Syn. *Lyngbya majuscula*), *Moorea bouillonii* (Syn. *Lyngbya bouillonii), and Lyngbya sordida* and exhibited potent cytotoxicity against various cancer cell lines (Luesch et al. 2001; Grindberg et al.

2011; Engene et al. 2012; Onda et al. 2017; Cai et al. 2021). Currently, APRs A and F have been discovered in *Perforatum tunneli* (Zimba et al. 2021).

The specific non-ribosomal antifungal glycolipopeptide hassallidin (HS), active against *Candida albicans*, was detected in terrestrial *Anabaena* sp. HAN7/1 from Finland and *Nostoc calcicola* 6sf Calc from the Czech Republic (Shishido et al. 2015). Before this the hassallidins A and B were found in *Tolypothrix* (basionim *Hassallia*) sp., collected from Bellano, Italy, with activity against human pathogenic fungi (Neuhof et al. 2005, 2006). In the genome of a freshwater *Anabaena* sp. 90 a cryptic inactive non-ribosomal peptide BGC was identified, and it was presumed as responsible for HS biosynthesis (Vestola et al. 2014). In the same study, the HS production and its relative BGC were reported for a *Nostoc* strain, symbiotic in a lichen. Other lipopetides, lyngbyabellins (LBLs) are cyclic and linear peptolides, most of which are with a strong potency to disrupt the cell microtubules, thus having cytotoxic effect against human breast cancer cells (MCF7) and other cancer cell lines (Fathoni et al. 2020). Some LBLs exhibit antimalarial and antifouling activities against *Amphibalanus amphitrite* barnacle larvae (Fathoni et al. 2020). Interestingly, the acyclic LBLs are less cytotoxic than the corresponding cyclic ones (Fathoni et al. 2020). The first LBLs and their variants were isolated from marine *Moorea producens* (Luesch et al. 2000; Han et al. 2005; Choi et al. 2012; Fathoni et al. 2020). Up to now, 27-deoxylyngbyabellin A was known only from Malaysian *Moorea bouillonii* and a Red Sea *Okeania* sp. (Fathoni et al. 2020), and therefore, its current finding in the sand-boring *Perforatum tunneli* (Zimba et al. 2021) is notable. Like the LBLs, malyngamides (MLGs) are also cytotoxic lipopeptides with activity against cancer cell lines, with the first compound, malyngamide (MLG) A, discovered in *Moorea producens* (Cardellina II et al. 1979). Later, more MLGs were discovered, all in marine cyanoprokaryotes from different strains of *M. producens* (Shaala et al. 2013; Kwan et al. 2010; Sueyoshi et al. 2017; Jiang et al. 2018). The single soil alga, in which MLGs (MGL O) have been found, is *Perforatum tunneli* (Zimba et al. 2021). A new lipopeptide, nostofungicidine, which contains a novel β-amino acid, 3-amino-6-hydroxy stearic acid (Ahs), was isolated from the methanolic extract of a soil *Nostoc commune*, field-grown at the campus of Okayama University (Kajiyama et al. 1998). Nostofungicidine is a cyclic peptide, strongly active against fungi like *Aspergillus candidus*, *Cladosporium herbarum*, *Botrytis fabae*, and *Cercospora beticala*; moderately active against *Glorruerella cingulata*, *Sclerotium rolfsii*, and *Rhizoctonia solani*; and cytotoxic against murine myeloid leukemic cell line NFS-60 (Kajiyama et al. 1998; Kobayashi and Kajiyama 1998; Chlipala et al. 2011). Puwainaphycins (PWPs) belong to the large group of ß-amino fatty acids that contain lipopeptides of unknown biosynthetic pathways, frequently occurring in Cyanoprokaryota, some of which are cytotoxic and cause strong disruption of human cell membranes (Mareš et al. 2014). Cytotoxic puwainaphycin (PWP), together with its F/G congeners and their BGC, was identified in the soil heterocytous filamentous *Cylindrospermum alatosporum* CCALA 988 (C24/1989) (Mareš et al. 2014). This is the first study in which a ß-amino lipopetide operon of Cyanoprokaryota was identified and which provided the first insight into the biosynthesis of ß-amino fatty acid lipopeptides in this group. Close to PWP is the calophycin, a cyclic decapeptide that contains the novel (2R,3R,4S)-3-amino-2-hydroxy-4-methylpalmitic acid unit (Moore 1996). It has been discovered in the soil *Calothrix fusca* EU-10-1 and has been proved as a potent broad-spectrum fungicide (Moon et al. 1992).

Two new cyclic undecalipopeptides, muscotoxins A and B, which possess the unique lipophilic residue 3-amino-2,5-dihydroxydecanoic acid (5-OH Ahdoa), were isolated from soil *Desmonostoc muscorum* NIVA-CYA 817 (Tomek et al. 2015). Muscotoxin A at molar range showed cytotoxicity to cancer cell lines (i.e., YAC-1, Sp/2, and HeLa) via membrane damaging and influx of calcium ions, and it had the ability to permeabilize liposomes, but toxicity of muscotoxin B was not investigated due to its poor yield (Tomek et al. 2015).

From the large group of the non-ribosomal/polyketide synthesized cytotoxins of Cyanoprokaryota, the cyclic depsipeptide cryptophycin (CRY) (in 27 congeners of different molecular weight) was isolated and identified in two strains of *Nostoc* sp., namely ATCC 53789 found as lichen symbiont in Scotland and *Nostoc* GSV224 (ATCC 55483) from India (Schwartz et al. 1990; Barrow et al. 1995; Golakoti et al. 1995, 2001; Trimurtulu et al. 1995; Fidor et al. 2019). Although being so distinct geographically, both strains demonstrated the same composition of cryptophycins (CRYs), and later the identity of their relevant BGC was demonstrated (Magarvey et al. 2006). Afterward, CRY was found in two more soil strains from Iran, *Nostoc* sp. ASN_M and *Nostoc* sp. BN KY303912 (Nowruzi et al. 2012, 2018). Today, more than 28 CRYs are known as strong "antimitotic" natural toxins with antiproliferative effect, suppressing tumors, but yet the first discovered CRY is considered as the most potent (Chlipala et al. 2011; Gärtner et al. 2021).

Macrolides belong to the structurally diverse family of polyketides, synthesized by multienzyme polypeptides (polyketide synthases) and include natural compounds with large macrocyclic lactone ring to which one or more deoxy sugars may be attached (Omura 2002). Currently, the class of macrolides comprises tolytoxin (TLT) and scytophycins (SCPs), isolated from terrestrial algae (Tomsickova et al. 2014; Delawská et al. 2021; Ivanov et al. 2021). The macrolide TLT (6-hydroxy-7-*O*-methylscytophycin B) was discovered in a *Tolypothrix conglutinata* var. *colorata*, collected from a soil sample at Fanning Island (Moore 1981). Its cytotoxicity, caused by actin disrupting with strong antiproliferative effects on human cells, was proved by Patterson et al. (1993). TLT from *Scytonema* sp. PCC 10023 had shown activity against tunneling nanotubes (Senol et al. 2019). These membrane protrusions contain actin and are significant in long-range intercellular communication, the role of which increases in pathological situations and diseases (Senol et al. 2019). The closely related to TLT acetogenic polyketide-derived macrolides, SCPs A-E, were isolated from *Scytonema pseudohofmannii* strain BC-1-2 (ATCC 53141) and were fully characterized (Carmeli et al. 1990a). The antifungal activity and cytotoxicity of SCPs and TLT were proved in five more strains of three other species of *Scytonema*, viz. *Scytonema burmanicum* DO-4-1 (from soil surface at Moon Beach, Okinawa), *Scytonema ocellatum* DD-8-1, FF-65-1, and FF-66-3, collected at Guam, Missouri, and Illinois, and in *Scytonema wolleanum* (strain *S. mirabile* BY-8-1) (Carmeli et al. 1990a), as well as in a crude extract of soil *Nostoc* sp. 5/96

(Tomsickova et al. 2014). Potent antifungal scytophycin SCP against *Candida albicans* and *Aspergillus flavus* was detected in *Nostoc* sp. HAN11/1, collected in Finland (Shishido et al. 2015). SCP was found also in a strain of *Cylindrospermum* (Jung et al. 1991; Patterson et al. 1994).

Among the nine toxins, produced by the sand-boring *Perforatum tunneli*, the polyketide δ-lactone (dihydropyranone) coibacin B was detected (Zimba et al. 2021). The first coibacin (coibacin A) was isolated from the marine non-heterocytous cyanoprokaryotes, close to *Oscillatoria*, collected from the Coiba National Park and exhibited selective antileishmanial activity and potent anti-inflammatory properties (Balunas et al. 2012; Carneiro et al. 2014). After the structure of coibacin A was unraveled, coibacin B was synthesized (Carneiro et al. 2014).

2.4 Toxic Ribosomally Synthesized and Post-Translationally Modified Peptides

Besides the non-ribosomal peptides, *Nostoc* can produce peptides that are ribosomally synthesized and post-translationally modified, such as the N-acetylated tricyclic dipeptides microviridins (MVDs) with a rare cage-like architecture that are protease and elastase inhibitors, and have been recorded also from other, mainly aquatic, cyanoprokaryotes (Micaleff et al. 2015; Fidor et al. 2019; Dehm et al. 2019; do Amaral et al. 2021; Gärtner et al. 2021; Uzunov et al. 2021a). In soil algae, according to Fidor et al. (2019), the new MVDs G and H were obtained from *Nostoc minutum* NIES-26 (Murakami et al. 1997), and the highly unusual microviridin (MVD) types N3–N9 were discovered in a symbiotic *Nostoc punctiforme* PCC 73102, isolated from *Macrozamia* in Australia (Dehm et al. 2019). By genome mining of another filamentous cyanoprokaryote, *Fischerella* sp. PCC 9339, an MVD (*mvd*) BGC of eight genes was identified (Micaleff et al. 2015). Although this BGC was highly similar to the *mvd* BGC from *Planktothrix agardhii* NIVA-CYA 126/8, it encodes two precursor peptides, the protein alignments of which revealed that the peptide sequences of the studied *Fischerella* strain were novel (Micaleff et al. 2015).

Cyanobactins (CBs) (including cyclamides) represent another group of small cyclic ribosomally synthesized and post-translationally modified peptides with more than hundred compounds, some of which are cytotoxic (Sivonen et al. 2010). Initially CBs were considered to consist only of oxazolines and thiazolines, or their oxidized derivatives oxazoles and thiazoles, but later the term was broadened in order to embrace also cyclic peptides that contain proteinogenic amino acids and some linear cyanobactins (Donia and Schmidt 2010; Sivonen et al. 2010; Leikoski et al. 2013; Micaleff et al. 2015). CBs are widespread in symbiotic and free-living Cyanoprokaryota from aquatic and terrestrial habitats, with an estimate that 10–30% of all blue-green algae can produce them (Sings and Rinehart 1996; Donia et al. 2008; Sivonen et al. 2010; Schmidt and Donia 2009; Donia and Schmidt 2011; Martins and Vasconselos 2011), and currently more than 100 their variants are known (Micaleff et al. 2015). For example, genome analysis of 126 strains of blue-green algae revealed the presence of 31 cyanobactin BGC in 24% of them (Leikoski et al. 2013). Although many of them have been supposed to be nonfunctional, within closely related strains their active versions have been identified (Leikoski et al. 2012; Micaleff et al. 2015).

Among CBs of soil algae, which contain heterocyclized amino acids, are the following hexapeptides: (1) the moderately toxic raocyclamides A and B, against sea urchin embryos, isolated from *Oscillatoria variabilis* (Syn. *O. raoi*; Admi et al. 1996) and tenuecyclamides, discovered in *Nostoc carneum* (as strain *N. spongiaeforme* var. *tenue* TAU IL-184-6) from a soil/litophytic sample collected at Bet Dagan, Israel (Banker and Carmeli 1998), and the BGC for the last was identified (Donia et al. 2008); (2) the mildly cytotoxic against colon cancer (LoVo) and human epithelial (KB) cell lines, and antibacterial westiellamide, isolated from *Westiellopsis prolifica* UH EN-3-1, collected from a mud sample at the island Oahu, Hawaii (Prinsep et al. 1992a; Patterson et al. 1994; Abed et al. 2019); (3) dendroamides (A, B, and C) from a lipophilic extract of the heterocytous soil *Stigonema dendroideum* UH-IA-45-3, isolated from a Maui rock sample, that have shown some MDR reversing activity via P-glycoprotein inhibition of a human breast carcinoma line (MCF-7/ADR), but their toxicity/nontoxicity has to be clarified (Ogino et al. 1995; Sivonen et al. 2010). Genome mining of heterocytous *Westiella intricata* UH HT-29-1 revealed that its cyanobactin BGC shared some homology to the tenuecyclamide BGC, but the analysis of its precursor gene supposed that it encodes two novel cyanobactins (Micaleff et al. 2015).

Bacteriocins (BCs), known for more than 100 years, represent a heterogeneous family of more than 200 small proteinaceous molecules that are ribosomally synthesized and have strong broad- or narrow-spectrum antimicrobial activity (Soltani et al. 2021). They are abundant and widely spread in nature, and different approaches have been used for their classification (e.g., Wang et al. 2011; Soltani et al. 2021). Since their activity is mainly directed against closely related bacteria, BCs are commonly considered to be nontoxic and safe with a broad range of applications in human life (Soltani et al. 2021). However, some BCs exerted cytotoxicity against eukaryotic cell-culture lines (Soltani et al. 2021). One of the first BCs produced by Cyanoprokaryota was discovered in *Nostoc* sp. 78-11-A-E during screening of 65 filamentous blue-green algae (Flores and Wolk 1986). BGC for four BCs was detected in *Brasilonema* CT11 (Saha et al. 2020). In the study by Micaleff et al. (2015) different numbers of bacteriocin BCG were identified in the strains of the following heterocytous cyanoprokaryotes: *Chlorogloeopsis fritschii* PCC 6912 (isolated from soil in Allahabad, India)—10, *Fischerella* sp. PCC 9339 (locality not indicated)—9, *Fischerella* sp. PCC 9605 (isolated from a limestone in Jerucham, Har Rahama, Israel)—12, *Fischerella* sp. JSC-11 (locality not indicated)—8, *Hapalosiphon welwitschii* UTEX B1830 (pointed by the authors as "also known as *Fischerella* sp. PCC 9431")—9, *Mastigocladopsis repens* PCC 10914 (isolated from soil in Tarragona, Spain)—6, and "*Mastigocladopsis testarum* BC008" (locality not indicated)—10 (Micaleff et al. 2015). The obtained gene clusters belong to the seven groups defined by Wang et al. (2011), and nine more BGC were identified but were not classified (Micaleff et al. 2015).

2.5 *Toxic Proteins*

The polypeptides lectins (LCs) comprise an elaborate collection of sugar-binding proteins without enzymatic activity, some of which are beneficial and some are toxic (e.g., Kocourek and Hořejši 1981; Mitchell et al. 2017; Mazur-Marzec et al. 2021). These diverse peptides are ubiquitous in nature and have been considered among the most interesting molecules with antibacterial, antiviral, and antifungal activities, found in micro- and macroalgae (Mazur-Marzec et al. 2021; Pagarete et al. 2021). The significant antimicrobial potential of LCs is caused by their ability to bind to the carbohydrates of outer membranes with high specificity (Mazur-Marzec et al. 2021; Pagarete et al. 2021). Two of five known from Cyanoprokaryota, cytotoxic at the nanomolar range LCs, have been discovered from soil algae (Mazur-Marzec et al. 2021). Since LCs are generally classified based on the glycan recognition of their carbohydrate recognition domains, here it is worth mentioning that they are specifically binding to oligomannoses (Mitchell et al. 2017). Cyanovirin-N, which has been obtained from a crude extract of *Nostoc ellipsosporum* Q68D170 collected in Hawaii, is a low molecular lectin (comprised of 101 amino acid residues) with strong antiviral activity (e.g., against human immunodeficiency virus—HIV, simian immunodeficiency virus—SIV, herpes simplex virus—HSV, Ebola, influenza viruses, and hepatitis C virus), but is not toxic for humans (Gustafson et al. 1997; Gandhi et al. 2000; Tsai et al. 2003, 2004; Lagenaur et al. 2015; Carpine and Sieber 2021). The lectin scytovirin, isolated from cultured *Scytonema varium* strain HG-24-1, is a novel protein (comprised of a single 95-amino acid chain with significant internal sequence duplication and 10 cysteines forming five intrachain disulfide bonds) with a potent activity against HIV-1, HSV, and Ebola virus (Bokesch et al. 2003; Xiong et al. 2006a, b; Moulaei et al. 2007; Takebe et al. 2013; Garrison et al. 2014; Mazur-Marzec et al. 2021; Pagarete et al. 2021).

2.6 *Toxic Alkaloids*

Alkaloids are diverse natural organic compounds, a class of widely occurring secondary metabolites which have at least one nitrogen atom, most of which are basic (Hesse 2002). They occur in different organisms and are of different origins (e.g., from amino acids, polyamines, peptides, and cyclopeptides, or from other compounds) with high structural diversity and therefore have diverse biological roles, but no common classification is applied to them (Hesse 2002).

Two types of well-known alkaloid neurotoxins, such as anatoxins (ANAs) and saxitoxins (SXTs), commonly spread in aquatic environments (Evangelista et al.

2007; Meriluoto et al. 2017), have been found also in soil algae (Gärtner et al. 2021). SXTs are blocking the sodium channels, while ANAs are nicotinic-acetylcholine mimics (Cox et al. 2009; Meriluoto et al. 2017). According to HPLC peaks, the presence of anatoxin (ANA) A was presumed for *Nostoc linckia* PACC 5085, while SXTs close to the detection limit of ELISA (0.010 ppb) in the same strain and in *N. punctiforme* PACC 8646 were detected (Teneva et al. 2012). By contrast, a significant production of SXTs (ca. 30 ppb, or 30 $\mu g\ L^{-1}$) was detected by ELISA in *Nostoc microscopicum* CCALA #124 (Batsalova et al. 2016). In both cases ELISA kits were used with 10–30% cross reactivity to dicarbamoyl saxitoxin and gonyatoxins II, III, B1, C1, and C2. Despite the first two strains being noted by the authors as freshwater, and for the third no habitat was indicated, considering the common occurrence of these three species in terrestrial habitats, they were enlisted among toxigenic aeroterrestrial and extremophilic algae (Gärtner et al. 2021). The first documentation of ANAs (ANA A) from the Arctic came from an investigation of soil crusts at Northwest coast of Spitsbergen, and it was supposed that the production of these toxins was related to non-heterocytous *Oscillatoria* and *Phormidium* (Chrapusta et al. 2015). In cultures obtained from soil crusts in deserts of northwest Qatar, the phosphorylated cyclic N-hydroxyguanine neurotoxin which inhibits the acetylcholine esterase, quanitoxin (formerly known as ANA-s, Fiore et al. 2020), was detected (Cox et al. 2009; Metcalf et al. 2012, 2015). It is noteworthy to mention that these "salt" crusts in dry, vegetation-free undrained desert basins were composed mainly of species from non-heterocytous filamentous genera *Microcoleus* and *Oscillatoria* and of the coccal genera *Anacystis*, *Gloeocapsa*, and *Synechocystis* (Cox et al. 2009).

Novel peptide alkaloids, the modified pentapeptides containing a unique array of four consecutive thiazoline/thiazole rings, tantazoles (A, B, F, and I), and mirabazoles (didehydromirabazole A, and mirabazoles B and C) as their related minor components, have been isolated from *Scytonema wolleanum* (as strain *S. mirabile* BY-8-1) (Carmeli et al. 1990c, 1991d, 1993; Welker and Döhren 2006). Among them, tantazole B (second major tantazole in *Scytonema*) and didehydromirabazole A exhibited murine solid tumor selective cytotoxicity (Carmeli et al. 1990c, 1991d).

A specific linear oxazole peptide alkaloid with two contiguous oxazoles and unusual prenyl groups, the antibacterial and cytotoxic muscoride A, was discovered in "terrestrial freshwater" *Nostoc muscorum* (Nagatsu et al. 1995). Currently, gene clusters important in biosynthetic pathways of muscoride A and B were identified in *Nostoc* spp. PCC 7906 and UHCC 0398 (Matilla et al. 2019).

Scytonemin is an indole alkaloid, which is a dimer of indolic and phenolic subunits, discovered in terrestrial *Scytonema* and *Chlorogloeopsis* sp. (e.g., Singh et al. 2017). The reduced scytonemin from *Nostoc* was proved cytotoxic against murine macrophage Raw 264.7 cells and as inducing autophagic cell death in human T-lymphoid line Jurkat cells (Itoh et al. 2013; Li and Guo 2018).

Specific tetracyclic indole alkaloids, hapalindoles (HPLs) A and B, were discovered in *Hapalosiphon pumilus* (Syn. *H. fontinalis*), collected from a soil sample at Marshall Islands (Moore et al. 1984). The study was carried out in an attempt to isolate the active ingredients responsible for the activity of *H. pumilus* against another blue-green alga, *Anabaena oscillarioides*. Further studies showed the antibacterial and antifungal properties of HPLs (Moore et al. 1987). Scheuer (1992) noted the HPLs, and the hapalindole (HPL) A in particular, as "noteworthy terrestrial isocyanides," a result of mixed biosynthesis of tryptophan and a linear monoterpene, but currently, more commonly they are classified among alkaloids (e.g., Bhat et al. 2014; Walton and Berry 2016; Carpine and Sieber 2021). After the discovery of HPLs, in the same strain, but in small amounts, other oxidized hapalindoles, such as hapaloxindoles, hapalonamides, and fontonamids (viz. anhydrohapaloxindoles B and M; hapalonamides G, H, V; and dechlorofontonamide) were detected (Moore et al. 1989; Walton and Berry 2016). From *Fischerella* UTEX LB1829 and *Westiellopsis* sp. SAG 20.93 along with the known hapalonamide H, anhydrohapaloxindole A, and HPLs A, C, G, H, I, J, and U, were isolated three new HPL-type alkaloids (namely 13-hydroxy dechlorofontonamide, HPL X, and deschloro HPL I) (Kim et al. 2012). At molar ranges, HPLs X and A and hapalonamide H demonstrated potent antifungal and antibacterial activity (i.e., against *Candida albicans* and *Mycobacterium tuberculosis)* (Kim et al. 2012).

From *Fischerella* sp. Java 94/2, chosen according to the results from screening for antibiotic activity of 189 cyanoprokaryote strains, newly isolated mainly from soils and freshwaters in Australia, Indonesia, Nepal, Thailand, and Vietnam (Schlegel et al. 1999), 12-epi-HPL E isonitrile and 12-epi-HPL H isonitrile were determined (Doan et al. 2000). In "*Hapalosiphon laingii* 89-785/4," isolated from dead coral debris at the ground of Laing Island (Papua New Guinea), three new HPLs (viz. 12-epi-HPL H, 12-epi-HPL G, and 12-epi-HPL Q isonitrile) were discovered along with the known 12-epi-HPL isonitriles E and C (Klein et al. 1995). From a *Hapalosiphon* sp. CBT1235, three known HPLs (A, D, and M—Moore et al. 1984, 1987; Schwartz et al. 1987) were extracted together with their two derivatives, HPL A-formamide and HPL J-formamide (Chilszuk et al. 2020). This was the first report of the extraction of these two formamide-bearing compounds from a natural source. All these HPLs, and HPL A in particular, demonstrated potent antiproliferation effects in the T-cell assay (Chilszuk et al. 2020). Considering that HPLs have been described to be neurotoxic metabolites as potent sodium channel modulators (Cagide et al. 2014), detailed further studies are necessary to better distinguish toxicity and immunomodulatory effects (Chilszuk et al. 2020).

Nowadays, more than 80 related alkaloids, named HPL-type alkaloids, which contain a polycyclic ring system, unique functional groups and various isomers (Hohlman and Sherman 2021), have been identified and most of them showed antibacterial and antifungal activities (e.g., Doan et al. 2000; Bhat et al. 2014; Walton and Berry 2016). Among them are structurally related fischerindoles (FSDs), ambiguines (AMBs), and welwitindolinones (WELs), found in strains of different species from the cyanoprokaryote order Stigonematales (Bhat et al. 2014; Walton and Berry 2016; Carpine and Sieber 2021). The first antifungal tetracyclic indole alkaloid, fischerindole (FSD) L, was discovered in the ethanol (70%) extract of the terrestrial *Fischerella muscicola* UTEX 1829 (Park et al. 1992) and later was again isolated from the similar (possibly similar) strain UTEX LB1829, and from *Westiellopsis* sp. SAG 20.93 as well (Kim et al. 2012). Four novel FSD congeners were recorded in *Hapalosiphon welwitschii* UH IC-52-3 along with seven variants of WELs, a novel class of terpenoid indole alkaloids, which possess fungicidal and insecticidal properties and are able to reverse MDR (Stratmann et al. 1994b). In the same study, some of them in small amounts were found also in *Westiella intricata* UH HT-29-1 (Stratmann et al. 1994b). A few years later, four novel, highly oxidized congeners of this class were characterized in soil strains of *Fischerella muscicola* HG-39-5 and *Fischerella major* HX-7-4 (viz. 3-hydroxy-*N*-methyl-WEL C isothiocyanate, 3-hydroxy-*N*-methyl-WEL C isonitrile, 3-hydroxy-*N*-methyl-WEL C formamide, and *N*-methyl-WEL D isonitrile) along with three already known compounds, such as WEL A isonitrile, *N*-methylWEL C isonitrile, and *N*-methylWEL C isothiocyanate (Jimenez et al. 1999; Bhat et al. 2014). However, one of them, the 3-hydroxy-*N*-methyl-WEL C formamide, was believed to be an artifact (Jimenez et al. 1999; Bhat et al. 2014). HPLs G-H along with AMBs A-F were isolated from *Fischerella ambigua* UTEX 1903, and AMBs D and E were isolated from *Hapalosiphon hibernicus* BZ-3-1 and *Westiellopsis prolifica* EN-3-1 (Smitka et al. 1992; Bonjouklian et al. 1993, 1994; Mo et al. 2009b). Ambiguine (AMB) D can inhibit mitosis indirectly, by inducing damage through oxidative stress (Bhat et al. 2014). The first nitrile-containing AMB, variant G, was found in the extract of *Hapalosiphon delicatulus* (Huber et al. 1998) with inhibitory effects against other cyanoprokaryotes (Bhat et al. 2014). In soil *Fischerella ambigua* IL-199-3-1 from Israel, nine isonitrile-containing alkaloids were identified, three of which were new AMBs (i.e., variants H-J with antimicrobial activity, Raveh and Carmeli 2007). Among them the AMB I was the most potent antibacterial, antifungal, and antiprotozoal compound (Raveh et Carmeli 2007; Mo et al. 2009b; Al-Yosef and Amina 2021). AMBs K, L, M, N, and O, isolated and characterized as novel compounds from *Fischerella ambigua* UTEX 1903, besides antifungal, antibacterial, and antiprotozoal activity, exhibited cytotoxicity against green monkey kidney (Vero) cells (Huber et al. 1998; Mo et al. 2009b; Bhat et al. 2014; Al-Yosef and Amina 2021) Afterward, the novel AMBs P and Q (along with the known AMB G nitrile) were extracted from the same strain (Mo et al. 2010).

The structural similarity between the HPL-type alkaloids suggested their common biosynthetic origin (Moore et al. 1984; Park et al. 1992; Stratmann et al. 1994b; Bhat et al. 2014). The BGC for WELs and AMBs were first identified from *Hapalosiphon welwitschii* UTEX B1830 and *Fischerella ambigua* UTEX 1903, respectively (Hillwig et al. 2013, 2014; Micaleff et al. 2015). Afterward, the BGC for HPL, WELs, and AMBs were found in terrestrial *Fischerella ambigua* UTEX 1903, *Fischerella muscicola* SAG 1427–1 (isolated from soil in Allahabad, India), *Fischerella* sp. ATCC 43239, *Fischerella* sp. PCC 9339, *Hapalosiphon welwitschii* UTEX B1830 (possibly identical *Fischerella* sp. PCC 9431), and *Westiella intricata* UH HT-29-1 (isolated from soil at Moen Island, Truk Atoll, Caroline Islands) (Micaleff et al. 2014, 2015).

Two novel hapalindole-related alkaloids, fischambiguines A and B, were discovered in the cultured *Fischerella ambigua* UTEX 1903 (Mo et al. 2010). They exhibit anti-inflammatory properties and fischambiguine B at molar rates showed strong inhibition of *Mycobacterium tuberculosis but* was not cytotoxic in a Vero cell line

(Mo et al. 2010). Most HPL-type compounds demonstrated antibacterial activity, as well as the possibility to inhibit cell proliferation and RNA polymerase, or to be sodium-channel moderators (Carpine and Sieber 2021). The biosynthetic pathways and encoding BGC for these alkaloids have been revealed in a series of studies (Carpine and Sieber 2021 and references therein).

Specific to their biological activity are the hapalindolinones, which were supposed to be derivatives of tricyclic hapalindoles (Walton and Berry 2016). This unique structural subclass, containing spiro-fused cyclopropane ring and oxindole ring, was first discovered from a soil *Fischerella* ATCC 53558, collected in Everglades, Florida (Schwartz et al. 1987). Later, hapalindonones A and B were isolated also from "*Hapalosiphon laingii* 89-785/4" (Klein et al. 1995). They inhibit the arginine vasopressin (antidiuretic hormone, ADH) binding to kidney tissue and are considered potential aid in the treatment of several diseases, such as congestive heart failure, edema, hypertension, and hyponatremia (Bhat et al. 2014).

By years the biosynthetic pathways of HPL-like alkaloids were clarified, and it was shown that almost all of these compounds express inhibiting activities against versatile organisms (e.g., bacteria, fungi, algae, insects) and cancer cells (Bhat et al. 2014). For example, a significant and distinct cytotoxicity associated with antiproliferative activity based on microtubule depolymerization was expressed by the indolinone welwistatin (formerly, WEL C isothiocyanate) isolated from *Hapalosiphon welwitschii* (Zhang and Smith 1996; Bhat et al. 2014).

From a strain of heterocytous *Tolypothrix tjipanasensis* DB-1-1, collected from soil in Vero Beach, Florida, a set of specific, novel halogenated natural indolealcazoles, named tjipanazoles (TPZs), was isolated (Bonjouklian et al. 1991). Some of them show mild nonselective cytotoxicity against leukemia and tumor cells, some (e.g., TPZs A1 and A2) are appreciable antifungicides (Bonjouklian et al. 1991; Bhat et al. 2014), and some can be used against MDR bacteria (Parsaeimehr and Lutzu 2016). In *Fischerella ambigua* EAWAG 108b, along with the already known variant A, a novel tjipanazole (TPZ), variant D, was found (Wright et al. 2005). Currently, from the same species three new TPZs (i.e., K, L, and M), which share the indolocarbazole core structure, were isolated together with the known TPZs D and I (Chilczuk et al. 2020). Among them, variant M had activity similar to that of arcyriaflavin A, an inhibitor of cell proliferation (Chilczuk et al. 2020).

Cytotoxic indolocarbazoles (i.e., 6-Cyano-5-methoxy-12-methylmdolo[2, 3-*a*]carbazole and 6-cyano-5-methoxyindolo[2, 3-*a*]carbazole), with potent antiviral (anti-HSV) and weak antitumor activity, were isolated from *Nostoc sphaericum* EX-5-1, collected from a mud sample in the region of the University of Hawaii campus grounds in Manoa and grown in mass culture (Knübel et al. 1990; Patterson et al. 1994). Three other novel indole alkaloids (9H-pyrido-[3,4-ß] indoles), the β-carbolines bauerines A-C, were extracted from the heterocytous filamentous cyanoprokaryote *Dichothrix baueriana* GO-25-2, collected soil of Hanakoa Valley at Hawaii and demonstrated antiviral activity against HSV-2 (Larsen et al. 1994; Pattersson et al. 1994). The first report on the production by prokaryotic algae of the simplest β-carboline, norharman(e), known from higher plants and some marine dinoflagellates, sponges, and bacteria, came from the study of different Nostocales

(Volk 2008). Among the cultured strains, which excreted this alkaloid, were *Anabaena inaequalis*, *Leptolyngbya foveolarum* (Syn. *Phormidium foveolarum*), *Nostoc carneum*, and *Nostoc minutum* that have been frequently recorded in both freshwaters and soils. Later, in a comparative study of four cyanoprokaryote strains, the highest production of this antimicrobial and antitumor compound was recorded from *Nostoc linckia* PACC 5085 (Karan et al. 2016).

In a strain of a soil *Nostoc commune*, field-grown in Okayama, Japan, the alkaloid cytotoxin nostodione A, considered a "mitotic poison," was discovered (Kobayashi et al. 1994; Gärtner et al. 2021).

Calothrixins A and B, pentacyclic indolophenanthridins, isolated from *Calothrix* sp., have a structural assembly of quinoline, quinone, and indole pharmacophores (Rickards et al. 1999; Doan et al. 2000; Xu et al. 2016). Almost along with their discovery their anticancer activity, potent cytotoxicity to HeLa and other cells, as well as their antimalarial and antibacterial activity, was proved (Rickards et al. 1999; Xu et al. 2016). Firstly, the mode of action was unknown until it was shown that calothrixin A can inhibit DNA replication and DNA-directed RNA synthesis (Doan et al. 2000).

Tubercidin is an N-glycosylpyrrolopyrimidine, an adenosine in which carbon replaces the not attached to the ribose moiety 5-membered ring (https://pubchem.ncbi.nlm.nih.gov/compound/Tubercidin). It was discovered in the Gram-positive bacteria *Streptomyces tubercidicus* (Anazi et al. 1957) and then was identified in *Hassalia byssoidea* (strain *Tolypothrix byssoidea* H-6-2), collected at the island of Oahu, Hawaii (Barchi Jr. et al. 1983). Subsequently it was found to be produced by Fijian *Plectonema radiosum* DF-6-1, *Tolypothrix distorta* BL-11-2, and *Scytonema saleyeriense* var. *indica* CV-14-1 (Stewart et al. 1988). This alkaloid exhibited strong antimicrobial and antitumor properties and is also a cytotoxin, nucleoside that inhibits DNA and RNA, and also the synthesis of protein in KB and NIH/3T3 cancer cell lines, and is effective against P-388 lymphocytic maurine leukemia (Anazi et al. 1957; Smith et al. 1973; Barchi Jr. et al. 1983; Demay et al. 2019). It has to be noted that in *P. radiosum* DF-6-1 and *T. distorta* BL-11-2, most of the antifungal activity was due to tubercidin 5′-α-D-glucopyranose (Stewart et al. 1988). Along with tubercidin, from the genus *Tolypothrix* was isolated another cytotoxin, the nucleoside toyocamycin, which was discovered earlier in *Streptomyces toyacaensis* (Stewart et al. 1988). Toyocamycin was the major constituent in *Tolypothrix tenuis* BN-7-4, but the most potent antifungal nucleoside in this strain was toyocamycin 5′-α-D-glucopyranose (Stewart et al. 1988).

2.7 *Toxic Polyphenols*

Ambigols (AMLs), a class of chlorinated phenols, were isolated from the terrestrial *Fischerella ambigua* in several variants and demonstrated diverse inhibitory activity against viruses (HIV-1), bacteria, fungi, plasmodia, trypanosomes algae, and mollusks, as well as cytotoxicity against versatile mammalian cell lines (Falch et al.

1993; Carpine and Sieber 2021). The first AMLs, A and B, characterized as highly halogenated aromatic compounds, were discovered in extracts from *F. ambigua* EAWAG 108b (Falch et al. 1993). Subsequently, after changing the cultivation medium from Z- to BG-11, from the same strain, ambigol (AML) C was extracted (Wright et al. 2005). One of the recent studies of AMLs, based on strain *F. ambigua* 108b, revealed two novel compounds (AMLs D and E), along with known congeners A and C, and the significant potential of AML C as quorum-sensing inhibitor which acted through induction of prodiogisin production in *Serratia marcesens* (Chilczuk et al. 2020). Since 1965 the strain *Fischerella ambigua* 108b has been frequently investigated as a promising source of novel biologically active compounds, and, therefore, it became a subject of a taxonomic revision (Jung et al. 2021). It was done by a polyphasic approach which combined genetic and morphological analyses, and the strain name was changed to *Symphyonema bifilamentata* (Jung et al. 2021).

A novel highly purified polyphenolic compound, the cyclopentenedione nostotrebin 6, with a fully substituted 2,2′-bis(cyclopent-4-en-1,3-dione) skeleton, was discovered in the soil *Nostoc* sp. strain Lukesová 27/97 (Zelík et al. 2010; Cheel et al. 2014). First its inhibitory potency against acetylcholinesterase and butyryl cholinesterase was demonstrated (Zelík et al. 2010), but later, due to its broader activities, it was presumed as a pan-assay interference compound (Kossack et al. 2020). Currently, nostotrebin 6 and two new related compounds, nostolactone 4 and nostotrebinlactone 7, with antibacterial activity, were isolated from *Nostoc* sp. CBT 1153 (Kossack et al. 2020; Carpine and Sieber 2021).

A simple intracellular polyphenol, anthraquinone with antimicrobial activity was found in *Nostoc commune* EAWAG 122b (Jaki et al. 2000a). According to Rojas et al. (2020) and Al-Yousef and Amina (2021) from a polar fraction of *Fischerella ambigua* a cytotoxic and antiprotozoal anthraquinone derivative with an indeno-anthracene-trione skeleton, eucapsitrione, was extracted. This toxin was named after the freshwater *Eucapsis*, in which it was discovered (Sturdy et al. 2010; Demay et al. 2019).

Alkylresorcinols are a large group of natural amphiphilic phenolic lipids, widely spread in bacteria, plants, fungi, and insects (Martins et al. 2019). Alkylresorcinols are involved in the membranes of their producers and have been related to some of their dermatotoxic, cytotoxic, genotoxic, antioxidant, or signaling properties (Martins et al. 2019). Among the alkylresorcinols of Cyanoprokaryota are the cylindrocyclophanes, a family of six cytotoxic natural products that contain a highly unusual [7.7] paracyclophane ring system (Nakamura et al. 2012). After the first isolation and structural characterization of the cylindrocyclophanes in the early 1990s (Moore et al. 1990), a BGC that is involved in cylindrocyclophane production was identified in *Cylindrospermum licheniforme* (Nakamura et al. 2012).

2.8 Toxic and Volatile Terpenes

Noscomin with its unique diterpenoid skeleton was discovered as a novel extracellular antibacterial metabolite of terrestrial *Nostoc commune* EAWAG 122b, active against *Bacillus cereus*, *Staphylococcus epidermidis*, and *Escherichia coli* (Jaki et al. 1999). Subsequently, from the same strain for the first time as a natural product was extracted an antimicrobial indane-derivative along with five more extracellular cytotoxic, antibacterial, and molluscicidal terpene compounds, named comnostines A-E (Jaki et al. 2000a, b; Carpine and Sieber 2021). Later, three other novel terpene compounds with strong antibiotic activity at a molar range have been isolated: (1) scytoscalarol of *Scytonema* sp. UTEX 1163 is a sesterterpene bearing a guanidine group, able to inhibit *Bacillus anthracis*, *Escherichia coli*, *Mycobacterium tuberculosis*, *Staphylococcus aureus*, and *Candida albicans* (Mo et al. 2009a); (2) cybastacines A and B from a cave *Nostoc* sp. BEA-0956 (collected in Montañon Negro, Canary Islands, Spain), possess a pentacyclic sesterterpenoid–alkaloid skeleton structure with a guanidinium group and are potent inhibitors of bacteria from genera *Enterococcus*, *Mycobacterium*, *Nocardia*, *Staphylococcus*, and *Streptococcus* (Cabanillas et al. 2018). Scytoscalarol was pointed also as cytotoxic against Vero cells at molar concentrations (Mo et al. 2009a). Tolypodiol, a novel diterpenoid, was identified from *Tolypothrix nodosa* HT-58–2, collected from soil in Nan Madol, Pohnpei, which along with its monoacetate derivative was responsible for the anti-inflammation in the maurine ear edema assay (Prinsep et al. 1996).

The specific abietic and dehydroabietic acids, occurring in high amounts in resin, long have been considered defense metabolites of coniferous plants that can serve as their biomarkers (Costa et al. 2016). These diterpenoids were detected spectroscopically and proved genetically in a study of 15 strains of cyanoprokaryotes (mostly from marine habitats), one of which was the terrestrial endolithic *Nodosilinea* sp. (=*Leptolyngbya antarctica*) LEGE 13457, collected from sandstones in McMurdo Valleys, Antarctica (Costa et al. 2016). In this strain dehydroabietic acid was the most abundant compound inside the cells. Both acids inhibited the growth of the coccoid cyanoprokaryote *Synechococcus* cf. *nidulans* LEGE 07171 (Costa et al. 2016).

In a search for the active antibacterial compound of Antarctic strain *Nostoc* CCC 537, an intracellular biomolecule similar to anthraquinone and indane-derivatives of a diterpenoid was characterized (Asthana et al. 2009).

Nostocionone (Nost) and its derivative NostD3 [(1E,4E)-1-(3,4-dihydroxyphenyl)-5-(2,6,6-trimethylcyclohex-1-enyl)penta-1,4-dien-3-one] were isolated from *Nostoc commune* and considered as potential beneficial agents for treatment of acne due to their strong inhibition of the growth of *Propionibacterium acne* (Itoh et al. 2014; Li and Guo 2018). These organic compounds are derivatives of ß-ionone (Li and Guo 2018), a member of the class of terpenes with three consecutive isoprene units, known as sesquiterpenoids (oodb.ca/compounds/FDB015469).

Cyanoprokaryotes can produce nontoxic terpenoid volatile compounds, which can cause taste and odor problems in aquatic environments, such as the monoterpene 2-methylisoborneol (MIB) and the irregular sesquiterpene geosmin (GSM) (Meriluoto et al. 2017). Up to now, MIB and its encoding BGC have not been identified in terrestrial cyanoprokaryotes (Micaleff et al. 2015). By contrast, the odoriferous GSM, well-known for its "earthy" smell, is more often associated with soils, with *Nostoc* pointed as its main causative agent (Churro et al. 2020). Recently, GSM production was associated also with *Microcoleus asticus*, newly described from the Lisbon city center (Churro et al. 2020; Gärtner et al. 2021). The GSM BGC (encoding GSM synthase and two cyclic nucleotide-binding domain proteins) was recorded in three more genomes of soil algae, specifically *Fischerella muscicola* SAG 1427–1, *Hapalosiphon welwitschii* UTEX B1830 (possibly identical *Fischerella* sp. PCC 9431), and *Westiella intricata* UH HT-29-1 (Micaleff et al. 2015). The GSM-encoding genes have been comprehensively investigated and a great similarity was found among the sequences of cyanoprokaryotes from soils and freshwaters (Giglio et al. 2008; Micaleff et al. 2015; Churro et al. 2020).

2.9 Toxic Carbohydrates

Nostoflan is a novel cytotoxic acidic polysaccharide, isolated from a culture of *Nostoc flagelliforme*, collected from soil at Alxa, China (Kanekiyo et al. 2005). This polysaccharide consists of similar parts of glucose, galactose, mannose, and glucorinic acid, contrasting to the composition of the field-grown alga, in which glucose, xylose, and galactose (in proportion 2:1:1) were detected along with mannose and arabinose in trace amounts (Huang et al. 1998). Nostoflan demonstrated a potent antiviral activity against a broad spectrum of enveloped viruses with carbohydrates as cellular receptors, such as influenza A virus, HSV-1, HSV-2, and human cytomegalovirus, but unlike sulfated polysaccharides showed no antithrombin activity (Kanekiyo et al. 2005, 2007). Here is worth noting that *Nostoc flagelliforme*, which commonly occurs in some arid and semiarid regions of China, is traditionally used for food in this country, and its safety for human consumption has been proved experimentally (Takenaka et al. 1998).

In *Nostoc commune* DRH-1, which showed tolerance to desiccation, a novel exopolyssacharide was found that possesses a 1-4-linked xylogalactoglucan backbone with D-ribofuranose and nosturonic acid pendant groups (Helm et al. 2000). This compound was investigated with regard to algal thermotolerance, but no for toxicity and it was proved that nosturonic acid was not detected in other strain of the same species (i.e., *N. commune* UTEX 584) even when it was cultured under the same conditions as the strain DRH-1 (Helm et al. 2000). Comparison of the structure of this exopolysaccharide with nostoflan showed their slight difference in pendant sugar residues (Kanekiyo et al. 2005).

2.10 *Toxic Sulfoglycolipids*

Cytotoxic sulfoglycolipids (sulfonic acid containing glycolipids) with antiviral activity against HIV were found to occur in thylakoid membranes and cell walls of heterocytes in some Cyanoprokaryota (Gustafson et al. 1989; Mazur-Marzec et al. 2021). The first data on these metabolites from lipophilic extracts of terrestrial algae were provided by Reshef et al. (1997), who proved the different extents of their inhibition of HIV-1 RT enzymatic activity. Two diacylglycerol glycosides were isolated from *Leptolyngbya tenuis* (as strain *Phormidium tenue* CN-2-1) collected from Palau Islands (Guistafson et al. 1989). From *Scytonema* sp. TAU SL-30-1-4 five new diacylated sulfoglycolipids were discovered and four novel acylated diglycolipids were isolated from *Oscillatoria variabilis* (as *O. raoi* TAU IL-76-1-2), while *Leptolyngbya tenuis* (as strain *Phormidium tenue* TAU IL-144-1) contained two glycolipids, known before (Reshef et al. 1997; Loya et al. 1998). At nanomolar range, most of these sulfoglycolipids inhibited the function of DNA polymerase but had no significant effect on ribonuclease H (Reshef et al. 1997; Loya et al. 1998; Mazur-Marzec et al. 2021).

2.11 *Miscelaneous Compounds*

Novel cyclic undeca- and dodecapeptides, laxaphycins A-E and other 15 minor laxaphycins, have been isolated from *Anabaena laxa* UH FK-1-2, collected from a mud sample in Manoa campus of the University of Hawaii, and were found to have antiproliferative effects against fungi (i.e., *Aspergillus oryzae*) and tumor cell lines (e.g., human epidermoid carcinoma and colorectal adenocarcinoma) (Frankmölle et al. 1992a, b). The most interesting is the synergistic action of these cytotoxic peptides (Patterson et al. 1994).

Tolybyssidins A and B are two novel cyclic tridecapeptides, comprised of non-natural amino acid dehydrohomoalanine (Dhha) and proteinogenic amino acids albeit with D- or L-configuration (Jaki et al. 2001). They were discovered from the culture medium of *Hassalia byssoidea* (strain *Tolypothrix byssoidea* EAWAG 195) and exhibited moderate antifungal activity against *Candida albicans* (Jaki et al. 2001).

In *Fischerella ambigua* PTCC 1635, isolated from a rice field sample collected in northern Iran, a novel cyclic polymer, parsiguine, was discovered (Ghasemi et al. 2004). This is an oxygenated ethylenic compound, unique and unrelated to known cyanoprokaryote metabolites, that exhibited antifungal and antibacterial activities (Ghasemi et al. 2004). In some cases, it can be used against MDR bacteria (Parsaeimehr and Lutzu 2016).

Tolyporphin, an unusual porphyrin from the lipophilic extract of *Tolypothrix nodosa* HT-58-2, accounted for most of the activity of this extract that reversed MDR of a human ovarian adenocarcinoma cell line (Prinsep et al. 1992c).

Cyanobacterin, "other linear non-peptide" (https://pubchem.ncbi.nlm.nih.gov/compound/ Cyanobacterin), is a natural product with strong algicidal and herbicidal activity caused by inhibition of the electron transport in all plants, isolated from *Scytonema hofmannii* UTEX 2349 (Mason et al. 1982; Inderjit and Dakshini 1994). This secondary metabolite was characterized by Pignatello et al. (1983) as a chlorine-containing γ-lactone, or butyrolactone (the simplest 4-carbon lactone, Schwarz et al. 2019). *Scytonema hofmannii* can produce also the hydroxy analogue of cyanobacterin, hydroxycyanobacterin, which is supposed to be its precursor and is a specific inhibitor of the photosynthetic electron transport (Lee and Gleason 1994; Singh et al. 2002).

An imide is a functional group, formed by two acyl groups linked to nitrogen, which occurs in some unique polychlorinated metabolites derived from amino acid precursors (Williard and De Laszlo 1984). The first naturally occurring imides, N-acylpyrrolinones (linear peptides which contain an N-methylated or acetylated N-terminal amino acid) were long associated with Indopacific marine sponges *Dysidea herbacea* and with marine cyanoprokaryote *Moorea producens* (Paik et al. 1994). Later studies proved that the production of such compounds (e.g., dysidin) in sponges is caused by their symbiotic blue-green algae, such as *Oscillatoria spongelliaea* (Paik et al. 1994). From *Scytonema wolleanum* (as strain *S. mirabile* BY-8-1) five cytotoxic N-acylpyrrolinones, mirabimides A-E, have been isolated, with mirabimide A being the major imide (Carmeli et al. 1991a; Patterson et al. 1994; Paik et al. 1994; Welker and Döhren 2006). Mirabimides were cytotoxic to KB cell lines, with mirabimide E displaying the strongest murine solid antitumor selective cytotoxicity (Patterson et al. 1994; Paik et al. 1994).

Dichlormethane/2-propanol with antifungal activity against *Aspergillus oryzae*, *Candida albicans*, *Penicillium notatum*, *Saccharomyces cerevisiae*, and *Trichophyton mentagrophytes* was isolated from *Fischerella ambigua* UTEX 1903 (Smitka et al. 1992, Al-Yosuf and Amina 2021)

A glycine-rich peptide toxin, isolated from *Scytonema* sp. MKU 106, had a biopesticidal effect against the common pests of cotton crops, the American boll worm (*Helicoverpa armegera*) and leaf roller (*Stylepta derogata*) (Sathyamoorthy and Shanmugasunduram 1996; Singh et al. 2002).

2.12 *Other Presumed Toxigenic Soil Algae, the Toxicity of Which Awaits Investigation*

Gärtner et al. (2021) supposed toxigenic potential of the genus *Neowestiellopsis*, discovered in dried paddy fields of Iran (Kabirnata et al. 2018), due to its close phylogenetic relations with *Westiellopsis* because in a strain of *W. prolifica*, collected from Tigris River (Iraq), both MCs and *mcyE* gene were proved (Abed et al. 2013, 2019), and in its strain, isolated from mud at Hawaii, westiellamide was detected (Prinsep et al. 1992a). On the other hand, in the strain *Westiellopsis* sp. NQAIF324, collected from a seasonal creek in tropical Australia, neither MCs nor their biosynthetic *mcy* genes were found (Cirés et al. 2014).

Apart from the genera, discussed in the text above, some other genera of soil algae, which have their toxic counterparts in aquatic environments, such as *Coleofasciculus* (Syn. *Microcoleus* p.p.), *Desertifilum*, and *Pseudanabena*, were included in the list of potentially toxigenic aeroterrestrial and extremophilic algae the toxicity of which awaits investigation (Gärtner et al. 2021).

3 Discussion and Conclusion

The conducted analysis revealed that toxic compounds are produced by more than 216 strains, belonging to at least 67 species and 3 varieties from 41 genera of soil algae: 5 species from 5 coccal genera, 10 species from 8 non-heterocytous genera, 45 species and 3 varieties of 20 heterocytous genera of Cyanoprokaryota, as well as 3 centric species from 3 genera and 4 species from 4 genera pennate diatoms (Bacillariophyceae, Ochrophyta). Four more genera are supposed to be toxigenic according to their phylogenetic relations, or because of existence of toxic aquatic counterparts. As it could be expected, most of the toxigenic soil algal species and genera are prokaryotic—90% (63/70) and 77% (31/41), respectively. However, it is possible to suggest an increase in these numbers because toxin production has been demonstrated in genera from other taxonomic groups, which have not been considered toxigenic before.

Despite the inequality of conducted research, it is possible to state that in the present state of the art in most of the strains low number of toxins was examined: less than 5 were found in 175 strains (81%), among which in 103 strains (48%) 2–5 toxins were identified, and from 72 (33%) just a single compound was reported (Fig. 1). Only in 2 strains (1%) 21 and 23 compounds, and in 4 strains (2%) 15–20 compounds have been recorded, while in 20 strains (9%) 6–10, and in 15 strains (7%) 11–15 toxins were identified.

Up to now, the most studied strains (196/216, or 91%) are from the group of heterocytous cyanoprokaryotes, and among them six genera can be outlined as best

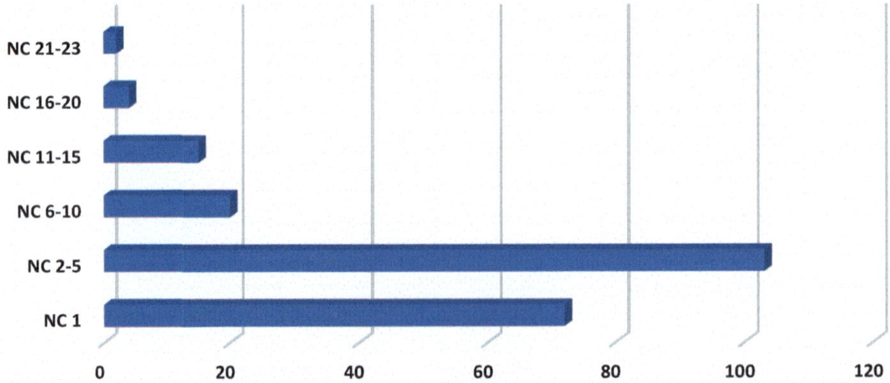

Fig. 1 Number of toxic compounds (NC) produced by relevant numbers of studied strains

investigated, such as *Nostoc* (107/216, or 50%), *Fischerella* (16/216, or 7%), *Scytonema* (14/216, or 6%), *Hapalosiphon* (9/216, or 4%), *Tolypothrix* (6/216, or 3%) and *Brasilonema* (5/216, or 2%). The highest number of main toxic chemical classes was investigated in heterocytous *Nostoc* (8) and *Scytonema* (7), *Fischerella* (5), *Hapalosiphon*, and *Tolypothrix* (4 in each of them), as well as in the non-heterocytous *Microcoleus* (4)—Fig. 2. The highest number of toxins was isolated or

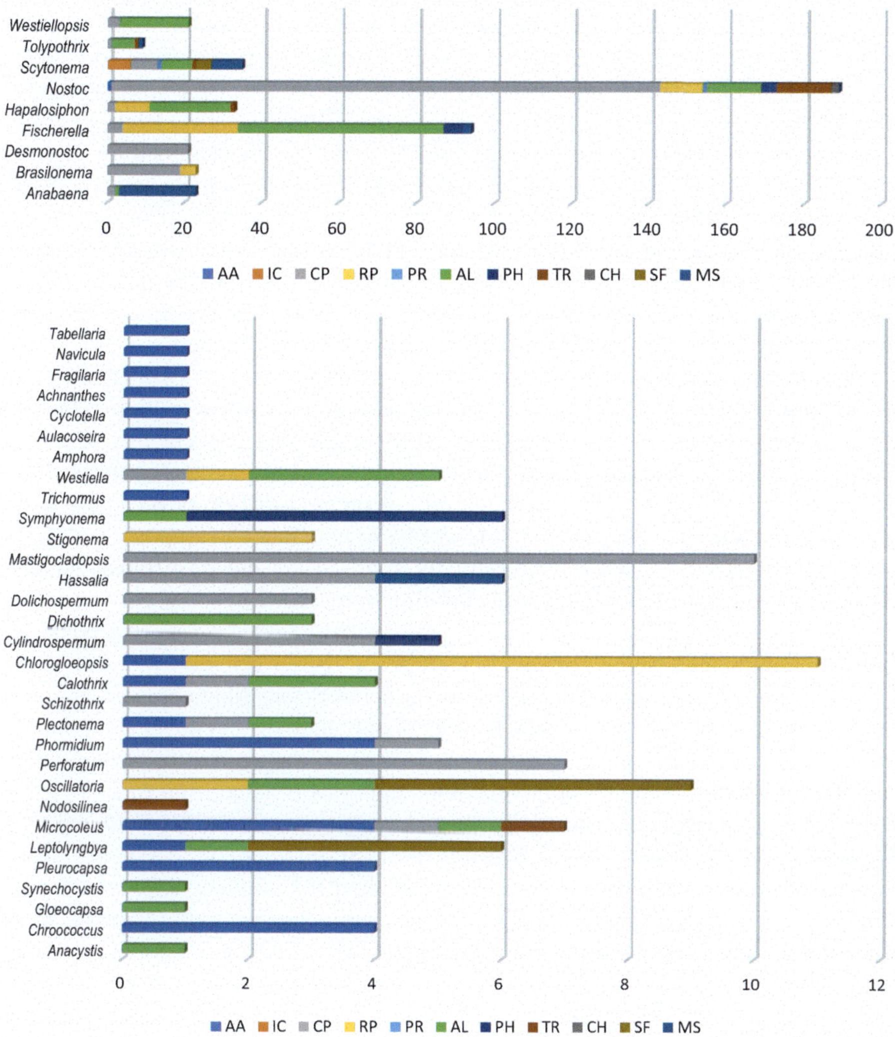

Fig. 2 Above: Number of toxic compounds by classes in the nine richest in toxins heterocytous genera of soil Cyanoprokaryota. Below: Number of toxic compounds by classes in the remaining 31 genera of Cyanoprokaryota and 7 of Bacillariophyceae. Legend: AA amino acids, IC isocyanides, CP non-ribosomally synthesized peptides (cyanopeptides) and polyketides, RP ribosomally synthesized and post-translationally modified peptides, PR proteins, AL alkaloids, PH polyphenols, TR terpenes, CH carbohydrates, SF sulfoglycolipids, MS miscellaneous compounds

genetically proved in the soil species of nine genera: *Nostoc* (189), *Fischerella* (94), *Scytonema* (35), *Hapalosiphon* (33), *Anabaena* and *Brasilonema* (23 in each of them), *Desmonostoc* (22) and *Westiellopsis* (21), and *Tolypothrix* (9)—Fig. 2. High number of compounds was isolated also from the non-heterocytous *Phormidium* (8), *Microcoleus* and *Perforatum* (7 in each of them), and *Oscillatoria* (5). Coccal cyanoprokaryotes practically have never been examined, and the presence of toxic compounds in their species was only presumed.

In total, 430 toxins and their variants, and the nontoxic odoriferous GSM as well, from 77 subclasses of the examined 11 classes have been reported from soil algae. The richest (59%) are peptides and polyketides, represented by 252 compounds, most of which are non-ribosomally synthesized (202/430, or 47% from all compounds)—Fig. 3. They have been found in 21 (51%) of the examined 41 genera, among which non-ribosomally synthesized peptides and polyketides were reported from 18 (44%), while ribosomal peptides were recorded only in 8 (20%) genera. By contrast, isocyanides and carbohydrates have been recorded only from *Scytonema* and *Nostoc*, respectively. The other classes have been investigated in genera as follows: amino acids—17, proteins—2, alkaloids—19, polyphenols—4, terpenes—6, sulfoglycolipids—3, and miscellaneous compounds—from 5 genera.

Most identified toxins (320/430, or 74%) are novel, and most of them showed potent antimicrobial or cytotoxic activity. The new compounds were discovered in all examined classes, except among the amino acids and polyphenols (Fig. 3). Among the newly characterized compounds peptides and polyketides comprised the biggest part (175/320, or 55%), and majority of them (146/175, or 83%) are non-ribosomal peptides and polyketides. It is worthy to note that all compounds (100%) from four classes (isocyanides, ribosomally synthesized and post-translationally modified peptides, carbohydrates, and sulfolipids) recorded in soil algae were new (Fig. 3).

Novel compounds were discovered in strains of 18 soil genera, which comprise 44% from all examined genera. Among them, heterocytous filamentous cyanoprokaryotes seem to be the most prolific producers of both toxic and novel compounds, with 14 genera (34% from all genera and 78% from the producers of new compounds). The highest number of novel compounds was isolated from *Nostoc* (157), *Fischerella* (38), and *Scytonema* (33) followed by *Desmonostoc* (22), *Anabaena* and *Hapalosiphon* (20 in each of them), *Brasilonema* (16), *Tolypothrix* (7), *Symphyonema* (6), *Hassalia* and *Westiellopsis* (4 in each of them), *Calothrix*, *Dichothrix*, and *Stigonema* (3 in each of them)—Fig. 4. Novel compounds were discovered also in non-heterocytous *Leptolyngbya* (4) and *Oscillatoria* (2), *Plectonema and Schizothrix* (1 in each of them)—Fig. 4. This result is not surprising when the number of investigated species or strains and number of examined classes and subclasses are considered (Fig. 5). For example, 7 classes and 48 subclasses of toxins have been examined in 107 soil strains of at least 16 species of *Nostoc*, which led to identification of 189 toxins, 157 (83%) of which are novel (Fig. 5). Interestingly, from genera with less studied number of strains the identified novel compounds belong to less examined toxin classes (Figs. 4 and 5).

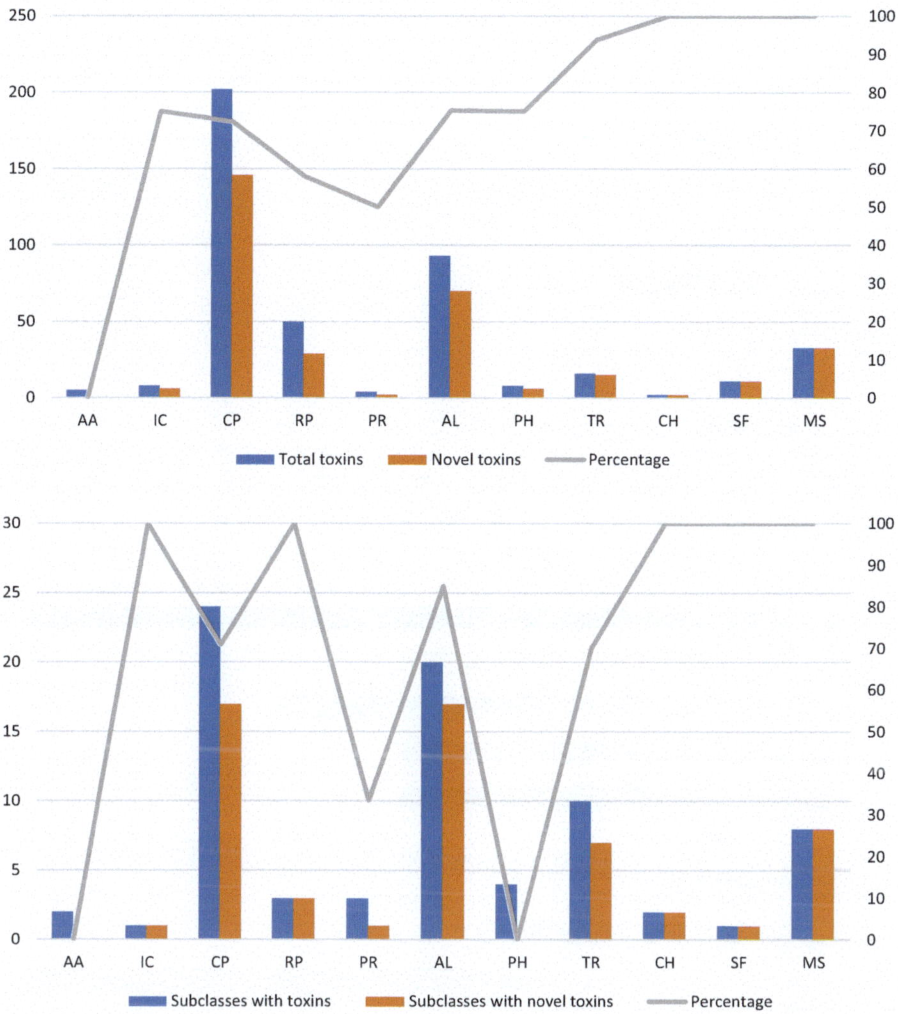

Fig. 3 Above Comparison of the number (right axis) and percentage (left axis) of total and novel toxic compounds in the 11 toxin classes. Below: Comparison of number (right axis) and percentage (left axis) of subclasses in which toxins were found with the number of subclasses from which novel toxic compounds were characterized in the 11 toxin classes, abbreviated as follows: AA amino acids, IC isocyanides, CP non-ribosomally synthesized peptides (cyanopeptides) and polyketides, RP ribosomally synthesized and post-translationally modified peptides, PR proteins, AL alkaloids, PH polyphenols, TR terpenes, CH carbohydrates, SF sulfoglycolipids, MS miscellaneous compounds

The percentage of novel compounds is quite high: in 10 genera (56%) the novel compounds comprise 80–100%, in 4 (22%) genera they account for 50–79%, and in the remaining 4 genera (22%) the percentage of novel compounds is 19–40% (Fig. 6).

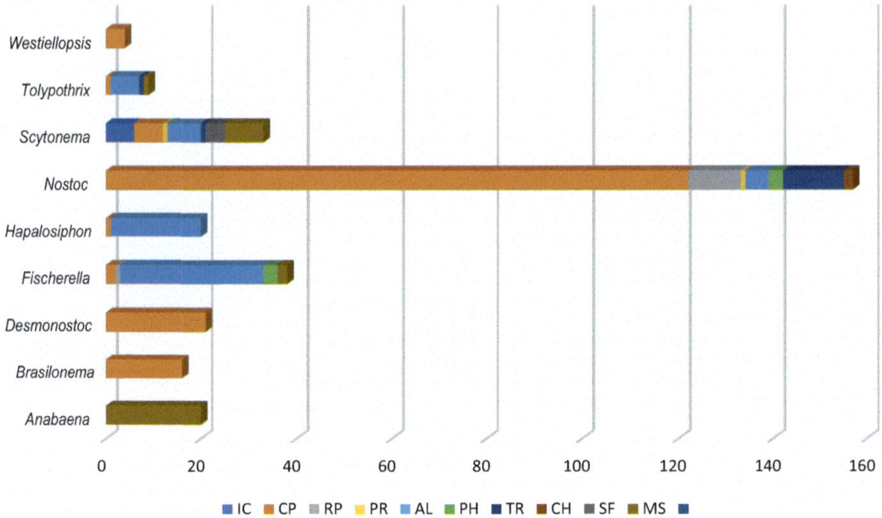

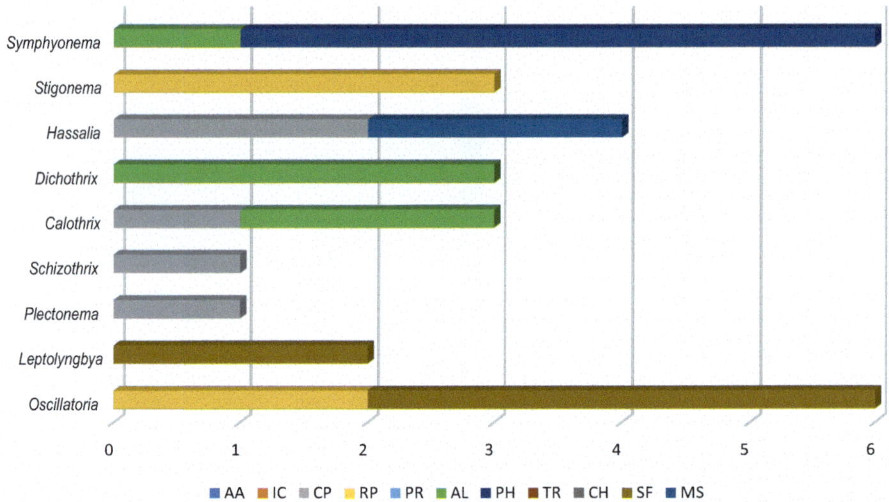

Fig. 4 Above: Number of novel toxic compounds by classes in the nine richest in toxins heterocytous genera of soil Cyanoprokaryota. Below: Number of novel toxic compounds by classes in the remaining 9 genera of Cyanoprokaryota. Legend: AA amino acids, IC isocyanides, CP non-ribosomally synthesized peptides (cyanopeptides) and polyketides, RP ribosomally synthesized and post-translationally modified peptides, PR proteins, AL alkaloids, PH polyphenols, TR terpenes, CH carbohydrates, SF sulfoglycolipids, MS miscellaneous compounds

In conclusion, soil algae of 41 genera (Fig. 7) have yielded a plethora of toxic compounds with remarkable structural diversity. Considering the high number of species in the peculiar ecological group of soil algae, it is to expect that further studies will result in higher number of compounds. Along with ecotoxicological

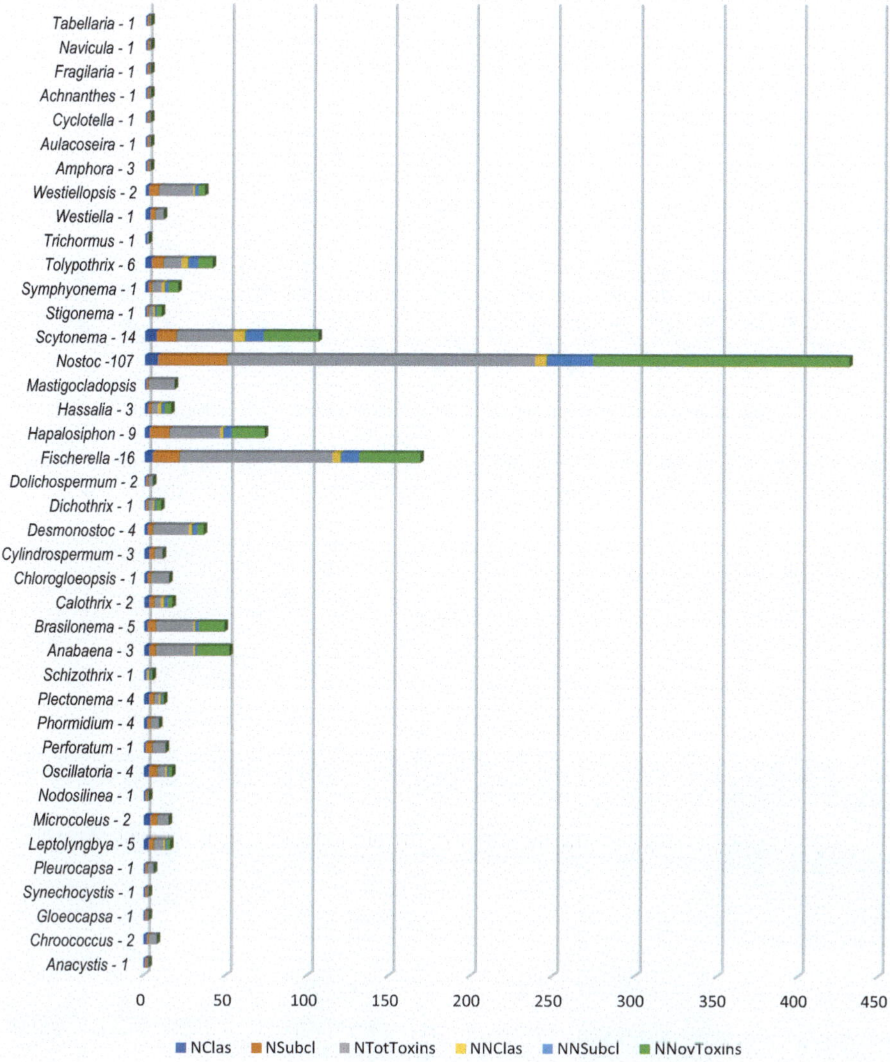

Fig. 5 Number of examined toxigenic classes (NClas) and subclasses (NSubcl), total number of toxic compounds (NTotToxins), and number of novel compounds (NNovToxins) with number of classes (NNClas) and subclasses (NNSubcl) from which novel toxins were identified in different soil genera, according to the number of studied strains indicated by number after the generic name

concern and potential hazardous effects on human and ecosystem health, some of these toxins are of potentially significant importance for application in pharmacology, medicine, cosmetics, and biotechnology. Considering the versatility of effects of these compounds, and the recent sharply increasing interest in antiviral compounds in conditions of COVID-19 pandemic, we believe that the summarized data provided in this chapter will enhance the work of future researchers and will give a

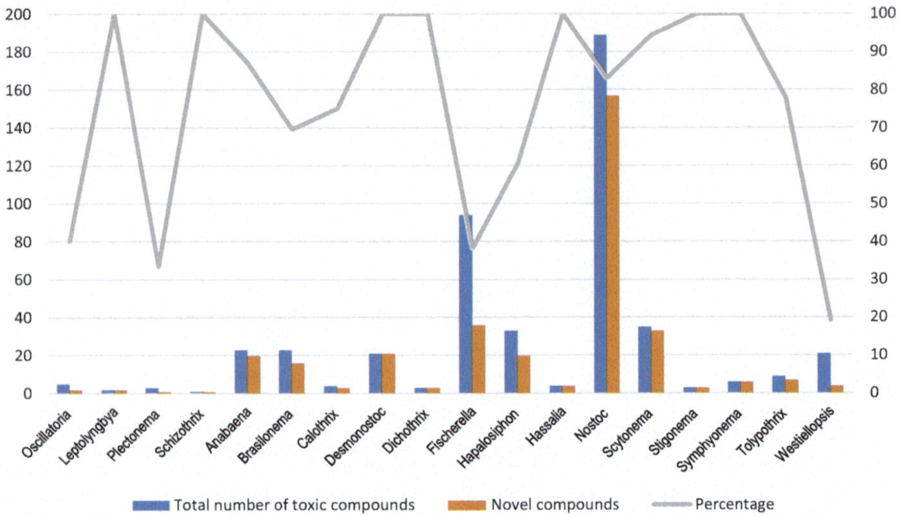

Fig. 6 Comparison of numbers of total and novel toxic compounds, and the percentage of novel compounds in 18 soil genera of Cyanoprokaryota, in which novel toxins were discovered

Fig. 7 The significance of each of the 39 soil algal genera investigated according to the total number of compounds and novel compounds produced in different classes, subclasses, and according to the number of examined strains

boost to further more intensive studies of the enormous diversity of secondary metabolites of soil algae.

Acknowledgments This study was supported financially by the National Science Fund of the Ministry of Education and Science (Bulgaria) through the grant DN13/9 from December 15, 2017.

References

Abed IJ, Abdulhasan GA, Al-Hussieny AA et al (2013) Molecular detection of toxigenic new record cyanobacterium: *Westiellopsis prolifica* in Tigris River. Int J Sci Tech Res 2(11):87–91

Abed IJ, Abdulhasan GA, Moushib LI (2019) Molecular and immunological methods to confirm toxiginicity (microcystin production) of *Westiellopsis prolifica* isolated from Tigris River -Iraq. Baghdad Sci J 16:2019978

Admi V, Afek U, Carmeli S (1996) Raocyclamides A and B, novel cyclic hexapeptides isolated from the cyanobacterium *Oscillatoria raoi*. J Nat Prod 59:396–399

Ahmad S, Saleem M, Riaz N et al (2020) The natural polypeptides as significant elastase inhibitors. Front Pharmacol 11. https://doi.org/10.3389/fphar.2020.00688

Al-Yousef HM, Amina M (2021) Phytoconstituents and pharmacological activities of cyanobacterium *Fischerella ambigua*. Arab J Chem 14(6):103153. https://doi.org/10.1016/j.arabjc.2021.103153

Anazi K, Nakamura D, Suzuki S (1957) A new antibiotic, tubercidin. J Antibiotic Ser A 10:201–204

Asthana RK, Deepali, Tripathi MK et al (2009) Isolation and identification of a new antibacterial entity from the Antarctic cyanobacterium *Nostoc* CCC 537. J Appl Phycol 21(81):81. https://doi.org/10.1007/s10811-008-9328-2

Balunas MJ, Grosso MF, Villa FA et al (2012) Coibacins A-D, antileishmanial marine cyanobacterial polyketides with intriguing biosynthetic origins. Org Lett 15:3878–3881. https://doi.org/10.1021/ol301607q

Banker R, Carmeli S (1998) Tenuecyclamide A-D, cyclic hexapeptides from the cyanobacterium *Nostoc spongiaeforme* var. *tenue*. J Nat Prod 61:1248–1251

Barchi JJ Jr, Norton TR, Furusawa E et al (1983) Identification of a cytotoxin from *Tolypothrix byssoidea* as tubercidin. Phytochemistry 22(12):2851–2852

Barrow RA, Hemscheidt T, Liang J et al (1995) Total synthesis of cryptophycins. Revision of the structure of cryptophycins A and C. J Am Chem Soc 117:2479–2490. https://doi.org/10.1021/ja00114a011

Bates SS (2000) Domoic acid producing diatoms: another genus added. J Phycol 36:978–983

Batsalova T, Moten D, Basheva D et al (2016) In vitro cytotoxicity and antioxidative potential of *Nostoc microscopicum* (Nostocales, Cyanobacteria). Toxicol Forensic Med Open J 1:9–17. https://doi.org/10.17140/TFMOJ-1-102

Becker JE, Moore RE, Moore BS (2004) Cloning, sequencing, and biochemical characterization of the nostocyclopeptide biosynthetic gene cluster: molecular basis for imine macrocyclization. Gene 325:35–42. https://doi.org/10.1016/j.gene.2003.09.034

Berry JP (2008) Cyanobacterial toxins as Allelochemicals with potential applications as algaecides, herbicides and insecticides. Mar Drugs 6(2):117–146. https://doi.org/10.3390/md6020117

Bhat V, Dave A, MacKay JA et al (2014) The chemistry of hapalindoles, fischerindoles, mbiguines, and welwitindolinones. Alkaloids Chem Biol 73:65–160. https://doi.org/10.1016/B978-0-12-411565-1.00002-0

Bokesch HR, O'Keefe BR, McKee TC et al (2003) A potent novel anti-HIV protein from the cultured cyanobacterium *Scytonema varium*. Biochemistry 42(9):2578–2584. https://doi.org/10.1021/bi0205698

Bonjouklian R, Smitka TA, Doolin LE et al (1991) Tjipanazoles, new antifungal agents from the blue-green alga *Tolypothrix tjipanasensis*. Tetrahedron 47(37):7739–7750

Bonjouklian R, Moore EE, Patterson GML et al (1993) Hapalindole alkaloids as antifungal and antitumor agents. Eur Patent Appl EP0543516 A1:EP19920309876
Bonjouklian R, Moore EE, Patterson GML et al (1994) Preparation of hapalindole-related alkaloids from Blue-green algae. U.S. Patent 5292650
Bouaïcha N, Corbel S (2016) Cyanobacterial toxins emerging contaminants in soils: a review of sources, fate and impacts on ecosystems, plants and animal and human health. In: Larramendy ML, Soloneski S (eds) Soil contamination - current consequences and further solutions. InTechOpen, Rijeka, Croatia. https://doi.org/10.5772/64940
Cabanillas AH, Perez VT, Corral SM et al (2018) Cybastacines A and B: antibiotic Sesterterpenes from a *Nostoc* sp. Cyanobacterium. J Nat Prod 81:410–413
Cagide E, Becher PG, Louzao MC et al (2014) Hapalindoles from the cyanobacterium Fischerella: potential sodium channel modulators. Chem Res Toxicol 27:1696–1706
Cai W, Ratnayake R, Wang M et al (2021) Inhibition of cotranslational translocation by apratoxin S4: effects on oncogenic receptor tyrosine kinases and the fate of transmembrane proteins produced in the cytoplasm. CRPHAR 2:100053. https://doi.org/10.1016/j.crphar.2021.100053
Campbell P (2011) The future prospects for global arable land. Future Directions International, 19 May 2011, 1–6
Cardellina JH II, Marner FJ, Moore RE (1979) Malyngamide A, a novel chlorinated metabolite of the marine cyanophyte *Lyngbya majuscula*. J Am Chem Soc 101(1):240–242. https://doi.org/10.1021/ja00495a051
Carmeli S, Moore RE, Patterson GML (1990a) Tolytoxin and new Scytophycins from three species of Scytonema. J Nat Prod 53(6):1533–1542. https://doi.org/10.1021/np50072a021
Carmeli S, Moore RE, Patterson GML et al (1990b) Isonitriles from the blue green alga *Scytonema mirabile*. J Org Chem 55(14):4431–4438
Carmeli S, Moore RE, Patterson GML et al (1990c) Tantazoles, unusual cytotoxic alkaloids from the blue-green alga *Scytonema mirabile*. J Am Chem Soc 112(22):8195–8197. https://doi.org/10.1021/ja00178a070
Carmeli S, Moore RE, Patterson GML (1991a) Mirabimides A–D, new N-acylpyrrolinones from the blue-green alga *Scytonema mirabile*. Tetrahedron 47:2087–2096. https://doi.org/10.1016/S0040-4020(01)96119-8
Carmeli S, Moore RE, Patterson GML (1991d) Mirabazoles, minor tantazole-related cytotoxins from the terrestrial blue-green alga *Scytonema mirabile*. Tetrahedron Lett 32(23):2593–2596. https://doi.org/10.1016/S0040-4039(00)78793-4
Carmeli S, Paik S, Moore RE et al (1993) Revised structures and biosynthetic studies of tantazoles A and B. Tetrahedron Lett 34(42):6681–6684. https://doi.org/10.1016/S0040-4039(00)61674-X
Carneiro VMT, Avila CM, Balunas MJ et al (2014) Coibacins A and B: Total synthesis and Stereochemical revision. J Org Chem 79(2):630–642. https://doi.org/10.1021/jo402339y
Carpine R, Sieber S (2021) Antibacterial and antiviral metabolites from cyanobacteria: their application and their impact on human health. CRBIOT 3:65–81. https://doi.org/10.1016/j.crbiot.2021.03.001
Chatziefthimiou AD, Richer R, Rowles H et al (2014) Cyanotoxins as a potential cause of dog poisonings in desert environments. Vet Rec 174:484–485. https://doi.org/10.1136/vr.g3176
Cheel J, Kučerová P, Garrard I et al (2014) Two-step separation of nostotrebin 6 from cultivated soil cyanobacterium (*Nostoc* sp.) by high performance countercurrent chromatography. Molecules 19:8773–8787. https://doi.org/10.3390/molecules19078773
Cheruku P, Plaza A, Lauro G et al (2012) Discovery and synthesis of namalide reveals a new anabaenopeptin scaffold and peptidase inhibitor. J Med Chem 55(2):735–742. https://doi.org/10.1021/jm201238p
Chilczuk T, Steinborn C, Breinlinger S et al (2020) Hapalindoles from the Cyanobacterium Hapalosiphon sp. Inhibit T Cell Proliferation. Planta Med 86(2):96–103. https://doi.org/10.1055/a-1045-5178
Chlipala GE, Mo S, Orjala J (2011) Chemodiversity in freshwater and terrestrial cyanobacteria - a source for drug discovery. Curr Drug Targets 12(11):1654–1673. https://doi.org/10.2174/138945011798109455

Choi H, Mevers E, Byrum T et al (2012) Lyngbyabellins K–N from two palmyra atoll collections of the marine cyanobacterium *Moorea bouillonii*. Eur JOC 27:5141–5150. https://doi.org/10.1002/ejoc.201200691

Chrapusta E, Węgrzyn M, Zabaglo K et al (2015) Microcystins and anatoxin-a in Arctic biocrust cyanobacterial communities. Toxicon 101:35–40. https://doi.org/10.1016/j.toxicon.2015.04.016

Christiansen G, Dittmann E, Via Ordorika L et al (2001) Nonribosomal peptide synthetase genes occur in most cyanobacterial genera as evidenced by their distribution in axenic strains of the PCC. Arch Microbiol 176(6):452–458. https://doi.org/10.1007/s002030100349

Churro C, Semedo-Aguiar AP, Silva AD et al (2020) A novel cyanobacterial geosmin producer, revising *Geo*A distribution and dispersion patterns in bacteria. Sci Rep 10:8679. https://doi.org/10.1038/s41598-020-64774-y

Cirés S, Alvarez-Roa C, Wood SA et al (2014) First report of microcystin-producing *Fischerella* sp. (Stigonematales, Cyanobacteria) in tropical Australia. Toxicon 88:62–66

Costa MS, Rego A, Ramos V et al (2016) The conifer biomarkers dehydroabietic and abietic acids are widespread in Cyanobacteria. Sci Rep 6:23436. https://doi.org/10.1038/srep23436

Cox PA, Banack SA, Murch SJ (2003) Biomagnification of cyanobacterial neurotoxins and neurodegenerative disease among the Chamorro people of Guam. PNAS 100:13380–13383. https://doi.org/10.1073/pnas.2235808100

Cox PA, Banack SA, Murch SJ et al (2005) Diverse taxa of cyanobacteria produce β-N-methylamino-l-alanine, a neurotoxic amino acid. PNAS 102:5074–5078. https://doi.org/10.1073/pnas.0501526102

Cox PA, Richer R, Metcalf JS et al (2009) Cyanobacteria and BMAA exposure from desert dust—a possible link to sporadic ALS among Gulf War veterans. Amyotroph Lateral Scler 10:109–117. https://doi.org/10.3109/17482960903286066

Dehm D, Krumbholz J, Baunach M et al (2019) Unlocking the spatial control of secondary metabolism uncovers hidden natural product diversity in *Nostoc punctiforme*. ACS Chem Biol 14:1271–1279. https://doi.org/10.1021/acschembio.9b00240

Delawská K, Divoká P, Sedlák D et al (2021) New insights into tolytoxin effect in human cancer cells: apoptosis induction and the relevance of hydroxyl substitution of its macrolide cycle on compound potency. Chem Bio Chem 23:e202100489. https://doi.org/10.1002/cbic.202100489

Demay J, Bernard C, Reinhardt A et al (2019) Natural products from Cyanobacteria: focus on beneficial activities. Mar Drugs 17(6):320. https://doi.org/10.3390/md17060320

Dhar BC, Cimarelli L, Singh KS et al (2015) Molecular detection of a potentially toxic diatom species. Int J Environ Res Public Health 12(5):4921–4941. https://doi.org/10.3390/ijerph120504921

Dillenius JJ (1741) Historia muscorum. Oxonii

Dittmann E, Gugger M, Sivonen K et al (2015) Natural product biosynthetic diversity and comparative genomics of the Cyanobacteria. Trends Microbiol 23(10):642–652. https://doi.org/10.1016/j.tim.2015.07.008

do Amaral SC, Monteiro PR, Neto JDSP et al (2021) Current knowledge on microviridin from cyanobacteria. Mar Drugs 19:17. https://doi.org/10.3390/md19010017

Doan TN, Rickards RW, Rothschild JM et al (2000) Allelopathic actions of the alkaloid 12-*epi*-hapalindole E isonitrile and calothrixin A from cyanobacteria of the genera *Fischerella* and *Calothrix*. J Appl Phycol 12:409–416. https://doi.org/10.1023/A:1008170007044

Donia MS, Schmidt EW (2010) Cyanobactins—ubiquitous cyanobacterial ribosomal peptide metabolites. In: Comprehensive natural products II chemistry and biology. Elsevier, Oxford, pp 539–558

Donia MS, Schmidt EW (2011) Linking chemistry and genetics in the growing cyanobactin natural products family. Chem Biol 18(4):508–519. https://doi.org/10.1016/j.chembiol.2011.01.019

Donia MS, Ravel J, Schmidt EW (2008) A global assembly line for cyanobactins. Nat Chem Biol 4(6):341–343

Draganov S, Stoyneva M (1992) Algal flora of the Danube river (Bulgarian sector) and adjoined water basins. III. Algae from some adjoined water basins. Ann Univ Sof 82(2):63–78

Engene N, Rottacker EC, Kaštovský J et al (2012) *Moorea producens* gen. nov., sp. nov. and *Moorea bouillonii* comb. nov., tropical marine cyanobacteria rich in bioactive secondary metabolites. Int J Syst Evol Microbiol 62(Pt 5):1171–1178. https://doi.org/10.1099/ijs.0.033761-0

Ersmark K, del Valle JR, Hanessian S (2008) Chemistry and biology of the aeruginosin family of serine protease inhibitors. Angew Chem Int Ed 47:1202–1223

Ettl H, Gärtner G (2014) Syllabus der Boden-, Luft- und Flechtenalgen. 2 ergänzte Auflage. Springer Spektrum, Berlin Heidelberg

Evangelista V, Barsanti L, Frassanito AM et al (eds) (2007) Algal toxins: nature, occurrence, effect and detection, proceedings of the NATO advanced study institute on sensor systems for biological threats: the algal toxins case, Pisa, Italy, 30 September—11 October 2007. Springer, Dordrecht

Facciponte DN, Bough MWD, Seidler JL et al (2018) Identifying aerosolized cyanobacteria in the human respiratory tract: a proposed mechanism for cyanotoxin-associated diseases. Sci Tot Environ 645:1003–1013. https://doi.org/10.1016/j.scitotenv.2018.07.226

Falch BS, König GM, Wright AD et al (1993) Ambigol A and B: new biological active polychlorinated aromatic compounds from the terrestrial blue-green alga *Fischerella ambigua*. J Org Chem 58:6570–6575

Fathoni I, Petitbois JG, Alarif WM et al (2020) Bioactivities of Lyngbyabellins from Cyanobacteria of *Moorea* and *Okeania* Genera. Molecules 25:3986. https://doi.org/10.3390/molecules25173986

Fidor A, Konkel R, Mazur-Marzec H (2019) Bioactive peptides produced by cyanobacteria of the genus *Nostoc*: a review. Mar Drugs 17:561. https://doi.org/10.3390/md17100561

Fidor A, Grabski M, Gawor J et al (2020) *Nostoc edaphicum* CCNP1411 from the Baltic Sea—a new producer of nostocyclopeptides. Mar Drugs 18:442. https://doi.org/10.3390/md18090442

Fidor A, Cekała K, Wieczerzak E et al (2021) Nostocyclopeptides as new inhibitors of 20S proteasome. Biomol Ther 11(10):1483. https://doi.org/10.3390/biom11101483

Fiore MF, de Lima ST, Carmichael WW et al (2020) Guanitoxin, re-naming a cyanobacterial organophosphate toxin. Harmful Algae 92:101737. https://doi.org/10.1016/j.hal.2019.101737

Flores E, Wolk CP (1986) Production, by filamentous, nitrogen-fixing cyanobacteria, of a bacteriocin and of other antibiotics that kill related strains. Arch Microbiol 145(3):215–219. https://doi.org/10.1007/BF00443648

Frankmölle WP, Knübel G, Moore RE et al (1992a) Antifungal cyclic peptides from the terrestrial blue-green alga *Anabaena laxa*. II. Structures of laxaphycins A, B, D and E. J Antibiot (Tokyo) 45(9):1458–1466. https://doi.org/10.7164/antibiotics.45.1458

Frankmölle WP, Larsen LK, Caplan FR et al (1992b) Antifungal cyclic peptides from the terrestrial blue-green alga *Anabaena laxa*. I. Isolation and biological properties. J Antibiot (Tokyo) 45(9):1451–1457. https://doi.org/10.7164/antibiotics.45.1451

Fujii K, Harada K, Suzuki M et al (1995) Novel cyclic peptides together with microcystins produced by toxic cyanobacteria, *Anabaena* sp. Symp Chem Nat Prod Symp Pap 37:445–450. https://doi.org/10.24496/tennenyuki.37.0_445

Gandhi MJ, Boyd MR, Yi L et al (2000) Properties of cyanovirin-N (CV-N): inactivation of HIV-1 by sessile cyanovirin-N (sCV-N). Dev Biol (Basel) 102:141–148

Garrison AR, Giomarelli BG, Lear-Rooney CM et al (2014) The cyanobacterial lectin scytovirin displays potent in vitro and in vivo activity against Zaire *Ebola* virus. Antivir Res 112:1–7

Gehringer MM, Adler L, Roberts AA et al (2012) Nodularin, a cyanobacterial toxin, is synthesized in planta by symbiotic *Nostoc* sp. ISME J 6:1834–1847. https://doi.org/10.1038/ismej.2012.25

Ghasemi Y, Yazdi MT, Shafiee A et al (2004) Parsiguine, a novel antimicrobial substance from *Fischerella ambigua*. Pharm Biol 42(4–5):318–322. https://doi.org/10.1080/13880200490511918

Gibbs HK, Salmon JM (2015) Mapping the world's degraded lands. Appl Geogr 57:12–21. https://doi.org/10.1016/j.apgeog.2014.11.024

Giglio S, Jiang J, Saint CP et al (2008) Isolation and characterization of the gene associated with geosmin production in cyanobacteria. Environ Sci Technol 42(21):8027–8032

Golakoti T, Ogino J, Heltzel CE et al (1995) Structure determination, conformational analysis, chemical stability studies, and antitumor evaluation of the cryptophycins. Isolation of new 18 analogs from *Nostoc* sp. strain GSV 224. J Am Chem Soc 117:12030–12049. https://doi.org/10.1021/ja00154a002

Golakoti T, Yoshida WY, Chaganty S et al (2000) Isolation and structures of nostopeptolides A1, A2 and A3 from the cyanobacterium *Nostoc* sp. GSV 224. Tetrahedron 56:9093–9102. https://doi.org/10.1016/S0040-4020(00)00764-X

Golakoti T, Yoshida WY, Chaganty S et al (2001) Isolation and structure determination of nostocyclopeptides A1 and A2 from the terrestrial cyanobacterium *Nostoc* sp. ATCC53789. J Nat Prod 64:54–59. https://doi.org/10.1021/np000316k

Gräbner P (1895) Studien über die norddeutsche heide. Bot Jahrb Syst 20:500

Grach-Pogrebinsky O, Carmeli S (2008) Three novel anabaenopeptins from the cyanobacterium *Anabaena* sp. Tetrahedron 64:10233–10238. https://doi.org/10.1016/j.tet.2008.08.015

Grindberg RV, Ishoey T, Brinza D et al (2011) Single cell genome amplification accelerates identification of the apratoxin biosynthetic pathway from a complex microbial assemblage. PLoS One 6(4):e18565. https://doi.org/10.1371/journal.pone.0018565

Gross EM, Wolk CP, Jüttner F (1991) Fischerellin, a new allelochemical from the freshwater cyanobacterium *Fischerella muscicola*. J Phycol 27:686–692

Guiry MD, Guiry GM (2021) AlgaeBase. Available online: http://www.algaebase.org/

Guljamow A, Kreische M, Ishida K et al (2017) High-density cultivation of terrestrial *Nostoc* strains leads to reprogramming of secondary metabolome. Appl Environ Microbiol 83:e01510–e01517. https://doi.org/10.1128/AEM.01510-17

Gustafson KR, Cardellina JH 2nd, Fuller RW et al (1989) AIDS-antiviral sulfolipids from cyanobacteria (blue-green algae). J Natl Cancer Inst 81(16):1254–1258. https://doi.org/10.1093/jnci/81.16.1254

Gustafson KR, Sowder RC 2nd, Henderson LE et al (1997) Isolation, primary sequence determination, and disulfide bond structure of cyanovirin-N, an anti-HIV (human immunodeficiency virus) protein from the cyanobacterium *Nostoc ellipsosporum*. Biochem Biophys Res Commun 238(1):223–228. https://doi.org/10.1006/bbrc.1997.7203

Gärtner G, Stoyneva-Gärtner M, Uzunov B (2021) Algal toxic compounds and their Aeroterrestrial, airborne and other Extremophilic producers with attention to soil and plant contamination: a review. Toxins 13:322. https://doi.org/10.3390/toxins13050322

Hagmann L, Jüttner F (1996) Fischerellin A, a novel photosystem-II-inhibiting alleolochemical of the cyanobacterium *Fischerella muscicola* with antifungal and herbicide activity. Tetrahedron Lett 37:6539–6542

Han B, McPhail KL, Gross H et al (2005) Isolation and structure of five lyngbyabellin derivatives from a Papua New Guinea collection of the marine cyanobacterium *Lyngbya majuscula*. Tetrahedron 61(49):1723–11729. https://doi.org/10.1016/j.tet.2005.09.036

Harada KI, Fujii K, Shimada T et al (1995) Two cyclic peptides, anabaenopeptins, a third group of bioactive compounds from the cyanobacterium *Anabaena flos-aquae* NRC 525-17. Tetrahedron Lett 36:1511–1514

Helm RF, Huang Z, Edwards D et al (2000) Structural characterization of the released polysaccharide of desiccation-tolerant *Nostoc commune* DRH-1. J Bacteriol 182(4):974–982. https://doi.org/10.1128/JB.182.4.974-982.2000

Hesse M (2002) Alkaloids: Nature's curse or blessing? Wiley-VCH. ISBN 978-3-906390-24-6

Hillwig ML, Zhu Q, Liu X (2013) Biosynthesis of ambiguine indole alkaloids in cyanobacterium Fischerella ambigua. ACS Chem Biol 9(2):372–377

Hillwig ML, Fuhrman HA, Ittiamornkul K et al (2014) Identification and characterization of a welwitindolinone alkaloid biosynthetic gene cluster in the stigonematalean cyanobacterium *Hapalosiphon welwitschii*. Chembiochem 15(5):665–669

Hofbauer WK (2021) Toxic or otherwise harmful algae and the built environment. Toxins 13:465. https://doi.org/10.3390/toxins13070465

Hofbauer WK, Gärtner G (2021) Microbial life on Façades. Springer Spectrum, Berlin Heidelberg

Hoffmann D, Hevel JM, Moore RE et al (2003) Sequence analysis and biochemical characterization of the nostopeptolide A biosynthetic gene cluster from *Nostoc* sp. GSV 224. Gene 311:171–180. https://doi.org/10.1016/S0378-1119(03)00587-0

Hohlman RM, Sherman DH (2021) Recent advances in hapalindole-type cyanobacterial alkaloids: biosynthesis, synthesis, and biological activity. Nat Prod Rep 38(9):1567–1588. https://doi.org/10.1039/d1np00007a

Huang Z, Liu Y, Paulsen BS et al (1998) Studies on polysaccharides from three edible species of Nostoc (Cyanobateria) with different colony morphologies: comparison of monosaccharide compositions and viscosities of polysaccharides from field colonies and suspension cultures. J Phycol 34(6):962–996. https://doi.org/10.1046/j.1529-8817.1998.340962.x

Huber U, Moore RE, Patterson GML (1998) Isolation of a nitrile-containing indole alkaloid from the terrestrial blue-green alga *Hapalosiphon delicatulus*. J Nat Prod 61(10):1304–1306. https://doi.org/10.1021/np9801561

Inderjit, Dakshini KMM (1994) Algal Allelopathy. Bot Rev 60(2):182–196

Itoh T, Tsuzuki R, Tanaka T et al (2013) Reduced scytonemin isolated from *Nostoc commune* induces autophagic cell death in human T-lymphoid cell line Jurkat cells. Food Chem Toxicol 60:76–82. https://doi.org/10.1016/j.fct.2013.07.016

Itoh T, Tsuchida A, Muramatsu Y et al (2014) Antimicrobial and anti-inflammatory properties of nostocionone isolated from *Nostoc commune* Vauch and its derivatives against *Propionibacterium acnes*. Anaerobe 27:56–63. https://doi.org/10.1016/j.anaerobe.2014.03.006

Ivanov D, Yaneva G, Potoroko I et al (2021) Contribution of cyanotoxins to the ecotoxicological role of lichens. Toxins 13:321. https://doi.org/10.3390/toxins13050321

Jaki B, Orjala J, Sticher O (1999) A novel extracellular diterpenoid with antibacterial activity from the cyanobacterium *Nostoc commune*. J Nat Prod 62(3):502–503. https://doi.org/10.1021/np980444x

Jaki B, Heilmann J, Sticher O (2000a) New antibacterial metabolites from the Cyanobacterium *Nostoc commune* (EAWAG 122b). J Nat Prod 63(9):1283–1285. https://doi.org/10.1021/np000033s

Jaki B, Orjala J, Heilmann J et al (2000b) Novel extracellular diterpenoids with biological activity from the cyanobacterium *Nostoc commune*. J Nat Prod 63(3):339–343. https://doi.org/10.1021/np9903090

Jaki B, Zerbe O, Heilmann J et al (2001) Two novel cyclic peptides with antifungal activity from the cyanobacterium *Tolypothrix byssoidea* (EAWAG 195). J Nat Prod 64(2):154–158. https://doi.org/10.1021/np000297e

Janssen EML (2019) Cyanobacterial peptides beyond microcystins—a review on co-occurrence, toxicity, and challenges for risk assessment. Water Res 151:488–499

Jiang L, Johnston E, Aberg KM et al (2013) Strategy for quantifying trace levels of BMAA in cyanobacteria by LC/MS/MS. Anal Bioanal Chem 405(4):1283–1292. https://doi.org/10.1007/s00216-012-6550-1

Jiang L, Eriksson J, Lage S et al (2014) Diatoms: a novel source for the neurotoxin BMAA in aquatic environments. PLoS One 9(1):e84578. https://doi.org/10.1371/journal.pone.0084578

Jiang W, Zhou W, Othman R et al (2018) A new malyngamide from the marine cyanobacterium *Moorea producens*. Nat Prod Res 32(1):97–104. https://doi.org/10.1080/14786419.2017.1338282

Jimenez JI, Huber U, Moore RE et al (1999) Oxidized welwitindolinones from terrestrial *Fischerella* spp. J Nat Prod 62(4):569–572. https://doi.org/10.1021/np980485t

Jokela J, Herfindal L, Wahlsten M et al (2010) A novel cyanobacterial nostocyclopeptide is a potent antitoxin against microcystins. Chem Bio Chem 11:1594–1599. https://doi.org/10.1002/cbic.201000179

Jokela J, Heinilä LMP, Shishido TK et al (2017) Production of high amounts of hepatotoxin nodularin and new protease inhibitors pseudospumigins by the Brazilian benthic *Nostoc* sp. CENA543. Front Microbiol 8:1963. https://doi.org/10.3389/fmicb.2017.01963

Joshi SK, Bajpai RK, Kumar P et al (2017) Long-term effect of fertilization and algalization on aggregate stability, aggregate associated carbon and nitrogen under rice-wheat cropping system. Int J Curr Microbiol App Sci 6(8):289–3295

Jung JH, Moore RE, Patterson GML (1991) Scytophycins from a blue-green alga belonging to the Nostocaceae. Phytochemistry 30:3615–3616

Jung P, D'Agostino PM, Büdel B et al (2021) *Symphyonema bifilamentata* sp. nov., the right *Fischerella ambigua* 108b: half a decade of research on taxonomy and bioactive compounds in new light. Microorganisms 9. https://doi.org/10.3390/microorganisms9040745

Kaasalainen U, Jokela J, Fewer DP et al (2009) Microcystin production in the tripartite cyanolichen Peltigera leucophlebia. Mol Plant-Microbe Interact 22(6):695–702. https://doi.org/10.1094/MPMI-22-6-0695

Kaasalainen U, Fewer DP, Jokela J et al (2012) Cyanobacteria produce a high variety of hepatotoxic peptides in lichen symbiosis. Proc Natl Acad Sci USA 109:5886–5891

Kabirnataj S, Nematzadeh GA, Talebi AF et al (2018) *Neowestiellopsis* gen. nov, a new genus of true branched cyanobacteria with the description of *Neowestiellopsis persica* sp. nov. and *Neowestiellopsis bilateralis* sp. nov., isolated from Iran. Plant Syst Evol 304:501–510. https://doi.org/10.1007/s00606-017-1488-6

Kajiyama S, Kanzaki H, Kawazu K et al (1998) Nostofungicidine, an antifungal lipopeptide from the field-grown terrestrial blue-green alga *Nostoc commune*. Tetrahedron Lett 39(22):3737–3740. https://doi.org/10.1016/S0040-4039(98)00573-5

Kampa A, Gagunashvili AN, Gulder TA et al (2013) Metagenomic natural product discovery in lichen provides evidence for a family of biosynthetic pathways in diverse symbioses. Proc Natl Acad Sci USA 110(33):E3129–E3137. https://doi.org/10.1073/pnas.1305867110

Kanekiyo K, Lee J-B, Hayashi K et al (2005) Isolation of an antiviral polysaccharide, nostoflan, from a terrestrial cyanobacterium, *Nostoc flagelliforme*. J Nat Prod 68(7):1037–1041. https://doi.org/10.1021/np050056c

Kanekiyo K, Hayashi K, Takenaka H et al (2007) Anti-herpes simplex virus target of an acidic polysaccharide, nostoflan, from the edible blue-green alga *Nostoc flagelliforme*. Biol Pharm Bull 30(8):1573–1575. https://doi.org/10.1248/bpb.30.1573

Kapuścik A, Hrouzek P, Kuzma M et al (2013) Novel aeruginosin-865 from *Nostoc* sp. as a potent anti-inflammatory agent. Chembiochem 14:2329–2337. https://doi.org/10.1002/cbic.201300246

Karan T, Daştan T, Baral İ et al (2016) Effects of differential time applications on some cyanobacterial norharman production rates. CSJ 37(4):398. https://doi.org/10.17776/CSJ.13491

Kim H, Lantvit D, Hwang CH et al (2012) Indole alkaloids from two cultured cyanobacteria, *Westiellopsis* sp. and *Fischerella muscicola*. Bioorg Med Chem 20(17):5290–5295. https://doi.org/10.1016/j.bmc.2012.06.030

Klein D, Daloze D, Braekman JC et al (1995) New hapalindoles from the cyanophyte *Hapalosiphon laingii*. J Nat Prod 58:1781–1785

Knübel G, Larsen LK, Moore RE et al (1990) Cytotoxic, antiviral indolocarbazoles from a blue-green alga belonging to the Nostocaceae. J Antibiot (Tokyo) 43(10):1236–1239. https://doi.org/10.7164/antibiotics.43.1236

Kobayashi A, Kajiyama S (1998) Secondary metabolites of Cyanobacteria *Nostoc* sp. Chin J Oceanol Limnol 16(Suppl):109–117

Kobayashi A, Kajiyama S, Inawaka K et al (1994) Nostodione A, a novel mitotic spindle poison from a blue-green alga *Nostoc commune*. Z Naturforsch C 49:464–470. https://doi.org/10.1515/znc-1994-7-812

Kociolek JP, Blanco S, Coste M, et al (2022) DiatomBase. *Halamphora coffeaeformis* (C. Agardh) Levkov, 2009. Accessed at: https://diatombase.org/aphia.php?p=taxdetails&id=624065 on 2022-01-05

Kocourek J, Hořejši V (1981) Defining a lectin. Nature 290:188

Kohler E, Grundler V, Häussinger D et al (2014) The toxicity and enzyme activity of a chlorine and sulfate containing aeruginosin isolated from a non-microcystin-producing *Planktothrix* strain. Harmful Algae 39:154–160. https://doi.org/10.1016/j.hal.2014.07.003

Komárek J (2013) Süsswasserflora von Mitteleuropa. Cyanoprokaryota: 3rd part: Heterocystous genera. Springer Spectrum, Berlin/Heidelberg

Kondratyeva NV (1962) Novyi vyd synio-zelenykh vodorostey—*Nostoc edaphicum* sp. n. [A new species of blue-green algae—*Nostoc edapicum* sp. n.]. Ukr Bot J 19:58–65

Kossack R, Breinlinger S, Nguyen T et al (2020) Nostotrebin 6 related cyclopentenediones and δ-lactones with broad activity spectrum isolated from the cultivation medium of the cyanobacterium *Nostoc* sp. CBT1153. J Nat Prod 83(2):392–400. https://doi.org/10.1021/acs.jnatprod.9b00885

Kwan JC, Teplitski M, Gunasekera SP et al (2010) Isolation and biological evaluation of 8-*epi*-malyngamide C from the Floridian marine cyanobacterium *Lyngbya majuscula*. J Nat Prod 73(3):463–466. https://doi.org/10.1021/np900614n

Lage S, Costa PR, Moita T et al (2014) BMAA in shellfish from two Portuguese transitional water bodies suggests the marine dinoflagellate *Gymnodinium catenatum* as a potential BMAA source. Aquat Toxicol 152:131–138. https://doi.org/10.1016/j.aquatox.2014.03.029

Lage S, Burian A, Rasmussen U et al (2016) BMAA extraction of cyanobacteria samples: which method to choose? Environ Sci Pollut Res 23:338–350. https://doi.org/10.1007/s11356-015-5266-0

Lagenaur LA, Swedek I, Lee PP et al (2015) Robust vaginal colonization of macaques with a novel vaginally disintegrating tablet containing a live biotherapeutic product to prevent HIV infection in women. PLoS One 10:e0122730

Larsen LK, Moore RE, Patterson GM (1994) beta-Carbolines from the blue-green alga *Dichothrix baueriana*. J Nat Prod 57(3):419–421. https://doi.org/10.1021/np50105a018

Lee ESJ, Gleason FK (1994) A second algicidal natural product from a cyanobacterium *Scytonema hofmanni*. Plant Sci 103:155–160

Leikoski N, Fewer DP, Jokela J et al (2012) Analysis of an inactive cyanobactin biosynthetic gene cluster leads to discovery of new natural products from strains of the genus *Microcystis*. PLoS One 7(8):e43002

Leikoski N, Liu L, Jokela J et al (2013) Genome mining expands the chemical diversity of the cyanobactin family to include highly modified linear peptides. Chem Biol 20(8):1033–1043

Li Z, Guo M (2018) Healthy efficacy of *Nostoc commune* Vaucher. Oncotarget 9(18):14669–14679. https://doi.org/10.18632/oncotarget.23620

Liaimer A, Helfrich EJN, Hinrichs K et al (2015) Nostopeptolide plays a governing role during cellular differentiation of the symbiotic cyanobacterium *Nostoc punctiforme*. PNAS 112:61862–61867. https://doi.org/10.1073/pnas.1419543112

Liaimer A, Jensen JB, Dittmann E (2016) A genetic and chemical perspective on symbiotic recruitment of cyanobacteria of the genus *Nostoc* into the host plant *Blasia pusilla* L. Front Microbiol 7:1693. https://doi.org/10.3389/fmicb.2016.01693

Liang ZX (2010) Complexity and simplicity in the biosynthesis of enediyne natural products. Nat Prod Rep 27(4):499–528. https://doi.org/10.1039/b908165

Liu L, Jokela J, Herfindal L et al (2014a) 4-Methylproline guided natural product discovery: co-occurrence of 4-hydroxy- and 4-methylprolines in nostoweipeptins and nostopeptolides. ACS Chem Biol 9(11):2646–2655. https://doi.org/10.1021/cb500436p

Liu L, Jokela J, Wahlsten M et al (2014b) Nostosins, trypsin inhibitors isolated from the terrestrial cyanobacterium *Nostoc* sp. strain FSN. J Nat Prod 77:1784–1790. https://doi.org/10.1021/np500106w

Loya S, Reshef V, Mizrachi et al (1998) The inhibition of the reverse transcriptase of HIV-1 by the natural sulfoglycolipids from cyanobacteria: contribution of different moieties to their high potency. J Nat Prod 61:891–895

Luesch H, Yoshida WY, Moore RE et al (2000) Isolation, structure determination, and biological activity of Lyngbyabellin A from the marine cyanobacterium lyngbya majuscula. J Nat Prod 63(5):611–615. https://doi.org/10.1021/np990543q

Luesch H, Yoshida WY, Moore RE et al (2001) Total structure determination of apratoxin A, a potent novel cytotoxin from the marine cyanobacterium *Lyngbya majuscula*. J Am Chem Soc 123:5418–5423

Luesch H, Hoffmann D, Hevel JM et al (2003) Biosynthesis of 4-methylproline in Cyanobacteria: cloning of *nosE* and *nosF* genes and biochemical characterization of the encoded dehydrogenase and reductase activities. J Org Chem 68(1):83–91. https://doi.org/10.1021/jo026479q

Magarvey NA, Beck ZQ, Golakoti T et al (2006) Biosynthetic characterization and chemoenzymatic assembly of the cryptophycins. Potent anticancer agents from *Nostoc cyanobionts*. ACS Chem Biol 1:766–779. https://doi.org/10.1021/cb6004307

Mareš J, Hájek J, Urajová P et al (2014) A hybrid non-ribosomal peptide/polyketide synthetase containing fatty-acyl ligase (FAAL) synthesizes the β-amino fatty acid lipopeptides puwainaphycins in the Cyanobacterium *Cylindrospermum alatosporum*. PLoS One 9(11):e111904. https://doi.org/10.1371/journal.pone.0111904

Martins A, Vasconcelos V (2011) Use of qPCR for the study of hepatotoxic cyanobacteria population dynamics. Arch Microbiol 193:615–627. https://doi.org/10.1007/s00203-011-0724-7

Martins TP, Rouger C, Glasser NR et al (2019) Chemistry, bioactivity and biosynthesis of cyanobacterial alkylresorcinols. Nat Prod Rep 36(10):1437–1461. https://doi.org/10.1039/c8np00080h

Mason CP, Edwards KR, Carlson RE et al (1982) Isolation of chlorine-containing antibiotic from the freshwater cyanobacterium *Scytonema hofmannii*. Science 215:400–402

Matern U, Oberer L, Falchetto RA et al (2001) Scyptolin A and B, cyclic depsipeptides from axenic cultures of *Scytonema hofmanni* PCC 7110. Phytochemistry 58(7):1087–1095. https://doi.org/10.1016/s0031-9422(01)00400-9

Matern U, Oberer L, Erhard M et al (2003a) Hofmannolin, a cyanopeptolin from *Scytonema hofmanni* PCC 7110. Phytochemistry 64(6):1061–1067. https://doi.org/10.1016/s0031-9422(03)00467-9

Matern U, Schleberger C, Jelakovic S et al (2003b) Binding structure of elastase inhibitor scyptolin A. Chem Biol 10(10):997–1001. https://doi.org/10.1016/j.chembiol.2003.10.001

Mattila A, Andsten RM, Jumppanen M et al (2019) Biosynthesis of the bis-prenylated alkaloids muscoride A and B. ACS Chem Biol 14(12):2683–2690. https://doi.org/10.1021/acschembio.9b00620

Mazor G, Kidron GJ, Vonshak A et al (1996) The role of cyanobacterial exopolysaccharides in structuring desert microbial crusts. FEMS Microbiol Ecol 21:121–130. https://doi.org/10.1016/0168-6496(96)00050-5

Mazur-Marzec H, Fidor A, Cegłowska M et al (2018) Cyanopeptolins with trypsin and chymotrypsin inhibitory activity from the Cyanobacterium *Nostoc edaphicum* CCNP1411. Mar Drugs 16:220. https://doi.org/10.3390/md16070220

Mazur-Marzec H, Cegłowska M, Konkel R et al (2021) Antiviral cyanometabolites - a review. Biomol Ther 11(3):474. https://doi.org/10.3390/biom11030474

Mehner C, Müller D, Kehraus S et al (2008) New peptolides from the cyanobacterium *Nostoc insulare* as selective and potent inhibitors of human leukocyte elastase. Chembiochem 9:2692–2703

Meriluoto J, Spoof L, Codd J (eds) (2017) Handbook of cyanobacterial monitoring and cyanotoxin analysis. Wiley, Chichester

Metcalf JS, Beattie KA, Purdie EL et al (2012) Analysis of microcystins and microcystin genes in 60–170-year-old dried herbarium specimens of cyanobacteria. Harmful Algae 15:47–52. https://doi.org/10.1016/j.hal.2011.11.004

Metcalf JS, Banack SA, Richer R et al (2015) Neurotoxic amino acids and their isomers in desert environments. J Arid Environ 112:140–144. https://doi.org/10.1016/j.jaridenv.2014.08.002

Metcalf JS, Banack SA, Wessel RA et al (2021a) Toxin analysis of freshwater cyanobacterial and marine harmful algal blooms on the West Coast of Florida and implications for Estuarine environments. Neurotox Res 39:27–35. https://doi.org/10.1007/s12640-020-00248-3

Metcalf JS, Tischbein M, Cox PA et al (2021b) Cyanotoxins and the nervous system. Toxins 13:660. https://doi.org/10.3390/toxins13090660

Metting B (1981) The systematics and ecology of soil algae. Bot Rev 47:195–312. https://doi.org/10.1007/BF02868854

Micallef ML, Sharma D, Bunn B et al (2014) Comparative analysis of hapalindole, ambiguine and welwitindolinone gene clusters and reconstitution of indole-isonitrile biosynthesis from cyanobacteria. BMC Microbiol 14(1):213. https://doi.org/10.1186/s12866-014-0213-7
Micallef ML, D'Agostino PM, Sharma D et al (2015) Genome mining for natural product biosynthetic gene clusters in the subsection V cyanobacteria. BMC Genomics 16:669. https://doi.org/10.1186/s12864-015-1855-z
Mitchell CA, Ramessar K, O'Keefe BR (2017) Antiviral lectins: selective inhibitors of viral entry. Antivir Res 142:37–54. https://doi.org/10.1016/j.antiviral.2017.03.007
Mo S, Krunic A, Pegan SD et al (2009a) An antimicrobial guanidine-bearing sesterterpene from the cultured cyanobacterium *Scytonema* sp. J Nat Prod 72(11):2043–2045. https://doi.org/10.1021/np900288x
Mo S, Krunic A, Chlipala G et al (2009b) Antimicrobial ambiguine isonitriles from the cyanobacterium *Fischerella ambigua*. J Nat Prod 72(5):894–899. https://doi.org/10.1021/np800751j
Mo S, Krunic A, Santarsiero BD et al (2010) Hapalindole-related alkaloids from the cultured cyanobacterium *Fischerella ambigua*. Phytochemistry 71(17–18):2116–2123. https://doi.org/10.1016/j.phytochem.2010.09.004
Monteiro PR, do Amaral SC, Siqueira AS et al (2021) Anabaenopeptins: what we know so far. Toxins 13:522. https://doi.org/10.3390/toxins13080522
Moon S-S, Chen JL, Moore RE et al (1992) Calophycin, a fungicidal cyclic decapeptide from the terrestrial blue-green alga *Calothrix fusca*. J Org Chem 57:1097–1103
Moore RE (1981) Constituents of blue-green algae. In: Scheuer PJ (ed) Marine natural products: chemical and biological perspectives, vol 4. Academic Press, New York, pp 1–52
Moore RE (1996) Cyclic peptides and depsipeptides from cyanobacteria: a review. J Ind Microbiol 16:134–143
Moore RE, Cheuk C, Patterson GML (1984) Hapalindoles: new alkaloids from the blue-green alga *Hapalosiphon fontinalis*. J Am Chem Soc 106:6456–6457
Moore RE, Cheuk C, Yang XQG et al (1987) Hapalindoles antibacterial and antimycotic alkaloids from the cyanophyte *Hapalosiphon fontinalis*. J Org Chem 52(6):1036–1043
Moore RE, Yang X-GG, Patterson GML et al (1989) Hapalonamides and other oxidized hapalindoles from *Hapalosiphon fontinalis*. Phytochemistry 28(5):1565–1567. https://doi.org/10.1016/S0031-9422(00)97798-7
Moore BS, Chen JL, Patterson GML et al (1990) [7.7]Paracyclophanes from blue-green algae. J Am Chem Soc 112(10):4061–4063. https://doi.org/10.1021/ja00166a066
Moulaei T, Botos I, Ziółkowska NE et al (2007) Atomic-resolution crystal structure of the antivirallectin scytovirin. Protein Sci 16:2756–2760
Mühlsteinová R, Hauer T (2013) Pilot survey of cyanobacterial diversity from the neighborhood of San Gerardo de Rivas, Costa Rica with a brief summary of current knowledge of terrestrial cyanobacteria in Central America. Braz J Bot 36:299–307. https://doi.org/10.1007/s40415-013-0030-5
Müller D, Kehraus S, Krick A et al (2005) Sponges and cyanobacteria: *Plectonema* sp. produces cyclic peptides related to sponge metabolites. 17. Irseer Naturstofftage, abstract
Murakami M, Ishida K, Okino T et al (1995) Aeruginosins 98-A and B, trypsin inhibitors from the blue-green alga *Microcystis aeruginosa* (NIES-98). Tetrahedron Lett 36:2785–2788
Murakami M, Sun Q, Ishida K et al (1997) Microviridins, elastase inhibitors from the cyanobacterium *Nostoc minutum* (NIES-26). Phytochemistry 45(6):1197–1202. https://doi.org/10.1016/S0031-9422(97)00131-3
Murch SJ, Cox PA, Banack SA (2004) A mechanism for slow release of biomagnified cyanobacterial neurotoxins and neurodegenerative disease in Guam. PNAS 101:12228–12231. https://doi.org/10.1073/pnas.0404926101
Nagatsu A, Kajitani H, Sakakibara J (1995) Muscoride A: a new oxazole peptide alkaloid from freshwater cyanobacterium *Nostoc muscorum*. Tetrahedron Lett 36:4097–4100. https://doi.org/10.1016/0040-4039(95)00724-Q

Nakamura H, Hamer HA, Sirasani G, Balskus EP (2012) Cylindrocyclophane biosynthesis involves functionalization of an unactivated carbon center. J Am Chem Soc 134(45):18518–18521
NCBI (2022) National Centre for Biotechnology Information https://www.ncbi.nlm.nih.gov/
Neuhof T, Schmieder P, Preussel K et al (2005) Hassallidin A, a glycosylated lipopeptide with antifungal activity from the cyanobacterium *Hassallia* sp. J Nat Prod 68(5):695–700. https://doi.org/10.1021/np049671r
Neuhof T, Schmieder P, Seibold M et al (2006) Hassallidin B - second antifungal member of the Hassallidin family. Bioorg Med Chem Lett 16(16):4220–4222
Nowruzi B, Khavari-Nejad R-A, Sivonen K et al (2012) Identification and toxigenic potential of a *Nostoc* sp. Algae 27:303–313
Nowruzi B, Blanco S, Nejadsattari T (2018) Chemical and molecular evidences for the poisoning of a duck by anatoxin-a, nodularin and cryptophycin at the coast of lake Shoormast (Mazandaran province, Iran). Int J Algae 20:359–376. https://doi.org/10.15407/alg28.04.409
Ogino J, Moore RE, Patterson GM et al (1995) Dendroamides, new cyclic hexapeptides from a blue-green alga. Multidrug resistance reversing activity of dendroamide A. J Nat Prod 59:581–586
Okino T, Qi S, Matsuda H et al (1997) Nostopeptins A and B, elastase inhibitors from the cyanobacterium *Nostoc minutum*. J Nat Prod 60:158–161
Oksanen I, Jokela J, Fewer DP et al (2004) Discovery of rare and highly toxic microcystins from lichen-associated cyanobacterium *Nostoc* sp. strain IO-102-I. Appl Environ Microbiol 70:5756–5763. https://doi.org/10.1128/AEM.70.10.5756-5763.2004
Omura S (ed) (2002) Macrolide antibiotics: chemistry, biology, and practice, 2nd edn. Academic Press. ISBN 978-0-1252-6451-8
Onda Y, Masuda Y, Yoshida M et al (2017) Conformation-based design and synthesis of apratoxin a mimetics modified at the α,β-unsaturated thiazoline moiety. J Med Chem 60(15):6751–6765. https://doi.org/10.1021/acs.jmedchem.7b00833
Pagarete A, Ramos AS, Puntervoll P et al (2021) Antiviral potential of algal metabolites-A comprehensive review. Mar Drugs 19(2):94. https://doi.org/10.3390/md19020094
Paik S, Carmeli S, Cullingham J et al (1994) Mirabimide E, an unusual N-acylpyrrolinone from the blue-green alga *Scytonema mirabile*: structure determination and synthesis. J Am Chem Soc 116(18):8116–8125. https://doi.org/10.1021/ja00097a019
Papke U, Gross EM, Francke W (1997) Isolation, identification and determination of the absolute configuration of *Fischerellin* B. A new algicide from the freshwater cyanobacterium *Fischerella muscicola* Thuret. Tetrahedron Lett 38(3):379–382. https://doi.org/10.1016/S0040-4039(96)02284-8
Park A, Moore RE, Patterson GML (1992) Fischerindole L, a new isonitrile from the terrestrial blue-green alga *Fischerella muscicola*. Tetrahedron Lett 33:3257–3260
Parsaeimehr A, Lutzu GA (2016) Algae as a novel source of antimicrobial compounds: current and future perspectives. In: Kon K, Rai M (eds) Antibiotic resistance mechanisms and new antimicrobial approaches. Academic Press/Elsevier, London, pp 377–396
Patil P, Ahmadian-Moghaddam M, Dömling A (2020) Isocyanide 2.0. Green Chem 22(20):6902–6911. https://doi.org/10.1039/D0GC02722G
Patterson GM, Smith CD, Kimura LH et al (1993) Action of tolytoxin on cell morphology, cytoskeletal organization, and actin polymerization. Cell Motil Cytoskeleton 24(1):39–48. https://doi.org/10.1002/cm.970240105
Patterson GML, Larsen LK, Moore RE (1994) Bioactive natural products from blue-green algae. J Appl Phycol 6:151–157. https://doi.org/10.1007/BF02186069
Pignatello H, Porwoll J, Carlson RE et al (1983) Structure of the antibiotic cyanobacterin, a chlorine-containing lactone from the freshwater cyanobacterium *Scytonema hofmanni*. J Org Chem 48:4035–4038
Ploutno A, Carmeli S. (2000) Nostocyclyne A, a novel antimicrobial cyclophane from the cyanobacterium Nostoc sp. J Nat Prod 63(11):1524–1526. https://doi.org/10.1021/np0002334. PMID: 11087597

Ploutno A, Carmeli S (2002) Modified peptides from a water bloom of the cyanobacterium *Nostoc* sp. Tetrahedron 58:9949–9995

Ploutno A, Carmeli S (2005a) Nostocyclyne A, a novel antimicrobial cyclophane from the cyanobacterium *Nostoc* sp. J Nat Prod 63:1524–1526

Ploutno A, Carmeli S (2005b) Banyasin A and banyasides A and B, three novel modified peptides from a water bloom of the cyanobacterium *Nostoc* sp. Tetrahedron 61:575–583

Prakash S, Nikhil K (2014) Algae as a soil conditioner. IJETR 2:68–70

Prinsep MR, Moore RE, Levine IA et al (1992a) Westiellamide, a bistratamide-related cyclic peptide from the blue-green alga *Westiellopsis prolifica*. J Nat Prod 55:140–142

Prinsep MR, Caplan FR, Moore RE et al (1992b) Microcystin LA from a blue-green alga belonging to the Stigonematales. Phytochemistry 31:1247–1248. https://doi.org/10.1016/0031-9422(92)80269-K

Prinsep MR, Caplan FR, Moore RE et al (1992c) Tolyporphin, a novel multidrug resistance reversing agent from the blue-green alga *Tolypothrix nodosa*. J Am Chem Soc 114:385–387

Prinsep MR, Thomson RA, West ML et al (1996) Tolypodiol, an antiinflammatory diterpenoid from the cyanobacterium *Tolypothrix nodosa*. J Nat Prod 59:786–788

Puddick J, Prinsep MR, Wood SA et al (2015) Further characterization of glycine-containing microcystins from the McMurdo dry valleys of Antarctica. Toxins 7:493–515. https://doi.org/10.3390/toxins7020493

Radau G (2005) Cyanopeptides: a new and nearly inexhaustible natural resource for the design and structure-activity relationship studies of new inhibitors of trypsin-like serine proteases. Curr Enzym Inhib 1(3):295. https://doi.org/10.2174/1573408057745805l0

Raveh A, Carmeli S (2007) Antimicrobial ambiguines from the cyanobacterium *Fischerella* sp. collected in Israel. J Nat Prod 70(2):196–201. https://doi.org/10.1021/np060495r

Reshef V, Carmeli S (2002) Schizopeptin 791, a new anabeanopeptin-like cyclic peptide from the cyanobacterium *Schizothrix* sp. J Nat Prod 65:1187–1189

Reshef V, Mizrachi E, Maretzki T (1997) New acylated sulfoglycolipids and digalactolipids and related known glycolipids from cyanobacteria with a potential to inhibit the reverse transcriptase of HIV-1. J Nat Prod 60:1251–1260

Riba M, Kiss-Szikszai A, Gonda S et al (2020) Chemotyping of terrestrial *Nostoc*-like isolates from alkali grassland areas by non-targeted peptide analysis. Algal Res 46:101798. https://doi.org/10.1016/j.algal.2020.101798

Rickards RW, Rothschild JM, Willis AC et al (1999) Calothrixins A and B, novel pentacyclic metabolites from *Calothrix* cyanobacteria with potent activity against malaria parasites and human cancer cells. Tetrahedron 55:13513–13520

Rojas V, Rivas L, Cárdenas C et al (2020) Cyanobacteria and eukaryotic microalgae as emerging sources of antibacterial peptides. Molecules 25:5804. https://doi.org/10.3390/molecules25245804

Rouhiainen L, Sivonen K, Buikema WJ et al (1995) Characterization of toxin-producing cyanobacteria by using an oligonucleotide probe containing a tandemly repeated heptamer. J Bacteriol 177(20):6021–6026. https://doi.org/10.1128/jb.177.20.6021-6026.1995

Rouhiainen L, Jokela J, Fewer DP et al (2010) Two alternative starter modules for the non-ribosomal biosynthesis of specific anabaenopeptin variants in *Anabaena* (*Cyanobacteria*). Chem Biol 17(3):265–273. https://doi.org/10.1016/j.chembiol.2010.01.017

Saha S, Esposito G, Urajová P et al (2020) Discovery of unusual cyanobacterial tryptophan-containing anabaenopeptins by MS/MS-based molecular networking. Molecules 25:3786

Sala SE, Sar EA, Ferrario ME (1998) Review of materials reported as containing *Amphora coffeaeformis* (Agardh) Kützing in Argentina. Diatom Res 13:323–336. https://doi.org/10.1080/0269249X.1998.9705454

Sanz M, Andreote APD, Fiore MF et al (2015) Structural characterization of new peptide variants produced by cyanobacteria from the Brazilian Atlantic Coastal Forest using liquid chromatography coupled to quadrupole time-of-flight tandem mass spectrometry. Mar Drugs 13:3892–3919

Sathyamoorthy P, Shansmugasundaram S (1996) Preparation of cyanobacterial peptide toxin as bipesticide against cotton pests. Appl Microbiol Biotechnol 46:511–513

Scheuer P (1992) Isocyanides and cyanides as natural products. Acc Chem Res 25(10):433–439. https://doi.org/10.1021/ar00022a001

Schlegel I, Doan NT, de Chazal N et al (1999) Antibiotic activity of new cyanobacterial isolates from Australia and Asia against green algae and cyanobacteria. J Appl Phycol 10:471–479

Schmidt EW, Donia MS (2009) Cyanobactin ribosomally synthesized peptides—a case of deep metagenome mining. Methods Enzymol 458:575–596. https://doi.org/10.1016/S0076-6879(09)04823-X

Schmidt EW, Harper MK, Faulkner DJ (1997) Mozamides A and B, cyclic peptides from a theonellid sponge from Mozambique. J Nat Prod 60:779–782

Schreuder H, Liesum A, Lönze P et al (2016) Isolation, co-crystallization and structure-based characterization of anabaenopeptins as highly potent inhibitors of activated thrombin activatable fibrinolysis inhibitor (TAFIa). Sci Rep 6:32958. https://doi.org/10.1038/srep32958

Schwartz RE, Hirsch CF, Springer JP et al (1987) Unusual cyclopropane-containing hapalindolinones from a cultured cyanobacterium. J Org Chem 52(16):3704–3706. https://doi.org/10.1021/jo00392a045

Schwartz RE, Hirsch CF, Sesin DF et al (1990) Pharmaceuticals from cultured algae. J Ind Microbiol 5:113–124. https://doi.org/10.1007/BF01573860

Schwarz W, Schossig J, Rossbacher R et al (2019) Butyrolactone. In: Ullmann's encyclopedia of industrial chemistry. Wiley-VCH, Weinheim. https://doi.org/10.1002/14356007.a04_495

Sedmak B, Carmeli S, Elersek T (2008) "Non-toxic" cyclic peptides induce lysis of cyanobacteria-an effective cell population density control mechanism in cyanobacterial blooms. Microb Ecol 56(2):201–209. https://doi.org/10.1007/s00248-007-9336-9

Senol AD, Pepe A, Grudina C et al (2019) Effect of tolytoxin on tunneling nanotube formation and function. Sci Rep 9:5741. https://doi.org/10.1038/s41598-019-42161-6

Shaala LA, Youssef DTA, McPhail KL et al (2013) Malyngamide 4, a new lipopeptide from the Red Sea marine cyanobacterium *Moorea producens* (formerly *Lyngbya majuscula*). Phytochem Lett 6(2):183–188. https://doi.org/10.1016/j.phytol.2013.01.002

Shen B, Yan X, Huang T et al (2015) Enediynes: exploration of microbial genomics to discover new anticancer drug leads. Bioorg Med Chem Lett 25(1):9–15. https://doi.org/10.1016/j.bmcl.2014.11.019

Shih PM, Wu D, Latifi A et al (2013) Improving the coverage of the cyanobacterial phylum using diversity-driven genome sequencing. PNAS 110(3):1053–1058

Shishido TK, Humisto A, Jokela J et al (2015) Antifungal compounds from cyanobacteria. Mar Drugs 13(4):2124–2140. https://doi.org/10.3390/md13042124

Shishido TK, Jokela J, Fewer DP et al (2017) Simultaneous production of anabaenopeptins and namalides by the cyanobacterium *Nostoc* sp. CENA543. ACS Chem Biol 12(11):2746–2755. https://doi.org/10.1021/acschembio.7b00570

Singh DP, Tyagi BM, Kumar A et al (2002) Bioactive secondary metabolites from Cyanobacteria. In: Anand N, Singh B, Singh MP (eds) Algological research in India, Dehra Dun, pp 275–292

Singh R, Parihar P, Singh M et al (2017) Uncovering potential applications of cyanobacteria and algal metabolites in biology, agriculture and medicine: current status and future prospects. Front Microbiol 8:515. https://doi.org/10.3389/fmicb.2017.00515

Sings HL, Rinehart KL (1996) Compounds produced from potential tunicate-blue-green algal symbiosis: a review. J Ind Microbiol Biotechnol 17:385–396

Sivonen K, Leikoski N, Fewer D et al (2010) Cyanobactins—ribosomal cyclic peptides produced by cyanobacteria. Appl Microbiol Biotechnol 86(5):1213–1225

Śliwińska-Wilczewska S (2019) Cyanobacteria and cyanometabolites used in the pharmaceutical and medical industry. Ann Univ Paedagog Crac Stud Naturae 4:180–190. https://doi.org/10.24917/25438832.4.12

Smith CG, Gray GD, Carlson RG et al (1973) Biochemical and biological studies with tubercidin (7-deazaadenosine), 7-deazainosine and certain nucleotide derivatives of tubercidin. Adv Enzym Regul 5:121–151

Smitka TA, Bonjouklian R, Doolin L et al (1992) Ambiguine isonitriles, fungicidal hapalindole-type alkaloids from three genera of blue-green algae belonging to Stigonemataceae. J Org Chem 57:857–861

Soltani S, Hammami R, Cotter PD et al (2021) Bacteriocins as a new generation of antimicrobials: toxicity aspects and regulations. FEMS Microbiol Rev 45(1):fuaa039. https://doi.org/10.1093/femsre/fuaa039

Srivastava A, Jüttner F, Strasser RJ (1998) Action of the allelochemical, fischerellin A, on photosystem II. Biochim Biophys Acta 1364(3):326–336. https://doi.org/10.1016/s0005-2728(98)00014-0

Srivastava VC, Manderson GJ, Bhamidimarri R (1999) Inhibitory metabolites production by the cyanobacterium *Fischerella muscicola*. Microbiol Res 153:309–317

Stewart JB, Bornemann V, Chen JL et al (1988) Cytotoxic, fungicidal nucleosides from blue green algae belonging to the Scytonemataceae. J Antibiot (Tokyo) 61:1048–1056

Stoyneva M (1995) Algal flora of the Danube river (Bulgarian sector) and adjoined water basins. V. Algal flora of the water bodies adjacent to the Lake of Srebarna. Ann Univ Sof 88(2):5–19

Stoyneva-Gärtner M, Uzunov B, Gärtner G (2020) Enigmatic microalgae from aeroterrestrial and extreme habitats in cosmetics: the potential of the untapped natural sources. Cosmetics 7:27. https://doi.org/10.3390/cosmetics7020027

Stratmann K, Burgoyne DL, Moore RE et al (1994a) Hapalosin, a cyanobacterial cyclic Depsipeptide with multidrug-resistance reversing activity. J Org Chem 59(24):7219–7226. https://doi.org/10.1021/jo00103a011

Stratmann K, Moore RE, Bonjouklian R et al (1994b) Welwitindolinones, unusual alkaloids from the blue-green algae *Hapalosiphon welwitschii* and *Westiella intricata*. Relationship to fischerindoles and hapalinodoles. Am Chem Soc 116(22):9935–9942. https://doi.org/10.1021/ja00101a015

Sturdy M, Krunic A, Cho S et al (2010) Eucapsitrione, an anti-mycobacterium tuberculosis anthraquinone derivative from the cultured freshwater cyanobacterium *Eucapsis* sp. J Nat Prod 73:1441–1443

Sueyoshi K, Yamano A, Ozaki K et al (2017) Three new malyngamides from the marine cyanobacterium *Moorea producens*. Mar Drugs 15:367. https://doi.org/10.3390/md15120367

Svirčev Z, Lalić D, Bojadžija Savić G et al (2019) Global geographical and historical overview of cyanotoxin distribution and cyanobacterial poisonings. Arch Toxicol 93(9):2429–2481. https://doi.org/10.1007/s00204-019-02524-4

Takebe Y, Saucedo CJ, Lund G et al (2013) Antiviral lectins from red and blue-green algae show potent in vitro and in vivo activity against hepatitis C virus. PLoS One 8:e64449

Takenaka H, Yamaguchi Y, Sakaki S et al (1998) Safety evaluation of *Nostoc flagelliforme* (Nostocales, Cyanophyceae) as a potential food. Food Chem Toxicol 36:1073–1077. https://doi.org/10.1016/S0278-6915(98)00089-1

Tan LT (2012) Chapter 4 - Marine Cyanobacteria: a treasure trove of bioactive secondary metabolites for drug discovery. In: Atta-ur-Rahman (ed) Studies in natural products chemistry, vol 36. Elsevier, pp 67–110

Teneva I, Stoyanov P, Belkinova D et al (2012) Production of cyanobacterial toxins from two *Nostoc* species (Nostocales) and evaluation of their cytotoxicity in vitro. J Bio Sci Biotechnol 1:33–43. http://www.oalib.com/paper/2762845#.V6BZ5vl97cs

Tomek P, Hrouzek P, Kuzma M et al (2015) Cytotoxic lipopeptide muscotoxin A, isolated from soil cyanobacterium *Desmonostoc muscorum*, permeabilizes phospholipid membranes by reducing their fluidity. Chem Res Toxicol 28(2):216–224. https://doi.org/10.1021/tx500382b

Tomsickova J, Ondrej M, Cerny J et al (2014) Analysis and detection of scytophycin variants by HPLC-ESI-MS. Chem Nat Compd 49:1170–1171

Trimurtulu G, Ogino J, Heltzel CE et al (1995) Structure determination, conformational analysis, chemical stability studies, and antitumor evaluation of the cryptophycins. Isolation of 18 new analogs from *Nostoc* sp. strain GSV 224. J Am Chem Soc 117:12030–12049. https://doi.org/10.1021/ja00154a002

Tsai CC, Emau P, Jiang YH et al (2003) Cyanovirin-N gel as a topical microbicide prevents rectal transmission of shiv89.6p in macaques. AIDS Res Hum Retrovir 19:535–541

Tsai CC, Emau P, Jiang YH et al (2004) Cyanovirin-N inhibits AIDS virus infections in vaginal transmission models. AIDS Res Hum Retrovir 20:11–18

Uzunov B, Stefanova K, Radkova M et al (2021a) First report on *Microcystis* as a potential Microviridin producer in Bulgarian waterbodies. Toxins 13:448. https://doi.org/10.3390/toxins13070448

Uzunov B, Stefanova K, Radkova M et al (2021b) Microcystis species and their toxigenic strains in phytoplankton of ten Bulgarian waterbodies. Botanica 27(1):77–94

Van Lanen SG, Shen B (2008) Biosynthesis of enediyne antitumor antibiotics. CTMCCL 8(6):448–459

Vega A, Bell EA (1967) α-Amino-β-methylaminopropionic acid, a new amino acid from seeds of *Cycas circinalis*. Phytochemistry 6(5):759–762. https://doi.org/10.1016/S0031-9422(00)86018-5

Venkataraman GS (1961) A method of preserving blue-green algae for seeding purposes. J Gen Appl Microbiol 7:96–99

Venkataraman GS (1972) Algal biofertilizers and rice cultivation. Today and Tomorrow's Printers and Publishers, New Delhi

Vestola J, Shishido T, Jokela J et al (2014) Hassallidins, antifungal glycolipopeptides, are widespread among cyanobacteria and are the end-product of a nonribosomal pathway. PNAS 111:1909–1917

Violi JP, Facey JA, Mitrovic SM et al (2019) Production of β-methylamino-L-alanine (BMAA) and its isomers by freshwater diatoms. Toxins 11:512. https://doi.org/10.3390/toxins11090512

Volk R-B (2008) Screening of microalgae for species excreting norharmane, a manifold biologically active indole alkaloid. Microbiol Res 163(3):307–313. https://doi.org/10.1016/j.micres.2006.06.002

Walton K, Berry JP (2016) Indole alkaloids of the Stigonematales (Cyanophyta): chemical diversity, biosynthesis and biological activity. Mar Drugs 14(4):73. https://doi.org/10.3390/md14040073

Wang H, Fewer DP, Sivonen K (2011) Genome mining demonstrates the widespread occurrence of gene clusters encoding bacteriocins in Cyanobacteria. PLoS One 6(7):e22384. https://doi.org/10.1371/journal.pone.0022384

Warren SD, Clair LLST, Stark LR et al (2019) Reproduction and dispersal of biological soil crust organisms. Front Ecol Evol 7:344. https://doi.org/10.3389/fevo.2019.00344

Welker M, von Döhren H (2006) Cyanobacterial peptides - nature's own combinatorial biosynthesis. FEMS Microbiol Rev 30(4):530–563. https://doi.org/10.1111/j.1574-6976.2006.00022.x

Whitford WG, Duval BD (2019) Ecology of desert systems, 2nd edn. Academic Press, London

Williard PG, De Laszlo SE (1984) Total synthesis of (.+−.)-dysidin, a marine metabolite containing an N-acyl-O-methyltetramic acid. J Org Chem 49(19):3489–3493. https://doi.org/10.1021/jo00193a011

Wood SA, Mountfort D, Selwood AI et al (2008) Widespread distribution and identification of eight novel microcystins in Antarctic cyanobacterial mats. Appl Environ Microbiol 74(23):7243–7251. https://doi.org/10.1128/AEM.01243-08

Wright AD, Papendorf O, König GM (2005) Ambigol C and 2,4-dichlorobenzoic acid, natural products produced by the terrestrial cyanobacterium *Fischerella ambigua*. J Nat Prod 68(3):459–461. https://doi.org/10.1021/np049640w

Xiong C, O'Keefe BR, Byrd RA et al (2006a) Potent anti-HIV activity of scytovirin domain 1 peptide. Peptides 27(7):1668–1675. https://doi.org/10.1016/j.peptides.2006.03.018

Xiong C, O'Keefe BR, Botos I et al (2006b) Overexpression and purification of scytovirin, a potent, novel anti-HIV protein from the cultured cyanobacterium *Scytonema varium*. Protein Expr Purif 46(2):233–239. https://doi.org/10.1016/j.pep.2005.09.019

Xu S, Nijampatnam B, Dutta S et al (2016) Cyanobacterial metabolite calothrixins: recent advances in synthesis and biological evaluation. Mar Drugs 14(1):17. https://doi.org/10.3390/md14010017

Yeager CM, Kornosky JL, Housman DC et al (2004) Diazotrophic community structure and function in two successional stages of biological soil crusts from the Colorado Plateau and Chihuahuan Desert. Appl Environ Microbiol 70:973–983. https://doi.org/10.1128/AEM.70.2.973-983.2004

Zelík P, Lukesová A, Cejka J et al (2010) Nostotrebin 6, a bis(cyclopentenedione) with cholinesterase inhibitory activity isolated from *Nostoc* sp. str. Lukesová 27/97. J Enzyme Inhib Med Chem 25(3):414–420. https://doi.org/10.3109/14756360903213481

Zhang X, Smith CD (1996) Microtubule effects of welwistatin, a cyanobacterial indolinone that circumvents multiple drug resistance. Mol Pharmacol 49(2):288–294

Zimba PV, Shalygin S, Huang I-S et al (2021) A new boring toxin producer—*Perforafilum tunnelli gen. & sp. nov.* (Oscillatoriales, Cyanobacteria) isolated from Laguna Madre, Texas, USA. Phycologia 60(1):10–24. https://doi.org/10.1080/00318884.2020.1808389

Nitrogen Recovery by Soil Algae

T. Purushothaman, M. B. Kavitha, and K. Irfana Mol

Abstract Algae is a photosynthetic microorganism, which can be found in the diverse ecosystem. Soil algae occur in different soil habitats such as tropical, subtropical, arid, semi-arid, and Polar Regions. Soil algae play a pivotal role in nitrogen fixation, and those found in the surface of the soil help in promoting the growth and development of plants. The soil algae also prevent the loss of nitrogen through the process of ammonia volatilization. The microalgal species such as *Anabaena sphaerica*, *Westiellopsis prolifica*, *Aulosira fertilissima*, *Nostoc hatei*, and *Cylindrospermum majus* increase the availability of nitrogen, thereby increasing the crop yield. The biofilm formed by various microalgal species also uplifted the nitrogen concentration in the soil. Hence, the nitrogen recovered by the soil algae is utilized by the plants for their growth and development. This chapter summarizes the general characteristics, ecology, as well as the beneficial role of soil algae in the recovery of nitrogen.

Keywords Soil algae · Nitrogen fixation · Nitrogen recovery · Soil fertility · Microalgae

1 Introduction

The term "algae" was first proposed by Carl Linnaeus in 1753, and the actual delimitation of algae from plant species was done by Antoine Laurent de Jussieu in 1789. The algae are considered a base for the aquatic food chain, over and above a vital contributor of photosynthetic activity (Day et al. 1999). The term "algae" comprises both microalgae and macroalgae, and most of the estimated numbers of species are microalgae (Barsanti and Gualtieri 2006). The microalgae are present in the littoral

T. Purushothaman (✉) · K. Irfana Mol
Department of Biotechnology and Research, Shri Nehru Maha Vidyalaya College of Arts and Science, Malumachampatti, Coimbatore, Tamil Nadu, India

M. B. Kavitha
Department of Microbiology, Shri Nehru Maha Vidyalaya College of Arts and Science, Malumachampatti, Coimbatore, Tamil Nadu, India

A. Mishra, A. Varma (eds.), *Soil Algae*, https://doi.org/10.1007/978-981-97-9873-5_11

and benthic zones; on the other hand, macroalgae are found in littoral zones (Rasul et al. 2017). Microalgae are considered the most primitive organisms residing on earth (Kulasooriya 2011).

Algae is a photosynthetic microorganism, which can be either prokaryotic or eukaryotic. It occurs in both terrestrial and aquatic regions and also can withstand extreme conditions. Algae lack major characteristic features found in plants such as vascular system, leaves, stem, or root. Hence, algae have an undifferentiated body called thallus, and it is included in the group thallophyta (Sahoo and Baweja 2015).

The algae and plants exhibit similarities in defense mechanism and production of some storage compounds. The algae occur in distinct forms relatively from simple microscopic cells to more complex structures. Most of the algae are photoautotrophs, which use sunlight to synthesize their necessary metabolites, whereas some of the algal division includes heterotrophic species that depend on the external environment for their nutritional requirements. But some of the algal species can utilize both strategies like photoautotrophy and heterotrophy, and those species are termed as "mixotrophy" (Barsanti and Gualtieri 2006).

Reproduction takes place sexually and asexually in algae. Sexual reproduction involves the fusion of gametes to form a zygote; conversely, asexual reproduction includes the production of spores and vegetative methods (Sahoo and Baweja 2015). Algae range from 1 μm to 600 meters in size and they possess several types of life cycles. Algae occur in different colors due to the existence of diverse chloroplast pigments such as chlorophyll, carotenoids, and phycobiliproteins (Rogers 2010). The algae can also play a symbiotic partnership with fungi to form lichens and also with coral reefs, thereby providing complex nutrients, color, and shelter to their partner and, in return, benefitting protection and nutrients (Barsanti and Gualtieri 2006).

2 Soil Algae

Soil algae usually occur on the surface or in the various soil habitats. The soil is one of the most preferred habitats for non-aquatic algae (Zenova et al. 1995). The soil algae can be isolated from distinct habitats such as polar, tropical, sub-tropical, arid, semi-arid, and coastal regions (Starks et al. 1981). The most prominent algae found in soil are green algae, blue-green algae (cyanobacteria), and diatoms, whereas the algal species of euglenophytes, xanthophytes, and rhodophytes occur occasionally (Metting 1981). Soil algae are considered primordial, and the first isolation of soil algae was reported by Beijerinck in 1893 (Johansen and Shubert 2001). So, algae are recognized as one of the vital parts of the soil microbial community. The important aspect of soil algae is the nitrogen fixation process, which contributes nitrogen content to the soil (Zancan et al. 2006). The algal populations found in the top layer of soil provide immense profitable services for agricultural activities via improving fertility as well as stability of the soil (Metting and Rayburn 1979).

3 Prokaryotic Soil Algae

The prokaryotic algal cell lacks the nucleus, organized chromosomes, and membrane-bound organelles such as the Golgi apparatus, mitochondria, and plastids (Lee 2018).

Phylum	Genera	Reference
Cyanophyta (blue-green algae)	*Anabaena* *Anacystis* *Calothrix* *Camptylonema* *Chlorogloeopsis* *Chroococcus* *Cylindrospermum* *Desmonema* *Fischerella* *Fremyella* *Hapalosiphon* *Hydrocoleum* *Hyphomorpha* *Gloeothece* *Gloeocapsa* *Lyngbya* *Mastigocladus* *Microcoleus* *Nostoc* *Nodularia* *Oscillatoria* *Phormidium* *Plectonema* *Schizothrix* *Stigonema* *Synechococcus* *Synechocystis* *Tolypothrix* *Westiella* *Westiellopsis* *Wollea*	Hunt et al. (1979), Metting (1981), Flechtner et al. (2008), Zancan et al. (2006), Thomas (2018), and Samolov et al. (2020)

4 Eukaryotic Soil Algae

The eukaryotic algae hold a nucleus enclosed with a double-walled nuclear membrane, chromosomes, and other organelles such as Golgi bodies, plastids, endoplasmic reticulum, lysosomes, peroxisomes, chloroplast, and the mitochondria. The nucleus contains the genetic material DNA (Lee 2018).

Phylum	Genera	Reference
Chlorophyta	*Ankyra* *Apatococcus* *Ascochloris* *Asterococcus* *Axilosphaera* *Borodinella* *Borodinellopsis* *Bracteacoccus* *Centrosphaera* *Characium* *Characiopsis Chlamydomonas* *Chlorella* *Chlorococcum* *Chlorogloea* *Chlorosarcina* *Chlorosarcinopsis* *Chlorozebra* *Coccomyxa* *Dictyochloris* *Dictyococcus* *Edaphochlorella* *Elliptochloris* *Fasciculochloris* *Friedmannia* *Gloeococcus* *Gleocystis* *Gloeotilopsis* *Halochlorococcum* *Hormotila* *Hormotilopsis* *Heterochlamydomona* *Ignatius* *Jaagichlorella* *Leptosira* *Lobosphaera* *Mesotaenium* *Muriella* *Myrmecia* *Nannochloris* *Nautococcus* *Neochloris* *Neocystis* *Neospongiococcum* *Oedocladium* *Oocystis* *Ourococcus* *Palmella* *Palmellococcus* *Planktosphaeria* *Planophila* *Pleurastrum* *Protoderma* *Protosiphon* *Pseudotrochiscia* *Pseudogloeocystis* *Radiosphaera* *Raphidonemopsis* *Rhizoclonium* *Rhopalocystis*	Hunt et al. (1979), Metting (1981), Flechtner et al. (2008), Ray and Thomas (2012), Zancan et al. (2006), and Samolov et al. (2020)

Phylum	Genera	Reference
	Scotiella *Spongiochloris* *Spongiococcum* *Sphaerellocystis* *Stichococcus* *Tetracystis* *Tetraedron* *Trichosarcina* *Trochisia* *Ulothrix* *Uronema* *Zygogonium*	
Euglenophyta	*Euglena* *Peranema*	Hunt et al. (1979) and Metting (1981)
Chrysophyta	*Bumilleria* *Bumilleriopsis* *Characiopsis* *Chlorellidium* *Ellipsoidion* *Eunotia* *Navicula* *Nitzchia* *Pleurochloris* *Tribonema*	Lund (1947), Metting (1981), and Maxwell (1991)
Ochrophyta	*Botrydiopsis* *Eustigmatos* *Vaucheria* *Vischeria*	Lund (1947), Flechtner et al. (2008), and Samolov et al. (2020)
Streptophyta	*Cylindrocystis* *Interfilum* *Klebsormidium*	Samolov et al. (2020)
Rhodophyta	*Cyanidium* *Gracilaria* *Porphyridium*	Metting (1981) and Subramanian et al. (2020)

5 Ecology of Soil Algae

Algae are considered ubiquitous organisms, which compensate for the distinct soil types, temperature, pH, moisture, availability of nutrients, etc. Since most of the algae are photoautotrophs, the density of the algal population is high on the surface of the soil and lowers with depth. Some species of green algae, cyanobacteria, and diatoms can exist in the absence of light (Petersen 1935). Some algal species survive in high temperatures and also in desiccated soils. *Prasiola crispa* and *Hormidium flaccidum* are reported desiccation-resistant soil algae, while *Nostoc*, *Nodularia*, *Cylindrospermum*, and *Anabaena* form resting spores in desiccated soils

(Hoffmann 1989). The soil algae such as *Vaucheria geminate*, *Botrydiopsis eriensis*, *Pleurochloris pyrenoidosa*, *Scenedesmus obliquus*, and *Tetradesmus* sp. have desiccation-tolerant capability (Terlova et al. 2020; El-Tablawy et al. 2020).

The species of Chlorophyceae, Cyanophyceae, Xanthophyceae, and Rhodophyceae are the notable soil algae present in the temperate soil (Akiyama 1961). The presence of *Chlorococcum*, *Hormidium (Klebsormidium)*, and *Chlamydomonas* in Enchanted Rock, Texas, was reported by Bischoff (1963). The green algae, blue-green algae, yellow-green algae, and diatoms were identified as species having active functions in the tundra region (Cameron et al. 1978). Soil algae such as *Anabaena*, *Schizothrix*, *Chlamydomonas*, and diatoms can also occur in the coastal areas (Dykstra et al. 1975). *Microcoleus vaginatus*, *Scytonema hofmanii*, *Schizothrix macbrideii*, *Microcoleus chthonoplastes*, *Nostoc muscorum*, *Protococcus grevillei*, *Plectonema nostocorum*, *klebsormidium* sp., and *Stichococcus subtilis* are the algal species distributed in the desert soils (Cameron 1963; Poulickova et al. 2001). The algal species belonging to Cyanophyceace are numerous in alkaline soil conditions, whereas members of Chlorophyceae take up the acidic environment (Starks et al. 1981). Consecutively, the algal species of Cyanophyceae, Bacillariophyceae, Chrysophyceae, Xanthophyceae, Eustigmatophyceae, and Chlorophyceae are present in the acid rain-impacted forest areas of the Krusné hory Mountains (Lukešová and Hoffmann 1996).

6 Importance of Soil Algae

- It helps in maintaining soil fertility in the ecosystem.
- It plays a vital role in the nitrogen fixation process.
- It produces various extracellular bioactive compounds, which influence the growth and development of plants.
- It can uptake heavy metals in soil (bioremediation).
- It is extensively used as a biofertilizer.
- It is an important atmospheric carbon contributor.
- It serves the organic matter in the soil thereby increasing the humus content and promoting plant growth.
- It aids in the reclamation of soil and also prevents soil erosion.
- Some species of soil algae are used as biological control agents (biopesticides) (Abdel-Raouf et al. 2012) (Fig. 1).

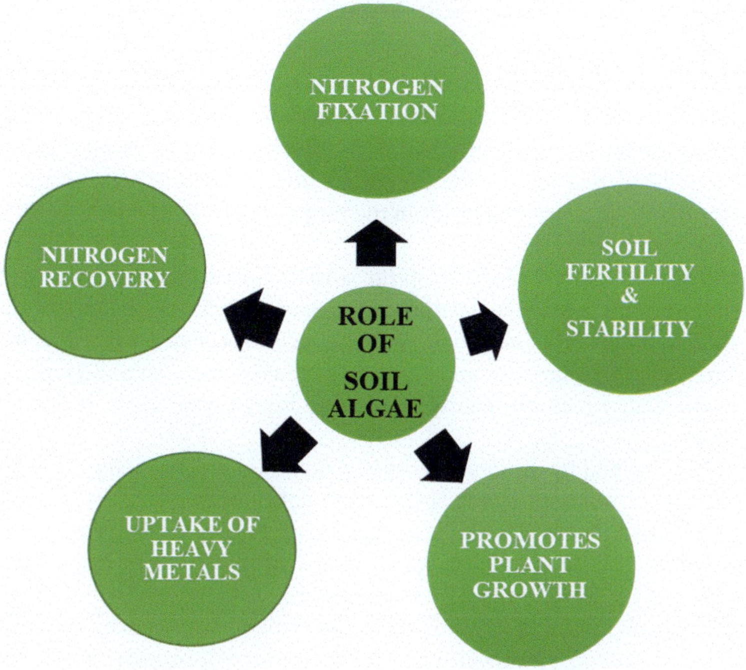

Fig. 1 Role of soil algae

7 Nitrogen Fixation

Algae are capable of fixing atmospheric nitrogen into the utilizable form of nitrogen, so it can be assimilated by the plants. The application of blue-green algae in the soil leads to the increased availability of nitrogen, thereby resulting in promoting the plants' growth (Singh 2021). The cyanobacteria found in the rhizosphere also promote nitrogen fixation. Notable species in cyanobacteria for nitrogen fixation are *Anabaena, Trichodesmium*, and *Nostoc* (Mishra et al. 2021).

8 Soil Fertility and Stability

Algal species belonging to Chlorophyta abet in soil stabilization. It also has a wide role in improving soil fertility. The exopolysaccharides produced by the algae help in creating a biofilm with other algal species or with other microbial populations. This microbial consortium benefits the higher availability of micronutrients as well as macronutrients (Prasanna et al. 2021).

9 Nitrogen Recovery by Soil Algae

The study done by de Souza et al. (2019) demonstrated that the presence of microalgal biofilm aids in preventing the loss of nitrogen in the soil via the process of ammonia volatilization. Nitrogen found in the blue-green algae *Nostoc* sp. incorporated in the soil was recovered by rice (*Oryza sativa*), resulting in the availability of nitrogen for the first and second crops in the range of 23–28% and 27–36%, respectively (Tirol et al. 1982).

The presence of microalgae *Chlorella sorokiniana* in the calcareous soil has remarkably elevated the concentration of nitrogen (Marks et al. 2019). The algal species such as *Anabaena sphaerica, Westiellopsis prolifica, Aulosira fertilissima, Nostoc hatei*, and *Cylindrospermum majus* increased the availability of nitrogen content in the soil, which ultimately resulted in higher crop yield in rice (Jha and Prasad 2006).

The biofilm formed by *Anabaena torulosa* demonstrated its potential by enhancing the amount of nitrogen in the soil (Swarnalakshmi et al. 2013). Interestingly, biofilm formed by other algal species such as *Anabaena doliolum, Nostoc carneum, Nostoc piscinale, Calothrix* sp., *Anabaena laxa*, and *Anabaena torulosa* has significantly improved the availability of nitrogen in the soil (Prasanna et al. 2014). The amount of nitrogen present in the soil can be effectively increased by the microalgal consortia of *Chlorella, Phormidium, Scenedesmus, Spirogyra, Chlorococcum, Anabaena, Chroococcus*, and *Fischerella* (Renuka et al. 2016). The work done by Usami et al. (2020) suggested the recovery of nitrogen from the microalgae *Fistulifera solaris* with the help of the 5 g/L carbon solid acid or a 0.02–0.50 M HCl as the catalyst.

10 Conclusion and Future Perspective

Soil algae play a crucial role in nitrogen fixation, nitrogen recovery, as well as in maintaining soil fertility and stability. Nitrogen can be recovered from a wide range of algal species such as *Chlorella, Anabaena, Spirogyra, Nostoc*, and *Calothrix* sp. These algal species also help in increasing the availability of nitrogen concentration in the soil. Furthermore, the use of these soil algal species in wastewater treatment could be promising for the recovery of nutrients such as nitrogen and phosphorous. Soil inoculation of algae benefits in enhancing the fertility of the soil, which paves the way for the efficient growth of plants. Therefore, further research studies are needed to identify and characterize the efficient strains of soil algae found in diverse ecosystems for the recovery of nitrogen.

References

Abdel-Raouf N, Al-Homaidan AA, Ibraheem IB (2012) Agricultural importance of algae. Afr J Biotechnol 11(54):11648–11658

Akiyama M (1961) Aerial and terrestrial algae in San-in region of Honshu, Japan. Bull Shimane Univ 10:75–89

Barsanti L, Gualtieri P (2006) Algae: anatomy, biochemistry, and biotechnology. Taylor & Francis, Boca Raton

Bischoff HC (1963) Some soil algae from Enchanted Rock and related algal species. Phycological studies IV. University of Texas Publ. No. 6318;6318:1–95

Cameron RE (1963) Algae of southern Arizona. Part I. Introduction-blue-green algae. Rev Algol NS 6:282–318

Cameron RE, Knox AD, Morelli FA (1978) The role of algae in tundra soils. In: Vegetation and production ecology of an Alaskan Arctic Tundra. Springer, New York, pp 207–227

Day JG, Benson EE, Fleck RA (1999) In vitro culture and conservation of microalgae: applications for aquaculture, biotechnology and environmental research. In Vitro Cell Dev Biol Plant 35(2):127–136

de Souza MH, Calijuri ML, Assemany PP, de Siqueira CJ, de Oliveira AC (2019 Feb) Soil application of microalgae for nitrogen recovery: a life-cycle approach. J Clean Prod 20(211):342–349

Dykstra RF, MacEntee FJ, Bold HC (1975) Some edaphic algae of the Texas coast. Tex J Sci 26:171–177

El-Tablawy NH, Mansour HA, Shaaban AE (2020) Antioxidant activities of some edaphic algae in Egypt. Beni-Suef Univ J Basic Appl Sci 9(1):1

Flechtner VR, Johansen JR, Belnap J (2008) The biological soil crusts of the San Nicolas Island: enigmatic algae from a geographically isolated ecosystem. West N Am Nat 68(4):405–436

Hoffmann L (1989) Algae of terrestrial habitats. Bot Rev 55(2):77–105

Hunt ME, Floyd GL, Stout BB (1979) Soil algae in field and forest environments. Ecology 60(2):362–375

Jha MN, Prasad AN (2006) Efficacy of new inexpensive cyanobacterial biofertilizer including its shelf-life. World J Microbiol Biotechnol 22(1):73–79

Johansen J, Shubert E (2001) Algae in soils. Nova Hedwig Beih 123:297–306

Kulasooriya SA (2011) Cyanobacteria: pioneers of planet earth. Ceylon J Sci 40(2):71–88

Lee RE (2018) Phycology. Cambridge University Press, p 3

Lukešová A, Hoffmann L (1996) Soil algae from acid rain impacted forest areas of the Krušné hory Mts. 1. Algal communities. Vegetatio 125(2):123–136

Lund JW (1947) Observations on soil algae II. Notes on groups other than diatoms. New Phytol 46(1):35–60

Marks EA, Montero O, Rad C (2019) The biostimulating effects of viable microalgal cells applied to a calcareous soil: increases in bacterial biomass, phosphorus scavenging, and precipitation of carbonates. Sci Total Environ 692:784–790

Maxwell CD (1991) Floristic changes in soil algae and cyanobacteria in reclaimed metal-contaminated land at Sudbury, Canada. Water Air Soil Pollut 60(3–4):381–393

Metting B (1981) The systematics and ecology of soil algae. Bot Rev 47(2):195–312

Metting B, Rayburn W (1979) Algal communities and soil microenvironments in an eastern Washington silt loam. Soil Sci 127(2):74–78

Mishra A, Rajput S, Gupta PS, Goyal V, Singh S, Sharma S, Shukla S, Singh A, Shukla K, Varma A (2021) Role of cyanobacteria in Rhizospheric nitrogen fixation. In: Soil nitrogen ecology. Springer, Cham, pp 497–519

Petersen JB (1935) Studies on the biology and taxonomy of soil algae. Dan Bot Ark 8(9):1–83

Poulickova A, Drimalova D, Novotny R, Válová P, Havránek P (2001) Morphological responses to varying environmental conditions in Klebsormidium flaccidum isolated from Moroccan desert soils. Nova Hedwig Beih 123:397–410

Prasanna R, Triveni S, Bidyarani N, Babu S, Yadav K, Adak A, Khetarpal S, Pal M, Shivay YS, Saxena AK (2014) Evaluating the efficacy of cyanobacterial formulations and biofilmed inoculants for leguminous crops. Arch Agron Soil Sci 60(3):349–366

Prasanna R, Renuka N, Nain L, Ramakrishnan B (2021) Natural and constructed cyanobacteria-based consortia for enhancing crop growth and soil fertility. In: Role of microbial communities for sustainability. Springer, Singapore, pp 333–362

Rasul I, Azeem F, Siddique MH, Muzammil S, Rasul A, Munawar A, Afzal M, Ali MA, Nadeem H (2017) Algae biotechnology: a green light for engineered algae. In: Rasul I, Azeem F, Siddique MH, Muzammil S, Rasul A, Munawar A, Afzal M, Ali MA (eds) Algae based, Polymers, blends, and composites. Elsevier, pp 301–334

Ray JG, Thomas TB (2012) Ecology and diversity of green-algae of tropical oxic dystrustepts soils in relation to different soil parameters and vegetation. Res J Soil Biol 4(3):42–68

Renuka N, Prasanna R, Sood A, Ahluwalia AS, Bansal R, Babu S, Singh R, Shivay YS, Nain L (2016) Exploring the efficacy of wastewater-grown microalgal biomass as a biofertilizer for wheat. Environ Sci Pollut Res 23(7):6608–6620

Rogers K (ed) (2010) Fungi, algae, and protists. Britannica Educational Publishing

Sahoo DJ, Baweja P (2015) The algae world, Cellular origin, life in extreme habitats and astrobiology, vol 26. Springer, Dordrecht, p 15. https://doi.org/10.1007/978-94-017-7321-8_1

Samolov E, Baumann K, Büdel B, Jung P, Leinweber P, Mikhailyuk T, Karsten U, Glaser K (2020) Biodiversity of algae and cyanobacteria in biological soil crusts collected along a climatic gradient in Chile using an integrative approach. Microorganisms 8(7):1047

Singh A (2021) Role of algae in soil nitrogen fixation. In: Soil nitrogen ecology. Springer, Cham, pp 483–496

Starks TL, Shubert LE, Trainor FR (1981) Ecology of soil algae: a review. Phycologia 20(1):65–80

Subramanian G, Nagaraj A, Sasikala J, Ravi P, Sona P (2020) Nutrient content of red algal species of a genus Gracilaria from the coastal areas of Rameswaram, Tamil Nadu, India. Int J Anal Exp Mod Anal 12(5):1721–1727

Swarnalakshmi K, Prasanna R, Kumar A, Pattnaik S, Chakravarty K, Shivay YS, Singh R, Saxena AK (2013) Evaluating the influence of novel cyanobacterial biofilmed biofertilizers on soil fertility and plant nutrition in wheat. Eur J Soil Biol 55:107–116

Terlova E, Holzinger A, Lewis L (2020) Desert green algae show higher desiccation tolerance than their aquatic sister-species. Authorea Preprints

Thomas BT (2018) In vitro culturing and biochemical properties of green and blue-green algae in the laterite soil. Vegetos 31(4):28–34

Tirol AC, Roger PA, Watanabe I (1982) Fate of nitrogen from a blue-green alga in a flooded rice soil. Soil Sci Plant Nutr 28(4):559–569

Usami R, Fujii K, Fushimi C (2020) Improvement of bio-oil and nitrogen recovery from microalgae using two-stage hydrothermal liquefaction with solid carbon and HCl acid catalysis. ACS Omega 5(12):6684–6696

Zancan S, Trevisan R, Paoletti MG (2006) Soil algae composition under different agro-ecosystems in North-Eastern Italy. Agric Ecosyst Environ 112(1):1–2

Zenova GM, Shtina EA, Dedysh SN, Glagoleva OB, Likhacheva AA, Gracheva TA (1995) Ecological relations of algae in biocenoses. Mikrobiologiya 64:121–133

Harnessing Nanopolysaccharides for Sustainable Agriculture and Environmental Protection

Dasarapu Santhosha, Kondi Vanitha, Swamykannu Dineshmohan, Sri Chandana Mavulati, and Onkar Nath Tiwari

Abstract There is an increasing demand for pharmaceuticals and cosmetics as additives and preservatives in the food industry for a variety of applications. However, global plant yields per hectare have significantly decreased due to aggressive cultivation practices aimed at feeding growing populations, as well as the impacts of climate change. There is, therefore, an urgent need to find alternative ways/smart ways of production, especially because the chemical synthesis of these valuables is complex and their production levels in nature are very low. Thus, accelerating plant growth and productivity by application of nanofertilizers can open new perspectives in agricultural practice. However, the response of plants to nanofertilizers varies with the type of plant species, their growth stages, and the nature of nanomaterials. This chapter deals with different nanomaterials/nanoparticles obtained from polysaccharides along with their applications in food and agriculture and industries. By following the techniques mentioned in this chapter, one can improve the food yield without the quality of food and fertility of the soil being affected.

D. Santhosha (✉)
Department of Pharmacognosy, Vishnu Institute of Pharmaceutical Education and Research, Narsapur, Medak, Telangana, India
e-mail: santhosha.d@viper.ac.in

K. Vanitha (✉) · S. Dineshmohan
Department of Pharmaceutics, Vishnu Institute of Pharmaceutical Education and Research, Narsapur, Medak, Telangana, India
e-mail: vanitha.k@viper.ac.in; dineshmohan.s@viper.ac.in

S. C. Mavulati
Department of Pharmacology, Vishnu Institute of Pharmaceutical Education and Research, Narsapur, Medak, Telangana, India

O. N. Tiwari
Centre for Conservation and Utilization of Blue-Green Algae, Division of Microbiology, ICAR, Indian Agricultural Research Institute, New Delhi, India
e-mail: chandana@svcp.edu.in

A. Mishra, A. Varma (eds.), *Soil Algae*, https://doi.org/10.1007/978-981-97-9873-5_12

Keywords Nanomaterials · Nanotechnology · Agriculture · Nanoparticles · Ecosystems

1 Introduction

The global population is projected to rise in number and affluence by the middle of this century, with projections of the world's population in 2050 ranging between 8 and 10 billion. Feeding the world's future population would put huge demands on agriculture (Wang et al. 2016). Therefore, there is a need to increase food production while reducing the negative impacts of agriculture on soil, water, and climate. Estimates suggest that the agricultural production will increase twofold to meet the projected food demands (Foley et al. 2011). More food is required and food quality needs to be improved, especially the nutrient content (Tester and Langridge 2010). The development of crops can be done simply by growing farmland to meet the needs of potential populations. Competition for land use, along with the loss of land due to salinization and desertification, will reduce the amount of land available for agricultural conversion (Fedoroff et al. 2010). Furthermore, the conversion of land into agriculture has significant environmental implications. Transformation of natural ecosystems into agricultural land impacts global carbon and hydrological cycles, habitat biodiversity, and soil conditions (Myers et al. 2014). Reproducing natural habitats for crop production may have significant effects on the emissions of greenhouse gases by releasing carbon that is stored in vegetation and soil biomass into the air (Matson et al. 1997).

Agricultural activity is the bridge that connects the environment and humans and is the leading cause of climate change and detriment of ecosystems. Over time, the increasing demand for food has encouraged the excessive use of agrochemicals, which has generated a harmful effect on the environment and health. The word "agrochemicals" refers to chemicals such as pesticides, herbicides, and fertilizers, among others (Campos et al. 2015). In this way, agriculture in the twenty-first century is facing significant challenges to meet the increasing demand for food of an increasing global population without negatively impacting the soil and human health. The main objective of the Food and Agriculture Organization in the 2030 Agenda is to implement sustainable agriculture practices that will eradicate hunger and improve food security and nutrition while maintaining ecosystems and improving soil quality. Agrochemicals play a significant role in maintaining the increasing demand for food to the growing population. However, the harmful effect of these chemicals on the environment and health limits their use. Hence, safe handling of these chemicals is very significant (Abhilash and Singh 2009). Nowadays, the primary concern is bridging the gap between agricultural poverty and the nutritional process of food production. Therefore, novel technologies should be adopted that focus on getting better agricultural yield (Yunlong and Smit 1994). Recently, food and nutritional security have been fully integrated into new knowledge. Agriculture development also depends on social inclusion, health, climate changes, energy, ecosystem processes, natural resources, moral supremacy, etc., and must also be documented in specific target-oriented goals. Therefore, sustainable agriculture strengthens the opportunity to prevent poverty and hunger. Agriculture on its way to recovery, needs environmental performance and at the same time participation of food chain ecosystems with agricultural food production (Thornhill et al. 2016). Currently, in the agricultural field, sustainable production and efficiency are unimaginable without the use of agrochemicals such as pesticides and fertilizers. However, every agrochemical has one or the other issues like contamination of water or presence of chemical residues on food products that threaten the human being and environmental health; thus, the precise management and control of inputs could allow reducing these risks (Kah 2015).

2 Nanotechnology and Polysaccharides for Agrochemical Delivery System

Nanotechnology is defined as the science of matter at dimensions of approx. 1–100 nm, where materials have unique physical properties that make novel applications possible. This definition is slightly rigid about size dimensions. Great emphasis could have been placed on the problem-solving capability of the materials. Other attempts to define nanoparticles in the context of agriculture include "particulate between 10 and 1,000 nm in size dimensions that are simultaneously colloidal particulate" (Mukhopadhyay 2014). In this context, scientific advances and the emergence of nanotechnology have allowed the development of "precision agriculture," which aims to reduce costs and resources and to increase yields with nanoscale control using nanosensors and intelligent delivery systems (Mukhopadhyay 2014). Nanotechnology has been widely exploited by mankind in various fields like pharmaceutical, medical, electronics, and cosmetic sectors, while agriculture, compared to other industrial sectors, has been relatively marginalized (Wang et al. 2016). In the last decade, there was a noticeable growth in the number of scientific publications and patents in agro-nanotechnology (Liu et al. 2008). Application of this tool in plant science is known as phytonanotechnology. It mainly scans systems that produce controlled release of biomolecules with specific temporal and spatial patterns (Mukhopadhyay 2014). Similarly, agriculture faces many challenges, such as reduced crop yield, nutrient deficiency, and environmental pollution, and the emergence of nanotechnology has offered great applications for precision agriculture (Fig. 1). The term "precision agriculture" is related to the site-specific crop management with a wide range of pre- and post-production aspects of agriculture, ranging from horticultural crops to field crops (Singh et al. 2016).

In the context of agriculture, nanotechnology has been used to increase the crop yield with quality enhancement by improving farming systems as shown in Fig. 2. The emergence of engineered nanomaterials and their actions within the range of sustainable agriculture has revolutionized the global agriculture canvass drastically

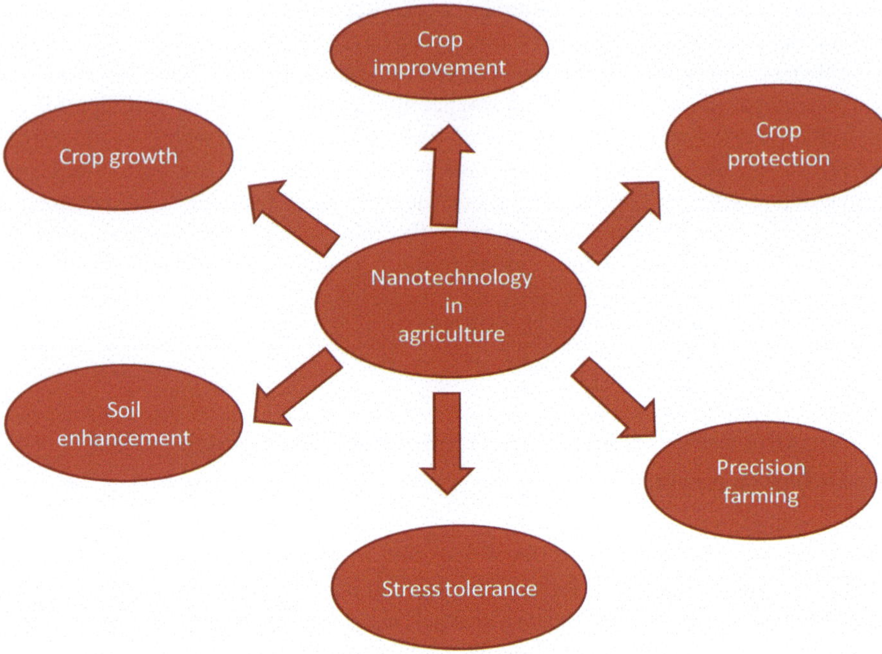

Fig. 1 Applications of nanotechnology in agriculture

by their novelty, fast growth, and enormity to meet the global food requirement. In sustainable agriculture, protecting the environment is a primary goal, and nanomaterials ensure better management and conservation of resources used in plant production.

Conventionally, agrochemicals are applied by spraying or soaking method. This technique consists of direct application of agrochemicals on plants and soil causing great environmental and economic disadvantages. A large percentage of the active substances applied are quickly lost by leaching, degradation, and volatilization, and only a small amount is left to fulfill their function and reach the site of action. Therefore, to maintain their biological efficiency and concentration at appropriate levels, periodic applications of agrochemicals in excessive concentrations are usually carried out, which implies high costs, low efficiency, and serious environmental and health problems (Nitta and Numata 2013). The application of engineered nanomaterials in the frame of sustainable agriculture has shown a different way of food production that could overcome the fluctuation in the crop sector with limited available resources (Prosekov and Ivanova 2018). The revolution of nanotechnology has drastically changed the global agriculture canvass and nanomaterials to meet the world's food demand and sustainable agriculture as well.

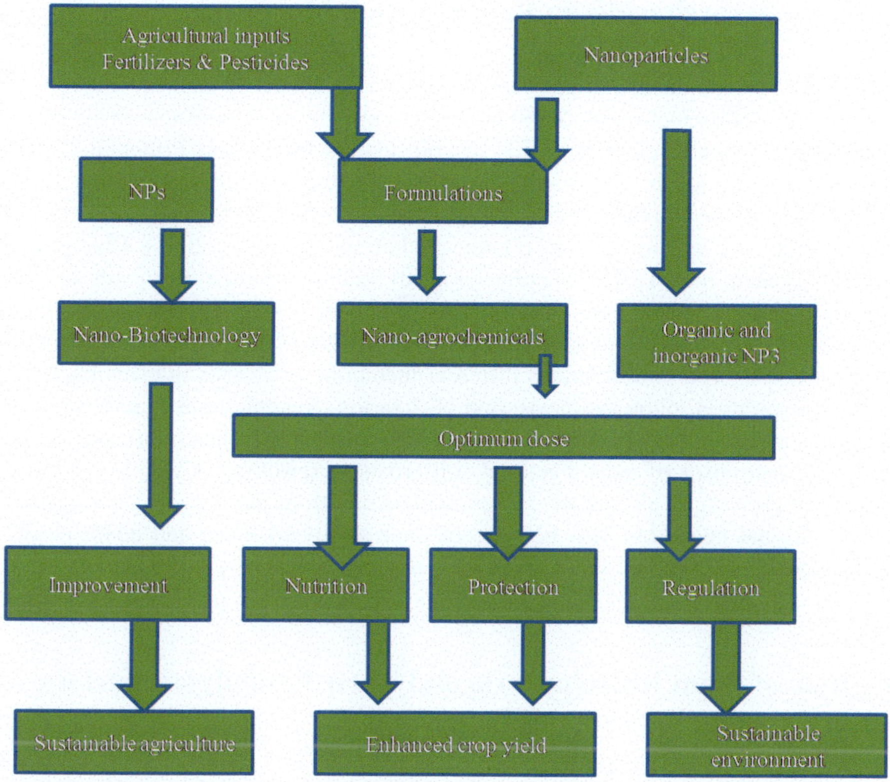

Fig. 2 Overview of applications of nanomaterials in sustainable agriculture production

2.1 *Polysaccharides for Agrochemical Delivery System*

Polysaccharides are macromolecules that are composed of monosaccharide units connected by glycosidic bonds (Raemdonck et al. 2013). These compounds are found in nature, including algae (alginate), plants (cellulose, pectin, cyclodextrin, and starch), microorganisms (dextran), and animals (chitosan) (Raemdonck et al. 2013). They are highly versatile since they have many reactive functional groups that can be used in chemical reactions to add new groups into their structure (Nitta and Numata 2013). The molecules can be linear or cyclic and differ in terms of their charge (Hassani et al. 2012). Advantages include their low cost and availability, which facilitates the large-scale production of products derived from them.

3 Exopolysaccharides in Retaining Soil Moisture

In drylands, the major factor that hinders soil microbial activity is the availability of moisture. Exopolysaccharides (EPS) are hygroscopic in nature and absorb the moisture present in and around the areas of soil. Garcia-Pichel and Pringault (2001) observed that cyanobacteria were able to migrate to the soil surface upon the availability of moisture and move back down the subsoil upon drying to avoid desiccation and damage due to sunlight (Garcia-Pichel and Pringault 2001). Changes in moisture content in soil affect the physiological function of crust organisms and the production of EPS. The early morning fog has been found to provide optimum moisture to stimulate activity (Thomas and Dougill 2007). The availability of moisture and its absorption accelerates different metabolic processes in cyanobacteria, such as respiration, photosynthesis, and N_2-fixation. Recent studies carried out in Kalahari sand soils (Thomas and Dougill 2007) have shown the complexity of the metabolic responses of cyanobacterial soil crusts to rewetting. Cyanobacteria are capable of growing in almost all environmental conditions not only because of their low nutritional needs but also because of their ability to sense and adapt to any environment.

4 Role of Plant Rhizobacteria and Their Exopolysaccharides in Drought Tolerance of Plants

Exopolysaccharides (EPS) constitute 40–95% of bacterial weight (Flemming and Wingender 2001). Bacteria generate EPS in two forms: (Abhilash and Singh 2009) slime EPS and (Achterbosch et al. 2014) capsular EPS. Important roles of EPS are as follows: protection, surface attachment, biofilm formation, microbial aggregation, plant-microbe interaction, and bioremediation. EPS-generating bacteria like *Pseudomonas* can survive even under drought conditions due to the presence of EPS (Sandhya et al. 2009). EPS of bacteria are hydrated with 97% of the water in polymer matrix protecting against desiccation (Hunter and Beveridge 2005) by improving the water retention and by regulation of organic carbon source's diffusion (Roberson and Firestone 1992; Chenu 1993; Chenu and Roberson 1996). EPS helps in the transformation of heavy metal and organic recalcitrant degradation (Van-Hullebusch et al. 2004; Pal and Paul 2008). The availability of moisture in the soil affects the soil structure (Roberson and Firestone 1992). This was proved by treating plants with EPS-producing bacteria, *Azospirillum*, which showed resistance to water scarcity (Bensalim et al. 1998) along with improvement in the soil structure (Sandhya et al. 2009) and soil aggregation. This was supported by another study by inoculating sunflower with EPS-producing bacterial strain YAAF34, which showed an increase in root tissue (Alami et al. 2000) (Fig. 3).

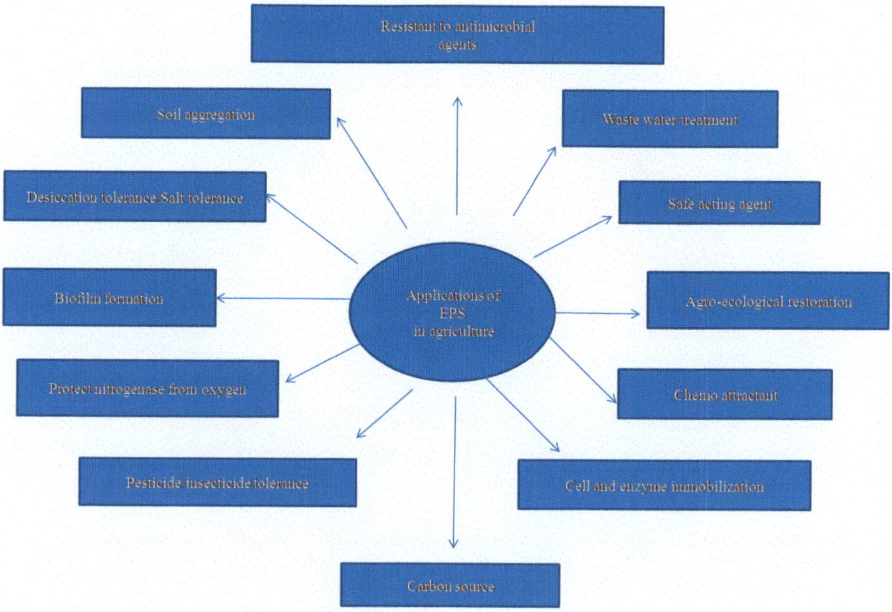

Fig. 3 Applications of EPS in agriculture

5 Applications of Nanoscale Materials in Agriculture

Nanoscale starch has many applications in agriculture:

1. For controlled release of fertilizer
2. To reduce fertilizer loss and environmental pollution
3. As a super absorbent in agriculture
4. To improve fertilizer utilization efficiency
5. To remediate of saline soil

Nanoscale cellulose materials, dispersed incompatible polymer matrices, can result in nanocomposites of enhanced mechanical/barrier properties (Zafar et al. 2016). Nanoscale cellulose plays a great role in packaging materials to protect the packaged foods due to their toxic nature (Maftoonazad and Ramaswamy 2018). Other studies have reported that antimicrobial particles incorporated onto nanocellulose materials, used for packaging food, have considerably increased the shelf life of different foods (Kargarzadeh et al. 2017). Nanocellulose has been used in agricultural practices as protective coatings for different parts like seeds, plants, and food materials to improvise stability (Iavicoli et al. 2017). The addition of plasticizer (e.g., glycerol) and surfactants (e.g., Tween 80 and Span 80) is required for optimal wettability, elasticity, and water resistance. Field application of cherry fruits with these nanocomposite coatings prevents cherry cracking without affecting the growth and quality of fruit. Nanostructured systems stabilized with carboxymethyl

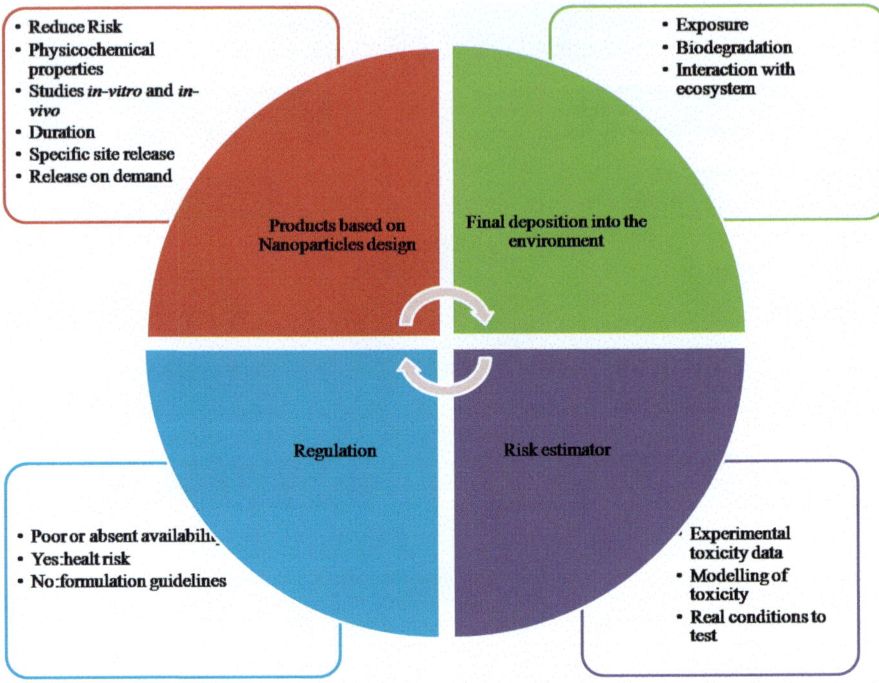

Fig. 4 Scheme of considerations related to the risk associated with nanoparticle-based products

cellulose have been studied for the removal of contaminants like heavy metals, dyes, or halogenated compounds in agricultural fields (Li et al. 2016a, b) (Fig. 4).

In the agricultural field, nanoscale lignin is used for the controlled release of fertilizers and herbicides (Norgren and Edlund 2014). Du et al. (2014) modified industrial softwood Kraft lignin by a process called amination, with and without phenolation pretreatments (Du et al. 2014). These aminated lignins produced stable aqueous colloidal suspensions with particle sizes ranging between 39 and 391 nm. Potential applications include surfactant chemicals and polycationic carrier for the sustain-release of fertilizers. Nanoscale pectin can improve stability in emulsions by interacting with protein. The proteins remain at the oil/water interface, while polysaccharides produce a denser stabilizing layer that prevents aggregation of droplets (Naqash et al. 2017). Nanoscale pectin materials are used in a prebiotic delivery vehicle, gelling agent, emulsion stabilizer, fat replacer, and so on. Nanoscale polysaccharide gums are abundantly used in industrial food and agriculture applications as stabilizing agents, thickeners, gelling agents, sugar/fat substitutes, prebiotics, encapsulation matrix, and so on (Jo et al. 2018). Guar gum is the important polysaccharide to form nanocomposites; according to Tang et al. (2018), the inclusion of nanocrystalline cellulose in chitosan/gum, guar/nanocrystalline cellulose nanocomposites improved their mechanical strengths (tensile strength) and reduced their air permeability, making them potential materials for food packaging.

Nanoscale cyclodextrins have been used to develop new agrochemicals such as herbicides, insecticides, fungicides, and growth regulators (Seglie et al. 2012; Villaverde et al. 2018). Cyclodextrins can be used for the removal of hydrophobic adulterants from soils (Morillo and Villaverde 2017). Pushpalatha et al. (2018) made use of cyclodextrin-based nanosponges (CD-NS) to deliver curcumin using diphenyl carbonate (DPC) and pyromellitic dianhydride (PMDA) as cross-linkers (Pushpalatha et al. 2018). They reported that PMDA resulted in CD-NS with higher solubility and curcumin release than those cross-linked by DPC. The former also exhibited higher cytotoxicity compared to the DPC cross-linker used in macro-scale materials. A wide variety of polysaccharide nanocomposites have been explored. These include polysaccharide nanocomposites containing metals, metal oxides and inorganic compounds, structured carbons, biomolecules, functional polymers and multiple polysaccharides, which have many potential applications. These potential applications can be divided based on the different overlapping groups. The major categories include drug delivery, tissue engineering and wound healing, blood and biocompatibility, antibacterial properties, applications in biomedicine, energy production and storage, industrial applications, electronic devices, affinity materials, and mechanical and thermal properties.

6 Applications of Controlled Released Nanofertilizers and Nanomaterials in Agriculture

Nano-based target delivery system is used for crop improvement. Nanopesticides play an important role in crop protection. Uses of nanosensors and computerized monitoring systems contribute to precision farming. Nanomaterials are used to promote tolerance toward plant stress and soil enhancement (Worrall et al. 2018). Recent studies have claimed the use of nanomaterials on germination as well as plant growth and development. For instance, the application of multiwalled carbon nanotubes (MWCNTs) affects seed germination of different crop species (Lahiani et al. 2013; Khodakovskaya et al. 2009). A few studies have claimed that nanomaterials have the capability to penetrate the seed coat and enhance absorption of water which in turn improves germination and seedling growth. Nonetheless, the mechanism of nanomaterial-induced water uptake is not clear. Nanomaterials are supposed to increase crop growth with quality enhancement in many crops. For example, carbon nanomaterials fullerols have shown positive effects on plant growth. It has been found that seed dressings with fullerol increase fruit number, fruit size, and final yield by up to 127% along with stimulating the content of bioactive compounds (Kole et al. 2013). Moreover, the studies of nutrients on controlled release of nanofertilizers in water and soil have confirmed the long-term availability of all the nutrients to the plant at all levels of growth cycle (Lateef et al. 2016).

7 Nanofarming in Agricultural Development

Nanoparticle engineering is one of the latest innovative technologies that explore targeted characteristics with elevated strength. Recent advancements in the manufacture of nanomaterials of various sizes and shapes have found their uses in environmental science, agriculture, medicine, and food processing areas. Agriculture always gets benefited from these innovations (Chen et al. 2016). In agriculture, there are many problems and unpredicted challenges, such as diminished crop yields due to biotic and abiotic stresses, nutrient deficiency, and environmental pollution. Nanotechnology has shown promising applications for precise agriculture (Fig. 1) meaning of development in farming or agriculture include wireless networking and miniaturization of the sensors for monitoring, assessing and controlling agricultural techniques. It is related to crop management with a wide range of pre- and post-production aspects of agriculture, ranging from horticultural crops to agriculture crops (Singh et al. 2016). Recent developments in nanomaterial-mediated targeted delivery of CRISPR (clustered regularly interspaced short palindromic repeats)/Cas (CRISPR-associated protein) mRNA and sgRNA for genetic modification (GM) is an important scientific achievement (Ran et al. 2017; Miller et al. 2017). Further, nanotechnology provides excellent solutions for a number of growing environmental challenges. For instance, nanosensors play excellent role in the observation of environmental stress and in enhancing the potentials of plants to fight against diseases (Afsharinejad et al. 2015; Kwak et al. 2017). Therefore, such continuous developments in nanotechnology with special preference on the identification of problems and development of collaborative solutions for sustainable agricultural growth have great potential to provide broad social and equitable benefits.

8 Nanoparticle-Incorporated Smart Delivery Options for Sustainable Agriculture

The delivery systems of agrochemicals and organic molecules are based on the transport of DNA molecules or oligonucleotides into the plant cells which is an important criterion for sustainable agriculture as well as precision farming (Joga et al. 2016). In conventional methods, agrochemicals are generally applied by spraying or broadcasting to crops. But, a very low amount of agrochemicals reaches the target areas of crops, which is much lesser than the minimum effective level of concentration necessary for the successful growth of the plant. The losses may be due to leaching of chemicals and degradation by photolysis, hydrolysis, and also microbial degradation (Nair et al. 2010). For example, more importance should be laid on the bioavailability of nutrients due to the chelation effects of soil, degradation by microorganisms, evaporation, over-application, hydrolysis, and run-off problems (Gogos et al. 2012). In case of application of pesticide, the efficacy can be enhanced with spray drift management (Ghormade et al. 2011). In developing eco-friendly agricultural practices, many advances in nanotechnology-based synthesis of slow or controlled release fertilizers, pesticides, and herbicides have been made (Singh et al. 2016). Nanoparticles are advantageous over agrochemicals due to their large surface area, quick attachment ability, and fast mass transfer capacity (Ghormade et al. 2011). For this, microcin- or submicronic-particle-based agrochemicals are prepared through absorption, capsulation, surface ionic attachments, and entrapment into the nanomatrix of active components (Nuruzzaman et al. 2016; Pandey et al. 2018). For example, encapsulation of potassium nitrate by graphene oxide films significantly extends the fertilizer release process; this process is relatively less expensive and can be carried out in large scale (Zhang et al. 2014). Nanomaterials improve the stability of agrochemicals and protect them from degradation and leakage into the environment, which increase the effectiveness and reduce the composition of agrochemicals. The recent advances in nanobiotechnology provide greater opportunities to the researchers to completely replace the genetic material of one species by another (Torney et al. 2007).

In genetic engineering, silicon dioxide nanoparticles have been designed to release DNA fragments to the target species, like tobacco and corn plants, without any side effects (Cheng et al. 2016; Galbraith et al. 2007). NP-based delivery systems are also used to develop insect-resistant crop varieties. For instance, DNA-coated NPs are used as bullets in gene-gun technology to transfer the desired genes to the targeted plants (Vijayakumar et al. 2010). The recent advancement in chitosan NPs-entrapped SiRNA delivery vehicles has provided a new platform by allowing the control of insect pests as chitosan has an efficient binding capacity with RNA as well as penetration capability through the cell membranes (Zhang et al. 2010). Advances in nanomaterial-incorporated specific delivery of CRISPR/Cas9 single-guide RNA (sgRNA) have set a new trend in genetic engineering. The CRISPR/Cas9 system, including CRISPR repeat-spacer arrays and Cas proteins, is an RNA-directed defense system in prokaryotes and is used for genome editing in plants (Miller et al. 2017). However, the low delivery efficiency is still a huge problem for its applications. Nanomaterials could possibly improve the efficiency and specificity of CRISPR/Cas systems. For instance, cationic arginine-incorporated gold nanoparticles (ArgNPs)-assembled Cas9En (E-tag)-RNP (ribonucleoproteins) delivery of sgRNA gives almost 30% effective cytoplasmic gene editing capacity in cultured cell lines, which facilitate crop development (Mout et al. 2017; Zhang et al. 2014) (Fig. 5).

9 Nanofertilizers for Balanced Crop Nutrition

Supplementation of essential (micro and macro) nutrients is necessary for improving crop productivity and soil fertility (Li et al. 2018). Nevertheless, fertilizer management is an important prerequisite for sustainable agricultural development (Huang et al. 2017). Food security is threatened due to the limitation of available

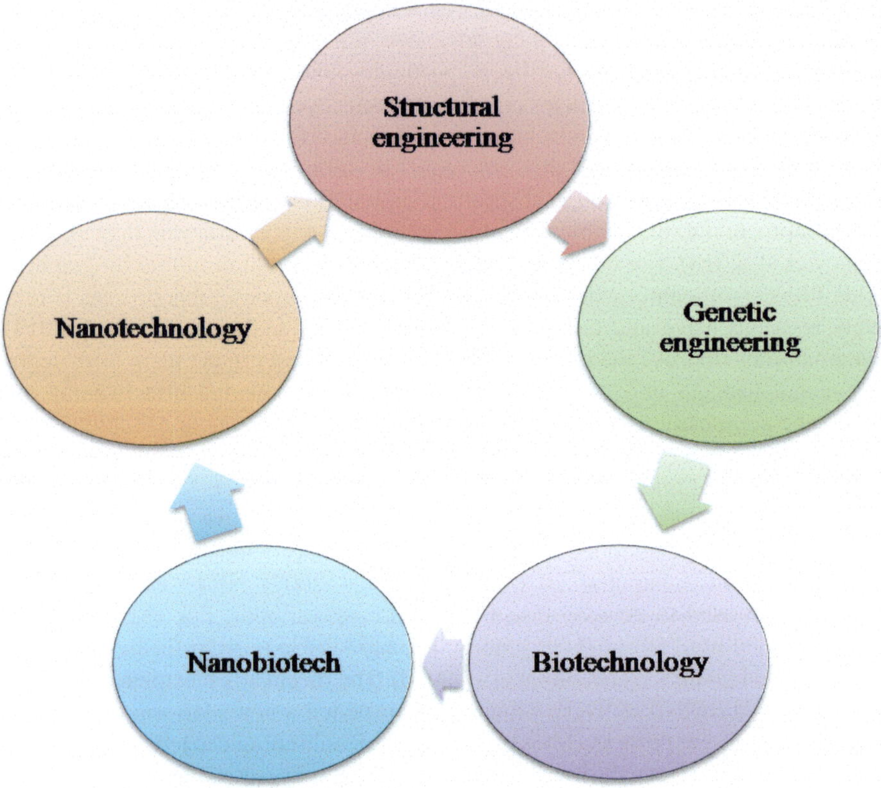

Fig. 5 An overview of nanobiotechnology

food resources. To feed the increasing population of India, almost 60% more food has to be produced (Achterbosch et al. 2014). For this, intensive farming is being practiced, which leads to the exhaustion of soil fertility and decline in the productivity. It has been surveyed that because of this practice, approximately 40% of the world's agricultural field has been degraded, leading to decrease in soil fertility. Consequently, a large amount of fertilizers or manures are used to improve soil fertility and crop yield. It has been observed that one-third of crop yield is pertaining to fertilizers and the rest depends on other agricultural inputs. The nutrient use efficiencies of traditional fertilizers exceed 35% approximately (Achterbosch et al. 2014). For instance, the use of fertilizers as nutrients, such as nitrogen (N) 30–35%, phosphorus (P) 18–20%, and potassium (K) 35–40%, remained constant for these years. But a very low amount reaches the targeted site due to leaching of chemicals, drift, hydrolysis, evaporation, and photolytic or even microbial degradation (Sabir et al. 2014). As a result, repeated use of fertilizers adversely affects the nutrient equilibrium of the soil. Apart from these, water bodies have been contaminated due to discharge of toxic materials into rivers and water reservoirs contaminating even drinking water (Rai et al. 2015). It has been noted that in the early 1970, only 27 kg

NPK was required to obtain 100 kg of grain, whereas in 2008, it increased to 109 kg of NPK to produce the same. Apart from these large amounts of traditional fertilizers, a large portion of the chemicals remain in the soil resulting in severe environmental pollution that can affect the health of flora and fauna. The utilization of engineered nanomaterials in the frame of sustainable agriculture has paved way for new food production methods that could effectively overcome the problem of limited food resources (Godfray et al. 2010).

The revolution of green nanotechnology has drastically changed the global scenario of agriculture purpose, and usage of nanomaterials as nanofertilizers has aroused to meet the global demand of food and sustainable agriculture. In order to reduce nutrient deficiency and to overcome the problem of eutrophication, nanofertilizers are the best choice (Shukla et al. 2019). Nanofertilizers are designed to regulate the release of nutrients depending on the need of the crops subsequently reducing the losses, have an intense potentiality. For example, natural nitrogen fertilizers are demonstrated by huge losses from the soil through leaching, evaporation, or even the degradation up to 60% approx, which reduces the efficiency of fertilizers and increases the production cost (Miao et al. 2015). In addition, nanoformulations of nitrogenous fertilizers monitor the release of fertilizer-N based on their demand by crops. Also, nano formulations prevent loss of nutrients via direct internalization by crops and avoid interaction between nutrients and soils, water, air, and microorganisms (Singh et al. 2016). For example, application of porous nanomaterials, such as zeolites, clay, or chitosan, effectively reduces the loss of nitrogen by monitoring the demand-based release and enhancing the uptake process. Ammonium-incorporated zeolites have the ability to accelerate the solubility of phosphate minerals and improve phosphorus availability and uptake by crops. Graphene oxide films, a carbon-based nanomaterial, can increase the time of potassium nitrate release, thereby extending the time of function and minimizing losses by leaching. Sabir et al. (2014) studied the efficacy of nanomaterials in crop production over conventional fertilizers. The study showed that the application of nanocalcite ($CaCO_3$-40%) with nano SiO_2 (4%), MgO (1%), and Fe_2O_3 (1%) improved the utilization of Ca, Mg, and Fe, as well as enhancing the intake of P with micronutrients (Singh et al. 2016; Lateef et al. 2016). Nanofertilizers are prepared by the encapsulation of nutrients with nanomaterials. Firstly, nanomaterials are produced by using both physical and chemical approaches, followed by encapsulation with targeted nutrients inside nanoporous materials or coated with thin film of polymer (Singh et al. 2016). Agricultural yield can be increased by balanced fertilizer application, irrigation, and quality seeds usage. It has been understood that nanoformulated fertilizers have a great capacity to improve productivity. For example, the use of carbon nanoparticles along with fertilizer can increase the production of rice (10.39%), maize (10.92%), soybean (16.84%), wheat (28.82%), and vegetables (12–20%). It was also found that combined application of chitosan-NPK fertilizer greatly improves the harvest index, crop index, and also mobilization index of the wheat yield as compared with control yield variables. The application of nanofertilizers can enhance nutrient uptake in plants by entering through pores, interacting with root exudates, or being absorbed through endocytosis (Rai et al. 2015).

Moreover, it has been observed that the size of nanomaterials is indirectly proportional to the surface mass ratio of particles leading to a huge number of nutrient ions get adsorbed and desorbed consistently for a long period of time (Monreal et al. 2016). Thus, the nanoformulations of fertilizers promise an equilibrium nutrition for crops throughout the growth process that would improve production. It is observed that the increased efficiency of a product may encourage the farmers to get more profits. Along with macronutrients, micronutrients like manganese and boron play an important role in stably increasing the crop yield. Other factors, such as soil pH and moisture levels, influence nutrient deficiencies in crops, especially with intensive farming practices (Ghormade et al. 2011). The deficiency of micronutrients affects the production, which in turn affects humans resulting in iron deficiency causing anemia, growth impairment, reproductive health issues, and impaired cognitive and physical performance (Monreal et al. 2016). The application of nano-micronutrients for the steady or controlled release of nutrients would trigger the uptake process, promote the productivity, and maintain soil health (Peteu et al. 2010). For instance, application of nano zinc oxide improves the growth and physiological responses, fresh dry weight, and photosynthesis in zinc-deficient soil. Kale et al. (2016) reported that application of zinc oxide nanoparticles either alone or in combination with other fertilizer in zinc-deficient soil promotes productivity by 91%, whereas conventional $ZnSO_4$ promotes productivity by 31%, both when compared to the control (Kale et al. 2016). Controlled release fertilizers are used to reduce the pollution of inputs in agriculture that act by forming a nano network which is formed upon contact with water in soil. Whereas the entrapped fertilizer nutrients are absorbed by the soil network through mechanisms like hydrogen bonds, surface tension, molecular force, etc. Eventually, their spatial scale increases in size so that they are blocked by soil filtration and remain intact in soil facilitating nutrient absorption throughout the growth cycle. This approach has been used to reduce the transfer rate of nitrogen into the environment (Cai et al. 2014). They also showed that application of control loss fertilizer not only decreases nitrogen and leaching loss by 21.7% and 24.6% but also increases soil residual mineral nitrogen by 9.8% together with 5.6% increased wheat production when compared to conventional fertilizers. Although extensive research has been carried out in this area, information and research on wider potentiality are still inadequate. Therefore, more research is needed to explore novel approaches that can control the migration of both macro- and micro-nutrients in micro-scale into the environmental matrix.

10 Role of Nanomaterials as Nanosensors

Nanomaterial engineering is the cutting-edge research area for sustainable agricultural growth. The use of nanomaterials in precision farming could reduce the cost and effort, increase efficiency, and lead environmentally sound development. The development of nanosensors would help assure soil and plant health, product quality, and safety for sustainable agriculture and environmental systems (Cheng et al.

2016). Naturally, biological organisms have the ability to identify the environmental conditions. However, the application of biology with nanomaterials into sensors has increased specificity, sensitivity, and rapid responses to sense the defects. For example, nanosensor-mediated global positioning system (GPS) is used for the monitoring of cultivated fields. Such networks of wireless nanosensors keep an eye on the controlled release mechanism using wireless signals located throughout the fields. This can assure a real-time regulation of the crop yield and effective high-quality data that provide opportunities for most excellent management practices by avoiding overdose of agricultural inputs (Singh et al. 2016). The automation of irrigation system by using sensor technology plays a significant role in maximizing water use efficiency. In context of water limitation, nanosensors detect soil–water tension in real time coupled with autonomous irrigation control. Similarly, a fast and precise detection of insects/pathogens would help in the application of pesticides or fertilizers to protect the crops from infestation. In this connection, a wireless nanosensor has been developed (Afsharinejad et al. 2015) for detecting insect attacks. This sensor differentiates the emitted volatile organics in many host plant species.

The nano-gold-based immunosensor is effective in detecting karnal bunt disease in wheat plants. Moreover, in the arena of nanobiotechnology, the development of bionic plants, by inserting nanoparticles into the cells and chloroplasts of living plants for detecting or imaging objects in their environment and for communicating as infrared devices or self-powering of plants as light sources, has huge importance in precision farming (Wong et al. 2017). For example, Giraldo et al. (2014) reported that the insertion of single-walled carbon nanotubes (SWNTs) accelerates electron transfer rate of light-adapted chloroplasts by 48% under in vivo conditions by augmenting photo absorption (Giraldo et al. 2014). They also showed that SWNTs are responsible for the light-harvesting capacity near-infrared fluorescence by suppressing the generation of RO species in chloroplast and could potentially influence the sensing process in plants; which in turn resulted in improved photosynthetic efficiency and quantum yield of plants. Therefore, advancements in nanobionic approaches for crops development and environmental monitoring pave way for research on plant nanomaterial hybrids (Giraldo et al. 2014; Wong et al. 2017). Plants react to stress through different physiological changes which are regulated by stress hormones. A modified gold electrode nanosensor with copper nanoparticles was developed to detect pathogenic fungal infestations by monitoring salicylic acid levels in oilseeds. Accordingly, multiwalled carbon nanotubes (MWCNTs) are also competent in plant growth study as they regulate hormones such as auxin, which may help scientists to explore how plant roots are acclimatized to their environment, particularly to marginal soils (McLamore et al. 2010). Ganesh Kumar et al. (2016) demonstrated that one-dimensional potassium niobate ($KNbO_3$) nanofibers are capable of detecting humidity because of their large surface-volume ratio. The $KNbO_3$ nanofiber-based humidity nanosensors provide a logarithmic–linear dependence behavior of the conductance with the relative humidity within couple of seconds, where the conductance increases from 10-10 M to 10-6 M for a relative humidity varying from 15% to 96% at room temperature. Recently, Fang et al. (2017) developed a simple, label-free glutathione-regulated dual-functional platform based on up-conversion nanoparticles (UCNPs)/AuNPs for the turn-on fluorescence detection of acetylcholinesterase (AChE) activity and toxic Cd2+ in real water samples. Besides, a remarkable advancement has been made toward the monitoring and quantification of small amount of pollutants like pesticides. For instance, biosensors containing photosystem II have the potential to bind to several molecular groups. Such nanosensors provide an easy and inexpensive technology for the identification of pesticides with a wide range of organic pollutants before their discharge into the environment. This provides the assurance of sustainable development by monitoring crop and soil health. However, considering extensive research in this area, the convincing applications of nanosensors, especially in field study, are surprisingly insufficient, suggesting a new era for future research.

11 Nanomaterials in Pesticide-Based Plant Protection

The application of nanotechnology in plant protection products has increased drastically to achieve higher crop production. Normally, traditional crop protection measures include applications of large-scale fungicides, herbicides, and insecticides. Among the applied pesticides, more than 91% are either lost in the environment or unable to reach the target areas important for effective pest control (Nuruzzaman et al. 2016). This not only increases the cost of crop production but also causes the depletion of environment. It is to be noted that the availability of active ingredients in minimum concentration of a formulation at the target sites is important for better protection of plants from pest invasion and eventual loss of crop. In this context, the development of new plant protection formulation has been a very cognitive field of agricultural research. One such technology is nanoformulation/encapsulation of pesticides, which is a breakthrough in the plant protection field. Nanoformulation of pesticides contains a very few numbers of particles that act as active ingredients of pesticides; however, other engineered nanostructures also have useful pesticidal properties (Prasad et al. 2019). Nanoencapsulation of pesticides is the coating of active ingredients of pesticides with another material of varying sizes at nano-range where encapsulated materials are referred to as internal phase of the core material and capsulation mate (Nuruzzaman et al. 2016). Nanoformulations or encapsulations of pesticides facilitate the persistence or controlled release of active ingredients in root zones or inside plants without compromising on its effectiveness. On the contrary, traditional formulations of pesticides/herbicides not only minimize water solubility of pesticides but also harm other living beings, leading to increased resistance to target organisms. Fascinatingly, in such circumstances, the use of nanoparticle is being reported as efficient alternative to directly suppress pathogen infection and activity, leading to increased crop growth and yield (Servin et al. 2015). Nanosilica is one such example that is hydrophobic in nature and can absorb into the cuticle layer of the insects upon contact, leading to the ultimate death of insects (Nair et al. 2010). For example, silica nanosphere formulations can increase the ability of pesticides to penetrate through the plant and reach the cell sap, thereby exerting systemic effect to control chewing- or sucking-type insects like aphid (Li et al. 2007). Such types of hollow formulation also protect pesticides from photodegradation due to direct exposure to sun rays (Singh et al. 2016).

The potential application of engineered nanomaterials in agriculture is also ascertained in disease and weed management as shown in Fig. 3. Inorganic NPs, such as ZnO, Cu, SiO_2, TiO_2, CaO, MgO, MnO, and AgNPs play important roles in various arena of plant protection including providing microbial activity and controlling bacterial diseases (Servin et al. 2015). For example, ZnO nanoparticles have recently been shown to provide effective growth control of *Fusarium graminearum*, *Penicillium expansum*, *Alternaria alternata*, *F. oxysporum*, *Rhizopus stolonifer*, *Mucor plumbeus*, and *A. flavus* as well as pathogenic bacteria *Pseudomonas aeruginosa* (Servin et al. 2015). Nano-Cu application was found to be more effective against *Phytophthora infestans* in tomato compared to currently available non-nano Cu formulations (Giannousi et al. 2013). Besides, Si, and TiO_2 have been found promising to suppress crop diseases directly through antimicrobial activity. MONPs inhibit the development of fungal conidia and conidiophores which cause ultimate death of fungal hyphae. Likewise, weeds are also considered serious threat to global agricultural production as they compete with crops for their nutrient, water, and light. However, the application of engineered nanomaterials containing herbicides provides a solution in ecofriendly manner. Usually, commercial herbicides control or kill the above-ground parts of the weeds without affecting the below-ground parts like rhizomes or tubers. As a result, the regrowth of weeds occurs; however, nanoherbicides prevent the regrowth of weeds. Thus, the nanomaterials in pesticides, fungicides, and herbicides have a tremendous scope in sustainable agricultural development (Patil et al. 2016).

12 Conclusions

In the field of agriculture, nanotechnology has been used to increase crop yield and quality by improving farming systems as shown in Fig. 6. The emergence of engineered nanomaterials and their applications in sustainable agriculture have drastically revolutionized global agriculture through innovation, rapid growth, and scalability, helping to meet the projected global food demand. In sustainable agriculture, protecting the environment from pollution is a key objective, and nanomaterials offer a reliable solution for better management and conservation of resources used in plant production. The potential of nanomaterials encourages a new green revolution with reduced farming risks. However, there are still huge gaps in our knowledge of the uptake capacity, permissible limit, and the ecotoxicity of different nanomaterials (Li et al. 2016a, b). Therefore, further research is extensively needed to untangle the fate of altered agriculture inputs and their interaction with biomacromolecules present in living systems and environment. The development of nanoscale materials has increased considerably over the past few years, as they offer significant advantages in comparison. The use of polysaccharide-based nanoparticles as smart delivery system

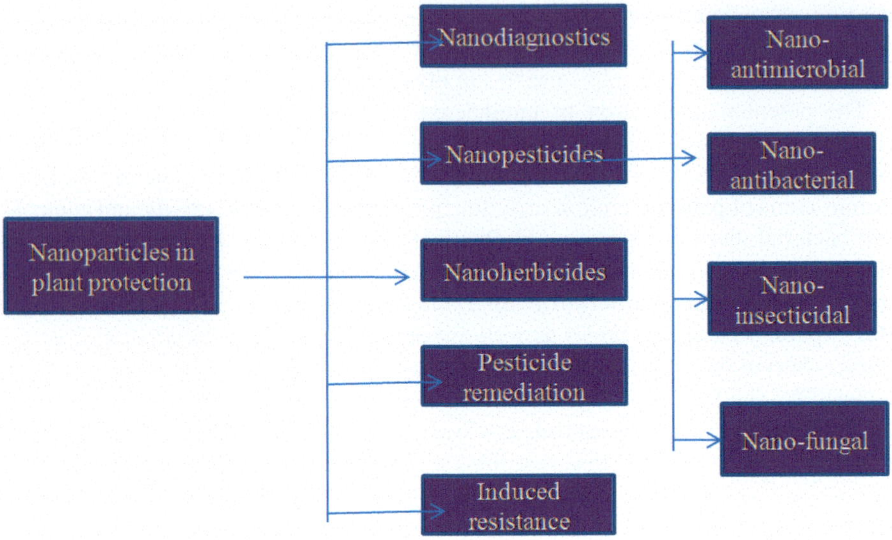

Fig. 6 Uses of nanoparticles in plant protection

in the agriculture field is still in budding stage. Significant outcomes have been reported on plant growth promotion and stress protection by chitosan- and alginate-chitosan-based nanomaterials at the laboratory level and green house scales. Since polysaccharide-based nanomaterials have shown favorable results for the controlled delivery of different agro compounds, including pesticides, plant growth regulators, herbicides, and nutrients at the laboratory scale, each of them may have great potential in future. Chitosan-based nanomaterials could be the most important polysaccharides for the development of controlled release systems for agricultural application. The combination of multiple bioactives (e.g., two herbicides or a herbicide and fungicide) with a matrix (such as chitosan/alginate) could enhance and expand the functionality of a chitosan-based nanomaterial. Despite these advantages, limited information on chitosan-based nanoparticles raises concerns about their potential environmental impact. To enhance our understanding, the following issues need to be addressed:

- Specificity versus universality of their biological properties
- Biophysical and chemical interactions with other components available in plants and the environment
- Mechanisms of action on specific parts of the plant or whole plants
- Interactions with biota they interact
- Life cycle assessments

Another important point to keep in mind is that several countries across the globe have been testing the appropriateness of their regulatory frameworks for dealing with nanotechnologies. Collaboration among countries is required to exchange information and to ensure a high level of protection for humans and the entire environment, without affecting the development of new beneficial products and their global marketing.

Acknowledgments The authors are very much thankful to the management of VIPER, Narsapur, Medak, Telangana, for providing facilities for carrying out this review work and also to the authority of ICAR-IARI, New Delhi, for moral support and encouragement.

References

Abhilash PC, Singh N (2009) Pesticide use and application: an Indian scenario. J Hazard Mater 165(1–3):1–12

Achterbosch TJ, Van Dorp M, van Driel WF, Groot JJ, Van der Lee J, Verhagen A, Bezlepkina I (2014) The food puzzle: pathways to securing food for all. Wageningen UR

Afsharinejad A, Davy A, Jennings B, Brennan C (2015) Performance analysis of plant monitoring nanosensor networks at THz frequencies. IEEE Internet Things J 3(1):59–69

Alami Y, Achouak W, Marol C, Heulin T (2000) Rhizosphere soil aggregation and plant growth promotion of sunflowers by an exopolysaccharide-producing Rhizobiumsp. Strain isolated from sunflower roots. Appl Environ Microbiol 66(8):3393–3398

Bensalim S, Nowak J, Asiedu SK (1998) A plant growth promoting rhizobacterium and temperature effects on performance of 18 clones of potato. Am J Potato Res 75(3):145–152

Cai D, Zhengyan W, Jiang J, Yuejin W, Feng H, Brown IG, Chu PK, Zengliang Y (2014) Controlling nitrogen migration through micro-nano networks. Sci Rep 4(1):1–8

Campos EV, Ramos JL, de Oliveira L, Fraceto F, Singh B (2015) Polysaccharides as safer release systems for agrochemicals. Agron Sustain Dev 35(1):47–66

Chen YW, Lee HV, Juan JC, Phang S-M (2016) Production of new cellulose nanomaterial from red algae marine biomass Gelidium elegans. Carbohydr Polym 151:1210–1219

Cheng HN, Klasson KT, Asakura T, Qinglin W (2016) Nanotechnology in agriculture. In: Nanotechnology: delivering on the promise, vol 2. American Chemical Society, pp 233–242

Chenu C (1993) Clay—or sand—polysaccharide associations as models for the interface between micro-organisms and soil: water related properties and microstructure. In: Soil structure/soil biota interrelationships. Elsevier, pp 143–156

Chenu C, Roberson EB (1996) Diffusion of glucose in microbial extracellular polysaccharide as affected by water potential. Soil Biol Biochem 28(7):877–884

Du X, Li J, Lindström ME (2014) Modification of industrial softwood Kraft lignin using Mannich reaction with and without phenolation pretreatment. Ind Crop Prod 52:729–735

Fang A, Chen H, Li H, Liu M, Zhang Y, Yao S (2017) Glutathione regulation-based dual functional upconversion sensing-platform for acetylcholinesterase activity and cadmium ions. Biosens Bioelectron 87:545–551

Fedoroff NV, Battisti DS, Beachy RN, Cooper PJM, Fischhoff DA, Hodges CN, Knauf VC et al (2010) Radically rethinking agriculture for the 21st century. Science 327(5967):833–834

Flemming H-C, Wingender J (2001) Relevance of microbial extracellular polymeric substances (EPSs)-part I: structural and ecological aspects. Water Sci Technol 43(6):1–8

Foley JA, Ramankutty N, Brauman KA, Cassidy ES, Gerber JS, Johnston M, Zaks DPM (2011) Solutions for a cultivated planet. Nature 478:7369

Galbraith DW (2007) Silica breaks through in plants. Nat Nanotechnol 2(5):272–273

Ganeshkumar R, Sopiha KV, Ping W, Cheah CW, Zhao R (2016) Ferroelectric KNbO3 nanofibers: synthesis, characterization and their application as a humidity nanosensor. Nanotechnology 27(39):395607

Garcia-Pichel F, Pringault O (2001) Cyanobacteria track water in desert soils. Nature 413(6854):380–381

Giannousi K, Avramidis I, Dendrinou-Samara C (2013) Synthesis, characterization and evaluation of copper based nanoparticles as agrochemicals against Phytophthora infestans. RSC Adv 3(44):21743–21752
Giraldo JP, Landry MP, Faltermeier SM, McNicholas TP, Iverson NM, Boghossian AA, Reuel NF et al (2014) Plant nanobionics approach to augment photosynthesis and biochemical sensing. Nat Mater 13(4):400–408
Godfray HC, Beddington JR, Crute IR, Haddad L, Lawrence D, Muir JF, Pretty J, Robinson S, Thomas SM, Toulmin C (2010) Food security: the challenge of feeding 9 billion people. Science 327(5967):812–818
Gogos A, Knauer K, Bucheli TD (2012) Nanomaterials in plant protection and fertilization: current state, foreseen applications, and research priorities. J Agric Food Chem 60(39):9781–9792
Hassani LN, Hendra F, Bouchemal K (2012) Auto-associative amphiphilic polysaccharides as drug delivery systems. Drug Discov Today 17(11–12):608–614
Huang M, Wang Z, Luo L, Wang S, Hui X, He G, Cao H et al (2017) Soil testing at harvest to enhance productivity and reduce nitrate residues in dryland wheat production. Field Crop Res 212:153–164
Hullebusch V, Eric D, Zandvoort MH, Lens PNL (2004) Nickel and cobalt sorption on anaerobic granular sludges: kinetic and equilibrium studies. J Chem Technol Biotechnol 79(11):1219–1227
Hunter RC, Beveridge TJ (2005) Application of a pH-sensitive fluoroprobe (C-SNARF-4) for pH microenvironment analysis in Pseudomonas aeruginosa biofilms. Appl Environ Microbiol 71(5):2501–2510
Iavicoli I, Leso V, Beezhold DH, Shvedova AA (2017) Nanotechnology in agriculture: opportunities, toxicological implications, and occupational risks. Toxicol Appl Pharmacol 329:96–111
Jo W, Bak JH, Yoo B (2018) Rheological characterizations of concentrated binary gum mixtures with xanthan gum and galactomannans. Int J Biol Macromol 114:263–269
Joga MR, Zotti MJ, Smagghe G, Christiaens O (2016) RNAi efficiency, systemic properties, and novel delivery methods for pest insect control: what we know so far. Front Physiol 7:553
Kah M (2015) Nanopesticides and nanofertilizers: emerging contaminants or opportunities for risk mitigation? Front Chem 3:64
Kale AP, Gawade SN (2016) Studies on nanoparticle induced nutrient use efficiency of fertilizer and crop productivity. Green Chem Technol Lett 2:88–92
Kargarzadeh H, Mariano M, Huang J, Lin N, Ahmad I, Dufresne A, Thomas S (2017) Recent developments on nanocellulose reinforced polymer nanocomposites: a review. Polymer 132:368–393
Khodakovskaya M, Dervishi E, Mahmood M, Yang X, Li Z, Watanabe F, Biris AS (2009) Carbon nanotubes are able to penetrate plant seed coat and dramatically affect seed germination and plant growth. ACS Nano 3(10):3221–3227
Kole C, Phullara Kole K, Randunu M, Choudhary P, Ramakrishna Podila P, Ke C, Rao AM, Marcus RK (2013) Nanobiotechnology can boost crop production and quality: first evidence from increased plant biomass, fruit yield and phytomedicine content in bitter melon (Momordica charantia). BMC Biotechnol 13(1):37
Kwak S-Y, Wong MH, Lew TTS, Bisker G, Lee MA, Kaplan A, Dong J et al (2017) Nanosensor technology applied to living plant systems. Annu Rev Anal Chem 10:113–140
Lahiani MH, Dervishi E, Chen J, Nima Z, Gaume A, Biris AS, Khodakovskaya MV (2013) Impact of carbon nanotube exposure to seeds of valuable crops. ACS Appl Mater Interfaces 5(16):7965–7973
Lateef A, Nazir R, Jamil N, Alam S, Shah R, Khan MN, Saleem M (2016) Synthesis and characterization of zeolite based nano–composite: an environment friendly slow release fertilizer. Microporous Mesoporous Mater 232:174–183
Li Z-Z, Chen J-F, Liu F, Liu A-Q, Wang Q, Sun H-Y, Wen L-X (2007) Study of UV-shielding properties of novel porous hollow silica nanoparticle carriers for avermectin. Pest Manag Sci 63(3):241–246

Li M, Ahammed GJ, Li C, Bao X, Jingquan Y, Huang C, Yin H, Zhou J (2016a) Brassinosteroid ameliorates zinc oxide nanoparticles-induced oxidative stress by improving antioxidant potential and redox homeostasis in tomato seedling. Front Plant Sci 7:615

Li Q, Chen X, Zhuang J, Chen X (2016b) Decontaminating soil organic pollutants with manufactured nanoparticles. Environ Sci Pollut Res 23(12):11533–11548

Li C, Li Y, Li Y, Guozhan F (2018) Cultivation techniques and nutrient management strategies to improve productivity of rain-fed maize in semi-arid regions. Agric Water Manag 210:149–157

Liu Z, Jiao Y, Wang Y, Zhou C, Zhang Z (2008) Polysaccharides-based nanoparticles as drug delivery systems. Adv Drug Deliv Rev 60(15):1650–1662

Maftoonazad N, Ramaswamy H (2018) Novel techniques in food processing: bionanocomposites. Curr Opin Food Sci 23:49–56

Matson PA, Parton WJ, Power AG, Swift MJ (1997) Agricultural intensification and ecosystem properties. Science 277(5325):504–509

McLamore ES, Diggs A, Marzal PC, Shi J, Blakeslee JJ, Peer WA, Murphy AS, Marshall Porterfield D (2010) Non-invasive quantification of endogenous root auxin transport using an integrated flux microsensor technique. Plant J 63(6):1004–1016

Miao Y, Wang Z, Li S (2015) Relation of nitrate N accumulation in dryland soil with wheat response to N fertilizer. Field Crops Res 170:119–130

Miller JB, Zhang S, Petra Kos H, Xiong KZ, Perelman SS, Zhu H, Siegwart DJ (2017) Non-viral CRISPR/Cas gene editing in vitro and in vivo enabled by synthetic nanoparticle co-delivery of Cas9 mRNA and sgRNA. Angew Chem Int Ed 56(4):1059–1063

Monreal CM, DeRosa M, Mallubhotla SC, Bindraban PS, Dimkpa C (2016) Nanotechnologies for increasing the crop use efficiency of fertilizer-micronutrients. Biol Fertil Soils 52(3):423–437

Morillo E, Villaverde J (2017) Advanced technologies for the remediation of pesticide-contaminated soils. Sci Total Environ 586:576–597

Mout R, Ray M, Tonga GY, Lee Y-W, Tay T, Sasaki K, Rotello VM (2017) Direct cytosolic delivery of CRISPR/Cas9-ribonucleoprotein for efficient gene editing. ACS Nano 11(3):2452–2458

Mukhopadhyay SS (2014) Nanotechnology in agriculture: prospects and constraints. Nanotechnol Sci Appl 7:63

Myers SS, Zanobetti A, Kloog I, Huybers P, Leakey AD, Bloom AJ, Carlisle E, Dietterich LH, Fitzgerald G, Hasegawa (2014) Increasing CO_2 threatens human nutrition. Nature 510(7503):139–142

Nair R, Varghese SH, Nair BG, Maekawa T, Yoshida Y, Sakthi Kumar D (2010) Nanoparticulate material delivery to plants. Plant Sci 179(3):154–163

Naqash F, Masoodi FA, Rather SA, Wani SM, Gani A (2017) Emerging concepts in the nutraceutical and functional properties of pectin—a review. Carbohydr Polym 168:227–239

Nitta SK, Numata K (2013) Biopolymer-based nanoparticles for drug/gene delivery and tissue engineering. Int J Mol Sci 14(1):1629–1654

Norgren M, Edlund H (2014) Lignin: recent advances and emerging applications. Curr Opin Colloid Interface Sci 19(5):409–416

Nuruzzaman MD, Rahman MM, Liu Y, Naidu R (2016) Nanoencapsulation, nano-guard for pesticides: a new window for safe application. J Agric Food Chem 64(7):1447–1483

Pal A, Paul AK (2008) Microbial extracellular polymeric substances: central elements in heavy metal bioremediation. Indian J Microbiol 48(1):49

Pandey G (2018) Challenges and future prospects of agri-nanotechnology for sustainable agriculture in India. Environ Technol Innov 11:299–307

Patil CD, Borase HP, Suryawanshi RK, Patil SV (2016) Trypsin inactivation by latex fabricated gold nanoparticles: a new strategy towards insect control. Enzym Microb Technol 92:18–25

Peteu SF, Oancea F, Sicuia OA, Constantinescu F, Dinu S (2010) Responsive polymers for crop protection. Polymers 2(3):229–251

Prasad R, Kumar V, Kumar M, Choudhary D (eds) (2019) Nanobiotechnology in bioformulations. Springer International Publishing

Prosekov AY, Ivanova SA (2018) Food security: the challenge of the present. Geoforum 91:73–77

Pushpalatha R, Selvamuthukumar S, Kilimozhi D (2018) Cross-linked, cyclodextrin-based nanosponges for curcumin delivery-physicochemical characterization, drug release, stability and cytotoxicity. J Drug Deliv Technol 45:45–53

Raemdonck K, Martens TF, Braeckmans K, Demeester J, De Smedt SC (2013) Polysaccharide-based nucleic acid nanoformulations. Adv Drug Deliv Rev 65(9):1123–1147

Rai M, Ribeiro C, Mattoso L, Duran N (eds) (2015) Nanotechnologies in food and agriculture. Springer, Cham/Heidelberg/New York/Dordrecht/London

Ran Y, Liang Z, Gao C (2017) Current and future editing reagent delivery systems for plant genome editing. Sci China Life Sci 60(5):490–505

Roberson EB, Firestone MK (1992) Relationship between desiccation and exopolysaccharide production in a soil Pseudomonas sp. Appl Environ Microbiol 58(4):1284–1291

Sabir A, Yazar K, Sabir F, Zeki Kara M, Yazici A, Goksu N (2014) Vine growth, yield, berry quality attributes and leaf nutrient content of grapevines as influenced by seaweed extract (Ascophyllum nodosum) and nanosize fertilizer pulverizations. Sci Hortic 175:1–8

Sandhya VZAS, Grover M, Reddy G, Venkateswarlu B (2009) Alleviation of drought stress effects in sunflower seedlings by the exopolysaccharides producing Pseudomonas putida strain GAP-P45. Biol Fertil Soils 46(1):17–26

Seglie L, Spadaro D, Trotta F, Devecchi M, Gullino ML, Scariot V (2012) Use of 1-methylcylopropene in cyclodextrin-based nanosponges to control grey mould caused by Botrytis cinerea on Dianthus caryophyllus cut flowers. Postharvest Biol Technol 64(1):55–57

Servin A, Elmer W, Mukherjee A, De la Torre-Roche R, Hamdi H, White JC, Bindraban P, Dimkpa C (2015) A review of the use of engineered nanomaterials to suppress plant disease and enhance crop yield. J Nanopart Res 17(2):92

Shukla P, Chaurasia P, Younis K, Qadri OS, Faridi SA, Srivastava G (2019) Nanotechnology in sustainable agriculture: studies from seed priming to post-harvest management. Nanotechnol Environ Eng 4(1):11

Singh DP, Singh HB, Prabha R (eds) (2016) Microbial inoculants in sustainable agricultural productivity, vol 2: Functional applications. Springer

Tang Y, Zhang X, Zhao R, Guo D, Zhang J (2018) Preparation and properties of chitosan/guar gum/nanocrystalline cellulose nanocomposite films. Carbohydr Polym 197:128–136

Tester M, Langridge P (2010) Breeding technologies to increase crop production in a changing world. Science 327(5967):818–822

Thomas AD, Dougill AJ (2007) Spatial and temporal distribution of cyanobacterial soil crusts in the Kalahari: implications for soil surface properties. Geomorphology 85(1–2):17–29

Thornhill S, Vargyas E, Fitzgerald T, Chisholm N (2016) Household food security and biofuel feedstock production in rural Mozambique and Tanzania. Food Secur 8(5):953–971

Torney F, Trewyn BG, Lin VS-Y, Wang K (2007) Mesoporous silica nanoparticles deliver DNA and chemicals into plants. Nat Nanotechnol 2(5):295–300

Vijayakumar PS, Abhilash OU, Khan BM, Prasad BLV (2010) Nanogold-loaded sharp-edged carbon bullets as plant-gene carriers. Adv Funct Mater 20(15):2416–2423

Villaverde J, Rubio-Bellido M, Lara-Moreno A, Merchan F, Morillo E (2018) Combined use of microbial consortia isolated from different agricultural soils and cyclodextrin as a bioremediation technique for herbicide contaminated soils. Chemosphere 193:118–125

Wang P, Lombi E, Zhao F-J, Kopittke PM (2016) Nanotechnology: a new opportunity in plant sciences. Trends Plant Sci 21(8):699–712

Wong MH, Giraldo JP, Kwak S-Y, Koman VB, Sinclair R, Lew TTS, Bisker G, Liu P, Strano MS (2017) Nitroaromatic detection and infrared communication from wild-type plants using plant nanobionics. Nat Mater 16(2):264–272

Worrall EA, Hamid A, Mody KT, Mitter N, Pappu HR (2018) Nanotechnology for plant disease management. Agronomy 8(12):285

Yunlong C, Smit B (1994) Sustainability in agriculture: a general review. Agric Ecosyst Environ 49(3):299–307

Zafar R, Zia KM, Tabasum S, Jabeen F, Noreen A, Zuber M (2016) Polysaccharide based bionanocomposites, properties and applications: a review. Int J Biol Macromol 92:1012–1024
Zhang X, Zhang J, Zhu KY (2010) Chitosan/double-stranded RNA nanoparticle-mediated RNA interference to silence chitin synthase genes through larval feeding in the African malaria mosquito (Anopheles gambiae). Insect Mol Biol 19(5):683–693
Zhang M, Gao B, Chen J, Li Y, Creamer AE, Chen H (2014) Slow-release fertilizer encapsulated by graphene oxide films. Chem Eng J 255:107–113

Soil Algae as a Feedstock for Biofuel Production and Value-Added Products

Vijayata Singh, Swati Mishra, and Deepjyoti Singh

Abstract Microalgae are a sustainable and economical source of biofuels and other valuable products. Algae are autotrophic organisms that can survive in aquatic or terrestrial forms. They are fast-growing microorganisms with a high potential for converting CO_2 from the atmosphere into valuable products such as polysaccharides, fats, and other bioactive metabolites. Fuel derived from algae is named "third-generation biofuel," which has several critical positives over plant-based (first) and animal waste feedstock (second). Biofuel companies are currently seeking to increase the commercial algal product and pursuing several engineering and biotechnological approaches to design an economic system for proper algal growth. However, microalgal research and production are mainly confined to aquatic forms, with specific limitations like water contamination and the high cost of dewatering feedstock during processing. Overexploitation of aquatic algae also leads to ecosystem imbalance as aquatic algae are the primary producers. Soil algae can be used as an alternative to counterbalance the usage of aquatic algae. The present chapter compiles some terrestrial algae that can be used as potential biomass to produce biofuels and various bioactive products.

Keywords Biofuels; Microalgae · Biorefinery · Photosynthetic microbial fuel cells (PMFCs) · Indole acetic acid (IAA) · Third-generation fuel

V. Singh (✉)
Ichan School of Medicine, Mount Sinai, New York, NY, USA

S. Mishra
Banaras Hindu University, Varanasi, India

D. Singh
North Carolina State University, Raleigh, NC, USA

A. Mishra, A. Varma (eds.), *Soil Algae*, https://doi.org/10.1007/978-981-97-9873-5_13

1 Introduction

Bioenergy production from biomass is in demand due to it's biodegradable nature and unlimited accessibility. Additionally, there is finite obtainability of first- and second-generation fuels and accelerating level of carbon dioxide (CO_2) and other greenhouses gases in atmosphere which makes bioenery production is must for future (Ullah et al. 2015). Microorganisms like bacteria, cyanobacteria, and microalgae are considered third-generation feedstock for renewable energy, nutraceutical, and pharmaceutical industries (Cuellar-Bermudez et al. 2015). The algal biomass mainly consists of major nutrients like carbohydrates, lipids and proteins by using sunlight, carbon dioxide and other greenhouse gases (Khan et al. 2018). Many microalgal species have been discovered and studied for their potential as value-added products with remarkable medicinal and biological qualities (Levasseur et al. 2020; Kumar et al. 2021; Siddiki et al. 2022).

Microalgae are primary producers, and their habitat ranges from aquatic to terrestrial forms. Due to its cosmopolitan and autotrophic nature, they can adopt a broad range of light intensities, temperatures, pH values, and different terrestrial forms (Zhang et al. 2011; Pushkareva et al. 2016; Krug et al. 2020; Lee and Ryu 2021). There are several advantage which makes algae as a potential candidate for the generation of biofuels and other biomaterials, like (1) oxygenic photosynthesis water is used as an electron donor; (2) more biomass productivity than crops; (3) nonfood feedstock; (4) no requirement of agricultural land or drinking water for cultivation; (5) adaption to grow in diverse climates such as marshy or deserted land; (6) production of byproducts like pigments, amino acids, many vitamins, bioactive compounds, and antioxidants; and (7) produce fewer waste materials (Nguyen and Hoang 2016; Khan et al. 2018; Shin et al. 2018; De Bhowmick et al. 2019; Kumar et al. 2020).

Microalgae also have applications for the treatment of wastewater and CO_2 mitigation in the atmosphere (Molazadeh et al. 2019; Shahid et al. 2020). Biorefineries focus more on microalgae as potential feedstock for biofuels production (Suganya et al. 2016; Chia et al. 2018). Growth enrichment techniques, new processing, extraction methods, new species composition and bioactivity, and genetic engineering may be used to get the most promising result as a future source of renewable bioproducts (Kostas et al. 2021). Microalgae can also be consumed directly as food or can be processed and use in pharmaceuticals, nutrition and cosmetics industries (Khan et al. 2018).

The algal population in the soil is far lesser than bacteria and fungi. It ranges from 100 to 10,000 numbers per gram of soil (Khan and Rao 2019). They concentrate in the topmost biological surface or beneath the soil's upper surface until 8 inches (Usharani et al. 2019). The soil algal flora includes members of the Chlorophyta (green algae), Cyanophyta (blue-green algae), Xanthophyta (yellow-brown algae), and Bacillariophyta (diatoms) (Adesalu and Olugbemi 2015). Generally, Chlorophyta and Bacillariophyta classes dominate temperate regions (Pentecost and Whitton 2012; Whitton 2012). Some commonly found algal genera in soil are *Chlorella*, *Chlamydomonas*, *Chlorococcum* (green algae), *Navicula*, and *Pinnularia* (Antonelli et al. 2017; Khan and Rao 2019; Foets et al. 2021)

1.1 Importance of Soil Algae

Soil algae can hold organic carbon for long periods of time because it can act as producer as well as component of carbon.

1. Recycling carbon by photosynthesis and sequestration (Brilman et al. 2013; Gayathri et al. 2021).
2. Increase the water retention quality of soil through algae-produced polysaccharides.
3. Works as a cementing agent to bind the soil particles to prevent soil erosion producing more aerated soil.
4. Soil algae also help in weathering rocks and helping maintain soil structure.
5. Some start-up companies like Gross-Wein Technologies are developing technology for algae production that can remove nitrogen and phosphorus from contaminated sources which can then produce fertilizer and bioplastics.
6. Soil algae, especially cyanobacteria, can take atmospheric nitrogen and convert it to usable nitrogen within the soil, which increases soil fertility (Lee and Ryu 2021).
7. Algal feedstock can be harvested all year round, which is a reliable source of continuous supply of oil and other products.

Considering the above-mentioned points, it is clear that soil algae can be a potential candidate to obtain bioenergy and other value-added products.

1.2 Soil Algae and Bioenergy

Due to the increased demand for energy for modernization and industrialization, fossil fuel is continuously required. However, fossil fuel is an unsustainable energy source that can create land degradation, climate change, and air and water pollution. Transportation, electricity generation, and agriculture are the main sectors where many fossil fuels are utilized. As the number of transportation facilities increases significantly, there is a need to find alternatives for petroleum-based fuels like biodiesel. Limited availability of fossil fuels also causes greenhouse gas emission. Biofuels are renewable, natural and ecofriendly than other petroleum-based fuels. It reduces carbon emission and highly biodegradable. These days, biofuel contributes around 10% of the total energy demand (Antoni et al. 2007). Biofuels are the primary sustainable source of energy that can be used in transportation; in 2018, around two million barrels of oil were consumed per day. In 2019, the United States consumed about 1.8 billion gallons of biomass-based diesel fuel with blending petroleum diesel (US Energy Information Administration 2020).

Biofuel is mainly produced from vegetable oils, plants, and certain species of algae. Among them, algae are a widely accepted source of bioenergy because of appreciable lipid contents (20–80% oil by dry weight biomass), high yields, and without impacting agricultural farms and crops. Biofuels can provide the public welfare of natural gas with net-zero carbon emissions. It can provide a renewable supply of fuel and electricity that can serve many developing countries, especially those who do not have oil reserves (Ahuja and Tatsutani 2009). Additionally, biofuels production from algae produces less GHG, decomposes the byproducts and waste no need for long transportation, also helps in nutrient recycling rural job creation (Bošnjaković and Sinaga 2020).

Microalgal biomass mainly consists of polysaccharides, fats, and proteins. These macronutrients are produced by the process of photosynthesis along with further metabolic processes. These polysaccharides and fats serve as raw materials for the production of different biofuels and biomaterials (Robak and Balcerek 2018). Algal biomass contains limited amount of hemicellulose and lignin which makes pretreatment uncomplicated further there is ease in downstream hydrolysis and fermentation process (Chandra et al. 2019). Apart from biofuel production, this biomass produces a broad number of nutrients, medicinal uses, laboratory reagents, biofertilizers, animal feed, and treatment for polluted soil and water (Behera et al. 2015; Mathimani and Mallick 2019; Ahmad et al. 2022). Algae can be converted into various types of fuels, depending on the technique and the part of the cells used. For example, the lipid content of biomass can be extracted and converted into biodiesel by transesterification or "drop-in" replacements for petroleum-based fuels. Further, the polysaccharide content of microalgae algae can be fermented and produce bioethanol or biobutanol; in contrast, the protein part of biomass can be utilized to produce bio-oil, bio-char, and gasoline by hydrothermal liquefaction or pyrolysis (Fig. 1) (Arun et al. 2018; Abomohra and Elshobary 2019).

2 Biofuels

Several biofuels like bioethanol, biodiesel, biogas, methane, jet fuel, biobutanol, and green diesel are produced from algae. Although most commercially available products are extracted from aquatic algae, in this chapter, we describe only those algae and their products that share aquatic and terrestrial habitats. Some of these biofuels and other value-added products are described in the following section.

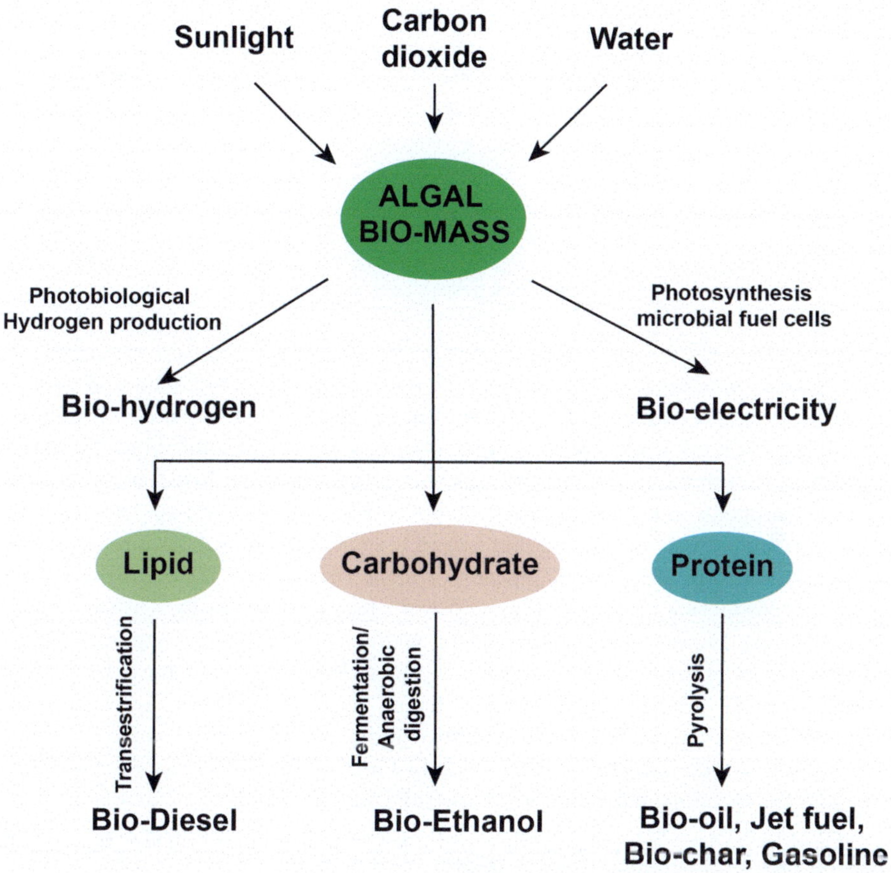

Fig. 1 Algae as energy: Schematic diagram of algal derived biofuels and their conversion processing

2.1 Biodiesel

Algal biodiesel can fulfill the excess demand of energy demand. Till now, the production and usage of algal biodiesel are tiny compared to conventional diesel oil. Biodiesel production from algae involves mainly two steps: (1) extraction of lipids, (2) transesterification. Several methods, such as physical, chemical, mechanical, and enzymatic (biological), can be applied to extract lipids from algal biomass (Li et al. 2014; Ranjith Kumar et al. 2015; Khan et al. 2018). Further, this lipid undergoes the process of transesterification in the presence of a chemical or a biological catalyst with the presence of alcohol and produces biodiesel and glycerol (Kumar and Thakur 2020; Ghedini et al. 2021). Biodiesel extraction from plants and aquatic algae is already in process for several decades. However, several marine species, which are also present in soil, can be used for biodiesel production. Some among

Table 1 The biomass and lipid yield from some common algal species

Microalgae	Lipid content (in % dry weight biomass)	Productivity of lipid	Biomass productivity	References
Chlorella vulgaris	32.5	0.837	0.2341–0.5322	Deng et al. (2018)
Chlamydomonas reinhardtii	–	0.08092	3.3	Shin et al. (2019)
Scendesmus quadricauda	38.61	0.13955	0.3614	Song and Pei (2018)

The table is adapted and modified from (Bošnjaković and Sinaga 2020)

them are *Chlorella vulgaris*, *Chlamydomonas reinhardtii*, *Euglena gracilis*, and *Scenedesmus quadricauda*. The biodiesel produced by them are saturated or low saturation products (Tucci et al. 2010; Furuhashi et al. 2015). Algal biodiesel production depends upon the yield of lipid obtained from particular algal biomass; some of the commonly used algal species and their content are listed in Table 1.

Some European countries (France, Ireland, Norway, and Iceland) and the Russian Federation are actively producing and researching algal biodiesel. As a result, there is a prediction to produce 41 megawatts of algae biomass by 2030 with less than $0.985/kg production costs (Bošnjaković and Sinaga 2020). In the United States, companies like Exxon Mobil and Synthetic Genomics are trying to organize large-scale biodiesel production at competitive prices. Several studies have been conducted whereby genetic engineering is trying to increase the amount of lipid compared to wild algae. The Government of India also introduced multiple initiatives to improve the indigenous production of biofuels. There is a target of production of 20% blending with ethanol in petrol and 5% blending of biodiesel with diesel products by 2030 (IEA Bioenergy 2017). The Indian government also reduced taxes on biodiesel supplied to oil marketing companies (OMCs) to enable a faster transition to green fuels and give an incentive to the biodiesel industry in the country.

2.2 *Bioethanol*

The green algae cell wall comprises large amounts of polysaccharides. Bioethanol production from algae is polysaccharides like starch, sugar, and cellulose. The production of bioethanol from these feedstocks requires microbial fermentation with prior pretreatment (with acid and alkali), saccharification (with amylase enzyme), followed by fermentation with a suitable microbial strain (Behera et al. 2015). Bioethanol production occurs mainly by yeast such as *Saccharomyces cerevisiae* (Phwan et al. 2019). Some aqatic algae like *Chlorococcum* sp., *Spirogyra* Sp., and *Chlamydomonas reinbardtii* and *Porphyidium cruentum* can also survive in terrestrial in nature, they can produce large amount to ethanol due to high carbohydrate content. *Chlorella vulgaris* contains about 55% carbohydrates of dry weight that is used as feedstock to produce bioethanol (Agwa et al. 2017).

Some efforts are already being made where genetically modified algae can produce ethanol. For example, Algenol Biofuels is a Florida USA-based company with a 36-acre integrated biorefinery that will produce around 100,000 gallons of algal-ethanol per year (Legere 2017).

Biobutanol produced from microalgae biomass has potential to replace bioethanol at some point. Biobutanol can be a substitute for gasoline and can be used directly in internal combustion engines without any engine alteration. As compared to bioethanol, biobutanol offers several inherent benefits like (1) high-energy density because of more carbon atoms, (2) a long burning time in the engines because of the higher boiling point, (3) low water miscibility, and (4) corrosion resistance (Hu et al. 2020). Some *Clostridium* species can produce biobutanol using acetone-butanol-ethanol fermentation (Jang et al. 2012).

2.3 *Biohydrogen*

Biohydrogen is basically produced from microbiological agents like bacteria, algae, and archaea from cultivations and organic byproducts (Wang and Wan 2009; Limongi et al. 2021). Biohydrogen is a clean alternative to fossil fuel, which can be used in transportation and electricity generation. There is certain soil green alga *Scenedesmus obliquus* and *Chlamydomonas reinhardtii*, which produce hydrogen naturally via photolysis.

Furthermore, microalgal carbohydrates, such as starch or cellulose, are ideal feedstocks for hydrogen production. The production of biohydrogen feedstock goes through pretreatment followed by dark microbial fermentation (Wang and Wan 2009). In addition, *Chlamydomonas moewusii* (Meuser et al. 2009), *Chlorococcum littorale* (Ueno et al. 1999), *Chlorella fusca* (Winkler et al. 2002), and *Mastigo claduslaminosus* (Miyamoto et al. 1979) can also be used as a potential hydrogen feedstock for biohydrogen production. Integrated biorefinery can be used for the production of biohydrogen from microalgae where it can be integrated with wastewater treatment and CO_2 mitigation, along with production of other value-added products (Kumar et al. 2018). Furthermore, biohydrogen has didtinct advantages as a potential future because of its high-density, zero carbon emission, and generates only water vapor after cobustion (Rathore et al. 2019).

2.4 *Biogas*

Lipids or oils work as a staple for biodiesel production from algal biomass. The remaining algal-enriched residue after lipid extraction, which consist of other nutrients such as carbohydrates, nitrogen, phosphorous, and proteins, can be further used as feedstock for biogas production by applying the anaerobic digestion method (Uggetti et al. 2014).

AD is a biochemical process that channels organic compounds to biogas under anaerobic conditions. Biogas comprises 50–70% methane; the remaining content is mostly carbon dioxide (30–45%) and a small portion of hydrogen sulfide, nitrogen, and hydrogen. Biogas can be used as fuel for electricity and thermal energy (Ojolo et al. 2007; Kvasauskas and Baltrėnas 2009). The anaerobic conversion process involves four basic steps: (1) hydrolysis, (2) acidogenesis, (3) acetogenesis, and (4) methanogenesis (Anukam et al. 2019). *Anabaena* sp., *Scenedesmus* sp., *Chlamydomonas rheinhardii*, *Euglena gracilis*, *Porphyridium cruentum*, *Chlorella vulgaris*, and *Tetraselmis* sp. (Legros et al. 1983; Hernández and Córdoba 1993) are some algal species that are already being used to produce biogas.

2.5 Biochar

Biochar is a form of charcoal that is produced from algal biomass by pyrolysis. Biochar is used as a soil conditioner, an electrode for lithium and sodium-ion batteries, electrochemical sensors, and wastewater management (Michalak et al. 2019; Binda et al. 2020). Pyrolysis temperature differs for different species. For example, pyrolysis temperatures 450–600 °C are good enough to prepare biochar for the biomass of *Chlorella vulgaris* (Palanisamy et al. 2017).

2.6 Syngas

Syngas consists of various gases such as carbon monoxide, hydrogen, carbon dioxide, and methane. Its chemical composition can be different depending upon which raw material has been used. Syngas is produced by gasification of carbon-rich feedstocks via partial oxidation; temperature can range from 100 to 1000 °C (Raheem et al. 2021). Microalgal strains like *Spirulina platensis* and *Chlorella vulgaris* are widely used as feedstock for syngas production (Raheem et al. 2017).

2.7 Green Diesel

Green diesel from algae can be produced through a hydrotreating refinery process in which larger molecules break into smaller hydrocarbon chains used in diesel engines. Most of the chemical properties of green diesel are similar to petroleum-based diesel. Therefore, it can be used in the same diesel engines and distributed through the same pipeline. Currently, green diesel is mainly produced through a catalytic reaction involving hydroprocessing, decarboxylation, and decarbonylation of triacylglycerols derived from the algal feedstock. The microalgae *Tetradesmus obliquus* is already in use to produce green diesel. *Tetradesmus obliquus* is mostly an aquatic form, but few isolates are also found in soil habitats that can be used to produce green diesel at the commercial level (Pragya and Pandey 2016).

2.8 Jet Fuel

Before the COVID-19 pandemic, around 1 billion tons of carbon was produced in 2019: about 3% of all emissions were released into the atmosphere worldwide. Due to reducing carbon emission, many companies such as Algenol, Sapphire Energy, Euglena, Solazyme, Alpha Biotech, and AlgaeTec. are successfully involved in the research and production of algae-based fuels. Two major airlines in Japan All Nippon and Japan's airlines, already operating their domestic flight by utilizing sustainable aviation fuel, which is produced from algal byproducts blended with traditional jet fuel. It used ASTM D7566 Annex 7; a blend of liquid hydrocarbon fuel obtained by hydrogenating crude oil (primary hydrocarbon) extracted from an aquatic microalga *Botryococcus braunii*. Another pioneering Japanese company Euglena is building a biorefinery that will produce jet fuel made from Euglena. *E. gracilis* synthesizes wax esters, mainly composed of C14:0 saturated fatty alcohols, myristyl alcohols, C14:0 saturated fatty acid, and myristic acid, which are considered high-value jet fuels. Furthermore, research should be done to make efficient jet fuel from soil algae, reducing fossil fuel use for aviation purposes (Bwapwa et al. 2019).

2.9 Bioelectricity

Photosynthetic microbial fuel cells are an efficient system that can generate bioelectricity. In this cell, algae accept electrons, act as a bio-catholyte, and produce oxygen and organic matters during the process of photosynthesis. These organic materials are further used as feedstock to produce bioelectricity. *Chlorella vulgaris* has already been used for the production of bioelectricity. It works as a cathode plate to use CO_2 and produce O_2, whereas another chamber that works as an anode consists of other microbes like Escherichia coli and sewage, which catalyze the organic materials into subatomic particles like electrons and protons (Enamala et al. 2020). Electricity will be produced by reducing oxygen at the cathode plate. So simply in PMFCs, with the help of algae and other microbes, organic matter's chemical energy is converted directly into electricity (Saba and Christy 2020).

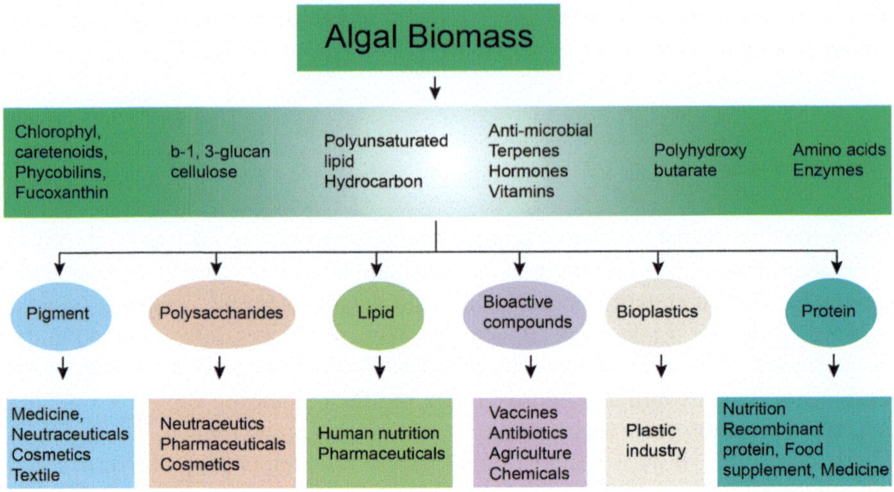

Fig. 2 Algae as potential feedstock for value added products

3 Soil Algae and Value-Added Products

Microalgal biomass consists of polysaccharides, lipids, proteins, different types of pigments, minerals, and vitamins (Spolaore et al. 2006; Safi et al. 2014; Ansari et al. 2021). However, the composition of biomass may differ according to species and culture environments, such as light condition, pH, temperature, salinity, and culture nutrient availability (Sharma et al. 2012; Vitova et al. 2015; Ru et al. 2020). In addition to biofuels, algal organic materials can be utilized as a staple crop for producing multiple commercial products like biofertilizers, pigments, medicines, and nutraceuticals (Fig. 2).

3.1 Biofertilizer

To increase the production of crops, chemical fertilizers have been used for a long time. However, these chemicals contain some harmful substances that cause soil pollution and create damage to our environment and health. These effects are cumulative and long term, including waterway pollution, chemical burn to crops, less nutrition value of food, increased air pollution, acidification of the soil, and mineral depletion. Therefore, biofertilizers are considered alternative and sustainable methods of fertilizing the soil. Biofertilizer contains living microalgae that can promote growth of plant through biogeochemical processes that increase the availability of nutrient by colonizing around rhizosphere and thus promoting plant growth (Vessey 2003; Herrmann and Lesueur 2013; Lesueur et al. 2016; Singh et al. 2019). Microalgae with omnipresent in all terrestrial environments are used in many ways to provide nutrition to the soil in the form of fertilizer and conditioner. In addition to fixing nitrogen, there are other benifits to soil algae: (1) excretion of growth-promoting hormones, vitamins, and amino acid; (2) solubilization of phosphate by secretion of organic acid; (3) reduction of the oxidizable matter of the soil and narrowing down the C: N ratio; (4) inhibition of weed growth; (5) increase in water holding capacity by mucous secretion; (6) increase in soil biomass by decomposition; (7) enhancement of soil porosity by their filamentous structure and secretions; (8) decrease in soil salinity; (9) attract other microorganisms to grow which further enrich the soil; and (10) gelatinous structure helps in chelation of molybdenum, iron, copper, zinc, and magnesium (Chatterjee et al. 2017).

Many blue-green algae, diatoms, Phaeophyceae, Chlorophyta, and Rhodophyta have been used as biofertilizers. Some most commonly found nitrogen-fixing cyanobacteria are *Anabaena variabilis*, *Nostoc linkia*, *Aulosira fertilisima*, *Microcoleus* sp., *Calothrix* sp., *Tolypothrix* sp., *Scytonema* sp., and *Phormidium* sp. (Moore 1969; Prasad and Prasad 2001). The *Azolla-Anabaena* association is a perfect example of symbiosis for nitrogen fixation as well as nutrient enrichment in the rice fields. Azolla is a water fern that assimilates nitrogen associated with nitrogen-fixing bacteria *Anabaena azollae*. Azolla is a good source of potassium when applied to the soil. A recent study showed that natural extract obtained from diatom *Navicula* sp. could stimulate the growth of *Salix viminalis*, *Helianthus tuberosus*, *Sida hermaphrodita* (Piotrowski et al. 2020). Coriander seeds inoculated with *Anabaena laxa* and *Calothrix elenkinii* showed increased germination (Kumar et al. 2013). Seeds socked with *Spirulina platensis* (2×10^4 cells/ml) show better crop plant growth, including rocket, Bayam red, and Pak choi (Wuang et al. 2016; Chiaiese et al. 2018). The filtrated extract of cyanobacteria *Nostoc* sp., *Calothrix* sp., *Westiellopsis* sp., and *Hapalosiphon* sp. shows better seed germination in wheat (Karthikeyan et al. 2009; Lee and Ryu 2021). In another study, green algae like *Chlorella* sp., *Nannochloris* sp., and *Scenedesmus* sp. have significant positive effects on soil percent organic matter (% OM), cation exchange capacity (CEC), pH, and microalgae species diversity (Hastings et al. 2014). Further, the cell extracts of *Chlorella vulgaris* added with *Nannochloropsis oculata* showed promoted plant growth in wheat, maize, sugar beet, and tomato (Shaaban 2001; Coppens et al. 2016; Barone et al. 2018).

There are specific vital steps that have to examine before applying any algae as biofertilizers. Some important concerns are mentioned as follows:

1. The particular biofertilizer can help in survival, establishment, colonization, and interaction with a particular crop.
2. Lab-to-field transition: specific algal strain with good in vitro performs poorly in the greenhouse and field trials.
3. Physiochemical property of soil (pH, moisture, texture, organic content).
4. Economic feasibility at the commercial rate (testing, scaling, storage, and application).
5. Farmer's accessibility and awareness.
6. Regulations and Intellectual Property Rights (Mitter et al. 2021).

7. Some soil algae also excrete phytohormones and amino acids, which enhance the beneficial microbial communities in soil (Singh et al. 2018).

3.2 Bioplastics

Bioplastics are plastics that are made from using biomaterials such as food crops or microalgae. They exhibit almost identical function as petroleum-based plastics (Chia et al. 2020). These include (1) hybrid plastics, which consist of microalgal biomass, petroleum-based polymers, and other additives; (2) biopolymers, such as polyhydroxy butyrate and starch, which cultivate intracellularly within microalgal cells. These bioplastics are remarkably biodegradable as compared to petroleum-based plastics. *Chlorella* sp. and *Spirulina* sp. have produced plastic similar to polyurethane (PE). By using genetic manipulation, we can generate the algal strains that synthesize compounds for bioplastics production. For example, *Chlamydomonas reinhardtii* having two expression vectors containing phbB and phbC genes from *Ralstonia eutropha*to can produce polyhydroxy butyrate (Chaogang et al. 2010). Although bioplastic production from soil algae is minimal due to its biodegradable nature, several attempts should overcome plastic pollution.

3.3 Extracellular Polymeric Substances (EPS)

Microalgae can eliminate numerous extracellular polymeric substances in response to their surroundings (Xiao and Zheng 2016). Exopolysaccharides are extracellular viscous polymers that accumulate outside the cell. These microalgal EPS mainly consist of carbohydrates, proteins, lipids and genetic materials. It retains matrix structure and forms a three-dimensional polymer network where the cell can interact and mediate their adhesion to surfaces. They are structurally and compositionally diverse biopolymers that possess unique bioactivities. These biopolymers are widely used in food, paper, textile, biotechnological, pharmaceutical, and cosmetics industries and for water retention. Cyanobacteria, red algae, and diatoms are the primary producers of EPS in response to stress conditions.

Nostoflan (NSF) is an EPS found in *Nostoc flagelliforme*, which has an antiviral effect on many viruses like HSV-1, HSV-2, human cytomegalovirus, and influenza A (Kanekiyo et al. 2007). Many red, brown and green algae produces Sulfated polymannuronate which combats HIV infection (Liu et al. 2005). Some more algal species and their useful EPS are listed in Table 2.

Table 2 EPS compounds and their functions

Functions	Type of EPS	Algal species	Reference
Soil particle texture, soil permeability	Polysaccharides, protein, and DHA	*Oscillatoria*, *Schizothrix delicatissima* AMPL0116, and *Lyngbya*	Sadeghi et al. (2017) and Mugnai et al. (2018)
Heavy metal absorption	Polysaccharides	*Kirchneriella aperta*, *Chlorella stigmatophora*, *Chlorella vulgaris*, *Anabaena spiroides*	Lombardi and Vieira (1999), Freire-Nordi et al. (2005), Zheng et al. (2019), Xie et al. (2020), and Babiak and Krzemińska (2021)
Water/nutrient retention and survival	Polysaccharides	*Nostoc calcicole*, *Phormidium* 94a	Vicente-García et al. (2004) and Bhatnagar et al. (2014)
Antiviral activity, anti-cancerous activity	Sulphated polysaccharides (sPS)	*Porphyridium cruentum*	Gardeva et al. (2009)

Adapted and modified from (Flemming and Wingender 2010)

3.4 Prebiotics

Prebiotics are different forms of polysaccharides such as galactans, arabinoxylans, and β-glucans.

Microalgae like *Chlorella* sp. produce an acidic polysaccharide that consists of rhamnose (52%), arabinose, and galactose, which can inhibit the proliferation of harmful pathogens (Gupta et al. 2017).

3.5 Nutraceuticals

Algae can be used as potential source to fight against world nutrition problem. Many countries like China, Korea, Chile, Ireland, North America, Ireland, British Columbia, and others are already using potential food sources. Vegetarian population cannot get docosahexaenoic acid (DHA) and eicosaoentaenoic acid (EPA) from vegetarian and dairy sources. Among them, algae emerged as a popular source of omega-3 fatty acids (Bucchini 2019).

Some terrestrial microalgae species, such as *Chlamydomonas* sp., *Chlorella* sp., *Spirulina* sp., and *Dunaliella* sp., have long been used as nutraceuticals. They can synthesize diverse bioactive compounds, such as carotenoids, essential amino acids, omega-3 and omega-6 fatty acids, and trace minerals (Paniagua-Michel 2015; Jha et al. 2017). They provide many vitamins and are rich in potassium, iodine, zinc, calcium, and magnesium. *Chlorella* and *Spirulina* are two widespread algae forms widely used in cuisine. *Chlorella* contains nine essential amino acids that the body cannot synthesize (Rani et al. 2018). *Spirulina* contains thiamine, copper, and 10% more proteins than *Chlorella*. *Euglena gracilis* can synthesize all 20 amino acids (Rodríguez-Zavala et al. 2010; Hasan et al. 2017). In addition to amino acids, *E. gracilis* has a high amount of vitamins C, vitamin E, PUFA, and β-1,3-glucan paramylon (Zeng et al. 2016). *Lobosphaera incisa* (*Parietochloris incisa*) is considered one of the richest sources of polyunsaturated fatty acid (PUFA) and arachidonic acid, which holds up to 47% of the triglycerides (Bigogno et al. 2002). *Nostoc flagelliforme*, commonly known as Fat choy, is widely consumed in traditional Chinese and Vietnamese cuisine. One of the *Clostridium* spp. can co-produce butanol and vitamin B_2 together so it can be used to produce both fuel and nutrition (Qureshi and Blaschek 2001; Sun et al. 2018).

3.6 Pharmaceuticals

Microalgae, such as *Nostoc flagelliforme* and *Porphyridium cruentum*, show antiviral and antibacterial activities against *Salmonella enteritidis* and some other viruses like African swine fever virus and Vaccinia virus (Liu et al. 2016). *Spirulina platensis* secreates sulfated polyscharide called spirulan which inhibits cancer metastasis (El-Baky et al. 2014). Strains like *Chroococcus disperses*, *Chlamydomonas reinhardtii*, and *Chlorella vulgaris* are emerging as prominent strains that exhibit a broad spectrum of antimicrobial activities (Ghasemi et al. 2007).

3.7 Pigments

All algae contain natural pigments, usually found in specialized plastids called chromophores, except in the members of Cyanophyceae (blue, green algae). Algal biomass produces many pigments like chlorophyll-a, b, c, carotenoid, and phycocyanin (Barkia et al. 2019). In addition to food and pharmaceutical uses, these natural pigments are being used as a substitute for coloring agents and chemical dyes. Carotenoids also have antioxidant effects that act as scavengers of harmful free radicals produced during inflammation inside the body (Young and Lowe 2018). Several carotenoids, such as α- and β-carotene, lutein, and zeaxanthin, are used as an anti-tumor drugs (Nishino et al. 2000). Several algal divisions and their commonly found pigments are listed in Table 3.

Spirulina and *Arthrospira* produce phycobilin pigments such as phycocyanin, which has several applications like anti-inflammatory agent, cosmetics, food colorant, fluorescent marker in histochemistry, and antibody labels (Manirafasha et al. 2018).

Table 3 Major pigments present in some soil algal species

Algal group	Algal species	Pigment
Chlorophyta (Green algae)	*Chlorella vulgaris*, *Dunalicella salina*, *Stichococcus* sp.	Chl-a, Chl-b, β-carotenoids, xanthophylls
Cyanophyta (Blue-green algae)	*Spirullina*, *Nostoc* sp.	Phycoerythrin, phycocyanin
Charophyta	*Spirogyra* sp.	Chl-b
Phaeophyta (Brown algae)	*Fucus vesiculosus*	Chl-a, Chl-c, fucoxanthin, β-carotenoid
Euglenophyta (Euglenoids)	*Euglena* sp.	Chl-b
Chrysophyta (Diatoms)	*Sellaphora nana*, *Hantzsochia amphioxys*	Chl-a, Chl-b, β-carotenoids, Fucoxanthin, diatoxanthin, diadonoxanthin, violaxanthin, zeaxanthin

3.8 Hydrocarbon

Hydrocarbon can include aromatic hydrocarbons, alkanes, alkenes, and alkyne-based organic compounds. These hydrocarbons are used as a biofuel and utilized for different pharmaceutical uses. Microalgae, like *Spirulina platensis*, consist 0.07% of total dry weight of hydrocarbon, with 84.9% of the 17:1 compound (Nevenzel 1989). *Spirulina* produces hydrocarbons like heptadecane and tetradecane, which are known as antibacterial agents and widely used in pharma industries (Seddek et al. 2019).

3.9 Animal Feed

Algae for animal feed can help improve an animal's health by providing micronutrients and antioxidants, which activate the immune system within the animal. As a result, it is being used as a feedstock in livestock, poultry, and other pet food. The most common algae for animal feed used currently are *Spirulina*, *Scenedesmus*, and *Chlorella*. They are added to whole foods, treats, and supplements for pets and are both rich in nutrients (Madeira et al. 2017; Saadaoui et al. 2021).

3.10 Enzymes

Enzymes are biological catalysts that accelerate chemical reactions for many industrial processes. Microalgal cells have a list of enzymes that include cellulases, galactosidases, proteases, phytases, lipases, laccases, and amylases; antioxidant enzymes like superoxide dismutases and peroxidases; and some carbon metabolic enzymes like carbonic anhydrases.

The coccoid green soil alga *Tetracystisaeria* Brown & Bold strain SAG 89.80 (Chlorophyta, Chlorococcales) can produce laccase-like phenoloxidases, which can degrade toxic phenolic substances in soil. In *Euglena gracilis*, thiolase and acyl-CoA-oxidizing enzymes are present in the peroxisomes and the mitochondria, which can purified and used for many industrial purposes (Graves Jr and Becker 1974; Winkler et al. 1988).

3.11 Phycoremediation

Many soil algae have shown appeared as better scavenger over bacteria and fungi in degrading different organic pollutants. Algae can degrade pollutants and use that carbon, contributing to CO_2 and oxygen production (Zhang et al. 2020). They can also fix N_2 from atmosphere and make them suitable for growth, maintenance, and adaptability to survive in polluted surroundings (Singh et al. 2016). Soil algae can be used to treat contaminants from soil infested with pesticides, hydrocarbon, phenol and catechol, heavy metals, and different xenobiotics (Al-Hasan et al. 2001; Singh et al. 2011).

Some blue-green soil algae such as *Nodularia*, *Synechococcus*, *Nostoc*, and *Cyanothece* degrade the pesticide named Lindane. Another organo-phosphorous herbicide, glyphosate, is metabolized by *Anabaena* sp. and *Nostoc* sp. and consumed as the phosphorus source (Forlani et al. 2008). Green algae like *Chlorella vulgaris* and *Scenedesmus bijugatus* can metabolize organophosphorus insecticides such as monocrotophos and quinalphos. DDT (1,1,1-trichloro-2,2-bis(p-chlorophenyl) ethane) residues have now become persistent environmental contaminants. Unicellular green algae *Chlorococcum* sp. works as an indicator of soil contamination from high-level DDT (Megharaj et al. 2000; Fu and Secundo 2016). Using several genetic manipulation methods, the biodegradation properties of soil algae can be enhanced many folds (Singh et al. 2016). However, regulatory bodies should provide ecological and environmental safety concerns under field conditions.

3.12 Biocontrol Agent

Soil algae produce a range of biological compounds that exhibit antibacterial, anti-algal, antifungal, and antiviral properties (Dahms et al. 2006; Yadav et al. 2017). These compounds are mainly alkaloids, amides, indoles, lipopeptides, and fatty acids (Abarzua et al. 1999; El-AAl and Azza 2013).

Nostoc muscorum, a genus of cyanobacteria, shows antifungal ability against soil-borne fungi, and it is mostly known for producing "damping off" (Chaudhary et al. 2012). *N. muscorum* extract can inhibit in vitro growth of the fungal plant pathogens, such as *Sclerotinia sclerotiorum* and *Rhizoctonia solini* (Singh et al. 2016). Two depsipeptide metabolites, scyptolin A and B, extracted from terrestrial cyanobacterium *Scytonema hofmanni* PCC 7110, have antimicrobial activity (Matern et al. 2003). Cyanobacteria can produce several enzymes that can degrade fungal cell wall. Some common enzymes are chitosanase, β-1,4-glucanase, β-1,3-glucanase, and benzoic acid, which can inhibit the growth of different fungal species like *Fusarium* sp., *Penicillium* sp., and *Candida* sp. (López et al. 2002; Chaudhary et al. 2012; Natarajan et al. 2012; Prasanna et al. 2013, 2016). Some blue-green soil algae, such as *Nodularia* sp. and *Nostoc* sp., produce antibacterial agents like 4,4′-dihydroxybiphenyl, norharmane, and diterpenoids, which act against *Escherichia coli*, *Pseudomonas aeruginosa*, *Bacillus subtilis*, and *Staphylococcus epidermis* (Jaki et al. 2000; Volk and Furkert 2006; Carpine and Sieber 2021).

Extensive research has to be done to establish soil algae as the potential biocontrol agent against various plant diseases (Singh et al. 2016).

4 Conclusion and Future Perspectives

Soil algae are an absolute biofuel feedstock, which could help reduce the requirement of petroleum-based fuel. Currently, production of algal biofuel is underrated and overpriced to be commercialized. To increase the production of different value-added products, including algal biofuel, integrated biorefinery should be established. The production of only one type of biofuel at a time is uneconomical production process; using a biorefinery approach, a wide range of products can be produced from a single feedstock (Milledge et al. 2019). Integrated biorefinery can help in many ways such as cutting down biofuels production costs, recycling of byproducts, and production and transportation of multiple value-added products (Spolaore et al. 2006).

Advanced gene-editing technology such as CRISPR and site-directed mutagenesis must be used to invent efficient algal systems to make the algae biofuel science work, say experts. In addition, by using the bioinformatics approach (i.e., genomics, transcriptomics, proteomics, and metabolomics data), data analysis can be done to find out more species genes of interest and their function. Furthermore, extensive research is required to explore more about terrestrial microalgae for using it efficiently such as (a). feedstocks for biofuels and other value products (b). selection of potential strains (c). efficiency of microalgalproductivity (d). efficient reactor design (e). optimum culture conditions for fermentation (f). strategies for upgrading strains for better yield (g). kinetics modeling (h) life cycle assessment. Most of the products, as mentioned earlier, are extracted from an aquatic form of the same species, which is also found in soil. So, an upgrade system of bioreactors should launch where soil algae can be grown in abundance. Specific modifications will be required at the processing and purification steps. Involving soil algae as a feedstock for biofuel and value-added production will help balance the ecosystem because there will not be the overconsumption of aquatic algae; on the other hand, it will help reduce water pollution.

Several governmental bodies worldwide are establishing research institutions for path-breaking discoveries and inventions to improve algal biology and culture-system engineering. National Renewable Energy Laboratory, USA, The Energy Resource and Institute (TERI) India, German Biomass Research Center (DBFZ) Germany, and Institute of Technological Research (IPT), Brazil, are some of the pioneer institutions that are involved in research and develop new methods and technologies to improve the algal production process, efficiency of commercial biofuel, and other value-added products (Farm Energy 2019; IEA 2020).

Acknowledgement We would like to thank Dr. Deepak K. Singh (Albert Einstein College of Medicine, New York) for his critical reading and suggestions for this article.

References

Abarzua S, Jakubowski S, Eckert S, Fuchs P (1999) Biotechnological investigation for the prevention of marine biofouling II. Blue-green algae as potential producers of biogenic agents for the growth inhibition of microfouling organisms. Bot Mar 42:459

Abomohra AE-F, Elshobary M (2019) Biodiesel, bioethanol, and biobutanol production from microalgae. In: Microalgae biotechnology for development of biofuel and wastewater treatment. Springer, Singapore, pp 293–321

Adesalu T, Olugbemi O (2015) Soil algae: a case study of two vegetable farmlands in Lagos and Ogun states, Southwest Nigeria. IFE J Sci 17:765–772

Agwa OK, Nwosu IG, Abu GO (2017) Bioethanol production from Chlorella vulgaris biomass cultivated with plantain (Musa paradisiaca) peels extract. Adv Biosci Biotechnol 8:478

Ahmad A, Banat F, Alsafar H, Hasan SW (2022) Algae biotechnology for industrial wastewater treatment, bioenergy production, and high-value bioproducts. Sci Total Environ 806:150585

Ahuja D, Tatsutani M (2009) Sustainable energy for developing countries. SAPIENS 2(1). http://journals.openedition.org/sapiens/823

Al-Hasan R, Khanafer M, Eliyas M, Radwan S (2001) Hydrocarbon accumulation by picocyanobacteria from the Arabian Gulf. J Appl Microbiol 91:533–540

Ansari FA, Guldhe A, Gupta SK, Rawat I, Bux F (2021) Improving the feasibility of aquaculture feed by using microalgae. Environ Sci Pollut Res 28:43234–43257

Antonelli M, Wetzel CE, Ector L, Teuling AJ, Pfister L (2017) On the potential for terrestrial diatom communities and diatom indices to identify anthropic disturbance in soils. Ecol Indic 75:73–81

Antoni D, Zverlov VV, Schwarz WH (2007) Biofuels from microbes. Appl Microbiol Biotechnol 77:23–35

Anukam A, Mohammadi A, Naqvi M, Granström K (2019) A review of the chemistry of anaerobic digestion: methods of accelerating and optimizing process efficiency. PRO 7:504

Arun J, Varshini P, Prithvinath PK, Priyadarshini V, Gopinath KP (2018) Enrichment of bio-oil after hydrothermal liquefaction (HTL) of microalgae C. vulgaris grown in wastewater: bio-char and post HTL wastewater utilization studies. Bioresour Technol 261:182–187

Babiak W, Krzemińska I (2021) Extracellular polymeric substances (EPS) as microalgal bioproducts: a review of factors affecting EPS synthesis and application in flocculation processes. Energies 14:4007

Barkia I, Saari N, Manning SR (2019) Microalgae for high-value products towards human health and nutrition. Mar Drugs 17:304

Barone V, Baglieri A, Stevanato P, Broccanello C, Bertoldo G, Bertaggia M, Cagnin M, Pizzeghello D, Moliterni V, Mandolino G (2018) Root morphological and molecular responses induced by microalgae extracts in sugar beet (Beta vulgaris L.). J Appl Phycol 30:1061–1071

Behera S, Singh R, Arora R, Sharma NK, Shukla M, Kumar S (2015) Scope of algae as third generation biofuels. Front Bioeng Biotechnol 2:90

Bhatnagar M, Pareek S, Bhatnagar A, Ganguly J (2014) Rheology and characterization of a low viscosity emulsifying exopolymer from desert borne Nostoc calcicola. Indian J Biotechnol 13:241–246

Bigogno C, Khozin-Goldberg I, Boussiba S, Vonshak A, Cohen Z (2002) Lipid and fatty acid composition of the green oleaginous alga Parietochloris incisa, the richest plant source of arachidonic acid. Phytochemistry 60:497–503

Binda G, Spanu D, Bettinetti R, Magagnin L, Pozzi A, Dossi C (2020) Comprehensive comparison of microalgae-derived biochar from different feedstocks: a prospective study for future environmental applications. Algal Res 52:102103

Bošnjaković M, Sinaga N (2020) The perspective of large-scale production of algae biodiesel. Appl Sci 10:8181

Brilman W, Alba LG, Veneman R (2013) Capturing atmospheric CO_2 using supported amine sorbents for microalgae cultivation. Biomass Bioenergy 53:39–47

Bucchini L (2019) Nutrition and health claims in Europe: oils & fats related claims, regulatory and labeling challenges. Ocl 26:48

Bwapwa JK, Akash A, Trois C (2019) Jet fuel blend from algal jet fuel and jet A1 in 50/50 volume ratio. Int J Low-Carbon Technol 14:234–240

Carpine R, Sieber S (2021) Antibacterial and antiviral metabolites from cyanobacteria: their application and their impact on human health. Curr Res Biotechnol 3:65–81

Chandra R, Iqbal HM, Vishal G, Lee H-S, Nagra S (2019) Algal biorefinery: a sustainable approach to valorize algal-based biomass towards multiple product recovery. Bioresour Technol 278:346–359

Chaogang W, Zhangli H, Anping L, Baohui J (2010) Biosynthesis of poly-3-hydroxybutyrate (phb) in the transgenic green alga chlamydomonas reinhardtii 1. J Phycol 46:396–402

Chatterjee A, Singh S, Agrawal C, Yadav S, Rai R, Rai L (2017) Role of algae as a biofertilizer. In: Algal green chemistry. Elsevier, pp 189–200

Chaudhary V, Prasanna R, Nain L, Dubey S, Gupta V, Singh R, Jaggi S, Bhatnagar AK (2012) Bioefficacy of novel cyanobacteria-amended formulations in suppressing damping off disease in tomato seedlings. World J Microbiol Biotechnol 28:3301–3310

Chia SR, Chew KW, Show PL, Yap YJ, Ong HC, Ling TC, Chang JS (2018) Analysis of economic and environmental aspects of microalgae biorefinery for biofuels production: a review. Biotechnol J 13:e1700618

Chia WY, Tang DYY, Khoo KS, Lup ANK, Chew KW (2020) Nature's fight against plastic pollution: algae for plastic biodegradation and bioplastics production. Environ Sci Ecotechnology 4:100065

Chiaiese P, Corrado G, Colla G, Kyriacou MC, Rouphael Y (2018) Renewable sources of plant biostimulation: microalgae as a sustainable means to improve crop performance. Front Plant Sci 9:1782

Coppens J, Grunert O, Van Den Hende S, Vanhoutte I, Boon N, Haesaert G, De Gelder L (2016) The use of microalgae as a high-value organic slow-release fertilizer results in tomatoes with increased carotenoid and sugar levels. J Appl Phycol 28:2367–2377

Cuellar-Bermudez SP, Garcia-Perez JS, Rittmann BE, Parra-Saldivar R (2015) Photosynthetic bioenergy utilizing CO2: an approach on flue gases utilization for third generation biofuels. J Clean Prod 98:53–65

Dahms H-U, Ying X, Pfeiffer C (2006) Antifouling potential of cyanobacteria: a mini-review. Biofouling 22:317–327

De Bhowmick G, Sarmah AK, Sen R (2019) Zero-waste algal biorefinery for bioenergy and biochar: a green leap towards achieving energy and environmental sustainability. Sci Total Environ 650:2467–2482

Deng X-Y, Gao K, Addy M, Li D, Zhang R-C, Lu Q, Ma Y-W, Cheng Y-L, Chen P, Liu Y-H (2018) Cultivation of Chlorella vulgaris on anaerobically digested swine manure with daily recycling of the post-harvest culture broth. Bioresour Technol 247:716–723

El-AAl A, Azza A (2013) Characterization of Anabaena azollae isolated from Azolla pinnata. Egypt J Agric Res 91:801–809

El-Baky AH, El Baz HK, El-Latife S (2014) Induction of sulfated polysaccharides in Spirulina platensis as response to nitrogen concentration and its biological evaluation. J Aquac Res Dev 5:1

Enamala MK, Dixit R, Tangellapally A, Singh M, Dinakarrao SMP, Chavali M, Pamanji SR, Ashokkumar V, Kadier A, Chandrasekhar K (2020) Photosynthetic microorganisms (algae) mediated bioelectricity generation in microbial fuel cell: concise review. Environ Technol Innov 19:100959

Farm Energy (2019) Algae for biofuel production

Flemming H-C, Wingender J (2010) The biofilm matrix. Nat Rev Microbiol 8:623–633

Foets J, Stanek-Tarkowska J, Teuling AJ, Van de Vijver B, Wetzel CE, Pfister L (2021) Autecology of terrestrial diatoms under anthropic disturbance and across climate zones. Ecol Indic 122:107248

Forlani G, Pavan M, Gramek M, Kafarski P, Lipok J (2008) Biochemical bases for a widespread tolerance of cyanobacteria to the phosphonate herbicide glyphosate. Plant Cell Physiol 49:443–456

Freire-Nordi CS, Vieira AAH, Nascimento OR (2005) The metal binding capacity of Anabaena spiroides extracellular polysaccharide: an EPR study. Process Biochem 40:2215–2224

Fu P, Secundo F (2016) Algae and their bacterial consortia for soil bioremediation. Chem Eng Trans 49:427–432

Furuhashi T, Ogawa T, Nakai R, Nakazawa M, Okazawa A, Padermschoke A, Nishio K, Hirai MY, Arita M, Ohta D (2015) Wax ester and lipophilic compound profiling of Euglena gracilis by gas chromatography-mass spectrometry: toward understanding of wax ester fermentation under hypoxia. Metabolomics 11:175–183

Gardeva E, Toshkova R, Minkova K, Gigova L (2009) Cancer protective action of polysaccharide, derived from red microalga Porphyridium cruentum—a biological background. Biotechnol Biotechnol Equip 23:783–787

Gayathri R, Mahboob S, Govindarajan M, Al-Ghanim KA, Ahmed Z, Al-Mulhm N, Vodovnik M, Vijayalakshmi S (2021) A review on biological carbon sequestration: a sustainable solution for a cleaner air environment, less pollution and lower health risks. J King Saud Univ Sci 33:101282

Ghasemi Y, Moradian A, Mohagheghzadeh A, Shokravi S, Morowvat MH (2007) Antifungal and antibacterial activity of the microalgae collected from paddy fields of Iran: characterization of antimicrobial activity of Chroococcus dispersus. J Biol Sci 7:904–910

Ghedini E, Taghavi S, Menegazzo F, Signoretto M (2021) A review on the efficient catalysts for algae transesterification to biodiesel. Sustain For 13:10479

Graves LB Jr, Becker WM (1974) Beta-oxidation in glyoxysomes from euglena. J Protozool 21:771–774

Gupta S, Gupta C, Garg A, Prakash D (2017) Prebiotic efficiency of blue green algae on probiotics microorganisms. J Microbiol Exp 4:00120

Hasan MT, Sun A, Mirzaei M, Te'o J, Hobba G, Sunna A, Nevalainen H (2017) A comprehensive assessment of the biosynthetic pathways of ascorbate, α-tocopherol and free amino acids in Euglena gracilis var. saccharophila. Algal Res 27:140–151

Hastings KL, Smith LE, Lindsey ML, Blotsky LC, Downing GR, Zellars DQ, Downing JK, Corena-McLeod M (2014) Effect of microalgae application on soil algal species diversity, cation exchange capacity and organic matter after herbicide treatments. F1000Research 3:281

Hernández ES, Córdoba LT (1993) Anaerobic digestion of Chlorella vulgaris for energy production. Resour Conserv Recycl 9:127–132

Herrmann L, Lesueur D (2013) Challenges of formulation and quality of biofertilizers for successful inoculation. Appl Microbiol Biotechnol 97:8859–8873

Hu Y, Bassi A, Xu CC (2020) Energy from biomass. In: Future energy. Elsevier, pp 447–471

IEA (2020) Outlook for biogas and biomethane: prospects for organic growth

IEA Bioenergy (2017) India Biofuel economy roadmap – challenges & prospects

Jaki B, Heilmann J, Sticher O (2000) New antibacterial metabolites from the cyanobacterium Nostoc c ommune (EAWAG 122b). J Nat Prod 63:1283–1285

Jang Y-S, Malaviya A, Cho C, Lee J, Lee SY (2012) Butanol production from renewable biomass by clostridia. Bioresour Technol 123:653–663

Jha D, Jain V, Sharma B, Kant A, Garlapati VK (2017) Microalgae-based pharmaceuticals and nutraceuticals: an emerging field with immense market potential. ChemBioEng Rev 4:257–272

Kanekiyo K, Hayashi K, Takenaka H, Lee J-B, Hayashi T (2007) Anti-herpes simplex virus target of an acidic polysaccharide, nostoflan, from the edible blue-green alga Nostoc flagelliforme. Biol Pharm Bull 30:1573–1575

Karthikeyan N, Prasanna R, Sood A, Jaiswal P, Nayak S, Kaushik B (2009) Physiological characterization and electron microscopic investigation of cyanobacteria associated with wheat rhizosphere. Folia Microbiol 54:43–51

Khan A, Rao TS (2019) Molecular evolution of xenobiotic degrading genes and mobile DNA elements in soil bacteria. In: Microbial diversity in the genomic era. Elsevier, pp 657–678

Khan MI, Shin JH, Kim JD (2018) The promising future of microalgae: current status, challenges, and optimization of a sustainable and renewable industry for biofuels, feed, and other products. Microb Cell Factories 17:36

Kostas ET, Adams JMM, Ruiz HA, Durán-Jiménez G, Lye GJ (2021) Macroalgal biorefinery concepts for the circular bioeconomy: a review on biotechnological developments and future perspectives. Renew Sust Energ Rev 151:111553

Krug L, Erlacher A, Markut K, Berg G, Cernava T (2020) The microbiome of alpine snow algae shows a specific inter-kingdom connectivity and algae-bacteria interactions with supportive capacities. ISME J 14:2197–2210

Kumar V, Thakur IS (2020) Biodiesel production from transesterification of Serratia sp. ISTD04 lipids using immobilised lipase on biocomposite materials of biomineralized products of carbon dioxide sequestrating bacterium. Bioresour Technol 307:123193

Kumar M, Prasanna R, Bidyarani N, Babu S, Mishra BK, Kumar A, Adak A, Jauhari S, Yadav K, Singh R (2013) Evaluating the plant growth promoting ability of thermotolerant bacteria and cyanobacteria and their interactions with seed spice crops. Sci Hortic 164:94–101

Kumar G, Nguyen DD, Sivagurunathan P, Kobayashi T, Xu K, Chang SW (2018) Cultivation of microalgal biomass using swine manure for biohydrogen production: impact of dilution ratio and pretreatment. Bioresour Technol 260:16–22

Kumar M, Sun Y, Rathour R, Pandey A, Thakur IS, Tsang DC (2020) Algae as potential feedstock for the production of biofuels and value-added products: opportunities and challenges. Sci Total Environ 716:137116

Kumar BR, Mathimani T, Sudhakar MP, Rajendran K, Nizami A-S, Brindhadevi K, Pugazhendhi A (2021) A state of the art review on the cultivation of algae for energy and other valuable products: application, challenges, and opportunities. Renew Sust Energ Rev 138:110649

Kvasauskas M, Baltrėnas P (2009) Research on anaerobically treated organic waste suitability for soil fertilisation. J Environ Eng Landsc Manag 17:205–211

Lee S-M, Ryu C-M (2021) Algae as new kids in the beneficial plant microbiome. Front Plant Sci 12:91

Legere E (2017) Algenol integrated pilot-scale biorefinery public version final report. Algenol

Legros A, di San Marzano CA, Naveau H, Nyns E-J (1983) Fermentation profiles in bioconversions. Biotechnol Lett 5:7–12

Lesueur D, Deaker R, Herrmann L, Bräu L, Jansa J (2016) The production and potential of biofertilizers to improve crop yields. In: Bioformulations: for sustainable agriculture. Springer, pp 71–92

Levasseur W, Perre P, Pozzobon V (2020) A review of high value-added molecules production by microalgae in light of the classification. Biotechnol Adv 41:107545

Li K, Liu S, Liu X (2014) An overview of algae bioethanol production. Int J Energy Res 38:965–977

Limongi AR, Viviano E, Luca MD, Radice RP, Bianco G, Martelli G (2021) Biohydrogen from microalgae: production and applications. Appl Sci 11:1616

Liu H, Geng M, Xin X, Li F, Zhang Z, Li J, Ding J (2005) Multiple and multivalent interactions of novel anti-AIDS drug candidates, sulfated polymannuronate (SPMG)-derived oligosaccharides, with gp120 and their anti-HIV activities. Glycobiology 15:501–510

Liu L, Pohnert G, Wei D (2016) Extracellular metabolites from industrial microalgae and their biotechnological potential. Mar Drugs 14:191

Lombardi A, Vieira A (1999) Lead-and copper-complexing extracellular ligands released by Kirchneriella aperta (Chloroccocales, Chlorophyta). Phycologia 38:283–288

López A, Ming DS, Towers GN (2002) Antifungal activity of benzoic acid derivatives from piper l anceaefolium. J Nat Prod 65:62–64

Madeira MS, Cardoso C, Lopes PA, Coelho D, Afonso C, Bandarra NM, Prates JA (2017) Microalgae as feed ingredients for livestock production and meat quality: a review. Livest Sci 205:111–121

Manirafasha E, Murwanashyaka T, Ndikubwimana T, Ahmed NR, Liu J, Lu Y, Zeng X, Ling X, Jing K (2018) Enhancement of cell growth and phycocyanin production in Arthrospira (Spirulina) platensis by metabolic stress and nitrate fed-batch. Bioresour Technol 255:293–301

Matern U, Oberer L, Erhard M, Herdman M, Weckesser J (2003) Hofmannolin, a cyanopeptolin from Scytonema hofmanni PCC 7110. Phytochemistry 64:1061–1067

Mathimani T, Mallick N (2019) A review on the hydrothermal processing of microalgal biomass to bio-oil-knowledge gaps and recent advances. J Clean Prod 217:69–84

Megharaj M, Kantachote D, Singleton I, Naidu R (2000) Effects of long-term contamination of DDT on soil microflora with special reference to soil algae and algal transformation of DDT. Environ Pollut 109:35–42

Meuser JE, Ananyev G, Wittig LE, Kosourov S, Ghirardi ML, Seibert M, Dismukes GC, Posewitz MC (2009) Phenotypic diversity of hydrogen production in chlorophycean algae reflects distinct anaerobic metabolisms. J Biotechnol 142:21–30

Michalak I, Baśladyńska S, Mokrzycki J, Rutkowski P (2019) Biochar from a freshwater macroalga as a potential biosorbent for wastewater treatment. Water 11:1390

Milledge JJ, Nielsen BV, Maneein S, Harvey PJ (2019) A brief review of anaerobic digestion of algae for bioenergy. Energies 12:1166

Mitter EK, Tosi M, Obregón D, Dunfield KE, Germida JJ (2021) Rethinking crop nutrition in times of modern microbiology: innovative biofertilizer technologies. Front Sustain Food Syst 5:29

Miyamoto K, Hallenbeck PC, Benemann JR (1979) Hydrogen production by the thermophilic alga Mastigocladus laminosus: effects of nitrogen, temperature, and inhibition of photosynthesis. Appl Environ Microbiol 38:440–446

Molazadeh M, Ahmadzadeh H, Pourianfar HR, Lyon S, Rampelotto PH (2019) The use of microalgae for coupling wastewater treatment with CO2 biofixation. Front Bioeng Biotechnol 7:42

Moore A (1969) Azolla: biology and agronomic significance. Bot Rev 35:17–34

Mugnai G, Rossi F, Felde VJMNL, Colesie C, Büdel B, Peth S, Kaplan A, De Philippis R (2018) Development of the polysaccharidic matrix in biocrusts induced by a cyanobacterium inoculated in sand microcosms. Biol Fertil Soils 54:27–40

Natarajan C, Prasanna R, Gupta V, Dureja P, Nain L (2012) Characterization of the fungicidal activity of Calothrix elenkinii using chemical methods and microscopy. Appl Biochem Microbiol 48:51–57

Nevenzel JC (1989) Biogenic hydrocarbons of marine organisms. In: Marine biogenic lipids, fats, and oils, vol 1. CRC Press, Boca Raton, pp 3–71

Nguyen MA, Hoang AL (2016) A review on microalgae and cyanobacteria in biofuel production. Hanoi, USTH

Nishino H, Tokuda H, Murakoshi M, Satomi Y, Masuda M, Onozuka M, Yamaguchi S, Takayasu J, Tsuruta J, Okuda M (2000) Cancer prevention by natural carotenoids. Biofactors 13:89–94

Ojolo S, Oke S, Animasahun K, Adesuyi B (2007) Utilization of poultry, cow and kitchen wastes for biogas production: a comparative analysis. J Environ Health Sci Eng 4:223–228

Palanisamy M, Mukund S, Sivakumar U, Sivasubramanian V (2017) Bio-char production from micro algal biomass of Chlorella vulgaris. Phykos 47:99–104

Paniagua-Michel J (2015) Microalgal nutraceuticals. In: Handbook of marine microalgae. Elsevier, pp 255–267

Pentecost A, Whitton BA (2012) Subaerial cyanobacteria. In: Ecology of cyanobacteria II. Springer, pp 291–316

Phwan CK, Chew KW, Sebayang AH, Ong HC, Ling TC, Malek MA, Ho Y-C, Show PL (2019) Effects of acids pre-treatment on the microbial fermentation process for bioethanol production from microalgae. Biotechnol Biofuels 12:1–8

Piotrowski K, Romanowska-Duda Z, Messyasz B (2020) Cultivation of energy crops by ecological methods under the conditions of global climate and environmental changes with the use of diatom extract as a natural source of chemical compounds. Acta Physiol Plant 42:1–13

Pragya N, Pandey KK (2016) Life cycle assessment of green diesel production from microalgae. Renew Energy 86:623–632

Prasad R, Prasad B (2001) Cyanobacteria as a source biofertilizer for sustainable agriculture in Nepal. J Plant Sci Bot Orientalis 1:127–133

Prasanna R, Chaudhary V, Gupta V, Babu S, Kumar A, Singh R, Shivay YS, Nain L (2013) Cyanobacteria mediated plant growth promotion and bioprotection against Fusarium wilt in tomato. Eur J Plant Pathol 136:337–353

Prasanna R, Kanchan A, Ramakrishnan B, Ranjan K, Venkatachalam S, Hossain F, Shivay YS, Krishnan P, Nain L (2016) Cyanobacteria-based bioinoculants influence growth and yields by modulating the microbial communities favourably in the rhizospheres of maize hybrids. Eur J Soil Biol 75:15–23

Pushkareva E, Johansen JR, Elster J (2016) A review of the ecology, ecophysiology and biodiversity of microalgae in Arctic soil crusts. Polar Biol 39:2227–2240

Qureshi N, Blaschek H (2001) ABE production from corn: a recent economic evaluation. J Ind Microbiol Biotechnol 27:292–297

Raheem A, Dupont V, Channa AQ, Zhao X, Vuppaladadiyam AK, Taufiq-Yap Y-H, Zhao M, Harun R (2017) Parametric characterization of air gasification of Chlorella vulgaris biomass. Energy Fuel 31:2959–2969

Raheem A, Abbasi SA, Mangi FH, Ahmed S, He Q, Ding L, Memon AA, Zhao M, Yu G (2021) Gasification of algal residue for synthesis gas production. Algal Res 58:102411
Rani K, Sandal N, Sahoo P (2018) A comprehensive review on chlorella-its composition, health benefits, market and regulatory scenario. J Pharm Innov 7:584–589
Ranjith Kumar R, Hanumantha Rao P, Arumugam M (2015) Lipid extraction methods from microalgae: a comprehensive review. Front Energy Res 2:61
Rathore D, Singh A, Dahiya D, Nigam PS-N (2019) Sustainability of biohydrogen as fuel: present scenario and future perspective. AIMS Energy 7:1–19
Robak K, Balcerek M (2018) Review of second generation bioethanol production from residual biomass. Food Technol Biotechnol 56:174
Rodríguez-Zavala J, Ortiz-Cruz M, Mendoza-Hernández G, Moreno-Sánchez R (2010) Increased synthesis of α-tocopherol, paramylon and tyrosine by Euglena gracilis under conditions of high biomass production. J Appl Microbiol 109:2160–2172
Ru ITK, Sung YY, Jusoh M, Wahid MEA, Nagappan T (2020) Chlorella vulgaris: a perspective on its potential for combining high biomass with high value bioproducts. Appl Phycol 1:2–11
Saadaoui I, Rasheed R, Aguilar A, Cherif M, Al Jabri H, Sayadi S, Manning SR (2021) Microalgal-based feed: promising alternative feedstocks for livestock and poultry production. J Anim Sci Biotechnol 12:1–15
Saba B, Christy AD (2020) Bioelectricity generation in algal microbial fuel cells. In: Handbook of algal science, technology and medicine. Elsevier, pp 377–384
Sadeghi SH, Kheirfam H, Homaee M, Darki BZ, Vafakhah M (2017) Improving runoff behavior resulting from direct inoculation of soil micro-organisms. Soil Tillage Res 171:35–41
Safi C, Zebib B, Merah O, Pontalier P-Y, Vaca-Garcia C (2014) Morphology, composition, production, processing and applications of Chlorella vulgaris: a review. Renew Sust Energ Rev 35:265–278
Seddek N, Fawzy M, El-Said W, Ahmed M (2019) Evaluation of antimicrobial, antioxidant and cytotoxic activities and characterization of bioactive substances from freshwater blue-green algae. Glob Nest J 21:328–336
Shaaban MM (2001) Green microalgae water extract as foliar feeding to wheat plants. Pak J Biol Sci 4:628
Shahid A, Malik S, Zhu H, Xu J, Nawaz MZ, Nawaz S, Asraful Alam M, Mehmood MA (2020) Cultivating microalgae in wastewater for biomass production, pollutant removal, and atmospheric carbon mitigation; a review. Sci Total Environ 704:135303
Sharma R, Singh P, Sharma K (2012) Effects of culture conditions on growth and biochemical profile of Chlorella vulgaris. J Plant Pathol Microbiol 3(5):1–6
Shin YS, Choi HI, Choi JW, Lee JS, Sung YJ, Sim SJ (2018) Multilateral approach on enhancing economic viability of lipid production from microalgae: a review. Bioresour Technol 258:335–344
Shin YS, Jeong J, Nguyen THT, Kim JYH, Jin E, Sim SJ (2019) Targeted knockout of phospholipase A2 to increase lipid productivity in Chlamydomonas reinhardtii for biodiesel production. Bioresour Technol 271:368–374
Siddiki SYA, Mofijur M, Kumar PS, Ahmed SF, Inayat A, Kusumo F, Badruddin IA, Khan TMY, Nghiem LD, Ong HC, Mahlia TMI (2022) Microalgae biomass as a sustainable source for biofuel, biochemical and biobased value-added products: an integrated biorefinery concept. Fuel 307:121782
Singh JS, Singh D, Dixit S (2011) Cyanobacteria: an agent of heavy metal removal. In: Bioremediation of pollutants. IK International Publisher, New Delhi, pp 223–243
Singh JS, Kumar A, Rai AN, Singh DP (2016) Cyanobacteria: a precious bio-resource in agriculture, ecosystem, and environmental sustainability. Front Microbiol 7:529
Singh V, Maharshi A, Singh DP, Upadhyay RS, Sarma BK, Singh HB (2018) Role of microbial seed priming and microbial phytohormone in modulating growth promotion and defense responses in plants. In: Advances in seed priming. Springer, pp 115–126

Singh M, Singh D, Gupta A, Pandey KD, Singh P, Kumar A (2019) Plant growth promoting rhizobacteria: application in biofertilizers and biocontrol of phytopathogens. In: PGPR amelioration in sustainable agriculture. Elsevier, pp 41–66

Song M, Pei H (2018) The growth and lipid accumulation of Scenedesmus quadricauda during batch mixotrophic/heterotrophic cultivation using xylose as a carbon source. Bioresour Technol 263:525–531

Spolaore P, Joannis-Cassan C, Duran E, Isambert A (2006) Commercial applications of microalgae. J Biosci Bioeng 101:87–96

Suganya T, Varman M, Masjuki HH, Renganathan S (2016) Macroalgae and microalgae as a potential source for commercial applications along with biofuels production: a biorefinery approach. Renew Sust Energ Rev 55:909–941

Sun C, Zhang S, Xin F, Shanmugam S, Wu Y-R (2018) Genomic comparison of Clostridium species with the potential of utilizing red algal biomass for biobutanol production. Biotechnol Biofuels 11:1–15

Tucci S, Vacula R, Krajcovic J, Proksch P, Martin W (2010) Variability of wax ester fermentation in natural and bleached Euglena gracilis strains in response to oxygen and the elongase inhibitor flufenacet. J Eukaryot Microbiol 57:63–69

Ueno Y, Kurano N, Miyachi S (1999) Purification and characterization of hydrogenase from the marine green alga, Chlorococcum littorale. FEBS Lett 443:144–148

Uggetti E, Sialve B, Trably E, Steyer J-P (2014) Integrating microalgae production with anaerobic digestion: a biorefinery approach. Biofuels Bioprod Biorefin 8:516–529

Ullah K, Ahmad M, Sharma VK, Lu P, Harvey A, Zafar M, Sultana S (2015) Assessing the potential of algal biomass opportunities for bioenergy industry: a review. Fuel 143:414–423

US Energy Information Administration (2020) Biofuels explained

Usharani K, Roopashree K, Naik D (2019) Role of soil physical, chemical and biological properties for soil health improvement and sustainable agriculture. J Pharmacogn Phytochem 8:1256–1267

Vessey JK (2003) Plant growth promoting rhizobacteria as biofertilizers. Plant Soil 255:571–586

Vicente-García V, Ríos-Leal E, Calderón-Domínguez G, Cañizares-Villanueva RO, Olvera-Ramírez R (2004) Detection, isolation, and characterization of exopolysaccharide produced by a strain of Phormidium 94a isolated from an arid zone of Mexico. Biotechnol Bioeng 85:306–310

Vitova M, Bisova K, Kawano S, Zachleder V (2015) Accumulation of energy reserves in algae: from cell cycles to biotechnological applications. Biotechnol Adv 33:1204–1218

Volk R-B, Furkert FH (2006) Antialgal, antibacterial and antifungal activity of two metabolites produced and excreted by cyanobacteria during growth. Microbiol Res 161:180–186

Wang J, Wan W (2009) Factors influencing fermentative hydrogen production: a review. Int J Hydrog Energy 34:799–811

Whitton BA (2012) Ecology of cyanobacteria II: their diversity in space and time. Springer Science & Business Media

Winkler U, Säftel W, Stabenau H (1988) β-Oxidation of fatty acids in algae: localization of thiolase and acyl-CoA oxidizing enzymes in three different organisms. Planta 175:91–98

Winkler M, Heil B, Heil B, Happe T (2002) Isolation and molecular characterization of the [Fe]-hydrogenase from the unicellular green alga Chlorella fusca. Biochim Biophys Acta, Gene Struct Expression 1576:330–334

Wuang SC, Khin MC, Chua PQD, Luo YD (2016) Use of Spirulina biomass produced from treatment of aquaculture wastewater as agricultural fertilizers. Algal Res 15:59–64

Xiao R, Zheng Y (2016) Overview of microalgal extracellular polymeric substances (EPS) and their applications. Biotechnol Adv 34:1225–1244

Xie Q, Liu N, Lin D, Qu R, Zhou Q, Ge F (2020) The complexation with proteins in extracellular polymeric substances alleviates the toxicity of Cd (II) to Chlorella vulgaris. Environ Pollut 263:114102

Yadav S, Rai S, Rai R, Shankar A, Singh S, Rai L (2017) Cyanobacteria: role in agriculture, environmental sustainability, biotechnological potential and agroecological impact. In: Plant-microbe interactions in agro-ecological perspectives. Springer, pp 257–277

Young AJ, Lowe GL (2018) Carotenoids—antioxidant properties. Multidisciplinary Digital Publishing Institute, p 28

Zeng M, Hao W, Zou Y, Shi M, Jiang Y, Xiao P, Lei A, Hu Z, Zhang W, Zhao L (2016) Fatty acid and metabolomic profiling approaches differentiate heterotrophic and mixotrophic culture conditions in a microalgal food supplement 'Euglena'. BMC Biotechnol 16:1–8

Zhang B, Zhang Y, Downing A, Niu Y (2011) Distribution and composition of cyanobacteria and microalgae associated with biological soil crusts in the Gurbantunggut Desert, China. Arid Land Res Manag 25:275–293

Zhang W, Zhao C, Cao W, Sun S, Hu C, Liu J, Zhao Y (2020) Removal of pollutants from biogas slurry and CO 2 capture in biogas by microalgae-based technology: a systematic review. Environ Sci Pollut Res 27:28749–28767

Zheng S, Zhou Q, Chen C, Yang F, Cai Z, Li D, Geng Q, Feng Y, Wang H (2019) Role of extracellular polymeric substances on the behavior and toxicity of silver nanoparticles and ions to green algae Chlorella vulgaris. Sci Total Environ 660:1182–1190

The manufacturer's authorised representative in the EU is Springer Nature Customer Service Centre GmbH, Europaplatz 3, 69115 Heidelberg, Germany. If you have any concerns regarding our products, please contact ProductSafety@springernature.com

Printed and bound by CPI Group (UK) Ltd, Croydon, CR0 4YY
30/09/2026
02225302-0014